ADVANCED TOPICS
IN SCIENCE AND TECHNOLOGY IN CHINA

ADVANCED TOPICS IN SCIENCE AND TECHNOLOGY IN CHINA

Zhejiang University is one of the leading universities in China. In Advanced Topics in Science and Technology in China, Zhejiang University Press and Springer jointly publish monographs by Chinese scholars and professors, as well as invited authors and editors from abroad who are outstanding experts and scholars in their fields. This series will be of interest to researchers, lecturers, and graduate students alike.

Advanced Topics in Science and Technology in China aims to present the latest and most cutting-edge theories, techniques, and methodologies in various research areas in China. It covers all disciplines in the fields of natural science and technology, including but not limited to, computer science, materials science, life sciences, engineering, environmental sciences, mathematics, and physics.

Zhong-Shan Gao Hua-Hao Shen
Min Zheng Lynn J. Frewer
Luud J.W.J. Gilissen

Multidisciplinary Approaches to Allergies

With 47 figures

ZHEJIANG UNIVERSITY PRESS
浙江大学出版社

Springer

Editors
Prof. Zhong-Shan Gao
Department of Horticulture/Allergy Research Center
Zhejiang University
Hangzhou 310058, China
E-mail: gaozhongshan@zju.edu.cn

Prof. Hua-Hao Shen
Institute of Respiratory Diseases
The Second Affiliated Hospital
School of Medicine, Zhejiang University
Hangzhou 310009, China
E-mail: hh_shen@yahoo.com.cn

Prof. Min Zheng
Department of Dermatology
The Second Affiliated Hospital
School of Medicine, Zhejiang University
Hangzhou 310009, China
E-mail: minz@zju.edu.cn

Prof. Lynn J. Frewer
Food and Society, Center for Rural Economy
School of Agriculture
Food and Rural Development
Newcastle University
Newcastle upon Tyne, UK
E-mail: lynn.frewer@newcastle.ac.uk

Dr. Luud J.W.J. Gilissen
Plant Research International
Wageningen University and Research Center
Wageningen, the Netherlands
E-mail: luud.gilissen@wur.nl

ISSN 1995-6819 e-ISSN 1995-6827
Advanced Topics in Science and Technology in China

ISBN 978-7-308-09349-1
Zhejiang University Press, Hangzhou

ISBN 978-3-642-31608-1 ISBN 978-3-642-31609-8 (eBook)
Springer Heidelberg New York Dordrecht London

Library of Congress Control Number: 2012941046

Printed on acid-free paper

Springer is part of Springer Science+Business Media (www.springer.com)

Preface

The world-wide increase in allergy is the consequence of drastic changes in people's lifestyles, living environments (both indoors and outdoors) and food. The allergy problem has become an important health issue in the past two decades, and valuable knowledge and strategies regarding prevention have been fully developed yet over this period. In China, the economy is growing at a very high speed. Currently, asthma, allergic skin diseases and several forms of food allergy are major recognized health problems. Although there are no exact national data to enable understanding of the prevalence of allergic diseases currently available, there is, non-the-less, a body of evidence available to suggest that the incidence of allergic disease is increasing. Allergy is caused by multiple factors, so prevention strategies should be based on data derived from medical, food and agricultural, environment-related, and consumer-related research, and should address human genetic and physiological/immunological as well as environmental aspects of allergic disease development and management. In addition, health organizations, food and pharmaceutical industries, and governmental authorities should be supplied with the relevant information. Preventing and managing allergy requires the simultaneous development of multidisciplinary and integrated strategies involving researches from medical, food, environment and societal approaches. Allergic disease should be researched from the perspective of the gene to molecular biology, cell biology, histopathology, symptomology, and social and environmental sciences. Impact on quality of life should also be considered. The prevention and treatment of allergy requires an interdisciplinary research strategy.

This book addresses a broad range of allergy issues, with chapters being contributed by leading scientists and experts, regarding the prevalence, basic mechanisms, allergenic sources and allergens, diagnosis, therapies and pharmacy, hypoallergenic products, environmental pollution, climate change and hygiene life style that are involved in the course of allergy development and its societal impact. We hope this book will stimulate more active collaboration in the common theme of multidisciplinary approaches to understand, manage and prevent the prevalent allergies. This book can be used as a reference by students, experts and end-users in education, research, and governmental administration of allergy.

Multidisciplinary approaches to reduce allergy was a cooperative initiative between China and the Netherlands in 2007. Now we are very glad to see that a special book on this theme will be published. We would like to thank firstly professor Jun Zhu at Zhejiang University (ZJU) and professor Evert Jacobsen at Wageningen University and Research Center (WUR), who played key roles in the starting of the project by their mutual visits. Zhejiang University has supported the idea and established a multidisciplinary Allergy Research Center at ZJU. As organized and financed by ZJU, WUR, HAL Allergy BV and Hangzhou Zheda Dixun Biological Gene Engineering Co. Ltd., Hangzhou Allergy Symposium in 2008 provided a platform for exchanges of broad knowledge on allergy and initiated a book publishing plan.

It was a challenging task to write allergy in such broad subjects, which is driven by great enthusiasm and strong belief of this new idea. Sincere thanks are addressed to all authors of this special book.

We thank Professors Shao-Heng He, Yi-Ping Xu, Xue-Jun Zhu, Chun-Di He for critical reading of this book. Also thanks go to editors of Zhejiang University Publisher and Springer, who encouraged us and with great assistance and management of the editing work.

This work was partly supported by the Natural Science Foundation of China (30971970) and the Special Research Fund for International cooperation with European Union (Contract No. 1114).

The editorial team experienced a real pleasure in the cooperation during the realization of the final volume on this interesting and challenging topic.

Finally, we would like to acknowledge the references and permissions to use text, tables, photos figures from publishers and authors which are indicated in the text through which the book gains additional value as a textbook and a reference study book.

Zhong-Shan Gao, Hua-Hao Shen, Min Zheng
Hangzhou, China
Lynn J. Frewer
Newcastle upon Tyne, UK
Luud J.W.J. Gilissen
Wageningen, the Netherlands
April, 2012

Contents

Contributors

Guan-Hao Bu
College of Food Science and Technology, Henan University of Technology, Zhengzhou, 450001, China

Hong-Bing Chen
State Key Laboratory of Food Science and Technology, Nanchang University, Nanchang, 330047, China

Zhi Chen
School of Medicine, Zhejiang University, Hangzhou, 310058, China

Hao Cheng
Department of Dermatology, Sir Run Run Shaw Hospital, School of Medicine, Zhejiang University, Hangzhou, 310016, China

Margrit Fooke
Dr. Fooke Laboratorien GmbH, Neuss, Germany

Lynn J. Frewer
Department of Marketing and Consumer Behaviour, Wageningen University, the Netherlands (current address: School of Agriculture, Food and Rural Development, Newcastle University, Agriculture Building NE17RU, Newcastle upon Tyne, UK)

Jin-Yan Gao
Department of Food Science, Nanchang University, Nanchang, 330047, China

Zhong-Shan Gao
Department of Horticulture / Allergy Research Center, Zhejiang University, Hangzhou, 310058, China

Luud J.W.J. Gilissen
Allergy Consortium Wageningen, Plant Research International, PO Box 619, 6700 AP Wageningen, the Netherlands

Rong-Xia Guan
Institute of Crop Sciences, Chinese Academy of Agricultural Sciences, Beijing, 100081, China

Lotte A.H. Hermsen
Faculty of Earth and Life Science, VU University of Amsterdam 1012 ZA Amsterdam, the Netherlands (VU medical Center, 1081 BT Amsterdam, The Netherlands)

Erina M. Lin
University of California, 10833 Le Conte Ave, Los Angeles, CA 90095, USA

Hong Lin
College of Food Science and Engineering, Ocean University of China, Qingdao, 266003, China

Xiu-Min Li
Pediatrics and Immunobiology, Division of Pediatric Allergy & Immunobiology, Mount Sinai School of Medicine; Box 1198, New York, NY 10029-6574, USA

Zhen-Xing Li
College of Food Science and Engineering, Ocean University of China, Qingdao, 266003, China

Yong-Kang Luo
College of Food Science and Nutritional Engineering, China Agricultural University, Beijing, 100083, China

Ying-Tao Ma
Deptartment of Horticulture, Zhejiang University, Hangzhou, 310058, China

Ingrid M. van der Meer
Plant Research International, P.O. Box 16, 6700AA Wageningen, the Netherlands

Huan Qian
Department of Dermatology, the Second Affiliated Hospital, School of Medicine, Zhejiang University, Hangzhou, 310009, China

Li-Juan Qiu
Institute of Crop Sciences, Chinese Academy of Agricultural Sciences, Beijing, 100081, China

Ronald van Ree
Academic Medical Center, Amsterdam University, 1105AZ Amsterdam, the Netherlands

Huub F.J. Savelkoul
Cell Biology and Immunology Group, Wageningen University, P.O. box 338, 6708 WD Wageningen, the Netherlands

Hua-Hao Shen
Department of Dermatology, the Second Affiliated Hospital, School of Medicine, Zhejiang University, Hangzhou, 310009, China

Marinus J. M. Smulders
Allergy Consortium Wageningen and Plant Research International, Wageningen UR, P.O. Box 16, NL-6700 AA Wageningen, the Netherlands

Hilde Tobi
Research Methodology Group- ECS, Wageningen University, P.O. Box 8130, 6700 EW Wageningen, the Netherlands

Yvonne M. Vissers
Cell Biology and Immunology Group, Wageningen University, P.O. box 338, 6708 WD Wageningen, the Netherlands (current address: Nestlé Research Centre, Allergy group, P.O. Box 44, 1000 Lausanne, Switserland)

Arnold J.H. van Vliet
Environmental Systems Analysis Group, Wageningen University, P.O. Box 47, 6700 AA Wageningen, the Netherlands

Hui-Ying Wang
Department of Allergy, the Second Affiliated Hospital, School of Medicine, Zhejiang University, Hangzhou, 310009, China

Xing-Jun Wang
High-Tech Research Center, Shandong Academy of Agricultural Sciences, Ji'nan, 250100, China

Ting-Huan Wen
Medical Acarology Laboratory, Shanghai Medical College, Fudan University, Shanghai, 200032, China

Harry J. Wichers
Food and Biobased Research, Wageningen University, P.O. Box 17, 6700 AA, Wageningen, the Netherlands

Gary W.K. Wong
Department of Pediatrics, Chinese University of Hong Kong, Hong Kong, China

Zhao-Wei Yang
Deptartment of Horticulture, Zhejiang University, Hangzhou, 310058, China

Zhuo-Hui Zhao
Department of Environmental Health, School of Public Health, Key Laboratory of Public Health Safety, Ministry of Education, Fudan University, Shanghai, 200032, China

Min Zheng
Department of Dermatology, the Second Affiliated Hospital, School of Medicine, Zhejiang University, Hangzhou, 310009, China

Cheng Zhou
School of Medicine, Zhejiang University, Hangzhou, 310058, China

Part I

General Allergy

1

Prevalence of Allergic Diseases in China

Huan Qian, Min Zheng*

Department of Dermatology, the Second Affiliated Hospital, School of Medicine, Zhejiang University, Hangzhou, 310009, China
* E-mail: minz@zju.edu.cn

Abstract: The world-wide increase of allergic diseases is the consequence of drastic changes in life style, living environment, food and medical care. In China, allergic rhinitis, asthma, skin diseases (for instance, eczema, urticaria and atopic and contact dermatitis), conjunctivitis and several forms of food allergy are major recognized problems. Although we do not have exact national data for the prevalence of each allergic disease in China so far, there is definitely a growing tendency of their increasing prevalence. The exact mechanisms and determining factors underlying the increase in prevalence remain unclear, although allergen sensitization has been found to be at least as common in the Chinese region as in the West. Meanwhile, allergic diseases affect the human quality of life with the same magnitude as cardio-vascular diseases. With the projected growth of the Chinese population over the next decade, the burden of allergic diseases is expected to increase considerably. There is a growing and world-wide need for strategies to identify, prevent and control the allergic diseases.

1.1 Introduction

Allergy refers to any exaggerated immune response to a foreign antigen regardless of its mechanism. It is a local or systemic disease, most commonly affecting the nose, eyes, skin and lungs. Common allergic reactions include atopic dermatitis, contact dermatitis, urticaria and angioedema, which may be primary skin disorders or symptoms of systemic disorders. Other allergies are latex allergies, allergic lung disorders such as asthma, allergic conjunctivitis, allergic reactions to venomous stings and food allergies. All atopic disorders are type Ⅰ hypersensitivity reactions involving an exaggerated IgE-mediated immune response. In accordance with the world-wide rising trends, increased prevalence of allergic diseases in China have been described through epidemiological surveys. At the beginning of the twentieth century, allergies were a mysterious and rare condition that affected only a tiny minority of people. In 1978, an epidemiological survey of 6,563 people was conducted in Beijing. About 37.7% of the people said they had experienced at least one kind of allergic disease during their life time (Ye and Qiao, 1984). It is estimated that the current number of people in China with an allergic history will amount to 10%–30% of the total population. To be more specific, it is estimated that there are at least 25 million asthma patients, 50 million allergic individuals suffering from rhinitis, and 45 million people with food allergies, many of which are children and middle-aged adults. With these figures, allergies become a new problem with regard to sustainable economic development, health issues and the quality of life in China.

However, at present, we lack sufficient widespread statistics of many allergic diseases. China is a big country in area and population. The geography of China stretches some 5,026 kilometres across the East Asian landmass in a changing configuration of broad plains, expansive deserts, and lofty mountain ranges. Traditionally, China is divided into 7 regions, including North, Northeast, East, South, Northwest, Central and Southwest China. Moreover, China represents a full 20% of the world's population, with dramatic demographic changes regarding age, gender, region, occupation, education and ethnic profiles. During the last several decades, the country has undergone enormous social, economic, and political changes. Consequently, the current status of allergic diseases in China needs to be extensively investigated and updated.

Since October 2009, the first nation-wide survey on the prevalence of allergic diseases has been started. This survey involves 180,000 persons in 18 districts, with different age-groups and socio-economical status. This study will also evaluate the importance of Chinese allergen features, both outdoors and indoors, such as from local pollens and pets, in order to understand their role in triggering immune responses. Through this survey, we will obtain an in-depth insight of allergic diseases in China and their associations with factors such as distribution, risk factors, meteorologic conditions and patients' socio-economic status. The wide variations in the lifestyle and environmental exposures among China offer attractive opportunities to identify the risk and protective factors for these allergic diseases. Epidemiologic and genetic research studies in China will increase our knowledge on disease prevalence, healthcare burden and pathogenesis, and will enable the establishment of prevention and treatment guidelines.

1.2 Common Allergic Diseases in China

In this section, the following common allergic diseases in China will be introduced: allergic rhinitis, allergic lung disorder and allergic dermatitis.

1.2.1 Allergic Rhinitis

Allergic rhinitis (AR) includes seasonal or perennial itching, sneezing, rhinorrhea, nasal congestion, and sometimes conjunctivitis, caused by exposure to pollens or other aeroallergens. AR exerts a major impact on the quality of life.

Data on the current prevalence of AR in China is scarce. To investigate the prevalence of self-reported AR among Chinese citizens, a cross-sectional, population-based study was conducted in 11 major cities in China from September 2004 to May 2005. The self-reported prevalence of AR was the lowest in Beijing (8.7%) and the highest in Urumqi (Northwest China) (24.1%). Among the subjects with self-reported AR, 25.6% were diagnosed with persistent AR and 74.4% suffered from intermittent AR. Less than half of the subjects with self-reported AR had visited a health clinic. In 37.3% of the cases, AR had previously been diagnosed by physicians, and 33.1% of the subjects with self-reported AR had been treated (Zhang *et al.*, 2009). Another similar study indicated that the self-reported prevalence of allergic rhinitis was the lowest in Xi'an (Central China) (8.0%), and again the highest in Urumqi (21.4%), with Nanjing (East China) having an intermediate value (11.5%). The gender-adjusted prevalence ranged from 8.5% in Xi'an to 21.3% in Urumqi, while the age-adjusted prevalence of self-reported allergic rhinitis ranged from 8.7% in Beijing to 24.1% in Urumqi (Han *et al.*, 2007).

Comorbidities rates reported were asthma (9.2%), rhinosinusitis (13.3%) and atopic dermatitis (16.4%) (Zhang *et al.*, 2009). A study in Beijing showed that almost half of the patients with autumnal pollen allergic rhinitis developed seasonal allergic asthma within 9 years (Yin *et al.*, 2006).

1.2.2 Allergic Lung Disorders

Allergic lung disorders include asthma, allergic bronchopulmonary aspergillosis, and hypersensitivity pneumonitis. It is widely accepted that asthma is the most common allergic lung-affecting disorder. Many epidemiologic studies have documented the prevalence of asthma in China. Surveys from 1997 to 1999, including 220,000 people from three representative areas showed an accumulative prevalence of asthma of 0.7%–1.5%, with a specific prevalence in Liaoning (North China) of 1.25%, in Guangdong (South China) of 0.94%, and in Shanghai (East China) of 0.41%; the overall prevalence in children was 0.11%–2.03%, And the male to female ratio was 1:1.64 in Liaoning Province (Chen, 2001). The onset of asthma occurring in children under the age of 14 accounted for 38.63% (Chen *et al.*, 2002).

In children, the prevalence rate in North China was relatively lower (1%) than that in South China (1.5%), and was the highest in the eastern areas of China (2.4%) and 4,670 children (70%) had their onset before 3 years of age (Ma *et al.*, 2009). A nation-wide survey among 0–14 years old children from 43 cities in 31 provinces showed that the average prevalence of asthma amounted to 1.97% and the current 2-year prevalence of Chinese urban children was 1.54%. There has been a significantly increasing trend of asthma prevalence during the last 10 years, especially in the older age group (Chen, 2003).

According to the World Health Organization (WHO), asthma alone affected 300 million people worldwide in 2005, killing 255,000 of them (He *et al.*, 2000). The death figures have yet to be determined in China. The trend in asthma mortality in Hong Kong rose annually by an average of 10.5% in male patients, which was estimated from published statistics for the years 1976–1985 (So *et al.*, 1990).

Asthma is the result of an interaction between the environment and genetic factors. The exact causes for the increase in asthma are generally attributed to the differences and changes in life-style and urbanization (Tillie-Leblond *et al.*, 2008). This phenomenon has also been seen in China. A recent study showed the prevalence rates of asthma and atopic sensitization in rural Chinese children were significantly lower than that in urban children, which may be caused by the different living environments leading to differences from early life onwards (Ma *et al.*, 2003). Hong Kong had a much higher rate of asthma and related allergic symptoms compared with those in mainland China. The use of foam pillows and gas as cooking fuel during infancy were identified as prominent environmental risk factors for current wheeze in children (Leung and Wong, 2008). A cross-sectional study of 3,945 children conducted in northeast China during April 2007 demonstrated that in-house environmental factors are particularly important for the development of respiratory morbidity among children. Boys may be more susceptible to these environmental factors than girls (Dong and Ma, 2008). Other factors such as pet keeping and parental atopy increased the risk of asthma and allergic respiratory diseases in children (Dong and Ding, 2008).

Despite the environment factors, genetic studies in Chinese patients evaluated the role of several candidate genes including the IL-13 gene, the CTLA4 gene, Mannose-binding lectin gene, and the prostaglandin- endoperoxidase synthase 2 genes in the pathogensis of asthma and allergies (Leung and Wong, 2008).

1.2.3 Allergic Dermatitis

Allergic Dermatitis includes atopic dermatitis, contact dermatitis, urticaria and angioedema, among which atopic dermatitis is of the most significant importance. Atopic dermatitis (eczema, or AD) is a chronic inflammatory skin disease with onset typically in early childhood and is the most common chronic inflammatory skin disease in children in industrialized countries (Schultz and Hanifin, 2002). In China, the incidence of atopic dermatitis has been increasing steadily over the past 2 decades and its progress is most common in children aged 6–7 years old (Tillie-Leblond *et al.*, 2008) , with severe impacts on the physical and mental health of these children.

From 1989 to 1990, the prevalence of AD in primary and middle school students was 0.46%, with 0.68% in 7–12 years olds and 0.12% in 13–18 years olds, respectively (Tian and Kang, 1992). In 1998, a study of more than 78,000 individuals from 6–20 years old in South, Middle and North China showed an AD prevalence of 0.69%. It should be noted that standardized prevalence of AD are higher in the urban areas than in the rural areas, higher in males than in females, and that this prevalence correlated negatively with age (Gu *et al.*, 2000). During 2002, another study including nearly 50,000 Children from 1–6 years old covering East, South and North China, demonstrated that the AD prevalence had been raised to 3.07%, with males to 3.86% and females to 2.20% (Gu *et al.*, 2004). In 2005, the crude AD prevalence in 1- to 6-year-old was 2.9% according to a study in Tianjin (North China). The prevalence in the city was 2.4%, against 3.5% in the rural was area (Zeng *et al.*, 2005). These figures add great weight to the rising tendency of AD prevalence.

1.3 Current Research on the Allergens in China

An allergen is a non-parasitic antigen capable of stimulating a type- I hypersensitivity reaction in atopic individuals (Goldsby *et al.*, 2003). Common allergens are dust mite excretion, pollen and certain foods, but it is also possible to be allergic to anything from chlorine to perfume. The study of allergens is the core of allergology, because of its significance in diagnosis and immunotherapy. Along with the rapid economical development and urbanization in China, more people will come to live in new environments and will be increasingly challenged by new and more potential allergens. Thus, investigation of allergens receives much more attention. The main focuses of current studies of allergens are the epidemic survey of pollen, fungal and other common allergens in China, future application of allergens in diagnosis and medical treatment, and basic research of allergens such as its separation, purification, recombinant production and standardization to allow for commercialization.

Admittedly, allergens are tied to regional environments. China is one of the largest countries in the world, with different climate zones and diversified geographic features. Currently, there are few nation-wide epidemiological investigations of allergen prevalence in China, and for many remote areas, basic statistic data are still lacking. Meanwhile, the diversity in environments and ethnic backgrounds would provide excellent opportunities for research into the environmental and genetic determinants of allergies. For instance, based on skin prick tests, the pattern of sensitization and allergic symptoms in an unselected agricultural Chinese population shows that atopic sensitization was common in this population, particularly to shellfish, peanut, dust mite, and cockroach. The prevalence of allergic symptoms, in contrast, was quite low (Kim *et al.*, 2008).

In conclusion, there is still insufficient statistics for various allergens, their concentrations, functional components or biological allergenic activity. Empirical results of skin prick testing have indicated the leading causes of indoor (house-dust mites, house dust, cockroaches, dogs and cats) and outdoor pollen allergens are the major triggering factors. In addition, food allergens and allergies to drugs are relevant, as well as some allergens specific to China. Below, most common allergens and their prevalence are discussed in more detail.

1.3.1 Pollen Allergy (Hay Fever)

Pollen is an important cause of allergic respiratory diseases, such as asthma and allergic rhinitis, also known as pollinosis or hay fever. The incidence of pollinosis is consistent to pollen peak times. In America, about 30% of all adults and 40% of all children have hay fever, mostly related to ragweed pollen, which torments sufferers from March to November (Ouyang *et al.*, 2007). Meanwhile, in Central China, the peak time of airborne pollen occurs in two seasons: spring (March and April) and autumn (August to October) (Zhu *et al.*, 2008). In Northeast China, the blooming period of pollen-allergenic ornamental trees in Beijing urban area was limited to springtime, while that of pollen-allergenic herbs ranged from July to September. Studies show that the cause of hay fever in Beijing urban area in the 1950's were artemisia pollen, humulus pollen during the summer-autumn period, and cypress, birch, *Fraxinus chinensis*, and-representatives of the *Sterculiaceae* family during spring. Autumnal pollens are very important causes because they especially induce asthma during autumn in Northern China (Yin *et al.*, 2005).

Limited to regional differences, the prevalence of pollen allergies all over China needs to be updated. Encouragingly, more and more regional pollinosis reports have emerged. In 1984, the genus and quantity of local pollen distributions, involving 28 provinces throughout China, were documented, via a 3-year's nation-wide investigation. From 1995 to 1998, an epidemiological survey among students in Nanjing (East China) showed that the total positive rate of orchard pollen scratch test to be 6.7%, and a prevalence of orchard pollinosis of 0.37% (Yin *et al.*, 1999). Another study showed incidences of ragweed pollinosis of about 1.04% in the Qingdao (Northeast China) (Lu *et al.*, 1994). The positive rate of cedar pollen scratch tests on 309 students in Wuhan (Central China) was 7.8% (Xu *et al.*, 2000). However, the prevalence and degree of symptoms-causing pollens are going to vary greatly between the ecological environment, the dynamic changes of the climate, as well as the socio-economical factors such as afforestation of cities. Consequently, much more detailed accounts of pollinosis are needed.

1.3.2 Fungal Allergens

Fungal allergens are considered to be a major source of airbourne allergens. From 1988 to 1993, a nation-wide epidemiological survey, involving 65 health institutions, on regional fungal allergens has been carried out. Despite huge geographic and climate differences from North to South, dominant airbourne fungal allergens are *Alternaria alternata* and *Cladosporium*. From this survey, knowledge of fungal allergens in China starts to accumulate. During mid-June, the wheat harvest season in Northern China, allergic symptoms among workers on the threshing floor are associated with fungal allergens. Other significant fungal allergens include *Aspergillus sp.* (ubiquitous fungi in the soil), which are linked to allergic bronchopulmonary aspergillosis (ABPA) (Ye and Qiao, 1995).

1.3.3 Dust Mite

The house dust mite, which feeds on organic detritus such as flakes of shed human skin, is a cosmopolitan guest in human habitats. Allergens produced by house dust mites are among the most common inhaled triggers of asthma and allergic symptoms worldwide. In 1987, studies showed that a mean of 1,328 mites/g of dust was detected in homes: *Dermatophagoides farinae* and *D. pteronyssinus* predominated. Mixed house dust mite extracts elicited a positive skin test in 78% of asthmatics but *D. pteronyssinus* and *D. farinae* allergens were found in only 40% of these persons (Chen *et al.*, 1987). In Southern China, the warm and dampened environment is agreeable for the dust mite. Meanwhile, Northern China is much colder and arid. Higher incidence of asthma in the South than in the North may thus result from different densities of the dust mites in both areas. However, with more and more enclosed-type buildings, wide use of lying carpets, and the increased application of air conditioning and heating systems in China, allergic diseases caused by dust mites may lead to more inconveniences.

1.3.4 Environment Pollution

China's rapid economic development has come at the cost of severe environmental degradations. Outdoor air pollution is associated with more than 300,000 deaths, 20 million cases of respiratory illness, and a health cost of more than 500 billion RMB annually, which is approximately 3% of the gross domestic product (Millman *et al.*, 2008). Exposure to environmental pollutants may partially account for the increased prevalence of allergic diseases as well.

1.3.4.1 Outdoor Pollution

Outdoor pollution is dominated by industrial emissions. China relies on coal for approximately 70%–75% of its energy needs. In addition to CO_2, a major greenhouse gas, coal burning emits vast quantities of dust particles, polycyclic aromatic hydrocarbons, sulfur dioxide, arsenic and mercury (Millman *et al.*, 2008). Due to the increased number of motor vehicles, particularly those that use diesel fuel, ambient air nitrogen oxide levels have been increasing. Additional diesel exhaust particles involve formaldehyde, carbon monoxide, transition metals, carbon particles and others.

Although several large epidemiological studies have demonstrated a strong association between exposure to motor vehicle traffic emissions and allergic symptoms and reduced lung function, the evidence for the development of allergic sensitization from diesel particulates is less abundant than that for the aforementioned associations (Polosa *et al.*, 2002). The role of particulate matter and ozone in triggering asthma or allergic rhinitis confirms their higher relevance (McCreanor *et al.*, 2007; Parker *et al.*, 2009). Pathophysiologic studies showed that pollutants tend to exacerbate allergic inflammatory processes.

Air pollutants not only have a direct or indirect effect upon the individual, but also exert important adjuvant actions upon aeroallergens. That is to say, pollution is a major potentiator of certain allergies. Recent comparisons of the prevalence of hay fever, as well as positive skin-prick tests, between citizens of former West and East Germany and between Hong Kong, China and mainland China civilians, have demonstrated marked differences (Polosa *et al.*, 2002). Pollen in heavily polluted zones can indicate a larger amount of proteins described as being allergenic, and therefore do not act solely through its allergens (Gruijthuijsen *et al.*, 2006). On the other hand, climate change in part gives rise to variations in the temperature pattern and the presence of higher CO_2 concentrations may also lead to both qualitative and quantitative variations of pollen output, with the subsequent risk of allergic sensitization among the exposed human population. Accordingly, investigation of allergens should take local environment factors into consideration.

1.3.4.2 Indoor Pollution

In effect, allergen exposure increases at home. Tobacco smoke is a major component of indoor air pollution in China. Exposure to environmental tobacco smoke leads to acute exacerbation of asthma resulting in hospitalization and possibly increased risk for asthma. Particularly, maternal smoking during pregnancy and exposure to environmental tobacco smoke early in life are associated with a greater risk of developing wheezing illnesses in childhood (Pedersen *et al.*, 2011). Other indoor pollutant can be linked to lifestyle, 70% of Chinese households' burn coal or biomass for cooking and heating, which contaminates indoor air. Also, allergies are rampant in the workplace. The proliferation of new chemicals and the poorly ventilated "sick buildings" have triggered new allergies.

1.3.5 Food Allergens

Current statistics from various western countries suggest that up to 5%–8% of children and about 2% of adults suffer from food allergies. While in China, it is conservatively estimated that about 45 million people (>3%) are allergic to common foods. For some food allergy sufferers, exposure to even a microscopic amount of an offending substance can cause an anaphylactic shock and even death.

Important food allergens in China are as the following, all of which sometimes are able to cause serious allergic reaction: (1) milk: dairy products are the most important food allergen sources in China; (2) egg: this is another common food allergen in China; (3) nuts: walnuts, hazelnuts, almonds, cashews, pistachios are relevant allergenic sources; (4) oil-bearing crops: peanuts, soybeans, sesame seeds contain several important allergens; (5) seafood: fish, shrimp, crab are allergenic to predisposed individuals; (6) fruits: peach, pear, apple, litchi, mango, pineapple; and (7) vegetables: beans, celery, tomato are mentioned to cause allergies; as well as (8) cereals: allergies to domestic oats, buckwheat, and wheat are reported from time to time (Zhang, 2008).

Next to the increased application of more and more food additives, the changing of diet habits may also trigger food allergies. In addition, anaphylactic shock caused by the consumption of a variety of protein-rich processed foods that contain hidden allergens has increased significantly. To emphasize this, the primary life-threatening food allergies in communities are caused by hidden food allergens (Yin, 2009). In recent years, serious allergic reactions related to the consumption of fruits containing allergenic pollen-related cross-antigens have been reported. We need to recognize that adolescent patients with pollinosis have an increasing chance for the development of plant food allergy later in their life (Wen *et al.*, 1990). Another special kind of lethal food allergies is food-dependant exercise-induced anaphylaxis, which is sometimes underestimated.

Nevertheless, China boasts of a long history of cuisine culture and also of a profundity for its variety. Although many people believe they have a food allergy, they sometimes may confuse a food allergy with food intolerance and identify the overall reaction to various cooking methods, seasonings, additives, even to Chinese medicine as a food allergy. As a matter of fact, food intolerance, although it is also an undesirable reaction to a kind of food, does not involve the immune system and therefore cannot be regarded as an allergy.

1.3.6 Drug

Drugs are the most common incentives for causing an anaphylactic shock. Currently, the overall Chinese data on the incidence of anaphylactic shocks caused by drugs remains unchecked. But we should pay special attention to the traditional Chinese medicine (TCM) related to the occurrence of anaphylactic shocks. The National Adverse Drug Reaction Monitoring Center reported several serious Chinese medicine injection events resulting in an allergic reaction. From 1988 to 2003, 272 cases of adverse reactions to Yuxingcao injection were reported, with 52 severe cases. In 2006, 44 persons died of a Yuxingcao injection-induced anaphylactic shock. Up to 2006, a retrospective study showed 228 cases of an anaphylactic shock from TCM injections, among which 90% of patients experienced shock symptoms within 30 minutes (Yin, 2009).

In addition, new drugs are spawning new allergies. About 3% of all hospitalized patients are expected to experience a severe allergic reaction to a new medication. In China, excessive use of antibiotics changes in fecal flora and thus affected the immunological balance towards increased sensitivity to allergies.

1.3.7 Specific Allergens in China

Specific allergens in China involve *Patinopecten yessoensis*, *Procambarus clarkii*, buckwheat, garouper and so on. Silk is also a highly potent allergen. A total of 64 children younger than 15 years of age with asthma caused by silk were studied. The first symptoms appeared on the average at 10 months after initial exposure to silk. In 61% of the patients, asthma was accompanied by allergic rhinitis but in only 14% of the cases by conjunctivitis. In most cases, asthma occurred in winter, due to seasonal use of bed quilts or clothes filled with silk. The average mean wheal diameter elicited by silk in skin prick testing was two times larger than the diameter of the histamine equivalent in the control. A cross reactivity exists among mulberry silk, and silkworm cocoons, batryticated silkworms, and silkworm chrysalis (Wen *et al.*, 1990). In this case, although we educated dust mites allergic patients to use cotton or silk quilts or cloths, we should also keep in mind that they also might be a possible allergen.

1.3.8 Other Allergens

In addition to pollen, food, medicines and the like, many more causes of allergies have emerged, including cockroach, pets, especially their furs, venomous stings, such as bee or wasp stings, detergents, cosmetics and others. With the improvement of people's living standards in China, more and more persons keep cats or dogs, potentially increasing risks of developing allergies. Household appliances produce electrostatic charges that would adsorb dust and micro-organisms in the air, thus leading to allergic skin diseases. Room and automobile interior decorations use paint, adhesives and plastics which will slowly release formaldehyde, benzene and other chemicals and jeopardize people's skin and respiratory tract. In conclusion, almost anything could envoke human's immune responses towards allergies.

1.4 Socioeconomic Burden of the Inexorable Rise in Allergies

Over the next decade in China, the potential risk of allergic diseases is expected to increase significantly. Recognition of the full health and economic costs of allergies, which is due to millions of children missing school and being hospitalized, and to adults staying away from their work, is crucial. Having an allergic reaction, such as a serious asthma exacerbation, which might be fatal in some cases, or will seriously compromise quality of life in others is a serious threat. For instance, more than a third of people with allergies cannot go to restaurants or have to avoid triggers such as perfumes, cleaning fluids and animals such as pets. Allergies among children are most worrying of all, and are responsible for the growing numbers suffering from asthma. In China, due to the limited number of trained allergists, many patients are seen by general physicians, and often, the appropriate diagnostic tests and treatments are not provided. In addition, the financial burden for health care may be prohibitive to a certain part of the population that would prevent obtaining a better understanding and treatment of allergies. Moreover, health organizations, food and pharmaceutical industries, and governmental authorities should be supplied with the relevant information necessary to understand the effect and risk of allergies. Therefore, a good control over a potential allergic epidemic requires the development of multidisciplinary and integrated strategies involving research based on medical, food, environment and societal approaches.

1.5 Conclusion

Allergies are a multifactorial disease that includes human genetic factors as well as environmental factors. The exact mechanism of each allergic disease is not yet very clear. It is thought that allergies belong to a systemic disease, but they are clinically very different. Also, there were many new developments in the field of pathogenesis research and clinical treatments. With the high speed economic growth and environmental changes in China, allergic diseases are currently of significant importance. We expect an increase of data on allergic triggers and risk factors, morbidity, mortality and the socio-economic burden of these different allergic diseases. Further epidemiologic research in China will apply knowledge on the disease prevalence and pathogenesis, and therefore enhance our diagnosis and treatment strategy. In conclusion, better identification of triggers and risk factors, increased surveillance of the burden of the disease, better public awareness, improved training of physicians and healthcare personnel, better access to essential medications, implementation of environmental controls, appropriate management and implementation of preventive measures are key to reducing the burden of these allergic diseases in China.

References

Chen, P., Yu, R., Hou, X., *et al.* (2002). Epidemiological survey on bronchial asthma in Liaoning province. Chinese J Tuberculosis Respir Dis, 25(10), 603-606 (in Chinese).

Chen, P. (2001). Asthma-results of epidermological survey in Liaoning and Guangdong Provinces and Shanghai Pudong China. Chin J Intern Med, 40(7), 433-434 (in Chinese).

Chen, Y.K., Yang, W.P., Xue, Z.L., *et al.* (1987). House dust mite asthma in China: a review. Ann Allergy, 59(2), 147-148.

Chen, Y.Z. (2003). A nationwide survey in China on prevalence of asthma in urben children. Chinese Journal of Pediatrics, 41(2), 123-127 (in Chinese).

Dong, G.H., Ding, H.L., Ma, Y.N., Jin, J., *et al.* (2008). Asthma and asthma-related symptoms in 16,789 Chinese children in relation to pet keeping and parental atopy. J Investig Allergol Clin Immunol, 18(3), 207-213.

Dong, G.H., Ma, Y.N., Ding, H.L., *et al.* (2008). Housing characteristics, home environmental factors and respiratory health in 3,945 pre-school children in China. Int J Environ Health Res, 8(4), 267-282.

Goldsby, R.A., Kindt T.K., Osborne B.A. *et al.* (2003). Immunology (5th ed.). San Francisco: W.H. Freeman.

Gruijthuijsen, Y.K., Grieshuber, I., Stöcklinger, A., *et al.* (2006). Nitration enhances the allergenic potential of proteins. Int Arch Allergy Immunol, 141(3): 265-275.

Gu, H., Chen, X.S., Chen, K. (2000). The evaluation of diagnostic criteria of atopic dermatitis. Chinese J Dermatol, 33(4), 379-382 (in Chinese).

Gu, H., You, L.P., Liu, Y.S., *et al.* (2004). Survey on the prevalence of childhood atopic dermatitis in ten cities of China. Chinese J Dermatol, 37(1), 29-31 (in Chinese).

Han, D.M., Zhang, L., Huang. D., *et al.* (2007). Self-reported prevalence of allergic rhinitis in eleven cities in China. Chinese Journal of Otorhinolaryngology Head and Neck Surgery, 42(5), 378-384 (in Chinese).

He, Q., He, B., Zhu, Y. (2000). Analysis of death of hospitalized patients with asthma in 16 hospitals in Beijing from 1988 to 1998. Chinese Journal of Tuberculosis and Respiratory Diseases, 23(11), 686-688 (in Chinese).

Kim, J.S., Ouyang, F., Pongracic, J.A., *et al.* (2008). Dissociation between the prevalence of atopy and allergic disease in rural China among children and adults. J Allergy Clin Immunol, 122(5), 929-935.e4.

Leung, T.F., Wong, G.W. (2008). The Asian side of asthma and allergy. Curr Opin Allergy Clin Immunol, 8(5), 384-390.

Lu, D., Zhau, P., Yu, Q., *et al.* (1994). Airway provocation test with ragweed pollen extract in Chinese asthmatics. Asian Pac J Allergy Immunol, 12(2), 125-129.

Ma, Y., Li, X.F., Zhao, J., *et al.* (2003). Relationships between the diameters of Bacille Calmette-Guerin scars and asthma, atopy in urban and rural Beijing children. Chinese J Tuberculosis Respir Dis, 26(9), 526-530 (in Chinese).

Ma, Y., Zhao, J., Han, Z.R., *et al.* (2009). Very low prevalence of asthma and allergies in schoolchildren from rural Beijing, China. Pediatr Pulmonol, 44(8), 793-799.

McCreanor, J., Cullinan, P., Nieuwenhuijsen, M.J., *et al.* (2007) Respiratory effects of exposure to diesel traffic in persons with asthma. N Engl J Med, 357(23), 2348-2358.

Millman, A., Tang, D., Perera, F.P. (2008). Air pollution threatens the health of children in China. Pediatrics, 122(3), 620-628.

Ouyang, Z.Y., Xin, J.N., Zheng, H., *et al.* (2007). Species composition, distribution and phenological characters of pollen-allergenic plants in Beijing urban area. Ying Yong Sheng Tai Xue Bao, 18(9), 1953-1958 (in Chinese).

Parker, J.D., Akinbami, L.J., Woodruff, T.J. (2009). Air pollution and childhood respiratory allergies in the United States. Environ Health Perspect, 117 (1), 140-147.

Pedersen, S.E., Hurd, S.S., Lemanske, R.F., *et al.* (2011). Global strategy for the diagnosis and management of asthma in children 5 years and younger. Pediatric Pulmonol, 46, 1-17.

Polosa, R., Salvi, S.D., Maria, G.U. (2002). Allergic susceptibility associated with diesel exhaust particle exposure: clear as mud. Arch Environ Health, 57(3), 188-193.

Schultz, L.F., Hanifin, J. (2002). Epidemiology of atopic dermatitis. Immunol. Allergy Clin North Am, 22, 1-24.

So, S.Y., Ng, M.M., Ip, M.S., *et al.* (1990). Rising asthma mortality in young males in Hong Kong, 1976-85. Respir Med, 84(6), 457-461.

Tian, R.M., Kang, K.F. (1992). Analysis of atopic dermatitis. Chinese Journal of Dermatology, 25, 183-184 (in Chinese).

Tillie-Leblond, I., Magnan, A., Pauli, G., *et al.* (2008). Asthma and allergy: short texts and recommendations of the expert conference of the French Speaking Pneumology Society (SPLF), in partnership with the French Society of Allergology and Clinical Immunology (SFAIC), the French Society of Occupational Medicine (SFMT) and the "Asthma-Allergy" association. Respir Med, 102(10), 1483-1493.

Wen, C.M., Ye, S.T., Zhou, L.X., *et al.* (1990). Silk-induced asthma in children: a report of 64 cases. Ann Allergy, 65(5), 375-378.

Wen, Z., Ye, S. (2002). A report of 50 patients with artemisia pollenosis and plant food allergy. National Med J China, 82(9), 626-629 (in Chinese).

Xu, L.J., Zhang, S.Y., Yang, Q., Cheng, L., *et al.* (2000). A preliminary study on the incidence of cedar pollinosis in district of Wuhan. J Clin Otorhinolaryngol, 14(11), 505-506 (in Chinese).

Ye, S.T., Pan, Y.Q., Zhang, H.Y. (1984). Prevalence of allergic diseases in 6563 individuals. Chinese Journal of Preventive Medicine, 1, 27-29 (in Chinese).

Ye, S.T., Qiao, B.S. (1995). Investigation of Chinese Airborne Fungal Allergens. Beijing: Science Press, pp.17-18 (in Chinese).

Yin, J., Yue, F.M., Wang, L.L., *et al.* (2005). The clinical study of the relationship between allergic rhinitis and allergic asthma in the patients with autumnal pollinosis. National Med J China, 85(24), 1683-1687 (in Chinese).

Yin, J., Yue, F.M., Wang, L.L., *et al.* (2006). Natural course from rhinitis to asthma in the patients with autumnal pollinosis: a clinical study of 1096 patients. National Med J China, 86(23), 1628-1632 (in Chinese).

Yin, I.J. (2009). A new challenge for chinese physicians: anaphylaxis. Chin J Allergy Clin Immunol, 3(1), 1-4 (in Chinese).

Yin, M., Cheng, L., Miyoshi, A., *et al.* (1999). Epidemiological survey on orchard pollinosis. J Clin Otorhinolaryngol, 13(11), 513-514 (in Chinese).

Zeng, W., Tang, N.J., Ji, J.M., *et al.* (2005). Prevalence of atopic dermatitis in infants and children in Tian Jin, China. J Investig Dermatol, 125, 854 (in Chinese).

Zhang, H.Y. (2008). Allergy study in China. China J Allergy Clin Immunol, 2(4), 255-260 (in Chinese).

Zhang, L., Han, D., Huang, D., *et al.* (2009). Prevalence of self-reported allergic rhinitis in eleven major cities in China. Int Arch Allergy Immunol, 149, 47-57.

Zhu, R., Li, W., Wang, Z., *et al.* (2008). A survey of airborne pollen in Wuhan and its relationship to pollinosis. J Clinl Otorhinolaryngol Head and Neck Surgery, 22(14), 647-650 (in Chinese).

2

Mechanism of Type Ⅰ Hypersensitivity

Cheng Zhou

School of Medicine, Zhejiang University, Hangzhou, 310058, China
E-mail: zhoucheng0113@zju.edu.cn

Abstract: The increasing prevalence of allergic disorders is currently a serious problem with public health. Understanding of the mechanisms underlying allergies (type Ⅰ hypersensitivity or immediate reactions) will enable us to improve our treatment with allergic diseases. In this chapter, the features of type Ⅰ hypersensitivity reactions and the major components involved as well as their potential roles in the induction and regulation of allergic responses are discussed. A half century ago, Gell and Coombs classified the hypersensitivity reactions into four types based on the immunologic mechanisms related to the symptoms in the organism (Gell and Coombs, 1963). Though this classification has many limitations since hypersensitive reactions always showed mixed pathological mechanisms, it is still widely accepted today. Here, we will present our insights on the basic mechanism of type Ⅰ hypersensitivity reactions.

2.1 Introduction

Type Ⅰ hypersensitivity reactions are characterized as IgE-mediated allergic reactions. The underlying mechanism of type Ⅰ hypersensitivity reactions is the switch from a physiological IgM/IgG antibody response to an IgE-dominated response against a specific allergen during sensitization (Pali-Scholl and Jensen-Jarolim, 2009). The development of type Ⅰ hypersensitivity consists of three phases: (1) the allergy-induction phase; (2) the allergy-stimulation phase; and

(3) the effector phase.

In the first phase, a specific allergen comes into contact with the skin, the mucosal epithelium of the eyes, the respiratory system, the digestive system, and even with the vascular system as a result of insect stings or when medication is administrated intravenously. Then the allergen is taken up by APCs, processed, combined with MHC II and presented to CD4+ helper T-cells. In allergic persons, the CD4+ T-cells differentiate into Th2 lymphocytes in the presence of IL-4, and Th2 lymphocytes release cytokines, particularly IL-4 and IL-13. In the presence of CD40L-CD40 interaction, isotype switching to IgE will occur in activated B-cells, resulting in the production of allergen-specific IgE which are excreted into the serum. Further, the Fc portion of IgE binds to high-affinity receptors (FcεRI) which are expressed on the surface of the mast cells and basophils. In the second phase, when the allergic person is re-exposed to the specific allergen, the bound IgE is cross-linked by the reintroduced allergen. In the third phase, the effector cells, like mast cells or basophils are activated by the crosslink of FcεRI, a series of cell signaling events will follow, and degranulation is activated causing the release of a large amount of inflammatory mediators from the mast cells, which leads to the development of the classical symptoms involving the skin, respiratory tract, circulation, or gastrointestinal tract (Pali-Scholl and Jensen-Jarolim, 2009) (Fig. 2.1)

Clinical and pathologic features of type I hypersensitivity are secondary to inflammatory mediators produced by mast cells in different tissues. The inflammatory mediators are divided into three groups: preformed mediators, newly synthesized mediators, and cytokines. Preformed mediators, namely histamine, serine proteases, tryptase, carboxypeptidase A, and proteoglycans, of which histamine is the principal and most important mediator. These are present within the granules of the mast cell and are released within minutes after activation, causing the early-phase or acute reaction. The newly synthesized mediators from the mast cells include prostaglandins (PGD 2), platelet-activating factor and leukotrienes (LTC4, LTD4, and LTB4). Activated mast cells can also release various cytokines, including IL-3, IL-4, IL-5, IL-6, IL-8, GM-CSF, and TNF-α. Preformed and newly synthesized mediators are the main components of the acute allergic reaction. In addition, numerous cytokines can be released from the mast cells recruiting leukocytes such as eosinophils, neutrophils, and Th2 cells, which in its turn are involved in the late phase reaction.

Despite the high occurrence of allergic disorders, only a few factors have been identified that are held responsible for the induction and maintenance of allergies. Some of the humoral components involved in type I hypersensitivity are discussed in the following paragraphs.

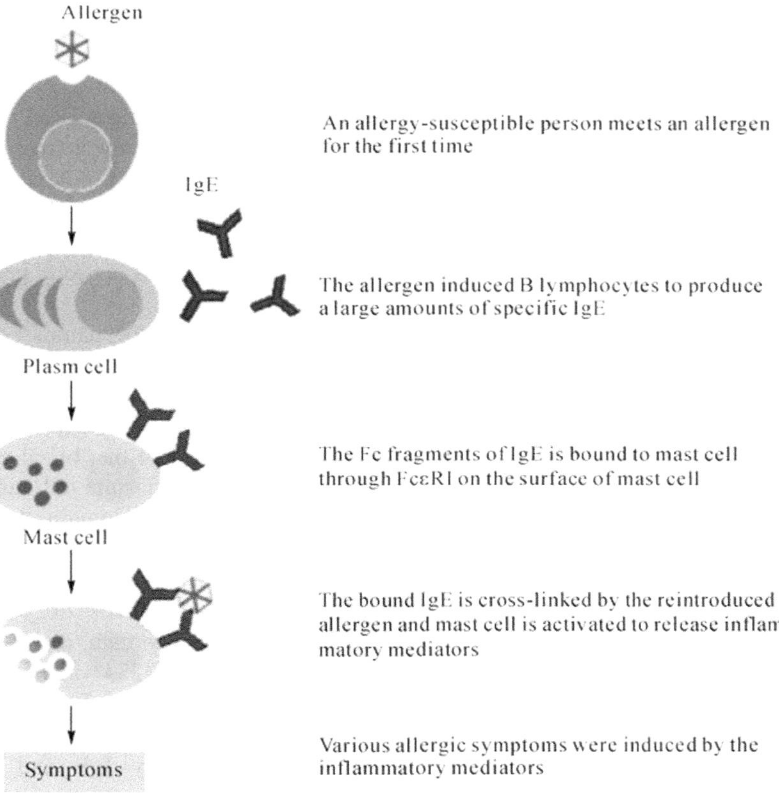

Fig. 2.1. The major steps of type Ⅰ hypersensitivity

2.2 IgE and IgE Receptors

IgE was discovered by Ishizake in 1966 as the primary allergy antibody. Its concentration in the serum is the lowest of the five immunoglobulin subtypes. IgE shares the same basic molecular architecture as the antibodies of the other classes, with 2 identical heavy chains and 2 identical light chains. The heavy ε-chain contains 3 domains, one more domain than the heavy γ-chain of IgG. Usually, a significant IgE response was induced only as a defense against parasitic helminth infections in humans (Jiz *et al.*, 2009). After an individual is exposed to the parasite, the IgE level increases in serum and remains high until the parasite is successfully cleared from the body. In atopic individuals (that are genetically predisposed to be able to produce relatively high amounts of IgE), non-parasitic allergens are able to stimulate inappropriate IgE production, leading to type Ⅰ hypersensitivity (Cooke and van der Veer, 1916; Yao *et al.*, 2010).

IgE synthesis begins when an allergen is taken up by the APC, and simultaneously caught by B-cell receptors through the cell-surface immunoglobulin receptors. In the APC, the allergen is packaged, processed, and presented to helper T-cells. The MHC-allergen complex is recognized by the Th2 cells, leading to the activation of the Th2 cells to produce IL-4, IL-13, and CD40L (CD154). IL-4 and IL-13 activate transcription, and the CD40L-CD40 interaction activates the DNA switch recombination, resulting in the isotype class switch to IgE (Stone *et al.*, 2010). The propensity toward Th2 development and IgE production has a strong genetic basis.

There are two receptors for IgE: the high-affinity IgE receptor, named FcεRI, primarily expressed on the surface of mast cells and basophils, and the low-affinity IgE receptor (FcεRII; CD23) expressed on the surface of B-cells, as well as other hematopoietic cells. FcεRI is made of four subunits, expressed as tetramers ($\alpha\beta\gamma_2$). The α-chain consists of an extracellular domain that binds to the Fc portion of IgE, a transmembrane domain, and a short cytoplasmic tail with no signaling motifs. The β-chain consists of four transmembrane domains with a single immunoreceptor tyrosine-based activation motif (ITAM) associated with Lyn kinase. The two γ-chains form a dimer, each of which contains an ITAM. FcεRII (CD23) is a Ca-dependent lectin. The receptor consists of a large extracellular domain that binds IgE, a single transmembrane domain, and a short cytoplasmic tail. Besides, CD23 also has a soluble form, sCD23, which can be shed from the membrane by proteases. CD23 activation mediates IgE regulation, differentiation of B-cells, activation of monocytes, and antigen presentation. The expressions of FcεRI and CD23 are both up-regulated by IgE and IL-4 (MacGlashan, 2005; Stone *et al.*, 2010).

2.3 Mast Cells

Mast cells are innate immune cells that were first described on the basis of their unique staining characteristics, displaying large granules. In human, there are two major subtypes of human mast cells: MC_T (characterized by the presence of tryptase) and MC_{TC} (containing tryptase and chymase). Their locations are different in the various body organs and tissues. MC_T cells predominate in the mucosa of the respiratory and gastrointestinal tract and increase with mucosal inflammation, whereas MC_{TC} cells are mainly localized in the connective tissues such as the dermis, submucosa of the gastrointestinal tract, heart, conjunctivae, and perivascular tissues (Stone *et al.*, 2010; Metcalfe, 2008). Both are originating from the bone marrow, and their precursor cells circulate in the blood and migrate into the various tissues, where they differentiate and mature. Mast cells are found throughout the connective tissues, particularly near blood and the lymphatic vessels. Some tissues, including the skin and the mucous membrane surfaces of the respiratory and gastrointestinal tracts, contain high concentrations of mast cells. The major survival and developmental factor for mast cells is stem cell factor (SCF), but many growth factors, cytokines and chemokines can also modulate their proliferation, differentiation and phenotypes. For example, IL-4 upregulates the expression of Fc *epsilon* RI (FcεRI) receptors, IL-5 promotes the proliferation in the presence of the stem cell factor (SCF), and IFN-γ decreases the number of mast cells (Stone *et al.*, 2010). Mast cells can function in both innate and adaptive immune responses and can have protective and pathogenic activity. In type I hypersensitivity reactions, such as rhinitis, urticaria, and asthma, mast cells increase dramatically in number. Mast cells are the main effector cells involved in the pathogenesis of allergic diseases, since most FcεRI are expressed on their surface. It is well known that mast cells play a critical role in the early response to allergen in the airways and skin, but it remains uncertain what roles they play in the development of late responses and in the following chronic inflammation. The signaling of mast cell activation is complicated. In IgE-mediated hypsensitivity, mast cells are activated through the crosslink of FcεRI by binding antigen-specific IgE. After FcεRI aggregation, Lyn phosphorylates tyrosine residues in the ITAMs of the β and γ subunits. Then Syk kinase, a central kinase in mast cell activation, is recruited by the tyrosine-phosphorylated γ subunit. Syk activates a series of downstream signaling events associated with mast cell activation. After being activated, the mast cell degranulates and a large amount of inflammatory mediators are released from the cell (Metcalfe *et al.*, 2009). Mast cells with Syk deficiency do not degranulate after FcεRI aggregation (Siraganian *et al.*, 2002). Mast cell activation through aggregation of FcεRI via the IgE-allergen complexes can be further enhanced by SCF-induced ligation of the mast cell growth factor receptor KIT (CD117) (shown in Fig. 2.2.).

2.4 Eosinophils

An eosinophil is a white blood cell type which takes up the red dye eosin when blood is examined under the microscope. Eosinophils are derived from CD34+ progenitor cells and mature in the bone marrow. The mature cells are released into the peripheral blood. The circulating eosinophils in the blood can be attracted into tissues by the effects of IL-4 and IL-13. Most eosinophils are located in the gut and lungs. Eosinophils can be primed by several mediators, including IL-3, IL-5, GM-CSF, CC chemokines, and platelet-activating factors. After activation, eosinophils release mediators from granules. Their natural role is to defend the body against parasites, and they accumulate wherever allergic reactions take place, like those in the lungs in the case of asthma. The mediators from the granules are important for killing parasites, but in asthma they are released inappropriately and damage the lining of the air passages.

Eosinophil granule proteins are thought to play important roles in the

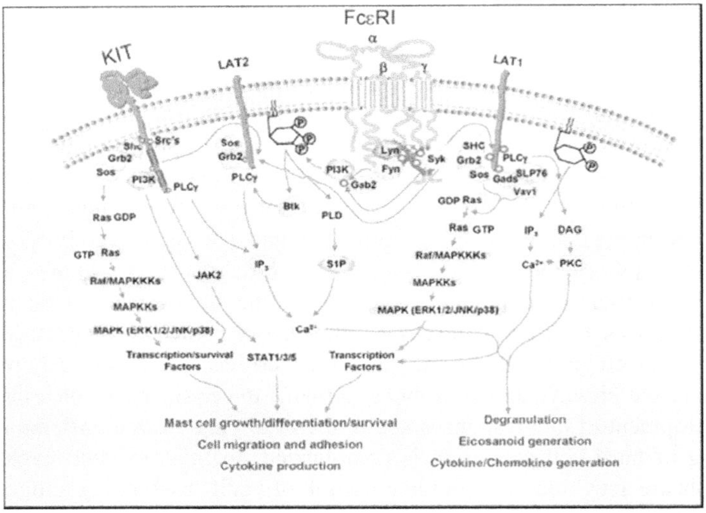

Fig. 2.2. Signaling pathways leading from activated KIT and aggregated FcεRI to mast cell responses. Antigen-induced aggregation of immunoglobulin E (IgE)-occupied Fc RI induces activation of the Src family tyrosine kinase, Lyn, whereas stem cell factor (SCF)-induced KIT dimerization induces activation of its intrinsic KIT kinase activation. Phosphorylation of tyrosine residues within the receptor chains thus allows recruitment of Src homology 2 (SH2) domain-containing signaling molecules. In the case of Fc RI, spleen tyrosine kinase (Syk) is recruited via immunoreceptor tyrosine-based activation motifs (ITAMs) contained in the c chain-cytoplasmic domains. Resulting activation of Syk, following its phosphorylation, leads to consequential phosphorylation of the transmembrane adapter molecules linker for activated T cells 1 (LAT1) and LAT2 [non T cell activation linker (NTAL) / linker for activation of B cells (LAB)]. Upon phosphorylation, these proteins serve as scaffolds for multimolecular signaling complexes comprising various cytosolic adapter molecules such as Gads, Grb2, SLP76, and SHC, guanosine triphosphate (GTP) exchangers including Sos and Vav1 and the signaling enzymes phospholipase C $_1$ (PLC $_1$) and PLC $_2$. PLC catalyzes the hydrolysis of PtdIns$_2$ to yield diacylglycerol (DAG) and inositol-1,4,5,-triphosphate (IP$_3$), which, respectively, result in the activation of protein kinase C (PKC) and the liberation of intracellular calcium. Following depletion of the intracellular calcium stores, the calcium signal is maintained by store operated calcium entry (not depicted). These signals lead to mast cell degranulation and eicosanoid generation and also contribute to activation of transcription factors required for cytokine and chemokine production. In parallel to this pathway, phosphoinositide 3-kinase (PI3K) is activated following binding to Gab2 upon the phosphorylation of this cytosolic adapter molecule by Fyn and / or Syk, phosphorylation of the p85 adapter subunit of PI3K, and activation of the catalytic subunit by small GTP-binding proteins. In the case of KIT, the p85a subunit directly binds to the phosphorylated molecule. The subsequent formation of membrane associated PtdInsP$_3$ results in the recruitment of pleckstin homology (PH) domain-containing signaling molecules such as Bruton's tyrosine kinase (Btk), PLD, and potentially others. PI3K-regulated pathways serve to enhance/maintain LAT/ PLC $_1$-regulated degranulation and, as depicted for KIT, regulate mast cell growth, differentiation, survival, migration, adhesion, and cytokine production. KIT- and Fc RI-mediated activation of the Ras-Raf-mitogen-associated protein kinase (MAPK) pathway following Sos- and Vav-regulated GDP–GTP exchange of Ras also contributes to these processes. The MAPK extracellular signal regulated kinase 1 / 2 (ERK1 / 2) also regulates phospholipase A$_2$ (PLA$_2$) activation, which leads to the liberation of arachidonic acid for the generation of eicosanoids (not depicted). The role of LAT2 in mast cell activation is still enigmatic; however, it has been proposed to both upregulate and downregulate antigen-mediated responses. It does appear to be required for the ability of KIT to enhance FceRI-dependent degranulation..(Gilfillan and Rivera, 2009). JAK, Janus kinase, and STAT, signal transducer and activator of transcription (With permission of John Wiley and Sons)

pathogenesis of allergic asthma, which include, for example, the major basic protein (MBP), the eosinophil-derived neurotoxin and the eosinophil cationic protein. Among them, major basic protein (MBP) is the major component of eosinophil granules. Usually, eosinophils comprise less than 5% of peripheral blood leukocytes. In patients with allergic diseases and helminth infections, eosinophils are always elevated in their peripheral blood and tissues. In addition, eosinophils also produce many kinds of pro-inflammatory cytokines, such as IL-1, IL-2, IL-3, IL-4, IL-5, IL-6, IL-8, IL-9, IL-10, IL-12, IL-13, IL-16, IL-17 and TNF-α, TGF-β, chemokines such as CCL5/RANTES and CCL11/eotaxin-1, and lipid mediators such as the platelet-activating factor (PAF) (Hogan *et al.*, 2008). However, the major signaling mechanism about how eosinophils are activated is still elusive.

2.5 Basophils

Basophils develop and mature in the bone marrow from CD34+ pluripotent stem cells under the effects of IL-3. Then they are released into peripheral blood, and comprise less than 1% of the peripheral blood leukocytes. Basophils share common features with mast cells and are well known for their role as effector cells of the Th2 immune response. They also express high level of FcεRI on their surface, and are activated by the crosslink of FcεRI. After activation, the same procedures as for mast cells occur, such as the production of Th2 cytokines (IL-4 and IL-13) as well as degranulation. Therefore preformed mediators such as histamine and eicosanoids are released, leading to the characteristic symptoms of the Th2 immune response (Sokol and Medzhitov, 2010).

It is generally regarded that basophils mainly participate in allergic reactions as effector cells, but recent studies identified important roles for basophils in many aspects of the immune response, from activation to memory, to the effector response (Schroeder, 2009; Sokol and Medzhitov, 2010). However, the detailed physiological role of basophils is not yet fully clear.

2.6 T Lymphocytes

T-cells play crucial roles in the orchestration of adaptive immunity. There are two major T-cell subsets in the periphery, divided by the expression of CD4 or CD8 antigens on their cell surface. CD4+ T-cells are activated by recognizing peptide fragments bound to MHC class II antigens. In 1986, it was reveallled that the activated CD4+ T-cells can be further classified into two subsets, Th1 and Th2, according to the secretions of specific cytokines (Mosmann *et al.*, 1986). Th1 cells characteristically secrete the cytokines IFN-γ, IL-2 and TNF-β that promote macrophage and other phagocyte activities to promote the intracellular killing of pathogens associated with protection against intracellular bacteria, as well as autoimmune diseases. On the contrary, Th2 cells secrete IL-4, IL-5 and IL-13 cytokines that promote antibody responses, and are critically important for allergic diseases. A number of factors determine the differentiation of naïve CD4+ T-cells to Th1 and Th2 subsets, and probably the most important factor is the cytokine milieu surrounding the T-cell. IL-12 is produced mainly by activated macrophages, causes antigen-primed naïve T-cells to differentiate into Th1 cells that secretes IFN-γ to kill the microbes. In contrast, IL-4 induces the differentiation of naïve T-cells into Th2 cells, which produce IL-4, IL-5 and IL-13 that stimulate IgE production and eosinophil activation.

It is thought that in allergic patients the T-cell response would be a Th2-dominated immune response. Th2 cells exert the crucial role in the pathogenesis of allergic diseases by secretion of Th2 associated cytokines, such as IL-4, IL-5 and IL-13, of which IL-4 and IL-13 are thought to be critical in promoting allergen-specific IgE production by B-cells. Thus, scientists have tried many ways to direct the immune response in atopy from the Th2-dominant response toward the Th1 response after antigen exposures. This Th1-Th2 model, to some extent, can explain the dis-regulation of the immune system in the case of allergies. However, more and more evidence indicates that this model is too simplistic to explain all the allergic responses (Durrant and Metzger, 2010).

The Th1-Th2 model dominated the allergy-science field for more than a decade, while in 2003 a third major Th cell type, derived from naive CD4+ T-cells, was discovered. This newly discovered Th cell type produces both IL-17A and IL-17F, and is therefore designated the Th17 cell (next to Th1 and Th2 cells) (Zhu *et al.*, 2010). Th17 subsets play important roles in neutrophilic inflammation in allergic diseases, since the Th17-induced cytokines IL-17A, IL-17F, and IL-22 induce numerous chemokines and growth factors to promote neutrophil and macrophage accumulation at the site of injury. The products of the neutrophils and macrophages, such as MMP9, promote inflammation, clearance of the infection, and remodeling of the airways (Fig. 2.3) (Alcorn *et al.*, 2010). The polarization of Th17 is controlled by antigen co-stimulation and cytokine-dependent signaling. It was shown that naïve CD4+ T-cell differentiation towards Th17 subsets was triggered by the presence of TGF-β and proinflammatory cytokines IL-1β and IL-6, signaling via Smad2/3, STAT3, nuclear factor-κB (NF-κB), and AP-1, inducing RORγT and RORα expressions and the production of IL-17A and IL-17F, as well as IL-21 and IL-22. Besides, IL-23 is important for TH17 cells, since loss of IL-23 leads to the decrease of IL-17. The signaling of IL-6 and TGF-β synergize to induce the expression of the IL-23 receptor, and IL-23 is responsible for the expansion and survival of Th17, although IL-23 is not necessary for Th17 polarization (Zhu *et al.*, 2010).

In addition to the above-mentioned effector Th cell subsets (Th1, Th2 and Th17), a fourth CD4+ T-cell subset with suppressive effect has been widely studied recently, which is called regulatory T-cell (Treg). These regulatory T-cells are further subdivided into two groups: natural regulatory T-cells (nTregs) and

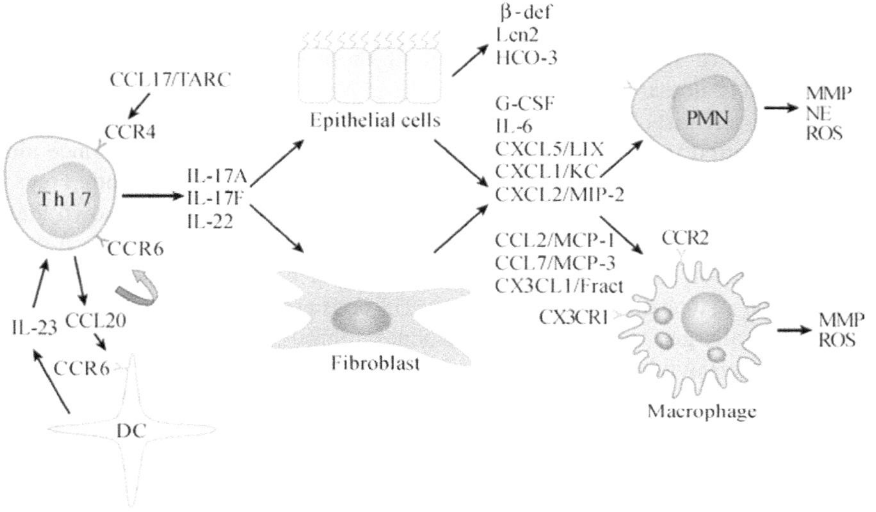

Fig. 2.3. Th17 regulation of inflammation. Th17-induced cytokines IL-17A, IL-17F, and IL-22 induce numerous chemokines and growth factors to promote neutrophil and macrophage accumulation at the site of injury. The neutrophil and macrophage products promote inflammation, clearance of infection, and airway remodeling. In addition, Th17 cytokines stimulate antimicrobial peptide production by the airway epithelium (Alcon *et al.*, 2010). β-def, β-defensin; DC, dendritic cell; Lcn2, lipocalin 2; MMP, matrix metalloproteinase; NE, neutrophil elastase; PMN, polymorphonuclear cell; ROS, reactive oxygen species (With permission of John Wiley and Sons)

induced regulatory T-cells (iTregs) according to their development and phenotype. The phenotypes and functions of nTregs were studied extensively during the last ten years. It is well known that nTregs develop in the thymus. FOXP3 acts as a mast switch transcription factor for the development and function of nTregs. The iTregs are differentiated from naive CD4+ T-cells in the presence of TGF-β and IL-2 (Zheng *et al.*, 2004). Both nTregs and iTregs play a crucial role in controlling and modifying the course of allergies via various signaling pathways on different cells. Tregs directly suppress allergen-specific Th2 cells to produce IL-4, IL-5, and IL-13, which are key cytokines in the development of allergic reactions. Tregs also directly act on inflammatory DC to suppress the activation of effector Th1/Th2/Th17 cells, whereas the function of tolerogenic DC is enhanced. In the effector phase, mast cells, basophils and eosinophils are directly or indirectly inhibited by Tregs, thus down-regulating allergic inflammation. In addition, the induction of Th0/Th1 cells is also suppressed by Tregs. As producers of allergen-specific antibodies, B-cells are directly acted upon by Tregs. The production of allergen-specific IgE is inhibited, while the induction of IgG4 is promoted. Tregs exert these various functions possibly through either a cell-cell contact-dependent or in a cytokine-dependent manner. They might employ a broad range of soluble and membrane-bound suppressor factors, such as IL-10, TGF-β, CTLA-4, programmed death-1 or histamine receptor 2. The exact molecular mechanism of how Tregs execute all of these functions remains to be fully elucidated (Palomares *et al.*, 2010).

Some scientists have also divided iTregs into two subgroups: IL-10-producing Tregs, referred to as Tr1, and TGF-β-producing Tregs named Th3. Since almost all Th cells can produce IL-10 under various circumstances and both Tregs can produce TGF-β, this classification is still under debate. It is well known that the Th3 lymphocytes play a major role in mucosal tolerance through antigen-specific IgA production via TGF-β signaling, but they are not thought to be important in protection against allergens. In contrast, Tr1 cells are thought to have a key role in reducing allergen-specific T-cell responses through the production of IL-10 (Lloyd and Hawrylowicz, 2009; Zhu and Paul, 2010).

In fact, the relationships between Tregs and other Th subsets (Th1/Th2/Th17) are still mysterious. Some studies indicate that master transcription factors T-bet, GATA3, and RORγt for Th1, Th2, and Th17 cells, respectively, can also be co-expressed in some Tregs. Therefore, Tregs with different combinations of transcription factors possibly can exert distinct regulatory functions. Additionally, Tregs may differentiate into effector Th cells (Th1/Th2/Th17) when the expressions of Th1, Th2, and Th17 master transcription factors are up-regulated and foxp3 is down-regulated (Zhu *et al.*, 2010).

2.7 B Lymphocytes

Approximately 15% of peripheral blood leukocytes are B lymphocytes (Chaplin, 2010). Like T lymphocytes, B-cells also play important roles in adaptive immunity. Since B-cells are the producers of IgE antibodies, the key immunoglobulin in allergic reaction, they are definitively crucial players in different stages of allergies.

Typically, the activation of B-cells requires two signals: the first are the specific antigen epitopes that are captured by the B-cell receptors (BCRs), leading to clustering of BCRs; the second is the co-stimulatory signal provided by CD4+ T-cells. The most important one is CD40L from the helper T-cell interaction with CD40 from the B-cell, the essential co-stimulatory signal for B-cell activation with T-cell help. After the activation of B-cells, they begin to produce IgM antibodies, which are the major immunoglobulin in the early immune response to most antigens, followed by an immunoglobulin class switch. This class switch is the result of a specialized DNA rearrangement that places a specific Ch (constant region of heavy chain) DNA sequence adjacent to a VDJ exon (Fig. 2.4). It should be pointed out that class switching does not affect antigen specificity as it does not change the structure of the V/D/J exon. The selection of the specific Ch isotype is mainly determined by cytokines. Exposure to IL-4 and IL-13 in the microenvironment promotes the activated B-cells to switch to IgE, while IFN-γ and IL-10 induces switching to IgG1 and TGF-β resulting in IgA production (Zhang, 2003). Once Ig class switching occurs, B-cells continue to divide and produce memory and plasma cells which express a new Ig isotype.

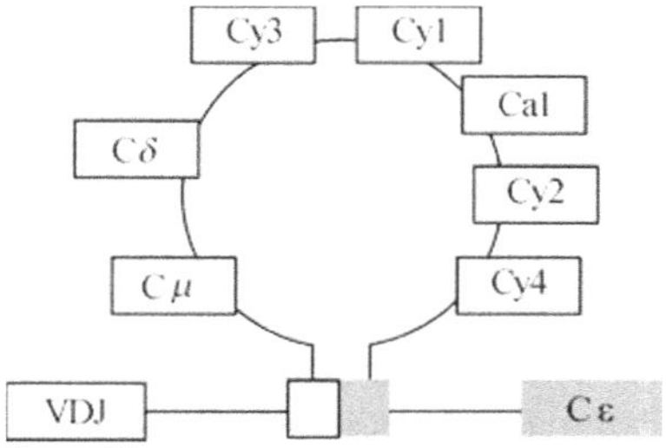

Fig. 2.4. The schematic diagram of Ig class switch to IgE

Therefore, the production of IgE requires three signals: The first is the antigen taken up by APCs (including B-cells); The second is the co-stimulatory signal via CD40L from T helper cells and the third is the soluble help via IL-4 or IL-13 produced by activated T-cells (Fig. 2.5) (Poulsen and Hummelshoj, 2007). Understanding the course of the activation of B-cells and their IgE production might enable to apply more effective ways to control allergic reactions.

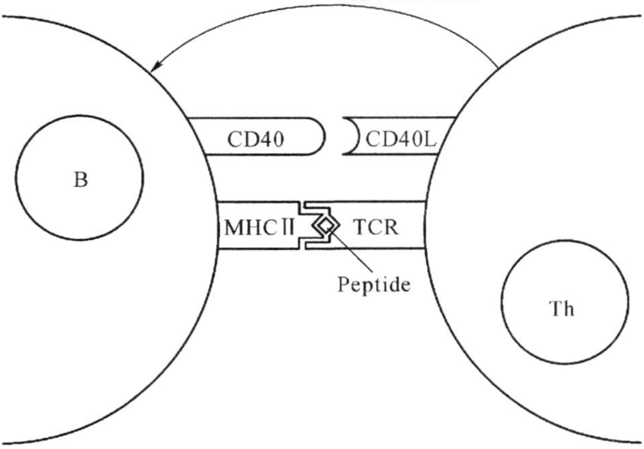

Fig. 2.5. Activation of B lymphocytes with T helper cells

Additionally, B-cells are also a kind of antigen-presenting cells as they express class II MHC proteins on their surface, produce numerous cytokines, and release exosomes (Samitas *et al.*, 2010). Recently, a novel subset of B cells, IL-10-producing regulatory B (Breg) was reported to have suppressive effects on allergen-induced airway hypersensitivity via a IL-10-dependent mechanism, which indicates that B-cells might have regulatory roles in the multiple aspects of immune responses via their production of inflammatory and regulatory cytokines (Amu *et al*., 2010). Up to date, it is not clear that how important the different subsets of B cells act in allergic diseases.

2.8 Conclusion

In this chapter we elucidated the roles of the various inflammatory and immune cells and their effective regulating compounds that are involved in allergies, ranging from sensitization to symptom development. Although these cells are important source of inflammatory mediators, the structural cells such as epithelial cells in the airways and skin also play a critical role in the secretion of inflammatory mediators and in maintaining chronic allergic inflammation (Paul and Zhu, 2010; Ziegler and Artis, 2010). This review clearly demonstrates the enormous complexity of cellular, physiological and biochemical reactions that occur coordinately, but nevertheless should not occur since allergies are an abnormal reaction of the immune system against antigens that should be tolerated. Although knowledge on the immune system is growing very rapidly, it appears that not all functions and their potential deviations are completely well understood. Therefore, further investigation is required when aiming at interruption (by medication or immunomodulation or changing of environmental factors) at any stage of this wrong sequence of cellular events that unfortunately lead to the development of allergic symptoms in a growing number of individuals' world-wide.

References

Alcorn, J.F., Crowe, C.R., Kolls, J.K. (2010). TH17 cells in asthma and COPD. Annu Rev Physiol, 72, 495-516.

Amu, S., Saunders, S.P., Kronenberg, M., Mangan, N.E., Atzberger, A., Fallon, P.G. (2010). Regulatory B-cells prevent and reverse allergic airway inflammation via FoxP3-positive T regulatory cells in a murine model. J Allergy Clin Immunol, 125(5), 1114-1124.

Chaplin, D.D. (2010). Overview of the immune response. J Allergy Clin Immunol, 125(2), S3-S23.

Cooke, R.A., van der Veer, A. (1916). Human sensitization. J Immunol, 1(3), 201-305.

Durrant, D.M., Metzger, D.W. (2010). Emerging roles of T helper subsets in the pathogenesis of asthma. Immunol Invest, 39(4-5), 526-549.

Gell, P.G.H., Coombs, R.R.A. (1963). The Classification of Allergic Reactions

Underlying Disease. Oxford: Blackwell Science.
Gilfillan, A.M., Rivera, J. (2009). The tyrosine kinase network regulating mast cell activation. Immunol Rev, 228(1), 149-169.
Hogan, S.P., Rosenberg, H.F., Moqbel, R., Phipps, S., Foster, P.S., Lacy, P., Kay, A.B., Rothenberg, M.E. (2008). Eosinophils: Biological properties and role in health and disease. Clin Exp Allergy, 38(5), 709-750.
Jiz, M., Friedman, J.F., Leenstra, T., Jarilla, B., Pablo, A., Langdon, G., Pond-Tor, S., Wu, H.W., Manalo, D., Olveda, R., *et al.* (2009). Immunoglobulin E (IgE) responses to paramyosin predict resistance to reinfection with Schistosoma japonicum and are attenuated by IgG4. Infect Immun, 77(5), 2051-2058.
Lloyd, C.M., Hawrylowicz, C.M. (2009). Regulatory T-cells in asthma. Immunity, 31(3), 438-449.
MacGlashan, D. Jr. (2005). IgE and FcepsilonRI regulation. Clin Rev Allergy Immunol, 29(1), 49-60.
Metcalfe, D.D. (2008). Mast cells and mastocytosis. Blood, 112(4), 946-956.
Metcalfe, D.D., Peavy, R.D., Gilfillan, A.M. (2009). Mechanisms of mast cell signaling in anaphylaxis. J Allergy Clin Immunol, 124(4), 639-646; quiz 647-638.
Mosmann, T.R., Cherwinski, H., Bond, M.W., Giedlin, M.A., Coffman, R.L. (1986). Two types of murine helper T-cell clone. I. Definition according to profiles of lymphokine activities and secreted proteins. J Immunol, 136(7), 2348-2357.
Pali-Scholl, I., Jensen-Jarolim, E. (2009). Basic aspects of allergy and hypersensity reations. Allergy Frontiers: Classification and Pathome- chanisms, 2, 3-17.
Palomares, O., Yaman, G., Azkur, A.K., Akkoc, T., Akdis, M., Akdis, C.A. (2010). Role of Treg in immune regulation of allergic diseases. Eur J Immunol, 40(5), 1232-1240.
Paul W.E., Zhu, J. (2010). How are T(H)2-type immune responses initiated and amplified? Nat Rev Immunol, 10(4), 225-235.
Poulsen, L.K., Hummelshoj, L. (2007). Triggers of IgE class switching and allergy development. Ann Med, 39(6), 440-456.
Samitas, K., Lotvall, J., Bossios, A. (2010). B cells: From early development to regulating allergic diseases. Arch Immunol Ther Exp (Warsz), 58(3), 209-225.
Schroede, J.T. (2009). Basophils beyond effector cells of allergic inflammation. Adv Immunol, 101, 123-161.
Siraganian, R.P., Zhang, J., Suzuki, K., Sada, K. (2002). Protein tyrosine kinase Syk in mast cell signaling. Mol Immunol, 38(16-18), 1229-1233.
Sokol, C.L., Medzhitov, R. (2010). Role of basophils in the initiation of Th2 responses. Curr Opin Immunol, 22(1), 73-77.
Stone, K.D., Prussin, C., Metcalfe, D.D. (2010). IgE, mast cells, basophils, and eosinophils. J Allergy Clin Immunol, 125(2 Suppl 2), S73-80.
Yao, W., Barbe-Tuana, F.M., Llapur. C.J., *et al.* (2010). Evaluation of airway reactivity and immune characteristics as risk factors for wheezing early in life. J Allergy Clin Immunol, 126(3), 483-488 e481.

Zhang, K. (2003). Accessibility control and machinery of immunoglobulin class switch recombination. J Leukoc Biol, 73(3), 323-332.

Zheng, S.G., Wang, J.H., Gray, J.D., Soucier, H., Horwitz, D.A. (2004). Natural and induced CD4+CD25+ cells educate CD4+CD25- cells to develop suppressive activity: The role of IL-2, TGF-beta, and IL-10. J Immunol, 172(9), 5213-5221.

Zhu, J., Paul, W.E. (2010). Heterogeneity and plasticity of T helper cells. Cell Res, 20(1), 4-12.

Zhu, J., Yamane, H., Paul, W.E. (2010). Differentiation of effector CD4+ T-cell populations. Annu Rev Immunol, 28, 445-489.

Ziegler, S.F., Artis, D. (2010). Sensing the outside world: TSLP regulates barrier immunity. Nat Immunol, 11(4), 289-293.

3

Multidisciplinary Approaches to Allergy Prevention

Luud J.W.J. Gilissen[1*], **Zhong-Shan Gao**[2], **Zhi Chen**[3]

[1] Allergy Consortium Wageningen, Plant Research International, PO Box 619, 6700 AP Wageniungen, the Netherlands
[2] Allergy Research Center, Zhejiang University, Hangzhou, 310058, China
[3] School of Medicine, Zhejiang University, Hangzhou, 310058, China
[*] E-mail: luud.gilissen@wur.nl

Abstract: Allergies are a complex medical and societal phenomenon. Its prevention requires well-elaborated actions. The classical avoidance rule is: no allergen, no allergy. However, with the high number of allergenic sources of allergens in the out-door and in-house environment and in foods on the one hand, and the many economic-related changes in life-style and food patterns that can bias the balance of the immune system towards allergic sensitivity in the general population or individuals on the other hand, allergen avoidance is a tough task. In this article, we focus on the development of an allergy prevention knowledge framework, running from fundamental research to societal implementation and education. Such a framework can be helpful to identify known and lacking scientific details. On the basis of this framework, adequate prevention strategies can be developed and implemented. It opens ways for cooperation in multi- and interdisciplinary settings at various societal (national and international) levels. Here, we present several examples of allergies and allergy prevention in different stages of the allergy prevention knowledge framework in which we distinguish five phases: (1) fundamental research; (2) designing research; (3) development of a prevention strategy; (4) implementation of the strategy into society; and (5) dissemination of acquired knowledge. Genetic and genomic approaches, also including the aspects of epigenetics, are generally at the level of fundamental research. The application of traditional medicines in allergy treatment and prevention has been implemented in Asian societies, and in many cases the scientific evidence (necessary for implementation in western societies) is beginning to accumulate. Regarding food allergies, labeling of allergens can result in the so-called "may contain" labeling that sometimes is an advantage regarding food safety, but mostly is a disadvantage to the individual allergic consumer as it may reduce the choice in food products unnecessarily. The gluten case, related to celiac disease (gluten intolerance) will be further elaborated on with the "oats issue". Regarding environmentally caused allergies, we focus on hay fever caused by pollen grains and on the development of "pollen calendars". The examples will be evaluated against the allergy prevention knowledge framework. In all, each allergen and each allergenic source requires its own prevention strategy.

3.1 Introduction

Allergy is a collective noun for aberrant and violent reactions of the immune system towards generally harmless compounds, mostly proteins, which should normally be tolerated by the immune system. Allergens from different origins in the outdoor and indoor environment (airborne and solid materials) and in foods can cause a variety of symptoms in the skin, the airway system and the intestine. Individuals can be sensitive to a single allergen, but many allergic patients react to more than one allergen. Asthma, hay fever, eczema, food and contact allergies, and sometimes combinations of these, are the most common symptoms. In a genetically predisposed individual, an allergy develops along two phases: (1) the individual becomes sensitized towards an allergen; (2) after repeated contacts, symptoms (inflammation reactions) occur.

Allergy is a multifactorial disease with high impact on the society. Several westernized countries now have a stabilized high allergy prevalence. The occurrence of allergies is often seen as the other side of the coin of prosperity. Reduced infections during childhood due to vaccinations, the application of antibiotics, and the highly increased hygiene status have saved the lives of many children on the one hand, however, the increase of smoking by pregnant women, the increase of stress and environment pollutions on the other hand make the immune system in a large part of the population, more susceptible to becoming allergic. Nowadays, as a result, over 25% of the industrialized population world-wide (Europe, USA, Japan) suffer from allergies. Since industrialization is rapidly growing in countries like China and other South-East Asian countries, India, Brazil and other South-American economies, these countries are expected to be on the eve of an allergic epidemic. The allergy-related economic burden in Europe is large and amounts up to many tens of billions Euros annually. The upcoming economies will not be safeguarded against such a burden. However, based on current knowledge about allergy development in Western countries, effective strategies can now be developed in upcoming economic countries to alleviate the large-scale allergy problems to be expected.

Allergy prevention strategies can start from two angles: from the sensitized individual and population on the one hand, and from the allergenic proteins on the other hand. Regarding symptom development, three strategies for prevention can be distinguished (adapted from the European Allergy White Paper, 1997, with additions):

Primary prevention addresses symptom-free children at risk. The risk of allergy can easily be assessed from the family history because genetics have a significant contribution. Sensitization can be counteracted by avoidance of potent allergens, such as hen's eggs, cow's milk, wheat bread, peanut butter, etc during the first months of life. Breast feeding is recommended. Exposure of infants to house dust mites, cats and tobacco smoke should be avoided. Smoking of women during pregnancy is a high risk factor, as is tobacco smoke to the newborn. Also vaccinations and the use of antibiotics can increase the risk of sensitization, as bacterial infections during childhood may naturally boost the robustness of the immune system.

Secondary prevention addresses sensitized individuals. Symptom development can be prevented by avoidance of the offending allergens in the environment and food. Allergen labeling on packaged foods is a straight forward procedure and currently obligatory by law in Europe. However, "may contain" labeling may pass the goal when applied to safeguard the producer and not to inform the consumer. Further, immunotherapy has been proven to be successful in cases of insect venom allergies, seasonal allergic rhinitis, and allergic asthma.

Tertiary prevention includes medical treatment, especially in cases of chronic symptoms.

Regarding primary and secondary prevention, the rule is clear: no allergen, no allergy. The cellular and immunological mechanisms by which an allergy develops in an individual are largely known, running from antigen presenting cells to Th2 helper cells, to B-cells and allergen-specific IgE production, to mast cell loading with IgE and mast cell degranulation upon IgE bridging by repeated allergen contact. In contrast, the individual (genetic, epigenetic and physiological) factors and the enhancing environmental factors that first lead to sensitization and then results in allergen-specific symptoms upon repeated contacts with the allergenic compound are still unclear to a large extent. However, recent developments in human genome research, the developments in novel sequencing and DNA chip array technology, genome-wide association studies (GWAS) and the rapid progress in bioinformatics have now had a revolutionizing stimulus on the discovery of disease-related genes and genotyping, also in the field of asthma and allergy (Vercelli, 2008; Weidinger *et al.*, 2008; Baye *et al.*, 2010). Nevertheless, in many cases and due to the complexity of measures to be taken, it will be too late for many individuals to apply primary prevention strategies: the mischief is done. Therefore, starting from daily practice, secondary prevention by contact avoidance with allergenic compounds from the environment and food will be in general the best-manageable strategy.

In this chapter, we will focus on recent research and industrial developments that have the potential to, or do already contribute to allergy prevention. From various widely divergent examples, we will demonstrate the complexity and multidisciplinarity of the allergy problem regarding the strategies for prevention. First we will deal with molecular genomics and genetics research going on in allergies. Then we will focus on allergy-related epigenetic issues and the application of traditional medicines in allergy prevention, care and cure and their potentials for application in Western countries. Regarding allergen-free foods, we give some details on current allergen labeling in Europe, including the issue of "may contain" labeling, and will elaborate on one of the most relevant "allergens" on the labeling list: gluten. Here we will explain about Celiac Disease (the largest food intolerance/allergy world-wide), the search for safe wheat and the potentials of oats as a safe food for the celiac patients. Then we move on to the allergen-free environment, and in the end we will elaborate on the dynamics in the allergy prevention knowledge framework, from fundamental science to communication and education. Based on these examples we will further discuss how to make a stand against the allergy epidemic by interdisciplinary cooperation based on multidisciplinary knowledge and its implementation in society. We have to bear in mind that each allergen will pose its own requirements in allergy prevention.

3.2 Genetic and Genomic Approaches

With the massive amount of genetic and genomic data becoming available, clinicians and researchers have the opportunity to complement and integrate their daily practice with these data to clarify the underlying causes of complex phenotypes and discover susceptibility genes in allergic diseases to make better assessments and predictions on the risks of an individual to become allergic. These researches considerably complement the data on family history, which is a predicting tool in allergy assessment and primary prevention. Allergic diseases have an undeniably genetic component, as can be observed from familial aggregation of skin allergies and asthma. A child has a 33% chance of developing an allergy if one parent is allergic, which increases to a 77% chance if both parents are allergic (Steinke, 2009).

There are two basic statistical approaches to identify genetic variants corresponding to a genetically determined disease. One approach involves linkage analysis with family data using molecular markers, and the other uses association analysis through linkage disequilibrium (LD). LD is based on the degree of association between alleles at two linked loci, which reflects, in part, their proximity and the corresponding probability of recombination breaking the haplotype on which they are found. LD exploits the recombination over many generations and does not require the collection of pedigrees (Collins, 2007). Mapping of candidate genes and analysis of allelic variation at the SNP and higher levels are commonly used in both approaches. Currently, LD is a major genetic approach to finding the genetic factors for various diseases. High density SNP chips facilitate the power and accuracy of LD.

3.2.1 Food Allergy

The genetics of food allergies were already reviewed in 1997, in which 5q31.1-33, 6p21.3, 11q13.14 genomic regions were suggested to be involved (Bonini and Ruffilli, 1997). Since then, only a few convincing reports have been published that pointed to specific genes related to food allergies. For specific populations, a clear understanding is necessary about the key allergens from a broad array of related and unrelated foods that are able to initiate IgE production and cross-reactivity. Some food allergies are restricted to a specific population, e.g., peanut allergy to Europe and North America, whereas others are universal, such as peach allergy. The various food allergic phenotypes should first be classified accurately before undertaking initiatives to dissect the underlying genetic human variations. Such phenotypes might become increasingly visible by new genomic tools for genomic and expression transcription analysis, like bead chip arrays, to survey genetic variation on a large scale. Through bioinformatics and computational biology, candidate genes and alleles can be extensively tested and verified further aiming at the development of functional genomics to better understand the complexity of the mechanism of the development of a certain food allergy.

3.2.2 Allergic Asthma

In contrast to food allergies, the genetics of allergic asthma has gained much interest. Numerous studies have been carried out worldwide since epidemiological data showed the consistent involvement of a genetic factor for asthma development. In the 1990s, researches on susceptibility genes/alleles of asthma-related traits through the candidate genes approach, resulted in the discovery of more than 25 genes (Ober and Hoffjan, 2005). By means of genome-wide linkage and association mapping, seven positional candidate genes ADAM33, PHF11, DPP10, GPR154, HLA-G, CYFIP2 (Laitinen, 2007) and ORMDL3 (Moffatt *et al.*, 2007) have been identified. The most important gene variants for asthma are those polymorphisms that exert their influence on the network system controlling biological responses to asthma-related exposures, e.g., to house dust mite (Martinez, 2007). Independent confirmation studies are rare but essential to unravel the real involvement of candidate genes and alleles in the key biological pathways of asthma development and to draw a clear picture of the genetic control of this complex disease. Recent new findings from researches on basic allergy mechanisms, such as the involvement of IL-17 (Schnyder-Candrian *et al.*, 2006), IL-4 (Yoshimoto *et al.*, 2009; Tachdjian *et al.*, 2009), IL33 (Pushparaj *et al.*, 2009), the TLR-4 factor (Hammad *et al.*, 2009) and the Th2 response to protease allergens (Sokol *et al.*, 2008; Perrigoue *et al.*, 2009; Charles *et al.*, 2009) will all contribute to the genetic analysis of susceptibility alleles in allergic asthma.

3.2.3 Skin Allergy

Skin allergy diseases are attributed to the occurrence of a general skin barrier defect to resist the specific allergens. Recent reports about the loss-of-function variants of filaggrin (filament aggregation protein) gene (FLG, 1q21) are found to be responsible for ichthyosis vulgaris and atopic dermatitis (AD) (Palmer *et al.*, 2006; McGrath and Uitto, 2008; O'Regan *et al.*, 2009). It is suggested now that FLG plays a key role in maintaining an effective skin barrier against the external environment. This breakthrough finding has greatly improved our understanding of the genetic predisposing factor of complex dermatitis and possibly related atopic asthma.

Future dissection of genetic and environmental factors for allergy will rely on further data mining of the genetic information and better definitions of phenotypes into specific or general clinical allergy symptoms, a challenging task that requires the analysis and comparison of huge amounts of DNA polymorphisms by powerful computers and sophisticated software. The identified SNP variants should then be checked for their exact physical location, in which gene with what biologic relevance, and in which population being present. We can expect breakthrough results to be discovered in coming years.

In what sense can the genetic study be applied in prevention? Knowledge on susceptibility alleles in combination with information of seasonal and geographic allergen load and distribution may become helpful as early warning systems and can contribute to taking intervening measures such as a proper choice for living and working place, travelling schedules, holiday areas, specific foods, etc.

3.3 The Potential of Epigenetic Approaches

"Genetics" times "Environment" determines the organism's phenotype. Each individual is specific regarding his genetic make-up. Humane genome variation includes all of the genetic characteristics within the human species, with genetic variations occurring within and among populations. The basis of genetic diversity became better understood from the recent full sequencing of individual genomes (Levy *et al.*, 2007). Variations in the genetic make-up at the level of individual cells as well as organisms include the coding and non-coding variability, the copy number variation, the gene expression profile and the gene-to-gene interaction networks. Among human beings, a genetic variation of 0.5% is estimated. This difference amounts to a significant number of distinct genetic traits that uniquely distinguish the genome of every person and contribute to unique and distinct responses to environmental exposures (including nutrition, skin contact and inhalation factors, and to risks for disease development and their (pharmacological) treatment. Regarding allergies, individual factors such as age and gender are relevant. Environmental factors include the physiological status (like body mass, microbiome) and physico-chemical factors (e.g., diet, exposures to tobacco smoke, toxins including drugs pollution, pathogens, allergens, stress and other physico-social factors) (Baye *et al.*, 2010). Upon and partly as a result of these factors, epigenetics plays a key role. Epigenetic variation does not affect the underlying DNA, but just modifies its expression through covalent modif—-cations such as DNA methylation, histone modification and through microRNAs. Such modifications are cell and tissue specific, and in the immune system, with its T-cell differentiation, epigenetic regulation occurs at many levels (Janson *et al.*, 2009). Epigenetic effects on gene expression can persist and be passed through mitosis and meiosis to the next generation as was seen in the 5'-CpG island in the ACSL3 allele associated with the risk of asthma, and in children born to mothers exposed to air pollutants (Perera *et al.*, 2009). The role of epigenetics in allergic disease is suggested to become increasingly important, but current knowledge is still limited (Baye *et al.*, 2010).

How can epigenetics be applied in allergy prevention? Vuillermin *et al.* (2009) mentioned that the period of immune programming during early life presents a critical window of opportunity, because epigenetic changes are known to be sensitive to environmental factors, and may therefore provide a mechanistic link between environmental cues, inappropriate immune programming, and the risk of allergic disease through preventive and therapeutic interventions. They focus especially on the microbial exposure, IFN-gamma gene demethylation in naïve T cells and the risk of allergic disease. They also suggest that low microbial exposure during early life increases the risk of allergy by activation of the IFN-gamma gene in these T cells. In this regard, the study of Seiskari *et al.* (2007) is relevant on allergic sensitization and microbial load in genetically comparable populations in Finland and Russian Karelia. From their results it is suggested that one possible way of allergy prevention, especially to prevent sensitization, may lead through the introduction (through oral vaccination in the same way as the mother "vaccinates" her child during the regular birth process) of microbial factors to the newborn.

3.4 The Application of Traditional Medicine

In the treatment of allergic symptoms, conventional Western drugs such as corticosteroids and β-agonists drugs are often used. Such treatments are not curative and withdrawal of these drugs is often accompanied with moderate or even severe relapse of disease symptoms. Many patients and doctors are also concerned about their side effects. Therefore, during the last two decades, complementary and alternative medicine (CAM) has gained much attention from both basic science research and in clinical trials. Hundreds of research papers have been dedicated to the potential application of traditional medicines in allergy prevention, care and cure, and their immunomodulatory effects. Traditional medicines have been used for thousands of years in China (TCM), India (Ayurveda), Japan and Korea. TCM is well known world-wide (Li *et al.*, 2007).

TCM is generally thought to be safe to most individuals of the Chinese population. There are different formulations to treat asthma, allergic rhinitis, atopic dermatitis and food allergy as described in old Chinese medicine literature. In 2000, a pioneer research was published to explore the potentials of the 14-herb MSSM-002 formulation to treat asthma and to provide evidences according to the standard Western scientific method (Li *et al.*, 2000). MSSM-0002 showed the down-regulation of Th2 response through inhibition of the Th2 specific transcription factor GATA-3 (Li *et al.*, 2000). Later on, a simplified formulation called antiasthma herbal medicine intervention (ASHMI) consisting of 3 herbs, Gan-Cao (*Glycyrrhiza uralensis*), Ku-Shen (*Sophorae flavescentis*) and Ling Zhi (*Ganoderma lucidum*), was derived from MSSM-002. The efficacy and tolerability of ASHMI was proved by an investigation on 91 patients in China compared with an oral prednisone therapy (Wen *et al.*, 2005). Further researches on the mechanism underlying ASHMI by using murine models were carried out in the last two years. It was demonstrated that its immunomodulatory effects included decreased numbers of peripheral blood eosinophils, reduced serum IgE and Th2 cytokines (IL-5, IL-13), blocked early phase airway reactions, and lowered histamine and leukotriene releases (Li and Brown, 2009). These effects were also associated with the secretion of T regulatory cytokines and the potent muscle relaxant prostaglandin PGI2 by smooth muscle cells in a murine model challenged with ovalbumin (Zhang *et al.*, 2010). This formulation modulates Th1 and Th2 responses, and induces long-lasting tolerance to allergen exposure. The key factor in the ASHMI effect is the production of interferon-γ (Srivastava *et al.*, 2010). From Gan-Cao (*Glycyrrhiza uralensis*), three more effective licorice flavonoids, liquiritigenin, isoliquiritigenin, and 7,4′-dihydroxyflavone, were identified that inhibited exotaxin-1 secretion in human fetal lung fibroblasts *in vitro* (Jayaprakasam *et al.*, 2009). In addition, ASHMI trials on 5-14 year-old children also showed greater symptom improvement (particularly regarding the nasal symptoms) as compared to the standard groups (Li, 2009).

The food allergy herbal formulation 1 (FAHF1) in the peanut allergy mouse model showed blocking effects on systemic anaphylactic signs and reduced mast cell degranulation and histamine release, reduced peanut-specific IgE levels and lower Th2 cytokine secretion (Li *et al.*, 2001). Further, a new FAHF-2 formulation (from which two herbs were eliminated) applied to a murine model of peanut allergy also completely blocked anaphylactic reactions, significantly reduced IL-4,IL-5and IL-13 production, enhanced IFN-γ production, and suppressed Th2 response after a repeated peanut stimulation (Srivastava *et al.*, 2005). FAHF-2 is also effective to treat multiple other food allergies, apart from peanut (Srivastava *et al.*, 2009), FAHF-2 applied to purified PBMCs, resulted in an antigen-dependent T-cell proliferation response, and selectively suppressed Th2-cytokine production. Phases II and III of this trial are in progress for both ASHMI and FAHF-2, and may become the first generation of commercial anti-asthma and anti-food allergy TCM products with sound science-based evidences.

Other reported formulations include modified Mai-dong-men-dong-tang (mMMDT) against persistent mild to moderate asthma (Hsu *et al.*, 2005), and Ding Chuan Tang (DCT) that reduces airway hyperreactivity (AHR) in stabilized asthmatic children (Chan *et al.*, 2006), STA-1 which is combined with mMMDT and LWDHW (Chang *et al.*, 2006), reduced total IgE and specific IgE. *Sophora flavescentis* Ait (Ku Shen) extract works as an excitatory modulator and is a safe and effective treatment against chronic asthma, which also could reduce or even eliminate the use of corticosteroids and β-agonists (Hoang *et al.*, 2007). *Citrus unshiu* powder demonstrated relief from seasonal allergic rhinitis to Japanese cedar pollen because of three effective flavonoids (Kobayashi and Tanabe, 2006). A TCM treatment of 5 herbs to children with moderate to severe atopic dermatitis resulted in a decrease of corticosteroid treatment by one third (Hon *et al.*, 2007). In an animal model, *Actinidia arguta* improved dermatitis skin lesions (Park *et al.*, 2007). Bu-zong-yi-qi-tang, composed of 10 herbs, was used to treat allergic rhinitis and atopic dermatitis in which it showed synergistic effects (Yang and Yu, 2008). Future trials should define the applicability in different allergic disease profiles caused by different allergenic sources and specific allergenic molecules to avoid the application of different complex mixtures to treat various disease severities and allergen exposures. In addition, the key effective ingredients should be identified in the near future. Kang *et al.* (2004) and Shen *et al.* (2008) reported a single-herb injection (*Astragalus membranaceus*) that prevented AHR in a mice model due to Th2 response inhibition. The formula Guo Min Kang was used to treat human male immune-related infertility with a success rate of 83% (35/42) in China (Wang *et al.*, 2005). This formulation was recently tested in an antigen (conalbumin) induced anaphylaxis mouse model to show its effects on reduction of symptoms, IgE level, histamine release and the number of degranulated mast cells (Li *et al.*, 2009).

A recent comprehensive review on complementary and alternative medicine(CAM) by Mainardi *et al.* (2009) stresses the importance for applications in the areas of allergy and immunology. CAM was recommended editorially in the field of allergy and clinical immunology (Sampson, 2009) and was highlighted as a new advance in 2009 (Peden and Bush, 2010). It is expected that more in-depth pharmacological and immunological ways of action of the selected TCM formulations or of their effective components will be elucidated in the near future.

3.5 Allergen-Free Foods

There are two methods to classify allergens. The first one is based on the biology of the protein itself positioned in a biological organism and related to a taxonomical context. For such classification, the Allergen Nomenclature Subcommittee of the International Union of Immunological Societies (IUIS) have implemented a set of rules. This list includes a long sub-list of allergens related to foods. Many proteins called allergens in this sub-list have been identified as being able to provoke an allergic reaction in one or more cases. Some allergens are very rare and almost harmless; others are highly frequent and severe. Based on this frequency of occurrence and severity in foods, the European Commission has made a separate classification, listing fourteen allergenic foods (compounds) (www.foodallergens.info/Legal/Labelling/FoodList.html). These are:

Cereals containing gluten (wheat, rye, barley, spelt, kamut or their hybridized strains; for oats that was initially also on this list, the new EC-Regulation 41/2009 came into force that allows oat products containing less than 20 ppm gluten to be sold as "gluten free" and to be labeled as such; see also below)

Crustaceans and products thereof

Eggs and products thereof

Fish and products thereof

Peanuts and products thereof

Soybeans and products thereof

Milk and products thereof (including lactose)

Nuts, i.e., almonds, hazelnuts, walnuts, cashews, pecan nuts, Brazil nuts, pistachio nuts, macadamia nuts and Queensland nuts and product thereof

Celery and products thereof

Mustard and products thereof

Sesame seeds and products thereof

Sulfur dioxice and sulfphites at concentrations of more than 10 mg/kg or 10 mg/L expressed as SO2

Lupin and products thereof

Molluscs and products thereof

3.5.1 "May Contain" Labeling

Because of the impact of (food) allergies on society, prevention is also a task of governments. The EU has made the prevention of food related allergies as one of its main concerns, resulting in the Allergen Directive (www.foodallergens.info/Legal/Labelling/FoodList.html), enabling consumers to make a healthy choice based on the information on the food product label offers. This information must be clear and must refer to the presence in the product of one or more of the allergenic foods or any product derived from them.

However, food allergic consumers are increasingly confronted with "may contain" indications referring to the residual or potential risk of allergen cross-contamination. The "may contain" labeling is unclear regarding the real presence of a specific allergen and can unnecessarily negatively affect the consumer's informed product choice. The food labeling policy involves two factors that can be clustered into two main groups: (1) the public-administrative chain, and (2) the agri-food chain. The first cluster is characterized by a hierarchical chain of delegated responsibilities, has policy-driven factors and consists of factors which are directly involved in the quality of the legal framework. The other cluster is characterized by a chain of factors that is product-driven with a shared responsibility regarding the quality of the food product. Each cluster has responsibilities for the implementation, execution and the compliance with the EU labeling policies and in subsequent Allergen Directives. A key issue for successful food safety policy is alignment and adjustment of the processes in both clusters. Because of the ambiguity in the management of the risk of allergen cross-contamination in food products, the food industry falters between two options. One is that the food industry acknowledges its responsibility and seeks to inform the consumer of a potential or residual risk. The other is that the food industry protects itself and applies the "may contain" labeling as a disclaimer for liability leaving the choice and the responsibility to the consumer.

Currently, European authorities still lack the legal instruments to enhance the food safety regime (Manders, 2007). In attempting to develop new national and international regulations for allergen labeling, e.g., in China, such ambiguous handling of labeling directives should be prevented.

3.5.2 Gluten

Gluten, causing Celiac Disease (CD), can serve here as a well-documented example. Celiac disease is a food-related inflammatory disorder of the small intestine caused by the ingestion of gluten proteins from wheat, barley and rye in genetically predisposed individuals. This "allergy" is not mediated by IgE, but by T-cells. In contrast to others, true food allergies that often have acute responses, the symptoms of CD are chronic. In children, the main symptoms are chronic bowel ache, diarrhea and retardation of growth. In adults, chronic fatigue, headache, bowel complaints, reduced fertility, miscarriage, dermatitis herpetiformis, osteoporosis and even, in rare cases, intestinal cancer (lymphoma) can occur. Currently, as a secondary preventative strategy, a life-long gluten-free diet is the only treatment.

Around 1% of the Western population suffers from CD, which makes CD one of the largest food-related diseases. Unfortunately, the vast majority of these people are not even aware of having the disease due to large-scale under-diagnosis or wrong diagnosis. These people continue with their daily unhealthy diet, which maintains or even worsens their problems. The prevalence and risk of death, measured in large numbers of undiagnosed individuals based on blood serum markers and epidemiological data, has increased four times during the last five decades (Rubio-Tapia *et al.*, 2009). The prevalence of CD in China, where wheat consumption is rapidly growing, is still unknown. The HLA DQ type of the T-cell receptors that responds to specific gluten protein fractions (epitopes) is crucial, since CD is only found in individuals that are DQ2 (95%) or DQ8 (5%). The first initiative to measure the prevalence of CD in China by testing blood serum markers in large numbers of young adults has recently been taken (Chen Hongbing, personal communication).

A gluten-free diet (as any allergen-free diet in general) is a form of secondary prevention. Such diet is not widely available because it is more expensive than a regular diet and many gluten-free foods (especially gluten-free bread products) appear to have a low palatability. In the Western world, it is very difficult to avoid gluten in food products, because wheat and gluten are detected in almost 30% of labeled products in supermarkets (Atchison *et al.*, 2010). In some of these products, the connection to wheat was visible, while in others invisible and even unexpected, such as canned vegetables, milk, meat and seafood. Gluten proteins have very specific characteristics due to the occurrence of repetitive domains, their high content of proline and glutamine, the presence of several cysteine residues with their specific sulphide groups through which large molecular networks can be formed, and their insolubility in water. These characteristics make gluten a highly useful and versatile group of proteins with numerous applications in the food industry, especially in bread making, and as a binder. In addition, gluten is a by-product from the wheat starch industry and, therefore, easy to obtain and very cheap. One of its disadvantages is the low nutritional value. They have a low digestibility, and if glutens are degraded, they end up in several peptides with immunogenic activity (with gluten- intolerance as a common inconvenience).

Two types of gluten proteins are distinguished: gliadins and glutenins, both with several subgroups consisting of various family members. A single wheat variety can express many gluten proteins. Different gluten proteins have different CD toxicities (some are even non-CD-toxic). Gliadins generally contain the highest number of CD-toxic fragments. The variation in CD-toxicity among the various gluten proteins opens ways to modulate the CD-toxicity of wheat, through selection of low-toxic varieties, through breeding, and through silencing of the genes coding for the toxic proteins. This is a huge task.

The increased gluten consumption, as well as the increased celiac disease toxicity of gluten proteins in modern wheat varieties compared to old varieties, is suggested to be a major current cause of its increased prevalence (van den Broeck *et al.*, 2010a). As a consequence, a vast reduction of the total CD-toxic gluten load in all food products could be a primary preventative method to reduce the initiation of the disease process. Within the framework of a large and multidisciplinary research consortium (the Dutch Celiac Disease Consortium, including medical and agricultural academic research partners, together with food and diagnostic industrial partners, and the Dutch Celiac Disease Patient Society) two strategies are now elaborated. The first strategy, to produce safe wheat varieties, is directed towards elimination of intrinsic CD-toxic sequences (epitopes) from gluten proteins in wheat. By selection, a few varieties with reduced CD-toxicity have now been identified (van den Broeck *et al.*, 2010b). These varieties will be tested for industrial quality and may contribute to the development of safer foods for celiac patients. Gluten proteins appear to be removed completely in several deletion lines of bread wheat lacking the chromosome parts of the coding gluten genes. Such deletions may have implications for the maintenance of the technological properties of the given wheat variety (van den Broeck *et al.*, 2009). Through gene silencing using the RNA-interference (RNAi) approach, the expression of gliadin genes can strongly be reduced more specifically (Becker *et al.*, 2006; Gil-Humanes *et al.*, 2010). In general, "low CD-toxicity" should become a new breeding target, either by classical breeding, or by genetic modification like the RNAi approach or other, more specific technologies, like the application of zinc finger nucleases. In all cases of low- or non-CD-toxicity, the need for wheat with a high technological quality (especially for bread making and for various applications in the production of processed foods) should be met.

The introduction of "low CD-toxicity" as a new breeding target in the agricultural setting meets two further challenges: the consumer's attitude towards genetically modified (GM) wheat, and the organization of agriculture. In Europe, GM foods are still a matter of debate, but there seems to be some movement in this field. Although conventional breeding is still the first choice of most consumers, acceptance of hypoallergenic products was found to be relatively high in consumers that perceive a personal benefit (Schenk *et al.*, 2008; van der Meer *et al.*, Chapter 24 of this book). Further, regarding agriculture, the less-or non-CD-toxic wheat varieties are required to be cultivated in complete separation and under strictly controlled (hazard assessments of critical control points, HACCP) production systems to avoid contamination with regular (high CD-toxic) wheat, rye and barley seeds or gluten.

All these facts summarized in the above paragraphs clearly indicate the complexity of gluten problems, and the large number of stakeholders involved in this gluten dilemma, starting with grain breeders, farmers, bakers and retail companies on the one hand, and then the patients and medical stakeholders on the other hand, with various research groups in immunology, food and agriculture as a third party. The major challenge for all these groups now is to answer the crucial question as to whether the high food industrial quality and applicability of gluten can be balanced to a low or absent CD-toxicity.

In the above paragraphs, we presented the prevention of gluten intolerance as an example. Similar considerations, for other occasionally less extensively occurring but in some cases more severe and acute allergens (e.g., peanut, seafood) need to be made with regard to the development of many other "allergen free" food production chains.

3.5.3 Oats

For CD patients, oats are a good replacement for wheat, rye and barley, and they serve as an important supplement of a patient's daily diet. In fact, the consumption of oats can be recommended for the general population, since they are a good source of slowly degrading carbohydrates, and contain lots of unsaturated fats (especially of the omega-6 and omega-9 type), β-glucan (which contributes to the reduction of cholesterol and can help in the case of heart and vascular diseases) and polyphenols with antioxidant activity. The slow degradation of its carbohydrates makes oats beneficial with respect to obesity and diabetes, as well (Andon and Anderson, 2008). Where the production of marketable CD-safe wheat may take at least ten years to develop, oats represent a short-term solution.

It has been thought for a long time that oats were about as similarly toxic to CD patients as wheat rye and barley. Indeed, in some very rare cases, oat intolerance was found (Arentz-Hansen *et al.*, 2004). It also appeared that some oat varieties have more immunogenic capacity than other varieties (Mujico *et al.*, 2010). During the last decade, numerous papers have been published, proving the safety of oats for the vast majority of CD patients (summarized by Pulido *et al.*, 2009). From these studies, it appeared that the most serious problem for CD patients in consuming oat products is the frequent occurrence of contamination with gluten and gluten-containing cereal material during cultivation, harvest, storage, milling, baking, etc. This requires the development of separate oat production chains. Such CD-safe oat production chains have now recently been established in Scandinavian countries and in the Netherlands (Gilissen *et al.*, 2008; 2010). Such production chains are under the continuous control of HACCP directives. This oats example clearly shows the impact of an indirect effect of an allergen (i.e., gluten) in the food production chain.

3.6 Allergen-Free Environment

The prevalence of allergies to environmental factors is a multitude of the food allergy prevalence. It includes indoor allergen sources, like cats, mites, cockroaches, fungi, and outdoor allergen sources, with pollen from trees, grasses and weeds, and insect stings as the most problematic. Whereas allergenic foods can be manipulated, environmental allergen sources are less easy to treat. Increased hygiene and cleaning, turning the cat out, reduction of carpets and improved ventilation may considerably reduce the indoor allergic problems. However, people suffering from hay-fever are confronted with an apparent elongation (starting time and end time) of the pollen season, as was observed in the Netherlands in the period between 1970 and 2000, probably related to climate change (van Vliet *et al.*, 2002). The severity of hay-fever complaints can be directly related to the amounts of pollen per allergenic species per day in the air. These numbers vary greatly due to annual fluctuations in pollen production by the given species and due to the prevailing daily weather conditions in the pollen season. Pollen allergy can be specific per person: some are highly sensitive to birch pollen and pollen from related fagales species (like hazel and alder), others are more sensitive to pollen of grasses, or weeds. Also combined sensitivities can occur. Especially the pollen from birch, most grass species, and Parietaria, Artemisia and Ambrosia are notorious in North-West Europe for their high allergenicity. In these groups of plants, different allergenic proteins are involved (D'Amato *et al.*, 1998; Radauer and Breiteneder, 2006)

In many European countries, but also in the USA, Canada and Australia, pollen concentrations in the air are counted daily. In 1989, several European organizations, EAACI (European Academy of Allergology and Clinical Immunology), EAN (European Allergy Network) and IAA (International Association for Aerobiology) have decided to jointly publish European pollen calendars. For this purpose, these organizations only use standardized pollen counts, executed according to specific procedures and using specific instruments. The procedure is standardized in three steps: (1) catching the pollen grains from a standardized volume of air using a Burkard or Lanzoni pollen sampler; (2) preparation of the microscopic slides and staining of the pollen grains; (3) determination and counting of the pollen grains on the slide. Pollen numbers per species and per cubic meter are classified as low (below 20 grains), moderate (from 21 to 80 grains), and high (more than 80 grains). Many sites on the Internet give additional information on the severity of allergenicity of the individual pollen species. Combining pollen counts and the weather forecast helps to make good predictions of the hay-fever risk on any given day. This gives hay-fever sufferers good opportunities to take adequate measures for pollen avoidance. The Internet provides many sites for pollen calendars and pollen counts.

3.7 The Allergy Knowledge Framework

The pre-medical allergy prevention strategies can be well organized within a knowledge framework. Such a framework runs from fundamental research to societal implementation and education in several phases (Gilissen *et al.*, 2006):

Fundamental research. It creates understanding of the phenomenon of allergy with regard to the etiology of the disease (sensitization and symptom development) and of the allergenicity of a given source in food and environment (what makes a protein an allergen).

Designing research. The fundamental knowledge can be further used in research for designing tools to better diagnose patients and to change their behavior aiming at allergen avoidance, as well as to better characterize allergenic sources and to design hypoallergenic products and environments.

Development of prevention strategies. Based on the designed tools and identified prerequisites and stakeholders, strategies can now be developed for allergy prevention and management: which efficient and adequate actions should be taken and by which stakeholder.

Implementation. At this phase, the stakeholders become actively involved, e.g., for the development of an allergen-free food production chain, a consumer-friendly database for product scanning in supermarkets, for the green management of city environments, for large-scale agriculture of allergy-safe crops, etc.

Dissemination of knowledge. The development of platforms for the dissemination and exchange of information, including education at different levels completes the knowledge framework to establish a solid infrastructure.

Table 3.1 visualizes the current position of the above examples in the knowledge framework. Such a matrix may be helpful in the identification of known and lacking scientific details, and is helpful to coordinate the various actions to be taken for allergy prevention, specifically or in general. The matrix shows that the phase of development of prevention strategies is on the edge of medicinal/natural sciences on the one hand, and social sciences, governmental responsibilities and daily practice on the other hand. This matrix clearly elucidates the multi-disciplinarity of allergy prevention and management and opens ways for cooperation in interdisciplinary settings. Such settings can be at the level of local initiatives (green management of a school yard; development website information on hypoallergenic foods) or the establishment of big national and international research consortia with industries.

Table 3.1 Position and guessed impact in the Allergy Prevention Knowledge Framework of the various allergies and their specific issues given as examples in this paper

Allergy type	Phase specific issue	Fundamental research	Designing research	Developing prevention strategies	Implementation in society	Dissemination of knowledge
Allergy and asthma – General	Genetics and genomics	+++	+	–	–	+
	Epigenetics	+++	+	+	–	+
	TCM	+	+	+	++	+
Food allergy	Labelling	–	+	+	+++	++
	CD and Gluten	+++	+++	+++	++	++
	Oats (safe chain)	+	++	+++	+	++
Pollen allergy	Pollen calendar	+	+	+++	+++	+++

3.8 Conclusion

Allergy is a big health problem. It seems to be stabalized in most Western countries with a high economic standard, with a prevalence of about 25% of the population. However, in upcoming economies like that of several countries in East and South-East Asia and South-America, a rapid increase in the allergy epidemic is expected. Timely measures towards allergy prevention and allergy management may be helpful to stem the tide and improve the quality of life or even guarantee safe life for many individuals.

Allergy includes a complex combination of immunological deregulations in the skin, the airways and the intestine. Cells of the immune system "see" harmful compounds (proteins) where they should not do so, and respond with the production of specific antibodies (IgEs in IgE-mediated allergies) or T-cells (in food-sensitivities like gluten intolerance) ultimately leading to inflammations. Symptoms can be mild and short-lasting, or chronic, or highly acute and life-threatening, depending on the genetic predisposition of the individual with its specific life-style in its specific environment and depending on the characteristics of the allergen. This determines the individual allergic response against a limited group of proteins in the environment and in food but being present in a large variety of sources and being immunologically active through cross-reactivity. The multifactorial nature of allergy demands for a multifactorial and multi- and interdisciplinary approach for prevention. Each allergen and its source require their own prevention strategy. Let's join hands across the world.

References

Arentz-Hansen, H., Fleckenstein, B., Molberg, Ø., *et al.* (2004). The molecular basis for oat intolerance in patients with celiac disease. PLoS Med, 1(1), e1.

Atchison, J., Head, L., Gates, A. (2010). Wheat as food, wheat as industrial substance; comparative geographies of transformation and mobility. Geoforum, 41, 236-246

Andon, M.B., Anderson, J.W. (2008). The oatmeal-cholesterol connection: 10 years later. Amer J Lifestyle Medicine, 2, 51-57.

Baye, T.M., Martin, L.J., Khurana, Hershey, G.K., *et al.* (2010). Application of genetic/genomic approaches to allergenic disorders. J Allergy Clin Immunol, 126, 425-436.

Becker, D., Folck, A., Knies, H., *et al.* (2006). Silencing the alpha-gliadins in hexaploid bread wheat. In: Gluten Proteins (Lookhart GL, Ng PKW, eds.). St Paul, MN, USA: AACC International.

Bonini, S., Ruffilli, A. (1997). Genetics of food allergy. Environ Toxicol and Pharmacol, 4, 71-78.

Chan, C.K., Kuo, M.L., Shen, J.J., *et al.* (2006). Ding Chuan Tang, a Chinese herb decoction, could improve airway hyper-responsiveness in stabilized asthmatic children: a randomized, double-blind clinical trial. Pediatric Allergy Immunol, 17, 316-322.

Chang, T.T., Huang, C.C., Hsu, C.H. (2006). Clinical evaluation of the Chinese herbal medicine formula STA-1 in the treatment of allergic asthma. Phytotherapy Res, 20, 342-347.

Charles, N., Watford, W.Q.T., Ramos, H.L., *et al.* (2009). Lyn kinase controls basophil GATA-3 trascription factor expression and induction of Th2 cell differentiation. Immunity, 30, 533-543.

Collins, A.R. (2007). Linkage disequilibrium and association mapping. Methods in Molecular Biology. 376: 1-15. In: Linkage Disequilibrium and Association Mapping, an Introduction. Totowa, NJ: Humana Press Inc.

D'Amato, G., Spieksma, F.T., Liccardi, G., *et al.* (1998). Pollen related allergy in Europe. Allergy, 53, 567-578.

Gilissen, L.J.W.J., van den Broeck, H.C., van't Westende, W.P.C., *et al.* (2010). The struggle against Celiac disease. Development of safe foods for Celiac patients–A multidisciplinary approach. Food for Thought, 3, 13-15.

Gil-Humanes, J., Piston, F., Tollefsen, S., *et al.* (2010). Effective shut down in the expression of celiac disease related wheat gliadin T-cell epitopes by RNA interference. PNAS, 107:17023-17028.

Gilissen, L.J.W.J., van der Meer, I.M., Smulders, M.J.M. (2008). Beyond Celeac Disease toxicity: Detoxified and non-toxic grains. In: Frontiers in Celiac Disease (Fasano, A., Troncone, R., Branski, D., eds.). Pediatr Adolesc Med. Basel, Karger, 12, 46-56.

Gilissen, L.J.W.J., Wichers, H.J., Savelkoul, H.F.J., *et al.* (2006). Future developments in allergy prevention: a matter of integrating medical, natural and social sciences. In: Allergy Matters-New Approaches to Allergy Prevention and Management (Gilissen, *et al.* eds.). Dordrecht, the Netherlands: Springer, pp. 3-10.

Hammad, H., Chieppa, M., Perros, F., *et al.* (2009). House dust mite allergen induces asthma via Toll-like receptor 4 triggering of airway structural cells, Nature Med, 15(4), 410.

Hoang, B.X., Shaw, D.G., Levine, S., *et al.* (2007). New approach in asthma treatment using excitatory modulator. Phytotherapy Res, 21, 554-557.

Hon, K.L., Leung, T.F., Ng, P.C., *et al.* (2007). Efficacy and tolerability of a Chinese herbal medicine concoction for treatment of atopic dermatitis: a randomized, double-blind, placebo-controlled study. British J Dermatol, 157, 357-363.

Hsu, C.H., Lu, C.M., Chang, T.T. (2005). Efficacy and safety of modified Mai-Men-Dong-Tang for treatment of allergic asthma. Pediatric Allergy and Immunology, 16, 76-81.

Janson, P.C., Winerdal, M.E., Winqvist, O. (2009). At the crossroads of T helper lineage commitment–epigenetics points the way. Biochem Biophys Acta, 1790: 906-919.

Jayaprakasam, B., Doddaga, S., Wang, R., *et al.* (2009). Licorice flavonoids inhibit eotaxin-1 secretion by human fetal lung fibroblasts *in vitro*. J Agri Food Chem, 57, 820-825.

Kang H., Ahn, K.S., Cho, C., *et al.* (2004). Immunomodulatory effects of Astragali radix extract on murine Thl/Th2 cell lineage development. Biological and Pharmaceutical Bulletin, 27, 1946-1950.

Kobayashi, S., Tanabe, S. (2006). Evaluation of the anti-allergic activity of Citrus unshiu using rat basophilic leukemia RBL-2H3 cells as well as basophils of patients with seasonal allergic rhinitis to pollen. Inter J Mol Med, 17, 511-515.

Laitinen, T. (2007). Gene mapping in asthma-related traits. Methods in Molecular Biology. 376: 213-234. In: Linkage Disequilibrium and Association Mapping. Totowa, NJ: Humana Press Inc.

Levy, S., Sutton, G., Ng, P.C., Feuk, L., *et al.* (2007). The diploid genome sequence of an individual human. PloS Biol, 5, e254.

Li, X.M. (2007). Traditional Chinese herbal remedies or asthma and food allergy. J Allergy Clin Immunol, 120, 25-31.

Li, X.M. (2009). Complementary and alternative medicine in pediatric allergic disorders. Current Opinion Allergy and Clinical Immunology, 9, 161-167.

Li, X.M., Brown, L. (2009). Efficacy and mechanism of action of traditional Chinese medicine for treating asthma and allergy. J Allergy Clin Immunol, 123, 297-306.

Li, X.M., Huang, C.K., Zhang, T.F., *et al.* (2000). The Chinese herbal medicine formula MSSM-002 suppresses allergic airway hyperreactivity and modulates TH1/TH2 responses in a murine model of allergic asthma. J Allergy Clin Immunol, 106, 660-668.

Li, X.M., Zhang, T.F., Huang, C.K., *et al.* (2001). Food allergy herbal formula-1 (FAHF-1) blocks peanut-induced anaphylaxis in a murine model. J Allergy Clin Immunol, 108, 639-646.

Li, X.M., Wang, Q.F., Schofield, B., *et al.* (2009). Modulation of antigen-induced anaphylaxis in mice by a traditional Chinese medicine formula, Guo Min Kang. The American Journal of Chinese Medicine, 37, 113-125.

Mainardi, T., Kapoor, S., Bielory, L. (2009). Complementary and alternative medicine: Herbs, phytochemicals and vitamins and their immunologic effects. J Allergy Clin Immunol, 123, 283-294.

Manders, J. (2007). From policy theory to actuality-an evaluation of the EU allergen labelling policy. MSc Thesis Wageningen Univerity, code PDQ 80439.

Martinez, F.D. (2007). Gene-environment interactions in asthma. Proc Ame Thoracic Society, 4, 26-31.

Mcgrath, J.A., Uitto, J. (2008). The filaggrin story: novel insights into skin-barrier function and disease. Trends in Molecular Medicine, 14, 20-27.

Moffatt, M.F., Kabesch, M., Liang, L.M., *et al.* (2007). Genetic variants regulating ORMDL3 expression contribute to the risk of childhood asthma. Nature. 448: 470-474.

Mujico, J.R., Mitea, C., Gilissen, L.J.W.J., *et al.* (2010). Natural variation in avenin epitopes among oat varieties: Implications for Celiac Disease. J Cereal Sci, 54(1): 8-12.

O'Regan, G.M., Sandilands, A., McLean, W.H., *et al.* (2009). Filaggrin in atopic dermatitis. J Allergy Clin Immunol, 124, R2-6.

Ober, C. and Hoffjan, S. (2005). Asthma genetics 2006: the long and winding road to gene discovery. Genes and Immunol, 7, 95-100.

Palmer, C.N., Irvine, A.D., Terron-Kwiatkowski, A., *et al.* (2006). Common loss-of-function variants of the epidermal barrier protein filaggrin are a major predisposing factor for atopic dermatitis. Nature Genetics, 38, 441-446.

Park, E.J., Park, K.C., Eo, H., *et al.* (2007). Suppression of spontaneous dermatitis in NC/Nga murine model by PG102 isolated from Actinidia arguta. J Investi Dermatol, 127, 1154-1160.

Peden, D.B., Bush, R.K. (2010). Advances in environmental and occupational respiratory diseases in 2009. J Allergy Clin Immunol, 125, 559-562.

Perera, F., Tang, W.Y., Herbstman, J., *et al.* (2009). Relation of DNA methylation of 5'-CpG island of ACSL3 to tranplacental to airborne polycyclic aromatic

hydrocarbons and childhood asthma. PLos One, 4, 2-e4488.
Perrigoue, J.G., Saenz, S., Siracusa, M.C., *et al.* (2009). MHC class II-dependent baophile-CD4 T cell interactions promote TH2 cytoline-dependent immunity. Nature Immunol.
Pulido, O.M., Gillespie, Z., Zarkadas, M., *et al.* (2009). Introduction of oats in the diet of individuals with celiac disease: a systematic review. Advances Food Nutrition Res, 57, 235-285.
Pushparaj, P.N., Tay, H.K., H'ng, S.C., *et al.* (2009). The cytokine interleukin-33 mediates anaphylactic shock. PNAS, 106(24), 9773-9778.
Radauer, C., Breiteneder, H. (2006). Pollen allergens are restricted to few protein families and show distinct patterns of species distribution. J Allergy Clin Immunol, 117, 141-147.
Rubio-Tapia, A., Kyle, R.A., Kaplan, E.L., *et al.* (2009). Increased prevalence and mortality in undiagnosed celiac disease. Gastroenterology, 137(1), 88-93.
Sampson, H.A. (2009). Role of complementary and alternative medicine in the field of allergy and clinical immunology. J Allergy Clin Immunol, 123, 317-318.
Schenk, M.F., Fischer, A.R.H., Frewer, L.J., *et al.* (2008). The influence of perceived benefits on acceptance of GM applications for allergy preention. Health, Risk and Society, 10, 263-282.
Schnyder-Candrian, S., Togbe, D., Couillin, I., *et al.* (2006). Interleukin-17 is a negative regulator of established allergic asthma. J Exp Med, 203(12), 2715-2725.
Seiskari, T., Kondrashova, A., Viskari, H., *et al.* (2007). Allergic sensitization and microbial load–A comparison between Finland and Russian Karelia. Clin Exp Immunol, 148, 47-52.
Shen, H.H., Wang, K., Li, W., *et al.* (2008). Astragalus membranaceus prevents airway hypersensitivity in mice related to Th2 response inhibition. Ethnopharmacology, 116, 363-369.
Sokol, C.L., Barton, G.M., Farr, A.G., *et al.* (2008). A mechanism for the initiation of allergen-induced T helper type 2 responses. Nature Immunol, 9(3), 310-318.
Srivastava, K.D., Kattan, J.D., Zou, Z.M., *et al.* (2005). The Chinese herbal medicine formula FAHF-2 completely blocks anaphylactic reaction in a murine model of peanut allergy. J Allergy Clin Immunol, 115, 171-178.
Srivastava, K., Sampson, H.A., Li, X.M. (2009). The traditional Chinese medicine formula FAHF-2 provides complete protection from anaphylaxis in a murine model of multiple food allergy [abstract]. J Allergy Clin Immunol, 123, S151.
Srivastava, K., Zhang, T., Yang, N., *et al.* (2010). Anti-asthma simplified herbal medicine intervention-induced long-lasting tolerance to allergen exposure in an asthma model is interferon-g, but not transforming growth factor-b dependent. Clin Exp Allergy, 40, 1678-1688.
Steinke, J.W. (2009). Genetics of hypersensitivity. In: Allergy Frontiers: Epigenetics, Allergens and Risk Factors (Pawankar R., *et al.* eds.). Berlin: Springer.

Tachdjian, R., mathias, C., Al Khatib, S., *et al.* (2009). Pathogenicity of a disease-assoicaited human IL-4 receptor allele in experimental asthma. J Exp Med, 206(10), 2191-2204.

van den Broeck, H.C., van Herpen, T.W.J.M., Schuit, C., *et al.* (2009). Removing celiac disease-related gluten proteins from bread wheat while retaining technological properties: A study with Chinese Spring deletion lines. BMC Plant Biology, 9, 41.

van den Broeck, H.C., De Jong, H.C., Salentijn, E.M.J., *et al.* (2010a). Presence of celiac disease epitopes in modern and old hexaploid wheat varieties: wheat breeding may have contributed to increased prevalence of celiac disease. Theor Appl Genet, 121:1527-1539.

van den Broeck, H.C., Chen, H.B., Lacaze, X., *et al.* (2010b). In search of tetraploid wheat accessions reduced in celiac disease-related gluten epitopes. Mol BioSyst, 6, 2206-2213. DOI 10.1039/c0mb00046a.

van Vliet, A.J.H., de Groot, R.S., Overeem, A., Jacobs, A.F.G., Spieksma, F.T.M. (2002). The influence of temperature and climate change on the timing of pollen release in the Netherlands. Int J Climatology, 22, 1757-1767.

Vercelli, D. (2008). Discovering susceptibility genes for asthma and allergy. Nature Reviews-Immunology, 8, 169-182.

Vuillermin, P.J., Ponsonby. A.L., Saffery R., *et al.* (2009). Microbial exposure, interferon gamma gene demethylation in naïve T-cells, and the risk of allergic disease. Allergy, 64, 348-353.

Wang, Q., Luo, Q.F., Zhao, H.W. (2005). Clinical observation on No. 2 Guominkang capsules for treatment of 42 cases of male immune infertility. J Traditional Medicine, 46, 119-121 (in Chinese).

Weidinger, S., Gieger, C., Rodriguez, E., *et al.* (2008). Genome-wide scan on total serum IgE levels identifies FCER1A as novel susceptibility locus. PloS Genetics, 4(8), e1000166.

Wen, M.C., Wei, C.H., Hu, Z.Q., *et al.* (2005). Efficacy and tolerability of anti-asthma herbal medicine intervention in adult patients with moderate-severe allergic asthma. J Allergy Clin Immunol, 116, 517-524.

Yang, S.H., Yu, C.L. (2008). Anti-inflammatory effects of Bu-zhong-yi-qi-tang in patients with perennial allergic rhinitis. J Ethnopharmacology, 115, 104-109.

Yoshimoto, T., Yasuda, K., Tanaka, H., *et al.* (2009). Basophils contribute to TH2-IgE responses *in vivo* via IL-4 production and presentation of peptide-MHC class II complexes to CD4+ T-cells. Nature Immunol, 10(7), 706-712.

Zhang, T.F., Srivastava, K., Wen, M.C., *et al.* (2010). Pharmacology and immunological actions of a herbal medicine ASHMITM on allergic asthma. Phytotherapy Res, 24, 1047-1055.

Part II

Allergenic Sources and Allergens

4

Overview of Allergen Sources in China

Zhao-Wei Yang[1], Zhong-Shan Gao[1,2*], Ying-Tao Ma[1], Shan-Dong Wu[2], Liya Qiao[1,2]

[1] Allergy Research Center/Department of Horticulture, Zhejiang University, Hangzhou, 310029, China
[2] Hangzhou Zheda Dixun Biological Gene Engineering Co. Ltd., Hangzhou, 310013, China
* E-mail: gaozhongshan@zju.edu.cn

Abstract: China is a large country both in area and population size, with diverse landscapes, climates and environments, and its domestic economy, international trade and tourism is rapidly growing. Under such conditions and in line with the above growths, the occurrence of allergy problems is unexpectedly increasing. A scarcity of public, governmental and food-industry knowledge and awareness of allergies and allergenic sources in China may cause a serious burden to the national population as well as to foreign consumers. Numerous allergenic sources have been reported in China, most of which are however only available in Chinese literature and have not yet been systematically reviewed and summarized in English. Here, we present an overview of the most relevant allergen types as well as their sources and the most prominent examples in China.

4.1 Introduction

Records and descriptions from clinical and individual reports of allergenic sources have accumulated substantially over the past 20 years. The identification and characterization of allergens is critical for management and prevention of allergies. In developed western countries, different types of allergenic sources have been well documented and frequently updated in pace with the progress in molecular biology (Burge and Rogers, 2000; Ong *et al.*, 1995; http://www. allergome.org). In parallel with the rapid increase in databases, most of the corresponding allergens in these developed countries have been extensively characterized and officially named by the WHO/IUIS (summary of allergens submitted to the allergen nomenclature subcommittee, http://www.allergen.org).

Current allergen database sources, however, do not provide definitive data or the clinical importance related to specific regions, e.g., China. This is important, since the complex nature of triggering and occurrence of an allergy can be specific for different human populations and individuals, also depending on age and specific conditions of target organs, physical status and a wide range of environmental factors. In addition, most reported pollen and mold allergen sources have not been precisely defined to a single allergen source but to mixtures from Allergo Pharma (Yang *et al.*, 2007).

This chapter, based on a survey of the current literature, presents a comprehensive summary of Chinese allergen sources, classified according to the route of exposure. Aeroallergens from plant and animal sources are classified in systematic order. Potential food allergens are listed according to a general classification rather than to species and cultivars.

4.2 Aeroallergens

Aeroallergens from the natural environment are known to cause various allergic responses, particularly rhinitis (pollinosis), asthma and atopic dermatitis. Aeroallergens that are active through inhalation are classified into three categories: pollen, mold and indoor allergens. The clinical relevance of various aeroallergenic species may vary from region to region, so regional surveys of aeroallergens and their aerobiology are necessary (Nilsson, 1990), as their management should take the geographic variation into consideration. A China-wide aerobiological survey of allergenic pollens and fungi was initiated in the mid-1980s, and continues in particular regions and major cities. Sources of aeroallergens specific to China are reviewed from aerobiological data and medical publications according to their clinical importance (Table 4.1).

Table 4.1 Clinically relevant aeroallergens in China

Family	Species	Common name (Chinese)	Dispersal months	City/ Province	References
Tree ollen					
Arecaceae	*Trachycarpus fortunei*	Palm (棕榈)	3–4	Zhaoqing/ Guangdong	Lü and Zhang, 2001
Betulaceae	*Alnus nepalensis*	Nepal alder (旱冬瓜)	9–12	Kunming/ Yunnan	Yang *et al.*, 2009; Gao,Y. *et al.*, 2007
Betulaceae	*Betula pubescens*	White birch (白桦)	4–6	Haerbin/ Heilongjiang	Li, L.P. *et al.*, 2009
Casuarina-ceae	*Casuarina equisetifolia*	Beach sheoak (木麻黄)	4–5	Haikou/ Hainan	Xie *et al.*, 2005
Cupressa-ceae	*Juniperus chinensis*	Chinese juniper (圆柏)	3–4	Hubei	Pei *et al.*, 2006
Cupressa-ceae	*Platycladus orientalis*	Chinese arbor-vitae (侧柏)	3–4	Hubei	Pei *et al.*, 2006
Fabaceae	*Albizia julibrissin*	Silk tree (合欢)	6–7	Kunming/ Yunnan	Xie *et al.*, 2005
Moraceae	*Broussonetia papyrifera*	Mulberry (构树)	5	Chongqing, Xi'an	Li *et al.*, 2007; Huang *et al.*, 2002
Oleaceae	*Fraxinus chinensis*	Chinese ash (白蜡)	3–4	Beijing, Xi'an	Li, Z.X. *et al.*, 2007
Phyllantha-ceae	*Bischofia javanica*	Bishop wood (重阳木)	4–5	Haikou/ Hainan	Xie *et al.*, 2005
Pinaceae	*Pinus massoniana*	Masson pine (马尾松)	4–5	Fuzhou/ Fujian	Zheng *et al.*, 2003
	Cryptomeria fortunei	Cedar (柳杉)	3	Central China	Xu *et al.*, 2000
	Cryptomeria japonica	Japanese cedar (日本柳杉)	3	Eastern China	Xu *et al.*, 2000
Pinaceae	*Pinus tabuliformis*	Chinese pine (油松)	4–5	Fuzhou/ Fujian	Zheng *et al.*, 2003
Platanaceae	*Platanus orientalis*	Oriental plane tree (梧桐)	4–5	Xi'an/ Shaanxi, Wuhan/ Hubei	Li, Z.X., 2007; Wu *et al.*, 2007
Rosaceae	*Rosa multiflora*	Rose (蔷薇)	4–6	Lijiang/ Yunnan	Zhu *et al.*, 2006; Li, L.F. *et al.*, 2008; Yang *et al.*, 2009

(To be continued)

(Table 4.1)

Family	Species	Common name (Chinese)	Dispersal months	City/ Province	References
Salicaceae	*Populus tomentosa*	Chinese white poplar (白杨)	4	Xi'an/ Shaanxi	Li, Z.X. *et al.*, 2007
Salicaceae	*Salix matsudana*	Hankow willow (旱柳)	4	Beijing	He and Liu, 2009; Zhang *et al.*, 2008;
Salicaceae	*Salix babylonica*	Weeping willow (垂柳)	4	Beijing	He and Liu, 2009; Zhang *et al.*, 2008;
Grass pollen					
Asteraceae	*Helianthus annuus*	Sunflower (向日葵)	7–9	Kunming/ Yunnan	Gao, Y. *et al.*, 2007
Euphorbia-ceae	*Ricinus communis*	Caster-oil plant (蓖麻)	6–10	Kunming/ Yunnan	Gao, Y. *et al.*, 2007
Poaceae	*Zea mays*	Maize (玉米)	6–8	Chongqing	Huang *et al.*, 2002
	Poe annua	Annual bluegrass (早熟禾)	3–4	N/A	He *et al.*, 2009
	Setaria viridis	Green bristlegrass (狗尾草)	6–8	N/A	He *et al.*, 2009
	Digitaria ciliaris	Henry's crab grass (升马唐)	6–8	N/A	He *et al.*, 2009
	Oryza sativa	Rice (水稻)	6–9	N/A	He *et al.*, 2009
	Sorghum bicolor	Sorghum (高粱)	7–8	N/A	He *et al.*, 2009
Weed pollen					
Amarantha ceae	*Amaranthus retroflexus*	Redroot pigweed (反枝苋)	7–8	N/A	He *et al.*, 2009
Amarantha-ceae	*Amaranthus viridis*	Slender amaranth (皱果苋)	7–8	N/A	He *et al.*, 2009
Asteraceae	*Artemisia vulgaris*	Mugwort (艾蒿)	9–10	Beijing, Urumqi	Li, J. *et al.*, 2008; Li, J. *et al.*, 2009
Asteraceae	*Artemisia sieversiana*	Sievers wormwood (大籽蒿)	7–8	Xi'an/ Shaanxi	He *et al.*, 2009; Wu *et al.*, 2007

(To be continued)

(Table 4.1)

Family	Species	Common name (Chinese)	Dispersal months	City/ Province	References
Asteraceae	*Artemisia annua*	Sweet wormwood (黄花蒿)	8–9	Shenzhen/ Guangdong	He *et al.*, 2009; Wu *et al.*, 2007
Asteraceae	*Artemisia apiace*	Sweet wormwood (青蒿)	7–8	Shenzhen/ Guangdong	Liang *et al.*, 2004
Asteraceae	*Ambrosia artemisiifolia*	Ragweed (豚草)	7–9	Beijing	Li, J. *et al.*, 2008
Asteraceae	*Parthenium hysterophorus*	Guayule (银胶菊)	4–10	Haikou/ Hainan	Xie *et al.*, 2005
Cannaba-ceae	*Humulus japonicus*	Scandent hop (葎草)	7–8	Beijing	Jia *et al.*,1995
Cannaba-ceae	*Cannabis sativa*	Cannabis (大麻)	7–8	Fuzhou/ Fujian	Zheng *et al.*, 2003
Chenopo-diaceae	*Chenopodium album*	Goosefoot (藜)	7–10	Beijing, Kunming/ Yunnan	Li, J. *et al.*, 2008; Gao, Y. *et al.*, 2007
Mold					
Dematia-ceae	*Cladosporium cladosporioides*	Bacillus (芽枝菌/枝孢霉属)	N/A	Beijing	Fang *et al.*, 2005; Zheng *et al.*, 2006
Pleospora-ceae	*Alternaria alternata*	*Alternata* (链格孢菌/交链孢霉)	N/A	Chongqing	Huang *et al.*, 2002; Fang *et al.*, 2005; Wang, C.Y. *et al.*, 2008;
N/A	*Ustilago*	Smut fungi (黑粉菌)	N/A	N/A	Wang, C.Y. *et al.*, 2008
N/A	*Uredinales*	Rust fungi (锈菌)	N/A	Changsha/ Hunan	Wang, C.Y. *et al.*, 2008; Xie *et al.*, 2007
N/A	*Penicillium*	Blue mold (青霉菌)	N/A	Beijing, Nanjing/ Jiangsu	Fang *et al.*, 2005; Sun *et al.*, 2000;
N/A	*Aspergillus*	*Aspergillus* (曲霉菌)	N/A	N/A	Fang *et al.*, 2005
N/A	*Paecilomyces*	*Paecilomyces* (拟青霉属)	N/A	Beijing, Nanjing/ Jiangsu	Sun *et al.*, 2000
N/A	Saccharomyces	Yeast (酵母属)	N/A	N/A	Sun *et al.*, 2000

(To be continued)

(Table 4.1)

Family	Species	Common name (Chinese)	Dispersal months	City/ Province	References
N/A	*Fusarium*	*Fusarium* (镰刀菌属)	N/A	N/A	Fang *et al*., 2005
N/A	*Helminthosporium*	*Helminthosporium* (蠕孢菌属)	N/A	N/A	He *et al*., 2009
Indoor allergens					
Pleosporaceae	*Alternaria alternata*	Mold	N/A	N/A	Kan *et al*., 2001
Trichocomaceae	*Aspergillus fumigatus*	Mold	N/A	N/A	Kan *et al*., 2001
Trichocomaceae	*Penicillium notatum*	Mold	N/A	N/A	Kan *et al*., 2001
Mycosphaerellaceae	*Cladosporium herbarum*	Mold	N/A	N/A	Kan *et al*., 2001
Felidae	*Felis domesticus*	Cat	N/A	N/A	He *et al*., 2009
Pyroglyphidae	*Dermatophagoides farinae*	American dust mite	N/A	N/A	Li, J. *et al*., 2009
Glycyphagidae	*Blomia tropicalis*	Mite	N/A	N/A	Li, J.*et al*., 2009
Pyroglyphidae	*Dermatophagoides pteronyssinus*	European dust mite	N/A	N/A	Li, J. *et al*., 2009
Muridae	*Mus musculus*	House mouse	N/A	N/A	He *et al*., 2009
Muridae	*Rattus norvegicus*	Brown rat	N/A	N/A	He *et al*., 2009
Blattellidae	*Periplaneta americana*	American cockroach	N/A	N/A	Jiang *et al*., 1991
Blattellidae	*Blattella germanica*	German cockroach	N/A	N/A	Jiang *et al*., 1991
Blattellidae	*Periplaneta fuliginosa*	Cockroach	N/A	N/A	Jiang *et al*., 1991
Others					
Bombycidae	*Bombyx mori*	Silkworm	N/A	N/A	Qiao *et al*., 1989; Li, 1996; Zhang, 2009
Chironomidae	*Tokunagayusurikataihuensis*	Chironomid midge	10–11	Taihu Basin	Wen *et al*., 1996

N/A: not applicable. (Some data from He and Liu, 2009) (With Courtesy of Dr. Shao-Heng He)

4.2.1 Pollen

Pollen from trees, weeds and grasses, which can be regarded as air pollutants, are a major cause of pollinosis. Pollen allergens are mainly from wind-pollinated plants, due to their highly efficient and abundant release of pollen. National investigations carried out in China revealed that about 60 species from 30 genera are responsible for the majority of clinically relevant airborne pollen (He *et al.*, 2009). In comparison to Europe, most of these sources belong to the same genera but involve different species. The prevalence of pollinosis among Chinese people has been reported to range from 0.5% to 1%, reaching 5% in high incidence areas (Ye *et al.*, 1988). Few reports on allergenic airborne pollens in China have been published in international peer-reviewed journals (Chen *et al.*, 1988; Li, J. *et al.*, 2008), the overwhelming majority being written in Chinese, covering most Chinese regions. Significant differences can be found in seasonal pollen diffusion and geographic distribution patterns, which is essential for effective diagnosis and management of pollinosis in any particular region. In general, the skin prick test (SPT) response in diagnosis is consistent with the pollen load in the air. The most common allergenic pollen sources in China and their clinical relevance are summarized in Table 4.1, with the representative regions indicated.

Pollen of mug-wort (*Artemisia vulgaris, Artemisia sieversiana*) and closely related species (Liang *et al.*, 2004), and Japanese hop (*Humulus japonicas*) (Fig. 4.1) is a major cause of allergenic reactions in late summer and autumn in northern China (Li, J. *et al.*, 2009) as well as other Asian countries. The alien invasive species, ragweed (*Ambrosia artemisiifolia*), proliferates in certain regions of central and eastern China and has had major impact as an allergen, deserving special attention. Other major airborne allergenic pollen sources in northern China include *Chenopodiaceae* and *Poaceae* (Li, J. *et al.*, 2008). Spring pollen, however, appears to be less sensitizing compared to summer or autumn pollens, which are primarily weed or grass.

Tree pollen accounts for the highest percentage of pollen produced during spring. Although exhibiting low allergenicity, tree pollens can cause considerable allergenic reactions due to their wide distribution as native species or as roadside plantings in urban areas, e.g., *Fraxinus* and *Salicaceae* in Beijing, *Platanus* in Wuhan and Nanjing, and *Alnus* in Kunming. Local cedar (*Cryptomeria fortunei*) and Japanese cedar (*C. japonica*) pollens are an important allergenic source causing rhinitis in spring, which is similar to the major pollen-related allergy in Japan, cedar pollinosis, caused by the same species (*Cryptomeria japonica*) (Xu *et al.*, 2000). Moreover, as a result of the high level of positive SPT responses, it has been reported that pollens from entomophilous plants also have high sensitizing potential, with *Phoenix roebelenii*, cultivated in tropical regions of China, being a major representative (Xie *et al.*, 2005). In addition, pollinosis as an occupational hazard for fruit growers is often caused by *Rosaceae* pollen during artificial pollination in orchards (Zhu *et al.*, 2006).

Fig. 4.1. Important allergenic pollens in China. (a) Cedar tree–*Cryptomeria fortunei*; (b) Mugword weed–*Artemisia sieversiana*; (c) Scandent hop weed–*Humulus japonicas*; (d) And annual grass–*Poe annua*

4.2.2 Molds

Molds, a well-known cause of allergies and asthma, are generally regarded as the second major class of clinically relevant aeroallergens (Waqar *et al.*, 2009). Molds primarily not only exist outdoors but also act as potent indoor allergens. Mold spores are smaller than pollen grains and substantially more abundant in the environment throughout the year. The quantity and diversity of the mold species and their spores varies from place to place due to different environmental conditions. Numerous studies, worldwide, have focused on the distribution of species composition and concentration of airborne fungi in outdoor environments. Although more than 80 mold genera have been recognized as aeroallergens (Simon *et al.*, 2008), there is evidence suggesting that several genera may be considered universally dominant, including *Cladosporium*, *Penicillium*, *Alternaria* and *Aspergillus* (de Ana *et al.*, 2006; Aimanianda *et al.*, 2009).

Unfortunately, we have found limited information on the composition and concentration of airborne fungi in China. In Beijing, *Cladosporium* has the highest concentration and is the most dominant fungal species followed by the non-sporulating isolates, *Alternaria*, *Pencillium* (the most abundant, comprising more than half of the total fungal species isolated), and *Aspergillus* (Fang *et al.*, 2005). Yan *et al.* (2000) found similar fungal species composition in another three cities (Shanghai, Chengdu, Guangzhou) in distinct regions of China. Conflicting results were found, however, in another survey conducted in Guangzhou (Zheng and Zhang, 2006), indicating that the most predominant fungal species was *Nigrospora*. The most common and predominant airborne mold spores are summarized in Table 4. 1, with the region of the survey indicated.

Mold is generally considered to be low in allergenicity, considering the relatively low percentage of positive SPTs compared to other (aero) allergenic sources. Recent studies have shown a barrier in immune recognition which is expanded by the existence of a blocking hydrophobic rodlet layer on the cell surface of the conidial spores (Aimanianda *et al.*, 2009).

4.2.3 Other Outside Aeroallergenic Sources

Besides pollen and mold, other areoallergenic sources in the outdoor environment are associated with respiratory disease or occupational allergies. One example is the chironomid midge (*tokunagayusurika taihuensis*) in the Taihu Basin of China, which resulted in a much higher positive SPT response in asthmatic subjects than that in normal controls (Wen and Zhou, 1994).

4.3 Indoor Allergens

Sensitivity to indoor allergens is generally associated with allergic asthma. The prevalence of asthma has markedly increased over the past 30 years, worldwide. The causative role of indoor allergens in the development of asthma has recently been reviewed (Gaffin and Phipatamakul, 2009). Evidence also suggests that indoor allergens, such as those produced by house dust mites, are the major allergens in patients with asthma and (or) rhinitis in China (Salo *et al.*, 2004; Li, J. *et al.*, 2009; Kong *et al.*, 2009).

Common indoor allergens are generated from house dust mites, pets (cats and dogs), molds and household pests such as cockroaches and mice. In contrast to allergenic pollens, indoor allergens are not particularly associated with specific regions in China. House dust mites are ubiquitous and the major source of allergens, particularly in Southern China, with a warm climate and relatively high humidity. More than 80% of asthmatic patients are sensitized to mite allergens (Wen and Xing, 2004). A standard mite preparation is still the only commercial allergen in China (Zhang, Y.J., 2008). Cockroaches, affecting about 30% of allergenic individuals in Shanghai, are one of the major indoor allergen producers (Wang *et al.*, 2001). In addition to the two well-documented allergenic cockroach species existing worldwide,

Periplaneta americana and *Blattella germanica*, *Periplaneta fuliginosa* has also been found in some regions of China, especially in the Yangtze river basin (Jiang *et al.*, 1991). The dominant molds in indoor environments are *Cladosporium* spp, *Aspergillus* spp, *Alternaria* spp and *Penicillium* spp (Kan *et al.*, 2001).

Another indoor, occupational allergenic source is silk waste and the allergen-bearing particles from the silkworm (during sericulture), where exposure can lead to respiratory diseases (Li, 1996; Qiao *et al.*, 1989; Zhang, 2009). Sensitization to silk has been identified as a strong predictor of rhinitis in Anqing, China (Celedon *et al.*, 2001).

4.4 Ingestible Allergenic Sources

Epidemiological studies have revealed that the frequency of food allergies in some Asian countries is low compared to developed countries, but apparently on the increase (Shek *et al.*, 2006; Chiang *et al.*, 2007). Although lacking definite evidence, a limited number of studies indicate a similar trend can be expected in China. An incidence of food allergies of 3.4%–5% has been estimated in three regions of China, including Beijing, Guangdong and the Sheng-li Oil Fields (Hill *et al.*, 1997). Milk is highest in the ranking of the most common sensitizing food groups, followed by eggs, soybean, fish, shellfish, peanuts and wheat. Rice allergies have been found to be relatively uncommon in the Sheng-li Oil Field population (Hill *et al.*, 1997). This ranking, however, varies when age and geographical differences are considered. Shellfish, for example, is a major sensitizing food source in children when it is a major component of their diet (Chiang *et al.*, 2007). The low rate of peanut allergies, despite a similar consumption (Sampson, 2004), may be associated to the different methods of preparation in China (mainly as a boiled food product) and the United States (mainly roasted) (Shaker *et al.*, 2009), or to a relatively low allergenicity of local cultivars (Cong *et al.*, 2008). Besides the most common food allergenic sources, everyday foods in Hong Kong, such as mangos, shrimps and eggs, can give rise to a relatively high rate of food allergies (Wong *et al.*, 2009).

China also has several unique food allergies due to its typical life styles and cultures. With an increasing incidence of anaphylaxis reported locally, allergies caused by unusual food sources, such as buckwheat in Japan, bird's nest in Singapore, chestnut in Korea, sesame in Israel and Japan, and chickpea in India (Shek *et al.*, 2006), is an area of significant concern and needs extensive attention. In a recent review (Ji *et al.*, 2009), anaphylactic shock and deaths were reported, which were caused by consumption of a panel of uncommon Chinese allergenic sources (Table 4.2). Local medical journals have reported 61 anaphylactic incidents from traditional Chinese foods, such as silkworm pupas, cicada pupas, grasshoppers, locusts and *Clanis bilineata*.

Another important type of allergy via ingestion is the adverse response to oral medication and to western and eastern medicines. Although these adverse events are extremely unusual, anaphylaxis is a life-threatening allergic reaction that may occasionally cause sudden death. Care must be taken when prescribing oral medication to patients. In a survey of local medical publications between 1995 and 2005, a total of 195 cases indicated antibiotics as the major allergenic source causing allergenic shock, followed by Chinese herbal medicines (Li *et al.*, 2006). Examples of medicines as representative of allergenic sources that may cause an allergic shock by oral administration are listed in Table 4.3. The list of ingestible allergenic sources available to date covers 12 classes and includes 82 kinds of medicine (Li *et al.*, 2006).

Table 4.2 Native, potentially allergenic sources causing anaphylaxis by ingestion

Category	Reported material
Foods	
Nuts	Pistachio, walnut, cashew nut, hazelnut
Fruits	Mango, pineapple, plum, mulberry, kumquat, lychee, apple, kiwifruit, *Shatian pomelo*, banana, pear, peach
Seafood	Soft-shelled turtle, shrimp, fish, sea fish, shellfish, crab
Meat	Duck meat, venison, beef, pigeon meat, hoptoad
Insect	Locust, grasshopper, silkworm pupa, cicada pupa, bee pupa, bee larva, *Clanis bilineata*
Vegetable and cereals	Millet congee, pepper, cabbage, cilantro, bamboo shoot, kidney bean, castor oil, sesame seed oil, corn, noodle (wheat and buckwheat)
Other	Honey, duck blood, sweet potato, tea, mushroom, garlic, Chinese prickly ash
Oral medication	
Antibiotic	Amoxicillin, penicillin V potassium, compound sulfamethoxazole tablets, pioneer IV capsule, pipemidic acid tablets, ciprofloxacin capsules, ethambutol hydrochloride tablets
Traditional Chinese medicine (TCM)	Huo xiang zheng qi shui, bezoar detoxicating tablet, guilong kechuanning capsules
Others	Astemizole tablets, ranitidine tablets, captopril tablets

4.5 Contactants

Contactants are defined as a group of allergenic sources eliciting symptoms of induced sensitivity via direct contact with the skin or mucosa. They trigger the two major clinical syndromes, contact urticaria and allergic (eczematous) contact dermatitis. Exposure and sensitization can stem from various kinds of substances, including metals, chemical agents, industrial products, food additives, medicines and even natural products (Table 4.3). Patch testing is usually carried out to reach a correct diagnosis for a specific substance causing inflammation of the skin. However, with a limited number of standardized screening allergens available, only a few essential contactants are included in a regular diagnosis. The most common contactants found in patch test data are listed in Table 4.4. This is consistent with the reported frequency of contact allergies that resulted from exposure to various allergenic sources, such as hair dyes, cosmetics, decorating materials in new homes, and medicines which contain allergenic ingredients.

Table 4.3 Allergenic contactants in China

Category	Sources*	References
Metal	Nickel (34.57%, 28/81), aluminum (23.45%, 19/81), chromium (21%, 17/81), mercury (18.52%, 15/81), cobalt (17.28%, 14/81), palladium (11.11%, 9/81)	Li, L.J. *et al.*, 2008
Chemical agent	Methanol (45.2%, 226/500), aniline (44.4%, (222/500)	Duan *et al.*, 2007
Industrial product	Detergent (54.8%, 285/520), hair dyes, pesticides (23.2%, 116/500), paint, cosmetics, cocamidopropyl betaine (CAPB) (9.8%, 42/429), rubber products (latex) (46.3%, 193/417)	Li, P.Y., 2008; Duan *et al.*, 2007; Zhang, H.Y., 2008; Zhang, Q.X. *et al.*, 2008
Food additives	Tartaric acid (48.6%, 243/500), perfume (40.8%, 204/500), nitrous acid (34.4%, 172/500), benzoic acid (33.6%, 168/500), citric acid (14.8%, 4/500), fructose (14.4%, 72/500)	Duan *et al.*, 2007
Medicine (Drug)	Medicated oil, Diduo, povidone iodine, piyan-ping ointment, moisture burn ointment	Cao, 2007; Chen, 2007; Zhang,X.Z., 2007; Yan *et al.*, 2007; Wang *et al.*, 2008
Natural product	Fresh cinnamon vine, smallage, propolis, mango, ginkgo seed, *toxicodendron vernicifluum* (leaf), *Ficus carica* (leaf), butter	Yang *et al.*, 2009; Liu *et al.*, 2008; Gao, X.P. *et al.*, 2007; Xie *et al.*, 2007; Li, L.P., 2009; Xu, 2009; Peng *et al.*, 2007; Wang, 2007

*Positive rate indicated as the percentage of sensitized individuals of all diagnosed individuals

Table 4.4 Most common contactants diagnosed by patch testing

Standard allergens	Cases reported*	References
Nickel sulfate	30.61% (75/245), 33.03% (107/312), 36.78% (25/68)	Ma *et al.*, 2009; Li and Chen, 2007; Luo *et al.*, 2009
Cobalt chloride	26.12% (64/245)	Ma *et al.*, 2009
Potassium dichromate	25.31% (62/245), 10.28% (33/312), 8.82% (6/68)	Ma *et al.*, 2009; Li and Chen, 2007; Luo *et al.*, 2009
Fragrance mix	22.04% (54/245), 14.95% (48/312)	Ma *et al.*, 2009; Li and Chen, 2007;
Rosin	15.1% (37/245), 2.8% (9/312), 11.76% (8/68)	Ma *et al.*, 2009; Li and Chen, 2007; Luo *et al.*, 2009
Formaldehyde	14.28% (35/245), 4.05% (13/312), 10.29% (7/68)	Ma *et al.*, 2009; Li and Chen, 2007; Luo *et al.*, 2009
Para-phenylenediamine (PPD)	13.06% (32/245), 6.23% (20/312), 13.24% (9/68)	Ma *et al.*, 2009; Li and Chen, 2007; Luo *et al.*, 2009
Benzocaine	4.05% (13/312), 11.76% (8/68)	Li and Chen, 2007; Luo *et al.*, 2009
White precipitate	30.88% (21/68)	Luo *et al.*, 2009
Thimerosal	40.98% (122), 9.66% (31/312)	Li and Chen, 2007; Feng *et al.*, 2006

* Positive rate indicated as a percentage of sensitized individuals of all diagnosed individuals

Nickel (Ni), for example, is well-known as the most common metal contactant in patch-tested patients. It is found in jewelry, metal spectacle frames, pant snaps and belt buckles. Other examples include heavy metals (e.g., Hg) and fragrances used in cosmetics, and paraphenylenediamine (PPD) used as a permanent hair dye (Lee *et al.*, 2009). It has been suggested that some contactants, as well as allergens, should also be included in routine patch tests. Cocamidopropyl betaine (CAPB), an amphoteric surfactant, is commonly used in rinse-off personal care products as well as in cosmetics. Reports indicate the prevalence of CAPB allergy in Beijing is higher than that previously reported elsewhere (Li, L.J. *et al.*, 2008). Some contactants are unusual, and little is known about their allergenic molecules, such as fresh cinnamon, vine and leaves of *Ficus carica*.

4.6 Injected Allergenic Sources

There are two major types of allergic reactions based on injection of a foreign substance: insect venom/sting allergies and drug-induced allergies. The majority of individuals who receive an insect sting or bite experience a local reaction, appearing as a swelling near to the site of the injection. A small number of highly sensitive allergic individuals may have a more severe response, such as potentially life-threatening anaphylaxis. In China, aggressive insects which may induce this response include wasps (*Vespidae*) (Chen and Zhu, 2009), bees (*Apidae*) (Sun, 2009), jellyfish (*Scyphozoa*) (Jiang *et al.*, 2008), ants (*Formicidae*, Zhang *et al.*, 2007), snakes (primarily the species *Naja naja*, *Bungarus multicinctus* and *Agkistrodon halys*) (Xie, 2009; Zhou *et al.*, 2009), poisonous spiders (Jiao *et al.*, 2004) and centipedes (Zeng, 2001).

In contrast, drug-induced allergies are a major source of anaphylaxis in China, with more than 80% of incidences being caused by intravenous administration (Li *et al.*, 2007). Symptoms usually appear rapidly, within 30 minutes after exposure, and can normally be relieved by administering adrenaline, a hormone that stimulates the heart and relaxes the airways. There are 12 classes of about 200 clinical drugs which can trigger anaphylaxis (Li, S.D. *et al.*, 2007), the most frequently reported being antimicrobial agents, Chinese medicinal herbs and biological and biochemical products (Table 4.5). The top 10 agents found in the literature include penicillin, low-molecular weight dextran, tetanus antitoxin, ciprofloxacin, aprotinin, compound Danshen injection, lincomycin, gentamicin, cefoperazone, cefradine and ofloxacin (Li, S.D. *et al.*, 2007). Penicillin is the most common trigger, cited both in China and worldwide, and the most well-known member of the beta-lactam antibiotic group characterized by a beta-lactam ring. Cephalosporin is another major antibiotic causing anaphylactic deaths as seen in forensic autopsies in Shanghai (Shen *et al.*, 2009). The relatively high rate of antibiotic allergies in China has been attributed to inappropriate use of antibiotics and to illegal medical practices (Shen *et al.*, 2009).

Traditional Chinese medicine (TCM) has become popular in other countries and is used for preventing and treating disease and illness, including asthma and allergies (Li *et al.*, 2009). With the growing application of TCM products, increasing numbers of severe adverse reactions have been reported (Sun, 2007). Intensive safety monitoring of TCM injections led to bitter debates on improvement of TCM injections (Li, Y. *et al.*, 2008; Science News, 2009). There are 109 TCM injections listed at the state level as standardized medicines (Wang, 2008) which, together with their complicated composition, make it difficult to evaluate and maintain the safety of TCM products. Comprehensive assessment of the potential allergenicity of the TCM raw materials and extracts with standard sera from different allergic populations will provide relevant knowledge for their administration. This, however, also requires international collaboration. Plant scientists, especially those involved in metabolomics and system biology, should play a leading role in this initiative.

Table 4.5 Top three classes of sources triggering drug-induced allergy

Category	Allergenic sources	References
Antimicrobial agent	β-lactam, cephalosporins, Quinolones	Li, S.D. *et al.*, 2007; Xu *et al.*, 2006
Traditional Chinese medicine	Shuanghuaglian, Qingkailing, Yuxingcao, Gegensu, Ciwujia, Mailuoning, Compound Danshen	Sun, 2007
Biological and biochemical product	Tetanus antitoxin, aprotinin, insulin, encephalitis vaccine, rabies vaccine	Li, S.D. *et al.*, 2007; Xu *et al.*, 2006

4.7 Future Perspective

This review provides an introduction and a starting point for further multidisciplinary collaborations for scientists focusing on the allergy problem in China, to continue investigating the molecular basis of the allergens, their possible cross-reactivity, the aetiology of the corresponding allergic symptoms and their treatment, cure and prevention. Further research is also needed to characterize the biochemical and immunological properties of novel and local allergens in China, for their accurate denomination following standard allergen nomenclature.

Acknowledgements

This work was supported by the Natural Science Foundation of China (No. 30971970).

References

Aimanianda, V., Bayry, J., Kniemeyer, O., *et al.* (2009). Surface hydrophobin prevents immune recognition of airborne fungal spores. Nature, 460(7259), 1117-1179.

Burge, H.A., Rogers, C.A. (2000). Outdoor allergens. Environ Health Perspectives, 108(4), 653-659.

Cao, J. (2007). Clinical analysis of 28 cases of clove oil contact-induced dermatitis. J Dermatol Venereol, 29(2), 15-16 (in Chinese).

Celedon, J.C., Palmer, L.J., Weiss, S.T., *et al.* (2001). Asthma, rhinitis, and skin test reactivity to aeroallergens in families of asthmatic subjects in Anqing, China. Am J Respir Critical Care Med, 163(5), 1108-1112.

Chen, H. (2007). Four cases of DiDou-induced contact dermatitis. J Dermatol and Venereol, 29(4), 54 (in Chinese).

Chen, K., Liao, Y.F., Zhang, J.T., *et al.* (1988). The major aeroallergens in Guangxi, China. Clinical Allergy, 18(6), 589-596.

Cheng, H., Zhu, C.Z. (2009). Clinical analysis of 59 cases of wasp sting. Clinical Medicine, 29(11), 85-86 (in Chinese).

Chiang, W.C., Kidon, M.L., Liew, W.K., *et al.* (2007). The changing face of food hypersensitivity in an Asian community. Clin Exp Allergy, 37(7), 1055-1061.

Cong, Y.J., Lou, F., Li, L.F., *et al.* (2008). Identification and quantification of major peanut allergens (in China) with IgE-binding property. J Food Biochem, 32(3), 353-367.

de Ana, S.G., Torres-Rodriguez, J.M., Ramirez, E.A., *et al.* (2006). Seasonal distribution of Alternaria, Aspergillus, Cladosporium and Penicillium species isolated in homes of fungal allergic patients. J Investig Allergology and Clin Immunol, 16(6), 357-363.

Duan, A.X., Liu, L.J., Wang, Q. (2007). Analysis of allergen test results of 500 cases with chronic urticaria. Journal of Ningxia Medical College, 29(3), 317-318 (in Chinese).

Fang, Z.G., Ouyang, Z.Y., Lin, X.Q., *et al.* (2005). Culturable airborne fungi in outdoor environments in Beijing, China. Science of the Total Environment, 350(1-3), 47-58.

Feng, P.Y., Zhu, G.X., Lu, C., *et al.* (2006). Clinical analysis of the patchtest in 122 cases of chronic eczema and dermatitis. J Clin Dermatology, 35(2), 81-83 (in Chinese).

Gaffin, J.M., Phipatanakul, W. (2009). The role of indoor allergens in the development of asthma. Curr Opin Allergy Clin Immunol, 9(2), 128-135.

Gao, X.P., Wang, R.Z., Wang, H., *et al.* (2007). One case of propolis-induced facial contact dermatitis. J Clin Dermatology, 36(4), 249 (in Chinese).

Gao, Y., Xie, S.Q., Liu, L., *et al.* (2007). Survey and clinical analysis of hay fever in summer and autumn in the Kunming area. Yunnan Medicine, 28(4), 416-417 (in Chinese).

He, S.H., Liu, Z.G. (2009). Fundamental Allergy, 1st ed. Beijing: Science Press.

Hill, D.J., Hosking, C.S., Chen, Y.Z., *et al.* (1997). The frequency of food allergy in Australia and Asia. Environ Toxicol and Pharmacol, 4(1-2), 101-110 (in Chinese).

Huang, J.J., Hong, S.L., Bai, Y., *et al.* (2002a). Survey on pollen allergy in the southern suburban area of Chongqing city. Acta Academiae Medicinae Militaris Tertiae, 24(4), 482-484 (in Chinese).

Huang, J.J., Yin, R., Liang, C.Y., *et al.* (2002b). Comparative investigation of airborne allergenic fungi b, etween Chongqing and Chengdu. Journal of Chongqing Medicine, 31(3), 222 (in Chinese).

Ji, K.M., Chen, J.J., Li, M., *et al.* (2009). Anaphylactic shock and l, ethal anaphylaxis caused by food consumption in China. Trends Food Sci Technol, 20(5), 227-231.

Jia, Y., Ye, S.T., Gu, R.J., *et al.* (1995). An important allergenic pollen in China – Humulus. Journal of Allergy and Clinical Immunology, 95(1), 164-164.

Jiang, X.F., Wang, L.J., Li, J.H., *et al.* (1991). Analysis of the results of surveillance of cockroach density in five cities. Chinese Journal of Vector Biology and Control, 2(1), 20-22 (in Chinese).

Jiang, Z.G., Wang, B. Xu, J.S., *et al.* (2008). 108 cases of jellyfish sting. Practical Journal of Medicine and Pharmacy, 25(12), 1457-1459 (in Chinese).

Jiao, J.H., Liu, S.S., Xi, Y.X. (2004). Two cases of spider bite-induced allergic toxic shock. Journal of Rare and Uncommon Diseases, 11(6), 57 (in Chinese).

Kan, H.D., Song, W.M., Jiang, R.F., *et al.* (2001). Distribution of indoor airborne fungi and the immune-suppressive effect of dominant fungal extracts. Practical Preventive Medicine, 8(1), 1-3 (in Chinese).

Kong, W.J., Chen, J.J., Zheng, Z.Y., *et al.* (2009). Prevalence of allergic rhinitis in 3- to 6-year-old children in Wuhan of China. Clin Exp Allergy, 39(6), 869-874.

Lee, P.W., Elsaie, M.L., Jacob, S.E., *et al.* (2009). Allergic contact dermatitis in children: common allergens and treatment: a review. Curr Opin Pediatr, 21(4), 491-498.

Li, J., Lu, Y., Huang, K., *et al.* (2008). Chinese response to allergy and asthma in Olympic athl, etes. Allergy, 63(8), 962-968.

Li, J., Sun, B., Zhang, C., *et al.* (2009). A multicentre study assessing the prevalence of sensitizations in patients with asthma and/or rhinitis in China. Allergy, 64(7), 1083-1092.

Li, L.F. (2008). A study of the sensitization rate to cocamidopropyl betaine in patients patch tested in a university hospital of Beijing. Contact Dermatitis, 58(1), 24-27.

Li, L.J., Yang, W.B., Li, X., *et al.* (2008). Clinical analysis of 81 cases of m, etal contact-induced dermatitis. Acta Academiae Medicinae Militaris Tartiae, 30(17), 1590-1591 (in Chinese).

Li, L.P. (2009) Report of 6 cases of ginkgo nuts-induced contact dermatitis. J Pract Clin Med, 6(3), 114 (in Chinese).

Li P.Y. (2008). The debate on the perspective of Chinese herbal injection: whether to be disabled. China Practical Medicine, 4(32), 199-199 (in Chinese).

Li, S.D., Wang, Q.F. (2007). Analysis of 608 cases of drug-induced allergic shock. Herald of Medicine, 26(2), 207-209 (in Chinese).

Li, W.J. (2007). Survey of airborne pollen and pollen hypersensitivity in Wuhan. Central China Medical Journal, 31(6), 493-495 (in Chinese).

Li, W.J., Chen, H. (2007). Analysis of the patch test results of 312 cases of contact dermatitis in Wuhan. Central China Medical Journal, 31(4), 257-258 (in Chinese).

Li, X.M., Brown, L. (2009). Efficacy and mechanisms of action of traditional Chinese medicines for treating asthma and allergy. J Allergy Clin Immunol,

123(2), 297-306.
Li, Y., Han, L.F., Qiao, H.Y. (2008). Etiological survey on occupational pollen allergy in fruit growers. Chinese Occupational Medicine, 35(3), 270 (in Chinese).
Li, Y., Li, H.S., Zeng, J., *et al.* (2006). Literature analysis of allergic shock in 195 cases induced by oral administration. China Pharmacy, 17(14), 1099-1100 (in Chinese).
Li, Z.X. (1996). Analysis of 52 cases of silk-based allergic asthma. Acta Univesitatis Scientiae Medicinae Chongqing, 21, 127-128 (in Chinese).
Li, Z.X., Zhang, G.S. (2007). Investigation and analysis of airborne pollen in central Xi'an in spring, 2006. Chinese Medicine of Factory and Mine, 20(3), 304-305 (in Chinese).
Liang, G.Z., Liu, Z.G., Yang, H., *et al.* (2004). Analysis of allergen composition in Artemisia argyi and Artemisia apiacea. Immunological Journal, 20(2), 120-124 (in Chinese).
Liu, H.F., Zheng, L., Lin, J.H. (2008). One case of wild celery-induced contact dermatitis. Central China Medical Journal, 32(5), 376 (in Chinese).
Liu R.J., Liu F.X., Wang Z.X., *et al.* (2009). Clinical analysis of 102 cases of birch pollen allergy patients. Symposium book of Chinese Allergy, Xiamen, 131-132 (in Chinese).
Lu, R.C., Cui, Z.J., Ma, S.L., *et al.* (2003). Analysis of 15 cases of oral drug-induced anaphylactic shock. Chinese J Clin Res, 79, 13128-13129 (in Chinese).
Luo, D.Y., Chen, H., Yu, H., *et al.* (2009) Analysis on the results of patch test in 68 cases with contact dermatitis. J Dermatology and Venereology, 31(2), 31-32 (in Chinese).
Lü, B., Zhang X.L. (2001). Allergen test of bronchial asthma in children. Guangdong Medical Journal, 22(1), 38-39 (in Chinese).
Ma, H.Y., Zong, Y.Y., Zhu, A.J., *et al.* (2009) Analysis on the results of patch test in 256 cases with contact dermatitis. Laboratory Medicine and Clinic, 6(3), 171 (in Chinese).
Nilsson, S. (1990). Regional and global distribution of aeroallergens. Review of Palaeobotany and Palynology, 64(1-4), 29-34.
Ong, E.K., Singh, M.B., Knox, R.B., *et al.* (1995). Aeroallergens of plant origin: Molecular basis and aerobiological significance. Aerobiologia, 11(4), 219-229.
Pei, B.F., Zhang, R.X., Yu, Y.H., *et al.* (2006). A survey of airborne pollen grains in Shiyan. Central China Medical Journal, 30(6), 509-510 (in Chinese).
Peng, G.H.,Zhang, L.R., Song, Q.J. (2007). Eight case of fig leaf-induced acute contact dermatitis. Journal of Qingdao Medical College, 43, 534 (in Chinese).
Qiao, B.S., Wen, Z.M., Zhou, L.X., *et al.* (1989). Clinical analysis of 65 cases of silk-induced respiratory allergy in children. Journal of Pediatrics in China, 27(1), 16-19 (in Chinese).
Salo, P., Xia, J., John, C.A., *et al.* (2004). Indoor allergens, asthma, and asthma-related symptoms among adolescents in Wuhan, China. Annals of

Epidemiol, 14(8), 543-550.

Sampson, H.A. (2004). Update on food allergy. J Allergy Clin Immunol, 113(5), 805-819.

Shaker, M., Woodmansee, D. (2009). An update on food allergy. Current Opinion in Pediatrics, 21(5), 667-674.

Shek, L.P.C., Lee, B.W. (2006). Food allergy in Asia. Curr Opin Allergy Clin Immunol, 6(3), 197-201.

Shen, Y., Li, L., Grant, J., *et al.* (2009). Anaphylactic deaths in Maryland (United States) and Shanghai (China): A review of forensic autopsy cases from 2004 to 2006. Forensic Science International, 186(1-3), 1-5.

Simon, N.B., Benk, U., Poll, V., *et al.* (2008). The spectrum of fungal allergy. Int Arch Allergy Immunol, 145(1), 58-86.

Sun, C.M. (2007). Literature analysis of Chinese herbal injection-induced adverse reactions. Chinese J Clin Med Res, 13(23), 3441-3442 (in Chinese).

Sun, J.Q., Zhai, J.H., Zhai, H., *et al.* (2000). Analysis and identification of Airborne Fungi in Beijing and Nanjing. China Public Health, 16(11), 1026 (in Chinese).

Sun, X. (2009). One case of anaphylactic shock caused by bee stings. Jiujiang Med J, 18(3), 149 (in Chinese).

Wang, C.S., Dong, F. (2008). Eight cases of applied MEBO-induced contact dermatitis. Modern Health Care: Medical Innovation Research, 5(21), 99 (in Chinese).

Wang, C.Y., Ding, X.Z., Wu, Q.Z., *et al.* (2008). A survey of airborne fungi in Jingzhou area. J Practical Med Techniques, 15(15), 1911-1913 (in Chinese).

Wang, H. (2008). Anaphylactic shock caused by traditional Chinese medicine injection: analysis of 448 cases in literature. Application and Monitoring of Drug in China, 5(4), 45-47 (in Chinese).

Wang, H.J. (2007). One cases of cream-induced facial contact dermatitis. J Dermatology and Venereology, 29(4), 57-58 (in Chinese).

Wang, W.C., Wang, J.D., Xu, Y.P., *et al.* (2001). Survey on reactivity of cockroach allergy in out-patients with allergic asthma through skin test. Acta Universitatis Medicinalis Secondae Shanghai, 21(1), 47-49 (in Chinese).

Waqar, M.A., Akhter, T., Waqar, A., *et al.* (2009). The role of fungi as aeroallergens in asthma and allergies. J Chem Society of Pakistan, 31(4), 688-704.

Wen, T.H., Xing, D.R. (2004). Prevalence of anaphylaxis to dust mite in the human population in Xuzhou. Chinese Journal of Parasitology and Parasitic Diseases, 22(5), 287-290 (in Chinese).

Wen, T.H., Zhou, C. (1994). Survey on hypersensitivity of Tokunagayusurika Taihuensis. Acta Parasitol and Medical Entomol, 1(2), 42-47 (in Chinese).

Wen, T.H., Zhou, C. (1996). Immunochemical characterization of chironomid midge tokunagayusurika Taihuensis. Chinese Journal of Parasitology and Parasitic Diseases, 14(2), 101-105 (in Chinese).

Wong, G.W., Ma, W., Chan, Y., *et al.* (2009). Comparative study of food allergy in rural and urban Chinese school children. J Allergy Clin Immunol, 123(2,

Suppl 1), S32-S32.

Wu, H., Zhang, G.S., Zhang, A.X., *et al.* (2007). Comparison of genera and content of the airborne pollens in 1986 and 2006 in Xi'an city. Chinese J Allergy Clin Immunol, 1(2), 167-171 (in Chinese).

Xie, C.M., Meng, G., Li, C.L., *et al.* (2005). Survey on pollen sensitization of tropical plants in Haikou. J Clin Otorhinolaryngol (China), 19(22), 1044-1045 (in Chinese).

Xie, L., Ding, M., Shan, Y.X. (2007). Analysis of 62 cases of mango-induced contact dermatitis. Chinese J Misdiagnosis, 7(28), 6863-6864 (in Chinese).

Xie, W.Q. (2009). Retrospective analysis of drug treatment for 66 cases of snakebite. Journal of Snake, 21(3), 215 (in Chinese).

Xu, L.J., Zhang, S.Y., Yang, Q., *et al.* (2000). A preliminary study on the incidence of cedar pollinosis in the district of Wuhan. Journal of Clinical Otorhinolaryngology (China), 14(11): 505-506 (in Chinese).

Xu, Z.L. (2009). Report of one case of paint leaves-induced contact dermatitis. Journal of Xianning College (Medical Science), 23(1), 12 (in Chinese).

Yan, E.P., Fu, C.L., He, W.H., *et al.* (2000). Special investigation of fungi from four cities in China. Bull etin of Microbiology, 27(4), 264-270 (in Chinese).

Yan, Z.H., Shi, X., Sun, X.D. (2007). Analysis of six cases of Piyanping ointment-induced contact allergy. Journal of Suzhou Medical, 30(1), 29-30 (in Chinese).

Yang L., Xu Y.P., Zou H.B., *et al.* (2007). Analysis of the allergens from 2475 patients with allergic disease. Labeled Immunoassays and Clinical Medicine, 14(3),141-143

Yang, L.M., He, Y.Q., Yang, S.S., *et al.* (2009). Investigation of airborne pollen in urban Lijiang area. Yunnan Medicine, 30(4), 473-475 (in Chinese).

Ye, S.T., Zhang, J.T., Qiao, B.S., *et al.* (1988). Airborne Allergic Pollen in China. 1st ed. Beijing: Science Press (in Chinese).

Zeng, G.Y. (2001). One case of centipede bite-induced anaphylactic shock. Journal of Practical Medicine, 17(10), 986 (in Chinese).

Zhang, H.Y. (2008). Chinese allergology. Chinese J Allergy Clin Immunol, 2(4), 255-260 (in Chinese).

Zhang, Q.L., Huang, X.K., Luo, X.B., *et al.* (2007). Clinical analysis of 18 cases of ant bite-induced allergy. Lingnan J Emergency Medicine, 12(5), 384-385 (in Chinese).

Zhang, Q.X., Wu, Y.Y., Zhou, H.N., *et al.* (2008). A survey of status quo of medical staff with lactoprene allergy. Chinese Nursing Res, 22(6B), 1519-1520 (in Chinese).

Zhang, X.Z. (2007). One case of povidone-iodine gauze-induced contact dermatitis. J Practical Medical Techniques, 14(10), 1362 (in Chinese).

Zhang, Y.J. (2008). Investigation on result of allergens test in allergic patients of Shandong province. Chinese Journal of General Practice, 6(8), 841-842 (in Chinese).

Zhang, Y.L. (2009). Analysis of allergen test results of 880 cases of chronic urticaria. J Chinese Rural Medicine, 16(11), 55 (in Chinese).

Zheng, X.Y., Liu, J.Y., Lin, G.Y. (2003). Research of allergenic pollen in patients with hay fever in urban Fuzhou. Strait J Preventive Medicine, 9(5), 76 (in Chinese).

Zheng, Z., Zhang, H. (2006). Distribution of airborne fungi and correlation with climate in Guangzhou. Journal of Sun Yat-sen University: Medical Science Edition, 2(6), 597-604 (in Chinese).

Zhou, Y.Z., Gong, G.G. (2009). Clinical analysis of 2150 cases of viper bite. Chinese J Rural Med and Pharmacy, 16(4), 30 (in Chinese).

Zhu, Y.H., Han, L.F., Guo, Z.M., *et al.* (2006). Clinical analysis of 65 cases of Rosaceae pollen allergy. Hebei Medical Journal, 28(10), 968-969 (in Chinese).

5

Allergen Protein Families and Cross-Reactivity

Luud J.W.J. Gilissen[1], Zhong-Shan Gao[2*]

[1] Allergy Consortium Wageningen, Plant Research International, PO Box 619, 6700 AP Wageniungen, the Netherlands
E-mail: luud.gilissen@wur.nl
[2] Allergy Research Center/Department of Horticulture, Zhejiang University, Hangzhou, 310058, China
[*] E-mail: gaozhongshan@zju.edu.cn

Abstract: Allergies are defined as an excessive reaction of the immune system to normally harmless proteins. What makes a protein an allergen, and why are some individuals affected but not others? This review deals with several issues which may provide answers to these crucial questions. From combined researches in allergen and protein databases, it appears that allergens represent only a narrow distribution regarding protein family membership and biological function. By far most allergies are caused by six plant food and pollen protein families, and by four food and airway-exposed protein families from animal sources. The majority of these allergens are cross-reactive: they are not capable of sensitization by themselves, but provoke allergic reactions via IgE binding in sensitized individuals. Analysis of 3D models of allergens representing the most important allergen protein families revealed that, in particular, highly exposed lysine residues on the surface of allergens were involved in IgE binding. Other amino acids that were frequently found in epitopes were alanine, serine, aparagine and glycine. In the birch pollen allergen Bet v 1, these amino acids accounted for 40% of the total amino acid content, with lysine on top (10%). The existence of a limited number of specific motifs in epitopes formed by these amino acids with a major role for alanine is suggested as a general cause of allergies in genetically predisposed individuals. Knowledge on true-sensitizing allergens is still largely lacking. The extended genomic and proteomic knowledge on the Bet v 1 in birch and its cross-reactive homolog Mal d 1 in apple may form a good basis to further elucidate the fundamental questions on allergic sensitization and cross-reactivity.

5.1 Introduction

In Europe, allergies have resulted in an economic burden of many billion Euros (van den Akker-van Marle *et al.*, 2005). The prevalence of allergies has rapidly increased over recent decades and is still increasing in countries associated with Western lifestyles and increasing urbanization (ISAAC, 1998). Urbanization takes place in countries with growing economies, such as China. Improvements in hygiene and health care (vaccinations and antibiotics), in daily diets, increased working stress, smoking by pregnant women, skin contacts with new materials, and so forth may all contribute to increases in allergies. Reduced microbial exposure in early life was found to be one of the critical factors (Janson *et al.*, 2007). All of these factors may influence the balance of the immune system, and may change the immune response towards (new) proteins as potential allergens.

Allergy is defined as the excessive reaction of the immune system to normally harmless proteins. In practice, a limited number of individuals will become allergic to a limited number of proteins. It is, however, difficult to predict who will become allergic to which protein(s). This makes an allergy a complex and multifactorial disease. With increasing prevalence, the scientific community and the governmental and regulatory authorities are becoming more and more interested in the specific human and protein factors involved, as only through this knowledge the epidemic can be halted and quality of life improved.

In this paper, we will firstly focus on protein-related issues. An important question involves which allergens are described in databases, and what we can learn from merging databases with regard to structural and functional characteristics of allergens, and their evolutionary relationships. Secondly, we will examine in more detail the phenomenon of cross-reactivity. Clinically, allergic symptoms may sometimes be provoked without prior exposure, or may result from sources that are unlikely to sensitize. The basis of this phenomenon deals with the ability of similar proteins from different sources to provoke an allergic reaction. Thirdly, allergens are specifically recognized by IgE, the central antibody in allergy. Several questions are relevant at this level. Can every protein become an allergen? Are specific protein characteristics involved in allergenicity? Can specific characteristics at the amino acid level of an allergen be attributed to IgE recognition? Lastly, we will elaborate on the birch-apple syndrome as a model to elucidate sensitization and cross-reactivity in greater detail.

5.2 Structural, Functional and Evolutionary Characteristics of Allergens

Several hundreds of allergens have been listed in databases in terms of their molecular, biochemical and clinical characteristics. These are amongst others, the Official List of Allergens issued by the International Union of Immunological Societies Allergen Nomenclature Subcommittee (http://www.allergen.org), the Allergome database (http://www.allergome.org), the Food Allergy Research Program Allergen database (http://www.allergenonline.com), and the InFormAll database (http://foodallergens.ifr.ac.uk). Records of most allergens in these databases are organized by type of allergen source, and route of exposure, according to the categories: inhalation, ingestion, sting/bite, contact, iatrogenic and auto-allergen. The allergen designation, according to official allergen nomenclature, is derived from the scientific name of the allergenic source species and a sequential number, e.g., Mal d 1 is the first identified allergens from *Malus domestica* (apple). This method of classification does not provide information on the type, the structure, the biological function and the evolutionary relationship of the allergens.

The Allergome database lists allergens on the basis of their molecular (gene and protein) sequence data. The extended information on protein families and their evolutionary and structural relationships is given in the Pfam database (http://pfam.sanger.ac.uk). To combine structural and evolutionary data of allergens, Radauer *et al.* (2008) constructed the allergen family (AllFam) database (http://www.meduniwien.ac.at/allergens/allfam) by merging the Allergome and Pfam databases. These data were subsequently matched with data in the Structural Classification of Proteins Database (http://scop.mrc-lmb.cam.ac.uk/scop) from which the structure of the various allergens became known. In order to classify allergens in terms of their biological function, the allergen sequences were compared with the Gene Ontology Annotation Database. These comparisons provide the basis for a novel classification of allergens based on their protein family relationship and distribution, which enables identification of common structural and biochemical allergen characteristics.

As a result, 707 allergens were classified into 134 AllFam families, containing 184 different Pfam domains, indicating that allergens were only found in 2% of the 9,318 Pfam families, and in 5% of the 3,012 structural protein families, respectively. This was about 4 times lower than that had been found in a random set of 707 proteins. In addition, the biological functions of allergens most frequently found were limited to hydrolysis (of proteins, polysaccharides and lipids), to binding of metal ions (Ca) and lipids, to storage, and to cytoskeleton associations. The number of allergens in the databases is still growing, and all newly added allergens and their related protein families have been found to contain minor allergens. This justifies the assumption that all major allergens from important sources have already been identified. In summary, the main conclusion from this research is that allergens represent only a narrow distribution regarding protein family membership and biological function (Radauer *et al.*, 2008). Earlier research already indicated that only four Pfam proteins (prolamins, PR10 proteins, cupins and profilins) accounted for more than 65% of all plant food allergens, indicating that conserved structures and biological activities may play a role in determining or promoting allergenic properties (Jenkins *et al.*, 2005).

Major allergen protein families were summarized and updated mainly by three researchers (Breiteneder and Radauer, 2004; Jenkins *et al.*, 2005; Radauer and Breiteneder, 2006, 2007; Jenkins *et al.*, 2007), and outlined as follows.

Plant food and pollen allergen families

- Prolamins (2S albumins, LTPs, proteinase inhibitors, seed storage proteins)
- Cupins (7S and 11S seed storage proteins)
- Pathogenesis-related (PR) proteins (e.g., PR10 proteins) in pollen and fruits
- Profilins (structural proteins)
- Expansins (related to cell growth and expansion; common in pollen)
- Polcalcins (EF-hand proteins involved in Ca-binding; common in pollen)

Animal food, airway and venom allergen families

- Tropomyosins (common throughout the animal kingdom)
- Parvalbumins (EF-hand proteins involved in Ca-binding; common in vertebrates)
- Caseins (milk proteins)
- Serpins (egg white proteins)
- Proteases, Phospholipases, Hyaluronidases (in insect stings and bites)

Jenkins *et al.* (2007) studied the allergenicity of animal food proteins with regard to the evolutionary distance to human homologs in order to identify how closely a protein must resemble a human protein to lose its allergenic potential. Using sequence homology methods, animal food allergens were classified into Pfam families and were further analyzed in silico for their evolutionary and structural relationships. Remarkably, proteins with a sequence identity above 62% to human homologs were rarely allergenic. Apparently, animal allergens challenge the capability of the human immune system to discriminate between foreign and self-proteins. Such immune responses run close to becoming autoimmune responses.

5.3 Cross-Reactivity

Two major questions remain from this study. The first question relates to the phenomenon of cross-reactivity. Many newly found allergens appear to be homolog of already identified allergy family members. Often, the source of these allergens is not known as a sensitizing agent. Cross-reactivity basically describes the relationship between three reagents: two allergens and one antibody (IgE). Aalberse (2007) distinguished between symmetric and asymmetric cross-reactivity. Symmetric cross-reactivity involves both allergens to be able to sensitize and to inhibit the binding to the same IgE equally. Good examples of symmetric cross-reactive allergens can be found among various grass families and among several mite families. In the case of asymmetric cross-reactivity, the non-sensitizing (incomplete) allergen has a lower affinity to IgE than the sensitizing allergen, but is nevertheless capable of provoking an allergic response. A good example is the birch-apple syndrome, where the birch pollen Bet v 1 allergen is able to sensitize, and the apple Mal d 1 cross reacts by eliciting symptoms through triggering mast cells loaded with IgE anti Bet v 1. Cross-reactivity is a common phenomenon in allergies. Currently, the sensitizing potency of most allergens is still unknown (Radauer *et al.*, 2008). It seems that the majority of the allergens currently filed in databases are asymmetrically cross-reacting allergens. Therefore, the number of true-sensitizing allergens will be limited to only a few representatives of the known allergen protein families.

The second question relates to the expected sequence similarity as the basis for high or low potential of allergens to cross-react. Sequence similarity was thought to be a suitable parameter to explain such cross-reacting potential. However, from several studies in different allergen protein families, no confirmation for this statement could be found (Radauer *et al.*, 2008, and various references therein). Even allergens with sequence homologies of less than 40% showed considerable cross-reactivity, e.g., the 2S albumins in peanut (Ara h 2) with allergens in almond and Brazil nut (De Leon *et al.*, 2007). In contrast, allergen isoforms with very high similarity (>95%) may reveal considerable differences in IgE binding, as was observed in birch (Schenk *et al.*, 2006). The characteristics of the true-sensitizing allergens, and the basic mechanism of cross-reactivity, are still not well understood.

5.4 Amino Acids and IgE Recognition

A start to a solution to this problem might be found in the work of Oezguen *et al.* (2008). On the basis of a previously created Structural Database of Allergenic Proteins SDAP (http://fermi.utmb.edu/SDAP), which facilitated rapid analysis of closely related allergens, continuous IgE epitope sequences could be compared in their structural setting. They obtained reliable 3D models for 433 allergen sequences, including the major allergens from peanuts, tree nuts, weed and tree pollen, fungi and insects, and used these 3D models only for those allergens of which the IgE epitopes had been mapped. From these epitopes, they determined which amino acids had the highest surface exposure. In this way they were able to identify on a structural basis the amino acids that are most likely to be involved in IgE binding. Data from 16 allergens, representing 9 Pfam protein families, were compared. It appeared that only a small subset of the amino acid residues in the epitopes had sufficient exposure to be involved in binding IgE in the intact protein. Overall, the IgE binding sites of the epitopes were considerably hydrophilic, and marked by highly exposed lysine side chains. Other amino acids that were found to occur with high frequency in the epitopes were alanine, serine, asparagine and glycine. Against this, the amino acids phenylalanine, tryptophane, methionine and isoleucine were significantly less often observed in IgE epitopes. This reflects the properties of the IgE binding and binding dynamics of allergens, suggesting that the binding process might be guided via electrostatic funneling (which would explain the net positive charge or at least the high density of lysine at the epitope region) (Oezguen *et al.*, 2008).

We counted the number of the amino acids with IgE binding potential in the consensus sequence of Bet v 1, including 160 amino acids. The results were noticeable: 16 lysine, 14 glycine, 13 alanine, 12 serine and 8 asparagine residues were counted, together representing about 40% of the total protein. This makes the chances of a high surface exposure of several of these amino acids in the Bet v 1 protein very high. In a preliminary analysis, we could identify the amino acid asparagine (residue 28) and the sequence motive glycine-glycine-serine (at position 110–112) as being involved in high allergenicity (Schenk *et al.*, 2011). From these results, allergenicity may be seen from a new perspective, with the existence of a limited number of specific amino acids, probably arranged in restricted sequence motifs, with especially lysine playing the major role.

Given these results, the phenomenon of cross-reactivity of proteins with low sequence homology, and the absence of cross-reactivity in highly homologous proteins, should be reconsidered.

5.5 The Birch-Apple Syndrome Model

Pathogenesis-related class 10 (PR10) proteins constitute the largest group of aeroallergens and are among the four most common food allergens (Breiteneder and Ebner, 2000). The main PR10 allergen is Bet v 1, which is expressed in high amounts (about 50% of the total protein content) in pollen of all birch species (Schenk *et al.*, 2006). The occurrence of sensitization to Bet v 1 is very high among atopic individuals, especially in regions (Eurasia) of higher latitude (above 45°), where birch is endemic. Pollen counts for several West-European cities show that birch pollen is highly abundant (Spieksma *et al.*, 2003). Individuals that are sensitized to Bet v 1 may experience oral allergic symptoms (OAS) upon consumption of nuts, vegetables and Roseceae fruits due to IgE cross-reactivity with Bet v 1 homologs in these foods. Given this prominent role of Bet v 1 allergens in the sensitization to hay fever and OAS, birch is a relevant target for research. Such research may focus on the development of allergy prevention strategies, as well as fundamental aspects of sensitization and cross-reactivity. Here, we will focus on the latter, with special emphasis on the birch-apple syndrome based on recent research into the genetics, genomics and proteomics that has been conducted on birch (Schenk *et al.*, 2006; 2009) and apple (Gao *et al.*,

2005; 2008). In birch, it has been found that at least seven different (but highly homologous) genes are encountered in pollen. The resulting proteins which showed differences in allergenicity, and pollen of a single tree contain a mixture of these. In apple, the Mal d 1 genes (16 loci) have been mapped on the genetic linkage map of apple. In particular, genes on linkage group 16 appear to have a role in apple allergy.

Using these extended sets of data now allow us to address in depth the following questions:

- Which of the Bet v 1 proteins has sensitizing potency, and which protein factors (T-cell epitopes and B-cell epitopes, but also expression levels) are particularly involved in the sensitization process?
- In what tertiary form (single or e.g., as stable multimers) are Bet v 1 proteins exposed to the immune system?
- Is sensitization a single event or does re-sensitization of an individual occur annually during the birch pollen season?
- Birch pollen-sensitive persons also suffer from hazel and alder pollen. Is this always the case, or just true for some individuals? Do the Bet v 1 homologous proteins of these species have equal sensitizing potency (symmetric cross-reactivity) to Bet v 1?
- Will comprehensive 3D analysis of the various Bet v 1 allergen isoforms and variants reveal the epitopes (amino acid motifs) with maximum IgE binding?
- Which Bet v 1 sensitization profile (IgE-profile) relates to Mal d 1 cross-reactivity (since only about 50%–70% of birch-allergic individuals have OAS to apple)?
- Which cross-reactive allergen profile in apple (according to expressed Mal d 1 proteins, their epitopes, and their relative amounts) causes OAS? Which sensitization IgE profile fits to which cross-reactive allergen profile?

It is likely that ongoing research will, in the near future, provide answers to these questions. These answers will reveal the fundamental causes of allergenicity. Comparisons with other cross-reactivity syndromes (e.g., in the LTP family with peach as a prominent example) may show molecular similarities (amino acid sequences and protein-surface exposition) in IgE recognition and general allergen epitope motifs. The keys to the questions mentioned above must be hidden somewhere in the complex regulatory activities of the immune system regarding T-cell differentiation and signaling, where allergenic proteins may mimic proteins from different pathogens, and where various protein structures are possible. This might be a reason why it is so difficult to ascribe allergic reactions to a single or a few protein structures. For example, recent research has reported that the mite allergen Der p 2 has structural homology and functional mimicry with MD-2 (also known as LY96), and the lipopolysaccharide (LPS)-binding component of the Toll-like 4 receptor (Trompette *et al.*, 2009). Further analysis in this research has indicated that about half of the allergens known are involved in lipid interactions, which suggests, at least for a large group of allergens, the occurrence of specific biological activity as a common mechanism underlying the phenomenon of allergenicity (Thomas *et al.*, 2005; Trompette *et al.*, 2009). And regarding signal pathways, it was found that the Ras-related protein Rab10, induced by LPS, facilitated TLR4 signaling by promoting replenishment of TLR4 onto the plasma membrane (Wang *et al.*, 2010).

Another finding, from research on the hyper- and hypo-allergenicity of Bet v 1 isoforms (Zaborsky *et al.*, 2010), indicated that particularly the hypoallergenic isoform, although not impaired in provoking an IgE response, was more potent to trigger the production of IgG and IgA antibodies, because of a more efficient activation of antigen presenting cells (APCs). This was thought to be caused by the formation of aggregates through disulfide linkages due to a specific serine to cysteine exchange in this hypoallergenic isoform. This aggregation might trigger the establishment of protective antibodies and not that of IgE. Allergen-mediated cross-linking of IgE antibodies bound to the FceRI receptors on the mast cell surface is the key feature of the type Ⅰ allergy. It was recently found that the majority are transient dimers that are formed through high protein concentrations that are obtained in cells by colocalization. They suggested that dimerization would be a very common and essential feature for allergens (Rouvinen *et al.*, 2010).

Whatever involvement in lipid metabolism, or occurrence of protein aggregation, or any other known biological (or mimicking) activity of a certain allergenic protein, this will have its particular consequences as well as potentials for specific immunotherapy.

References

Aalberse, R.C. (2007). Assessment of allergen cross-reactivity. BMC Clin Mol Allergy, 5, 2.

Breiteneder, H., Ebner, C. (2000). Molecular and biochemical classification of plant-derived food allergens. J Allergy Clin Immunol, 106, 27-36.

Breiteneder, H., Radauer, C. (2004). A classification of plant food allergens. J Allergy Clin Immunol, 113, 821-830.

De Leon, M.P., Drew, A.C., Glaspole, I.N., *et al.* (2007). IgE cross-reactivity between the major peanut allergen Ara h 2 and tree nut allergens. Mol Imunol, 44, 463-471.

Gao, Z.S., van de Weg, W.E., Schaart, J.G., *et al.* (2005). Genomic cloning and linkage mapping of the Mal d 1 (PR-10) gene family in Apple (Malus domestica). Theor Appl Genet, 111, 171-183.

Gao, Z.S., van de Weg, W.E., Matos, C.I., *et al.* (2008). Assessment of allelic diversity in intron-containing Mal d 1 genes and their association to apple allergenicity. BMC Plant Biology , 8, 116.

International Study of Asthma and Allergies in Childhood (ISAAC) Steering Committee. (1998). Worldwide variation in prevalence of symptoms of asthma, allergic rhinoconjunctivitis, and atopic eczema. Lancet, 351,

1125-1132.

Janson, C., Asbjornsdottir, H., Birgisdottir, A., *et al.* (2007). The effect of infectious burden on the prevalence of atopy and respiratory allergies in Iceland, Estonia, and Sweden. J Allergy Clin Immunol, 120, 673-679

Jenkins, J.A., Griffiths-Jones, S., Shewry, P.R., *et al.* (2005). Structural relatedness of plant food allergens with specific reference to cross-reactive allergens: An in silico analysis. J Allergy Clin Immunol, 115, 163-170.

Jenkins, J.A., Breiteneder, H., Mills, E.N.C. (2007). Evolutionary distance from human homologs reflects allergenicity of animal food proteins. J Allergy Clin Immunol, 120, 1399-1405.

Oezguen, N., Zhou, B., Negi, S.S., Ivanciuc, O., *et al.* (2008). Comprehensive 3D-modeling of allergenic proteis and amino acid composition of potential conformational IgE epitopes. Mol Immunol, 45, 3740-3747.

Radauer, C., Breiteneder, H. (2006). Pollen allergens are restricted to few protein families and show distinct patterns of species distribution. J Allergy Clin Immunol, 117, 141-147.

Radauer, C., Breiteneder, H. (2007). Evolutionary biology of plant food allergens. J Allergy Clin Immunol, 120: 518-525

Radauer, C., Bublin, M., Wagner, S., Mari, A., *et al.* (2008). Allergens are distributed into few protein families and possess a restricted number of biochemical functions. J Allergy Clin Immunol, 121, 847-852.

Rouvinen, J., Janis, J., Laukkanen, M.L., *et al.* (2010). Transient dimers of allergens. PLoS ONE, 5(2), e9073.

Schenk, M.F., Gilissen, L.J.W.J., Esselink, G.D., *et al.* (2006). Seven different genes encode a diverse mixture of isoforms of Bet v 1, the major birch allergen. BMC Genomics, 7, 168.

Schenk, M., Cordewener, J., America, T., *et al.* (2009). Characterization of PR-10 genes from eight Betula species and detection of Bet v 1 isoforms in birch pollen. BMC Plant Biology, 9, 24.

Schenk, M.F., Cordenener, J.H., America, A.H., *et al.* (2011). Proteomic analysis of the major birch allergen Bet v 1 predicts allergenicity for 14 birch species. J Proteomics, 74(8) : 1290-1300.

Spieksma, F.T.M., Corden, L.M., Detandt, M., *et al.* (2003). Quantitative trends in annual totals of five common airborne pollen types (Betula, Quercus, Poaceae, Urtica, and Artemisia), at five pollen-monitoring stations in Western Europe. Aerobiologica, 19, 171-184.

Thomas, W.R., Hales, B.J., Smith, W.A., (2005). Structural biology of allergens. Curr Allergy Asthma Rep, 5, 388-393.

Trompette, A., Divanovic, S., Visintin, A. *et al.* (2009). Allergenicity resulting from functional mimicry of a Toll-like receptor complex protein. Nature, 457 (7229), 585-U91.

van den Akker-van Marle, M.E., Bruil, J., Detmar, S.B. (2005). Evaluation of cost of disease: Assessing the burden to society of asthma in children in the European Union. Allergy, 60, 140-149.

Wang, D., Lou, J., Ouyang, C., *et al.* (2010). Ras-related protein Rab10 facilitates

TLR4 signaling by promoting replenishment of TLR4 onto the plasma membrane. PNAS, doi/10.1073/pnas.1009428107.

Zaborsky, N., Brunner, M., Wallner, M., *et al.* (2010). Antigen aggregation decides the fate of the allergic immune response. J Immunol, 184, 725-735.

6

Seafood Allergens in China and Anti-allergenic Property of Seaweeds

Zhen-Xing Li*, Hong Lin

College of Food Science and Engineering, Ocean University of China, Qingdao, 266003, China
* E-mail: lizhenxing@ouc.edu.cn

Abstract: Seafood is composed of diverse organisms and humans are allergic to many types of seafood. Although many proteins have been identified as allergens, tropomyosin is recognized as the major allergen of Crustacean and Mollusc, while parvalbumin is a major allergen of fish. These allergens are resistant to food processing such as heating and enzyme hydrolysis. Currently, there are no effective methods to produce hypoallergenic foods. In China, brown seaweed has been reported to have some anti-allergic components, identified as polyphenols. These findings will provide safer medicines for more effective treatment of patients allergic to seafood.

6.1 A Brief Introduction to Seafood Allergy

Seafood consumption is increasing rapidly due to their nutrients and flavour, with a world trade of more than $60 billion a year (FAO, 2008). Almost 200 countries supply seafood products to the global marketplace, consisting of more than 800 commercially important species of fish, crustaceans and molluscs, including 30 species of shrimp alone. The products take hundreds of forms, ranging from canned tuna to fresh boneless salmon fillets, from salted herring roe to dried shark fins, frozen pollock blocks, individually quick-frozen breaded cod portions, smoked mackerel, clam juice, live lobster, and fish meals and oil. The market is supplied by a global network of thousands of fishing vessels and marine and inland aquaculture establishments, thousands of processors, and tens of thousands of wholesalers and brokers.

Food allergies are an abnormal immunological response caused by sensitisation to foods or food components. In the past few years, food allergies have become more and more widespread especially in developed countries. In the European Community Respiratory Health Survey administered to 17,280 adults in 15 countries, 12% of respondents reported a food allergy or intolerance, ranging from 4.6% in Spain to 19.1% in Australia (Woods *et al.*, 2001). In a similar study with 33,110 people in France, 3.5% of those surveyed reported having a food allergy (Kanny *et al.*, 2001). The main foods responsible for an allergic reaction in childhood were eggs and milk, with atopic dermatitis as a frequent manifestation. In adulthood, seafood, fruits, and vegetables were the most common causes, with angioedema or urticaria, or both, as the most frequent symptoms.

A more recent survey (Sicherer *et al.*, 2004) found that 1 in 50 Americans had shellfish allergies. Shellfish is the number one cause of food allergies in adults in the United States and is responsible for the majority of emergency visits requiring medical treatment (Clark *et al.*, 2004). According to the announcement by FAO, 8 kinds of major foods are responsible for 90% of food allergy reactions. Among them, 2 kinds of seafood were identified-fish and its products, and crustaceans and its products. Mollusc allergies have also been treated as an important issue (Ishikawa *et al.*, 1998). Higher prevalence of seafood allergies is found in countries where seafood is a staple part of the diet, as in Scandinavian countries, Spain, east China and Japan. In comparison to milk or egg allergies, seafood allergies are a life-long allergy and most allergic patients are adults. This current review gives an overview of the potential allergens present in seafood.

6.2 Seafood Allergens

Seafood is a potent allergen in sensitised individuals and can cause life-threatening adverse reactions that are usually life-long. Extreme sensitivity to minute quantities of fish is occasionally noted, and even exposure to fumes of fish being cooked is enough to precipitate reactions in certain individuals. Three groups of seafood causing allergic response can be identified. The Mollusc group includes three different classes of seafood with species such as abalone, oysters, mussels, and squid (calamari). The second group, the crustacean, includes rock lobsters (crayfish), prawns, crabs and shrimps; and the third important group of seafood includes all the common edible fish, such as hake, cod and snook. Cod is the most frequently reported cause of fish allergies, but reactions to other fish such as haddock, herring, sprat, halibut, plaice, mackerel, trout and salmon are also well recognized. Very often patients are only allergic to certain species and are able to eat other seafood species without problems.

6.2.1 Fish Allergens

Parvalbumin exists in various species of fish, such as Atlantic cod, carp, Atlantic salmon, Pacific mackerel and tuna, and is the major and sole allergen for 95% of fish-allergic patients suffering from IgE-mediated hypersensitivity to fish (Arif *et al.*, 2007). Fish muscle parvalbumin is a small, stable acidic Ca^{2+}-binding protein, resistant to heat, chemical denaturation, and proteolytic enzymes (Arif, 2009). Based on amino acid sequence data, the parvalbumin protein family can be subdivided into two evolutionary distinct lineages: the α-group, consisting of less acidic parvalbumins with isoelectric points at or above pI 5.0, and the β-group, consisting of more acidic parvalbumins with isoelectric points at or below pI 4.5 (Lehrer *et al.*, 2003). Most fish muscle contains parvalbumins of either α- or β-lineage (van Do *et al.*, 1999). Since parvalbumins belong to a family of Ca^{2+}-binding proteins, it has recently been demonstrated that IgE binding to certain Ca^{2+}-binding plant allergens requires protein-bound calcium. Bugajska-Schretter *et al.* (1998) reported a significant reduction of IgE binding to parvalbumin after Ca^{2+} depletion. Structural studies with different Ca^{2+}-binding proteins have demonstrated dramatic conformational changes not only in the Ca^{2+}-binding regions but also in distant regions of these proteins.

Other fish allergens, such as type-I collagen which is present in many fish species, a 41 kDa minor allergen in cod, 35−90 kDa allergens in snapper, and 94−105 kDa allergens in tuna and marlin (Nakamura *et al.*, 2009), are not as important as parvalbumin in causing allergic responses.

6.2.1.1 Atlantic Cod (*Gadus Morhua* or *Gadus Callarias*) (Fig. 6.1)

The Atlantic cod (*Gadus morhua*) is a well-known demersal food fish belonging to the family Gadidae. It is also commercially known as cod, codling or Haberdine. The allergen in this fish was first identified in 1969, and called Allergen M, Gad m 1 and finally Gad c 1. This 113 amino-acid protein has a molecular weight of 12.3 kDa molecular weight and an isoelectric point of 4.75. Several following studies focused on its characterizations and the complete primary structure (Aas and Elsayed *et al.*, 1969; Elsayed and Bennich, 1975). Then the epitopes were illustrated using tryptic hydrolysis and synthetic peptides in radioallergosorbent assay (RAST). It was demonstrated that Gad c 1 contains at least 5 IgE-binding sites along its polypeptide chain at residues 13–32, 33–44, 49–64, 65–74, and 88–96 (Elsayed *et al.*, 1983; 1991). Sequence comparison showed that this protein had high sequence similarities (more than 60%) with other parvalbumins.

6.2.1.2 Atlantic Salmon (*Salmo Salar*) (Fig. 6.2)

The major allergen of Atlantic salmon is designated Sal s 1. Two distinct full-length parvalbumin cDNAs of Atlantic salmon were characterised and designated 14.1 and 24.1, with an amino acid identity of 51%–58% compared to the cod Gad c 1. Both clone 24.1 and 14.1 include the entire coding region and some nucleotides upstream. The nucleotide sequence similarity between clone 24.1 and 14.1 for the coding region is 69%, 57% for the first 45 bp in the upstream, and 51% for the 30-untranslated first 200 bp. The translated AA sequences of clones 24.1 and 14.1 were found to be identical in 65% of the residues (LindstrØm *et al.*, 1996).

Fig 6.1. Atlantic cod

Fig. 6.2. Atlantic salmon

6.2.1.3 Alaska Pollack (*Theragra Chalcogramma*) (Fig. 6.3)

In Alaska pollack, the muscle calcium-binding parvalbumin is the potent major allergen, designated c1, with a molecular weight of 11.5 kDa. There are two parvalbumin isotypes, P1 and P2, present in similar concentration and with an amino acid sequence identity of 67%. Both proteins belong to the β-lineage of parvalbumins, with isoelectric points of 4.39 for P1 and 4.60 for P2. They have 72% identical residues with Atlantic cod T1, while P2 was somewhat more similar to Atlantic cod T2 (77%) than P1 (65%) (van Do *et al.*, 2005). The fish allergen with the highest capability of inhibiting IgE-reactivity to Gad c 1 in patients' sera was c 1.

Fig. 6.3. Alaska pollack

6.2.1.4 Tuna (Fig. 6.4)

The purified allergens from big-eye tuna (*Thunnus obesus*), with a MW of 10–12 kDa, have been designated as Thu o 1.01 and Thu o 1.02 (Shiomi *et al.*, 1999). The parvalbumin was found in the white muscle of tuna and was absent in the red muscle (Lim *et al.*, 2005). An allergen of about 46 kDa detected in yellowfin tuna did not belong to the parvalbumin group, with an indication of the weak allergenicity of tuna.

6.2.1.5 Pacific Pilchard (*Sardinops Sagax*) (Fig. 6.5)

The most allergenic pilchard isoform has an estimated MW between 11.5 kDa and 11.9 kDa. Nucleotide sequencing of the gene revealed a 327 bp open reading frame encoding 109 amino acids, which has been designated to be Sar sa 1.0101. This protein was identified as a β-type parvalbumin of 109 residues by the positions of the following 6 amino acids: A-14, L-16, C-19, F-67, Q-69 and T-79. Sar sa 1.0101 displayed the highest homology to the European chub and zebrafish allergens as well as to the β-type parvalbumin from common carp, which shares the same lineage (Beale *et al.*, 2009).

Fig. 6.4. Tuna

Fig. 6.5. Pacific pilchard

6.2.1.6 Mackerel (Fig. 6.6)

There are three species of mackerel: Pacific mackerel (*Scomber japonicus*), spotted mackerel (*Scomber australasicus*) and Atlantic mackerel (*Scomber scombrus*). Their parvalbumins are called Sco j 1, Sco a 1 and Sco s 1, respectively, according to Hamada *et al.* (2003). They all have a MW of 11 kDa, and none has an exposed N-terminus. In the case of *S. scombrus*, one fragment (AAGSFDHK) is identical with segment 20–27 of the *S. japonicus* parvalbumin, and the other two fragments (SGYIEEEELK and IGVDEFAAK), while having one or two alterations, correspond well to the segments 55–64 and 97–105, respectively, which supports the homogeneity of these parvalbumins. Sco j 1, sco a 1 and Sco s 1 have been judged to have almost the same potency in IgE binding ability from their reactivity with patient serum. This suggests that the three species of mackerels are almost equally allergenic. The deduced amino acid sequence (108 residues) of Sco j 1 shows high identity (58%–76%) with those of parvalbumins from cod, β1 and β2 from Atlantic salmon and Cyp c 1.01 and Cyp c 1.02 from carp, with especially high identities (more than 80%) in the two calcium-binding regions (segments 49–62 and 90–101). Similar to Atlantic salmon and carp parvalbumins, Sco j 1 is a β-type parvalbumin.

6.2.1.7 Carp (*Cyprinus Caris*) (Fig. 6.7)

The major allergens of carp are two distinct, highly homologous parvalbumin isoallergns, Cyp c 1.01 and Cyp c 1.02 (Swoboda *et al.*, 2002). The open reading frames of both variants encode mature proteins of a size typical for parvalbumins of the β-lineage, with a calculated molecular mass of 11.5 kDa and isoelectric points of 4.41 (Cyp c 1.01) and 4.77 (Cyp c 1.02).

Fig. 6.6. Mackeral

Fig. 6.7. Carp

6.2.1.8 Cross-Reactivity

It has been demonstrated that parvalbumin is present in the white muscle of many fish species and the fact that almost 70% of the patients tested reacted exclusively to parvalbumin indicates common cross-reactivity (Hilger *et al.*, 2004). An amino acid homology of 60%–90% has been observed between parvalbumin from fish species, but results from studies on cross-reactivity are conflicting and the degree of allergenicity between species has been found to vary significantly. Patients with an allergy to cod were also particularly sensitised to salmon, and other fish species. Cod, salmon, pollack, herring, and wolffish contained the most potent cross-reacting fish parvalbumins, whereas halibut, flounder, tuna, and mackerel were the least allergenic (Kuehn *et al.*, 2010). Recent studies with Cyp c 1 and Sco j 1 suggest the importance of conformational-type IgE epitopes for fish parvalbumins. It is therefore likely that IgE cross-reactivity among fish parvalbumins largely depends on conformational-type rather than linear-type IgE epitopes (Yoshida *et al.*, 2008).

6.2.2 Shellfish Allergens

Shellfish is a broad term for all aquatic animals that have a shell or shell-like exoskeleton. In general, shellfish are separated into crustaceans and molluscs. In coastal countries, shellfish are among the most common causes of food allergies. It has been reported that hypersensitivity reactions to crustaceans and molluscs can cause clinical symptoms such as urticaria, asthma and diarrhoea. Among Spanish children, crustaceans cause 3.8% and molluscs 1.6%, of type 1 food-induced allergic reactions and up to 33% of food allergies in adults (Woods *et al.*, 2001). A recent survey by Sicherer *et al.* (2004) found that 1 in 50 Americans had shellfish allergy.

Studies on shellfish allergies indicate that tropomyosin, a 35–38 kDa myofibrillar protein involved in muscle contraction, which was first identified as a major allergen in shrimp, is a major cross-reactive allergen in crustaceans and molluscs. In recent years, tropomyosin has been demonstrated to be a major allergen for various shellfish, including shrimp, lobster, crab, squid, oyster and octopus.

6.2.2.1 Crustacea

It is well-known that crustaceans are highly cross-reactive, and usually avoidance of all crustaceans is recommended by individuals who suffer from related allergies. Tropomyosin has been identified as the main cross-reactive molecule. There is considerable *in vitro* cross-reactivity between molluscs and invertebrates such as dust mites and cockroach. The cross-reactivity among crustaceans, cockroach, and dust mites seems to be based on the sequence similarities of tropomyosin IgE-binding epitopes. Arginine kinase has also been described as a cross-reacting allergen within crustaceans and between crustaceans and insects. Table 6.1 shows the allergens found in crustaceans.

Table 6.1 Identified allergen in Crustaceans

Species	Allergen	MW (kDa)	Protein family	References
Shrimp (*Metapenaeus ensis*)	Met e 1	34	Tropomyosin	Leung *et al.*, 1994
Northern brown shrimp (*Penaeus aztecus*)	Pen a 1	36	Tropomyosin	Daul *et al.*, 1994
Indian white shrimp (*Penaeus indicus*)	Pen i 1	38	Tropomyosin	Naqpal *et al.*, 1989
Giant tiger shrimp (*Penaeus monodon*)	Pen m 1 Pen m 2	38 39.9	Tropomyosin Arginine kinase	Yu *et al.*, 2003
Pacific white shrimp (*Litopenaeus vannamei*)	Lit v 1 Lit v 2 Lit v 3 Lit v 4	34 40 20 22	Tropomyosin Arginine kinase Myosin light chain Sarcoplasmic calcium-binding protein (SCP)	García-Orozco *et al.*, 2007; Ayuso *et al.*, 2008; Ayuso *et al.*, 2009
American lobster (*Homarus americanus*)	Hom a 1	34	Tropomyosin	Mykles *et al.*,1998
Spiny lobster (*Panulirus stimpsoni*)	Pan s 1	34	Tropomyosin	Leung *et al.*, 1998a
Crab (*Charybdis feriatus*)	Cha f 1	34	Tropomyosin	Leung *et al.*, 1998b

6.2.3 Shrimp Allergens

6.2.3.1 Tropomyosin

To date, the only major shrimp (Fig. 6.8) allergen identified is the muscle protein tropomyosin, termed Pen a 1 (Daul *et al.*, 1994), Pen m 1 (Yu *et al.*, 2003), Met e 1 (Leung *et al.*, 1994), Pen i 1 (Naqpal *et al.*, 1989), or Lit v 1, depending on the species . At least 80% of shrimp-allergic subjects react to tropomyosin (Morgan *et al.*, 1990). The nucleotide sequence cDNA clone of Met e 1 reveals an open reading frame of 281 amino acid residues, coding for a protein of 34 kDa. Based on the frequency and intensity of IgE reactivity, 5 major IgE-binding regions of Pen a1 have been identified (Reese *et al.*, 1997). All 5 major IgE-binding regions were 15–38 amino acids long, with no substantial differences in amino acid group composition compared to the whole molecule. The 5 major IgE-binding regions identified are at regular intervals of approximately 42 amino acids (7 heptads), suggesting a relationship with the repetitive coiled-coil structure of the tropomyosin molecule. The high degree of similarity between Pen a 1 IgE-binding regions and homologous sequences in invertebrate tropomyosins, and the lower percentage similarity with homologous regions of vertebrate tropomyosins supports a structural basis for cross-reactivity of allergenic tropomyosins. Sequence identities and similarities of the Pen a 1 IgE-binding regions with homologous regions of allergenic arthropod tropomyosins have been found to be as high as 100%, whereas identities with homologous vertebrate sequences ranged from 36% to 76% and similarities from 53% to 85% (Emoto *et al.*, 2009).

Comparison of the amino acid sequences of the antigens Pen i 1, Pen a 1 and Met e 1 showed that they are similar or identical. Pen a 1 and Lit v 1 are 100% identical. An 86% homology has been found between Met e 1 and Pen i 1.

Fig. 6.8. Shrimp

6.2.3.2 Arginine Kinase

In addition to tropomyosin, a minor shrimp allergen, arginine kinase, like Pen m 2 (Yu *et al.*, 2003) and Lit v 2 (García-Orozco *et al.*, 2007), has been reported. The cDNA of Pen m 2 has a 1,071 bp open reading frame encoding a 356-amino-acid protein with a theoretical molecular weight of 39.9 kDa. The sequence of this protein showed similarity (60%) to crustacean arginine kinase (Yu *et al.*, 2003). Pen m 2 exhibited arginine kinase activity and reacted with IgE from shrimp-allergic patients. The reactivity of purified arginine kinase from shrimp (*Metapenaeu- sensis*), lobster (*Homarus gammarus*), crawfish (*Metanephrops thomsoni*), and crab (*Scylla serrata*) with anti-Pen m 2 antibody and sera from shrimp-sensitive patients indicates that arginine kinase is a common allergen among crustaceans.

6.2.3.3 Myosin Light Chain (MLC)

Muscle myosin has two heavy chains. The globular motor domains interact with actin, whereas the tails dimerise in a coiled-coil structure. Two light chains, each 20 kDa, wrap around the neck region of each myosin heavy chain. An MLC protein, named Lit v 3, has been identified as a new major shrimp allergen (Ayuso *et al.*, 2008). It was recognized by more than 50% of subjects with shrimp allergies. Lit v 3 has 177 amino acids, a molecular weight of 20 kDa, and a calculated isoelectric point of 4.2. IgE binding to Lit v 3 is detected in both raw and boiled shrimp extracts, sometimes with stronger IgE-binding to the raw protein extract. IgE binding to the boiled form of MLC appears to be greater in adults, while children tend to react to the MLC in the raw extract with higher intensity. Overall, MLC is a predominant allergen particularly in children. The amino acid sequence of MLC is 66% similar and 51% identical to cockroach Bla g 8, the allergenic MLC of *B. germanica*. Sequence similarity between MLCs can be implicated with *in vitro* and possibly clinical cross-reactivity of shrimp and cockroach and possibly also with dust mites. In contrast, sequence identity with other invertebrate MLCs, such as *Schistosoma* (13% identity) and *Aedes* (17% identity) species, was low (Ayuso *et al.*, 2008).

6.2.3.4 Sarcoplasmic Calcium-Binding Protein (SCP)

In a recent study (Ayuso *et al.*, 2009), SCP was characterised as a new shrimp allergen, and named Lit v 4.0101. SCP has very similar molecular weight and isoelectric point to MLC, with 194 amino acids, a molecular weight of 22 kDa, and a calculated isoelectric point of 4.7. It has high sequence identity with the α-B and α-A chains (93.8%) of *Penaeus* spp (P02636), and 80% with the β chain. SCPs are acidic, cytosolic, EF-hand type Ca^{2+}-binding proteins (20–22 kd). In shrimp, SCPs are dimers with 2 polypeptide chains, with 3 calcium-binding sites in each chain. Although sensitisation to tropomyosin has been implicated in cross-reactivity between crustaceans and molluscs, and also with other arthropods, sensitisation to SCP appears to be involved only in cross-reactivity among crustaceans. Seventeen of 23 (74%) children recognized SCP compared with 3 of 29 (10%) adults, suggesting that SCP is a major allergen in the young population (Shiomi *et al.*, 2008).

Parvalbumin, tropomyosin, MLC, and SCP are all EF-hand type Ca^{2+}-binding proteins. Amino acid sequence identity of shrimp SCP with other EF-hand type Ca^{2+}-binding proteins is low (12% sequence identity with cod parvalbumin Gad m 1 and shrimp MLC Lit v 3), but it has been suggested that they are all derived from a common ancestral protein because of the common structure of Ca^{2+}-binding sites (Wopfner *et al.*, 2007).

6.2.4 Lobster Allergens

Allergens of two species of lobster have also been identified as tropomyosin. The first to be identified was the fast muscle tropomyosin protein from the American lobster (*Homarus americanus)*, Hom a 1 (Mykles *et al.*, 1998).The allergen in spiny lobster (*Panulirus stimpsoni*) is Pan s 1 (Leung *et al.*, 1998a). The amino acid compositions of Pan s 1 and Hom a 1 are very similar, having an open reading frame of 274 and 284 amino acid residues, respectively, with an identical molecular weight of 34 kDa. Comparison of amino acid sequences of Pan s 1 and Hom a 1 with the shrimp tropomyosin Met e 1 indicates that these proteins are very similar, with 98.2% identity between Pan s 1 and Met e 1, 94.7% identity between Hom a 1 and Met e 1, and 97.5% identity between Pan s 1 and Hom a 1.

Fig. 6.9. American lobster

6.2.5 Crab Allergens

Leung *et al.* (1986b) first reported a 34 kDa protein, designated Cha f 1, as the major crab allergen, and identified it as tropomyosin based on nucleotide and amino acid sequence comparison. Liang *et al.* (2008) compared tropomyosin genes among three kinds of crab (Chinese mitten crab, mud crab, and swimming crab). The results demonstrated that, although Chinese mitten crab (*Eriocheir sinensis*) (Fig. 6.10) live in fresh water while the mud and swimming crab live in salt water, all three tropomyosin genes are 855 bp in size, encoding 284-amino-acid residues. Nucleotide sequence analysis revealed that Chinese mitten crab tropomyosin has extensive similarity in amino acid composition and peptide sequence identity with that from other crab species. It is 100% identical with that of *P. sanguinolentus*, and 92.4% homologous to the Homerianus fast muscle tropomyosin, Hom a 1, and 91.4% to *Panulirus stimpsoni* tropomyosin, Pan s 1. This high sequence identity established the molecular basis for the immunological cross-reactivity among crab and shrimp tropomyosins.

The 34 kDa tropomyosin is the major crab allergen identified so far, while there are notably other potential crab allergens, such as the 40 kDa arginine kinase in Chinese mitten crab (Liang *et al.*, 2008).

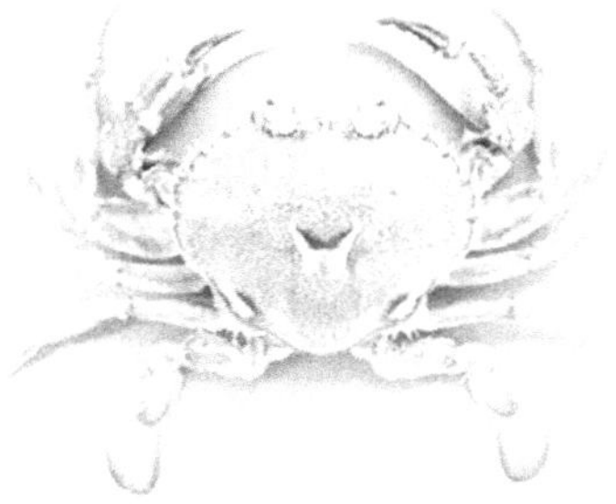

Fig. 6.10. Chinese mitten crab

6.2.6 Mollusca Allergens

Molluscs include cephalopods such as squids, gastropods such as abalone, and bivalves such as oysters. The major allergens of several kinds of mollusc that have been reported are shown in Table 6.2.

Table 6.2 Mollusca allergens

Species	Allergen	MW (kDa)	Protein family	References
Squid (*Todarodes pacificus*)	Tod p 1	38	Tropomyosin	Miyazawa *et al.*, 1996
Octopus (*Octopus vulgaris*)	Oct v 1	31–34	Tropomyosin	Ishikawa *et al.*, 2001
Mussel (*Perna viridis*)	Per v 1	38	Tropomyosin	Marsh *et al.*, 1988
Sea scallop *Placopecten magellanicus*)	Mim n 1	30	Tropomyosin	Patwary *et al.*, 1999
Scallop (*Chlamys nobilis*)	Chl n 1	38	Tropomyosin	Chu *et al.*, 2000
Pacific oyster (*Crassostrea gigas*)	Cra g 1.01 Cra g 1.02 Cra g 1.03	35 35 31	Tropomyosin	Leung *et al.*, 2001
Abalone (*Haliotis midae*)	Hal m 1	49	–	Lopata *et al.*, 1997
Abalone (*Haliotis diversicolor*)	Hal d 1	38	Tropomyosin	Chu *et al.*, 2000

6.2.7 Cephalopod Allergens

Cephalopods include decapods (cuttlefish and squid) (Fig. 6.11) and octapods (octopus). In 9 species of cephalopods, tropomyosin has been shown to be the major allergen in common (Motoyama *et al.*, 2006), with an open reading frame of each tropomyosin cDNA coded for 284-amino-acid residues. The 5 enzymatic peptides of Japanese flying squid tropomyosin were completely consistent with some fragments of tropomyosin as 37–42, 50–61, 62–76, 129–146 and 252–264.

Cephalopod tropomyosins are highly homologous (92%–96%). This high sequence identity is the molecular basis for the cross-reactivity among cephalopod tropomyosins. In contrast, cephalopod tropomyosins share only 70%–82% sequence identities with other mollusc tropomyosins, 63%–64% with crustacean tropomyosins and 51%–55% with human α- and β-tropomyosins.

Cephalopod tropomyosins have the same amino acid sequence in the region 249–259 as Pen a 1 and only one replacement in the regions 88–101, 137–141, 187–197 and 273–281 compared to Pen a 1. It is assumed that these 5 regions are IgE-binding epitopes for both cephalopod and crustacean tropomyosins, accounting for their cross-reactivity. In contrast, there are more than 2 alterations in the regions 43–55, 144–151 and 266–273 between cephalopod and crustacean tropomyosins.

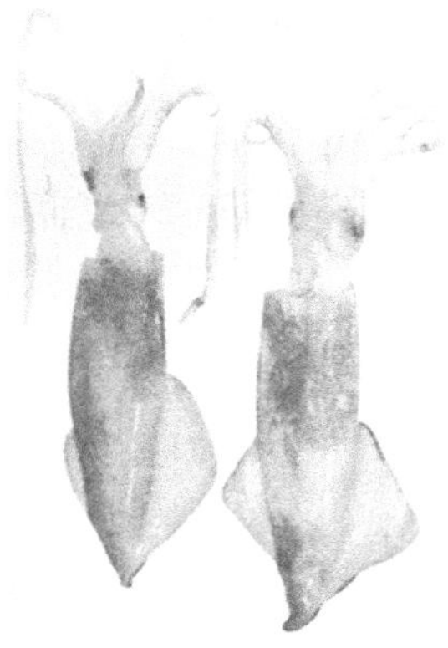

Fig. 6.11. Squid

6.2.8 Gastropod and Bivalve Allergens

Mussel, scallop and whelk (Fig.6.12) tropomyosins have an IgE-binding epitope in the region 249–259, common to cephalopod and crustacean tropomyosins. The amino acid sequences of gastropod and bivalve tropomyosins share only 60% sequence identities with crustacean tropomyosins and fairly low identities (mostly 70%–80%) with cephalopod tropomyosins (Leung *et al.*, 1996). They are highly homologous (with more than 90% identity) with one another within the same group (the same family or the same order) except between the gastropod and bivalve groups (with 70%–80% identity). Cephalopod tropomyosins have the same sequence as Pen a 1 in the region 249–259. In the gastropod and bivalve tropomyosins, the sequence of the region 249–259 is also completely conserved (Leung *et al.*, 2001).

Allergenic cross-reactivity has been reported between shrimp and oyster, shrimp and squid, and between shrimp and other crustaceans, suggesting chemical and immunological similarities between crustacean and mollusc allergens (Reese *et al.*, 1999). All allergens from shrimp, squid and oyster have been found to be chemically identical with one another.

Fig. 6.12. Whelk

6.3 Effect of Food Processing on Allergenicity

Food allergies are receiving increased attention and seem to be increasing, especially in western countries, and consumers are becoming more aware and educated about food allergies. Although several reports have demonstrated that seafood allergens are resistent to food processing, some, such as thermal processing, high intensity ultrasound, irradiation and enzyme hydrolysis, can affect the structure and properties, and change the allergenicity (Shimakura *et al.*, 2005).

6.3.1 Thermal Processing

Food processes such as baking, boiling and microwaving produce thermal effects. Some scientists have reported that seafood allergens can maintain allergenicity even after being boiled in water for 4 h (Leung *et al.*, 1994). The mechanism for this persistence is due to the epitopes of seafood allergens being linear epitopes, with about 8 amino acids forming one epitope. The short peptides are very stable and retain activity even after destruction of the allergens (Nakamura *et al.*, 2006).

Another report pointed out that canned tuna has hypoallergenicity compared with raw tuna muscle (Kelso *et al.*, 2003).

6.3.2 High Intensity Ultrasound

High-intensity ultrasound is an efficient food processing and preservation technology, used successfully for homogenising emulsions, deactivating enzymes, enhancing extraction processes, and accelerating dehydration, ageing and ripening processes. Application of high-intensity ultrasound causes chemical and physical changes in a viscous medium by cyclic generation and collapse of cavities. Increased pressure and temperature in the vicinity of these cavities cause the observed chemical and mechanical effects, converting the native protein structure into a molten globule state and even causing degradation. It has been reported that shrimp allergen treated with high intensity ultrasound exhibited decreased allergenicity, as measured by ELISA, with pooled serum of shrimp allergy patients and polyclonal antibodies (Li *et al.*, 2006). This decrease in allergenicity was confirmed by an immunoblot assay with human sera from shrimp allergy patients. The decrease in allergenicity determined by immunoblotting was higher than the decrease measured by ELISA for an equal treatment time. The decrease in allergenicity of allergen extracted from shrimp treated with high intensity ultrasound was less than that of treated pure allergen. A linear relationship between the decrease in allergenicity and treatment time was observed. The results obtained indicate that high intensity ultrasound may be an efficient way to reduce food allergenicity.

6.3.3 Irradiation

Food irradiation is used primarily as a preservation method, but it can also be used to produce specific changes in food materials. Irradiation processing under various conditions could have unpredictable effects on the allergenicity of allergens in food matrices. The allergenicity of irradiated shrimp allergen extracts, measured by immunoblot and Ci-ELISA, was found to be significantly decreased, but at an irradiation dose less than 10 kGy, there was an increase in allergenicity of shrimp muscle. Above this level, the allergenicity of shrimp muscle began to decrease (Li *et al.*, 2007). These changes in the allergenicity of treated products show the need to control the effects of irradiation on shrimp muscle, especially at doses close to 10 kGy. A combination of irradiation and heat can have a major effect on the integrity and structure of shrimp allergen, with a decrease in the overall IgE-binding ability. This could prove highly beneficial for reducing the allergic response in various raw and processed foods.

6.4 Analytical Methods for Seafood Allergens Detection in Food Matrices

Seafood may be mistakenly ingested due to shared production lines, where cross-contamination may take place between seafood containing and seafood-free food materials, or inappropriately labelled food products, especially prepared food and certain Asian dishes containing seafood. In the past few years, several methods have been established to detect allergens in food matrix (Poms *et al.*, 2004), but compared with other allergens, only a few have been developed to detect seafood allergens. An ELISA assay with LOD of 4 ng/L extracts has been developed to standardise shrimp tropomyosin concentrations in skin testing solutions for patients (Jeoung *et al.*, 1997). Other ELISA assays for tropomyosin detection in foods has been developed with similar LOD (Marianne *et al.*, 2007). Only one quantitative ELISA kit for shellfish protein in food matrices is currently commercially available, based on a prawn (*Penaeus latisulcatus*) tropomyosin polyclonal antibody. An effective sandwich protein chip for shrimp tropomyosin detection in food has recently been successfully developed and validated with 4 different typical food matrixes (Li *et al.*, 2010). This protein chip could be used not only for raw materials and end products, but also for safety control during food processing and inspection. Though methods suitable for food analysis have been reported for the detection of tropomyosins, this sandwich protein chip method has the potential to detect most major allergens in foods at the same time.

6.5 Anti-allergenic Compounds from Seaweeds

It has been frequently reported that food allergies induced by ingestion can cause a severe hypersensitive reaction in humans (Daul *et al.*, 1990; 1993). Food allergy is considered as type I allergies out of 4 general categories, based on the mechanism of immunological involvement. It has been reported that anti-allergic agents had a strong inhibitory effect on the activation of hyaluronidase (Fujitani *et al.*, 2001). Some metals, metallic salts, polyphenols, flavonoids, polysaccharides, and clinical drugs have been reported as anti-allergic agents or inhibitors of hyaluronidase (Jeong *et al.*, 1999; 2000; Akhtar and Bhakuni, 2003). Some food materials tested have also been demonstrated to be anti-allergic (Sanbongi *et al.*, 2004; Yamamoto *et al.*, 2004). Marine algal polyphenols, which contain phlorotannin that is only present in brown algae and restricted to phloroglucinols (1, 3, 5-Tri-hydroxi-benzene) polymers, have been tested for their antihyaluronidase activity (Shibata *et al.*, 2002).

In the past, several reports mentioned anti-allergenic compounds. Meyer and Rapport (1951) reported hyaluronidase inhibition by iron, copper and zinc salts, heparin, polyphenols and flavonoids. Due to the structural similarity of heparin and heparin sulphate to hyaluronic acid, these oligosaccharides were investigated as possible inhibitors of hyaluronidase, but inhibition was achieved only at concentrations much higher than physiological levels (Mio and Stern, 2002). Asada *et al.* (1997) examined the effect of various types of alginic acid, consisting of L-glucuronic and D-mannuronic acids, on the bovine testicular hyaluronidase. Inhibition by sodium alginate was dependent on the molecular weight: the higher the molecular weight, the stronger the inhibition. Based on these results, Toida *et al.* (1999) investigated O-sulphated glycosamino-glycans of which the fully sulphated compounds showed the highest inhibitory effect. Flavones and flavone analogues such as apigenin and kaempferol also inhibit hyaluronidase (Kakegawa *et al.*, 1992) but not selectively and only at millimolar concentrations (Salmen *et al.*, 2003). Other compounds with a similarly weak inhibitory activity have also been detected, e.g. aescin, disodium cromoglycate, tranilast, traxanox, hederagenin, norlignane and urolithin B (Jeong *et al.*, 1999; 2000). Recently, vitamin C, 96 L-arginine derivatives (Akhtar and Bhakuni, 2003) and fatty acids (Suzuki *et al.*, 2002) have been reported to inhibit a streptococcal hyaluronidase with IC50 values at (sub) millimolar concentrations.

Seaweeds, and photosynthetic marine macro algae, supposedly primitive, live in sea or brackish water. They are classified on the basis of their pigment constituents into 4,500 species of red algae (*Rhodophyta*), 3,000 species of brown algae (*Phaeophyta*) and 7,000 species of green algae (*Chlorophyta*). Brown seaweed is the most common type of seaweed found on rocky beaches, normally having a method to securely attach themselves to rock surfaces. A group of brown algae, whose initial development is along tropical shore line, later break free and drift in the open ocean where they reproduce vegetatively, growing at a depth of 0–3 metres. The brown colour of the seaweed is due to the brown fucoxanthin pigment overriding the green chlorophyll pigment. Both pigments are used in photosynthesis, fucoxanthin improving the process when the algae are covered by water.

It was found that phlorotannin might be one of the compounds responsible for the antihyaluronidase activity. However, the inhibitory effect on hyaluronidase of samples from different regions might be affected by various factors, including high molecular weight phlorotannin or the degree of sulphation of the compounds present in the crude extract. Asada *et al.* (1997) reported that the inhibition of hyaluronidase by sodium alginate was dependent on molecular weight; the higher the molecular weight, the stronger the inhibition. Toida *et al.* (1999) investigated O-sulphated glucosaminoglycan as a fully sulphated compound inhibitor. The research illustrated that anti-allergic activities of Sargassum tennerimum is as potent as catechin, the natural hyaluronidase inhibitor, and more potent than DSCG, a clinically used anti-allergic medicine (Haider, 2010). This suggests a promising future for development of natural anti-allergic medicines or functional foods.

Acknowledgments

This work was supported by the grant from "National Natural Scientific Foundation of China" (No. 30800859). Permission was obtained to use the photographs from *A Picturial Series of Seafood in International Trade*, edited by Hong Lin and Yong Liu, Ocean University of China University Press, October 2008.

References

Aas, K., Elsayed, S. (1969). Characterization of a major allergen (cod): Effect of enzymic hydrolysis on the allergenic activity. J Allergy, 44, 333-343.

Akhtar, M.S., Bhakuni, V. (2003). Streptococcus pneumoniae hyaluronate lyase contains two noncooperative independent folding/unfolding structural domains: characterization of functional domain and inhibitors of enzyme. J Biol Chem, 278(28), 25509-25516.

Arif, S.H. (2009). A Ca^{2+}-binding protein with numerous roles and uses: parvalbumin in molecular biology and physiology. BioEssays, 31(4), 410-421.

Arif, S.H., Jabeen, M., Hasnain, A. (2007). Biochemical characterization and thermostable capacity of parvalbumins: The major fish-food allergens. J Food Biochem, 31, 121-137.

Asada, M., Sugie, M., Inoue, M., *et al.* (1997). Inhibitory effect of alginic acids on hyaluronidase and on histamine release from mast cells. Biosci Biotechnol Biochem, 61(6), 1030-1032.

Ayuso, R., Grishina, G., Bardina, L., *et al.* (2008). Myosin light chain is a novel shrimp allergen, Lit v 3. J Allergy Clin Immunol, 122(4), 795-802.

Ayuso, R., Grishina, G., Ibanez, M.D., *et al.* (2009). Sarcoplasmic calcium-binding protein is an EF-hand-type protein identified as a new shrimp allergen. J Allergy Clin Immunol, 124 (1), 114-120.

Beale, J.E., Jeebhay, M.F., Lopata, A.L. (2009). Characterization of purified parvalbumin from five fish species and nucleotide sequencing of this major allergen from Pacific pilchard, Sardinops sagax. Mol Immunol, 46, 2985-2993.

Bugajska-Schr A., Elfman, A.L., Fuchs, T., *et al.* (1998). Parvalbumin, a cross-reactive fish allergen, contains IgE-binding epitopes sensitive to periodate treatment and Ca^{2+} depl etion. J Allergy Clin Immunol, 1, 67-74.

Chu, K.H., Wong, S.H., Leung, P.S. (2000). Tropomyosin is the major mollusk allergen: reverse transcriptase polymerase chain reaction, expression and IgE reactivity. Mar Biotech, 2, 499-509.

Clark, S., Bock, S.A., Ga T.J., *et al.* (2004). Multicenter study of emergency department visits for food allergies. J Allergy Clin Immunol, 113, 347-352.

Daul, C.B., Morgan, J.E., Lehrer, S.B. (1990). The natural history of shrimp hypersensitivity. J Allergy Clin Immunol, 86(1), 88-93.

Daul, C.B., Morgan, J.E., Lehrer, S.B. (1993). Hypersensitivity reaction to crustacean and mollusks. Clin Rev Allergy, 11(2), 201-222.

Daul, C.B., Slattery, M., Reese, G., *et al.* (1994). Identification of the major brown shrimp (*Penaeus aztecus*) as the muscle protein tropomyosin. Int Arch Allergy Clin Immunol, 105, 49-55.

Elsayed, S., Bennich, H. (1975). The primary structure of allergen M from cod. Scand J Immunol, 4(2), 203-208.

Elsayed, S., Ragnarsson, U. N., Etteland, B. (1983). Solid-phase synthesis of the non-calcium-binding loop of cod allergen M. Direct evidence of the reactivity of the amino-terminal segment. Scand J Immunol, 17(3), 291-295.

Elsayed, S., Apold, J., Holen, E., *et al.* (1991). The structural requirements of epitopes with IgE binding capacity demonstrated by three major allergens from fish, egg and tree pollen. Scand J Clin Lab Invest (Suppl 204), 17-31.

Emoto, A., Ishizaki, S., Shiomi, K. (2009). Tropomyosins in gastropods and bivalves: Identification as major allergens and amino acid sequence features. Food Chem, 114, 634-641.

Fishery and Aquaculture Statistics (2008). FAO year book.

Fujitani, N., Sakaki, S., Yamaguchi, Y., *et al.* (2001). Inhibitory effects of microalgae on the activation of hyaluronidase. J Appl Phycol, 13(6), 489-492.

García-Orozco, K.D., Aispuro-Hernández, E., *et al.* (2007). Molecular characterization of arginine kinase, an allergen from the shrimp Litopenaeus-vannamei. Int Arch Allergy Immunol, 144, 23-28.

Haider, S., Li, Z.X., Lin, H., *et al.* (2009). *In vivo* study of antiallergenicity of , ethanol extracts from Sargassum tenerrimum, Sargassum cervicorne and Sargassum graminifolium turn. European Food Res Technol, 229(3), 435-441.

Hamada, Y., Tanaka, H., Ishizaki, S., *et al.* (2003). Purification, reactivity with IgE and cDNA cloning of parvalbumin as the major allergen of mackerels. Food Chem Toxicol, 41, 1149-1156.

Hilger, C., Thill, L., Grigioni, F., *et al.* (2004). IgE antibodies of fish allergic patients cross-react with frog parvalbumin. Allergy, 59(6), 653-660.

Ishikawa, M., Nagashima, Y., Shiomi, K. (1998). Identification of the oyster allergen Cra g 2 as tropomyosin. Fish Sci, 64, 854-855.

Ishikawa, M., Suzuki, F., Ishida, M., *et al.* (2001). Identification of tropomyosin as a major allergen of the octopus *Octopus vulgalis* and elucidation of its IgE-binding epitopes. Fisheries Science, 67, 934-942.

Jeong, S.J., Ahn, N., Kim, Y., *et al.* (1999). Norlignans with hyaluronidase inhibitory activity from Anemarrhena asphodeloides. Planta Med, 65(4), 367-368.

Jeong, S.J., Kim, N.Y., Kim, D.H., *et al.* (2000). Hyaluronidase inhibitory active 6H-dibenzo[b,d]pyran-6-ones from the feces of Trogopterus xanthipes. Planta Med, 66(1), 76-77.

Jeoung, B.J., Reese, G., Hauck, P., *et al.* (1997). Quantification of the major brown shrimp allergen Pen a 1 (tropomyosin) by a monoclonal antibody-based sandwich ELISA. J Allergy Clin Immunol, 100, 229-234.

Kakegawa, H., Matsumoto, H., Satoh, T. (1992). Inhibitory effects of some natural products on the activation of hyaluronidase and their antiallergic actions. Chem. Pharm Bull., 40(6), 1439-1442.

Kanny, G., Moner, et-Vautrin, D.A., Flabbee, J., *et al.* (2001). Population study of food allergy in France. J Allergy Clin Immunol, 108, 133-140.

Kelso, J.M., Bardina, L., Beyer, K. (2003). Allergy to canned tuna. J Allergy Clin Immunol, 111(4), 901.

Kuehn, A., Scheuermann, T., Hilger, C., *et al.* (2010). Important variations in parvalbumin content in common fish species: A factor possibly contributing to variable allergenicity. Int Arch Allergy Immunol, 153(4), 359-366.

Lehrer, S.B., Ayuso, R., Reese, G. (2003). Seafood allergy and allergens: a review. Mar Biotechnol, 5, 339-348.

Leung, P.S., Chow, W.K., Duffey, S., *et al.* (1996). IgE reactivity against a cross-reactive allergen in crustacea and mollusca: evidence for tropomyosin as the common allergen. J Allergy Clin Immunol, 98, 954-961.

Leung, P.S.C., Chu, K.H. (2001). cDNA cloning and molecular identification of the major oyster allergen from the Pacific oyster Crassostrea gigas. Clin Exp Allergy, 31, 1287-1294.

Leung, P.S.C., Chen, Y.C., Gershwin, M.E., *et al.* (1998b). Identification and molecular characterization of Charybdis feriatus tropomyosin, the major crab allergen. J Allergy Clin Immunol, 102, 847-852.

Leung, P.S.C., Chen, Y.C., Mykles, D.L., *et al.* (1998a). Molecular identification of the lobster muscle protein tropomyosin as a seafood allergen. Mol Mar Biol Biotechnol, 7, 12-20.

Leung, P.S.C., Chu, K.H., Chow, W.K., *et al.* (1994). Cloning, expression, and primary structure of M, etapenaeus ensis tropomyosin, the major heat-stable shrimp allergen. J Allergy Clin Immunol, 94(5), 882-890.

Leung, P.S.C., Chu, K.H., Chow, W.K., *et al.* (1994). Cloning, expression, and primary structure of M, etapenaeus ensis tropomyosin, the major heat-stable shrimp allergen. Allergy Clin Immunol, 94(5), 882-890.

Li, Z.X., Lin, H., Cao, L.M., *et al.* (2006). Reduction of allergenic properties of shrimp (Penaeus Vannamei) allergens by high intensity ultrasound. European Food Reasearch & Technology, 223(5), 639-644.

Li, Z.X., Lin, H., Cao, L.M., *et al.* (2007). The influence of gamma irradiation on the allergenicity of shrimp (*Penaeus vannamei*). J Food Engin, 79, 945-949.

Li, Z.X., Zhang, Y.Q., Lin, H., *et al.* (2010). Quantitative analysis of shrimp allergen in food matrices using a protein chip based on sandwich immunoassay. Euro Food ResTechnol, (231), 47-54.

Liang, Y.L., Cao, M.J., Su, W.J., *et al.* (2008). Identification and characterisation of the major allergen of Chinese mitten crab (*Eriocheir sinensis*). Food Chem, 111, 998-1003.

Lim, D.L., Neo, K.H., Goh, D.L., *et al.* (2005). Missing parvalbumin: Implications

in diagnostic testing for tuna allergy. J Allergy Clin Immunol, 5, 874-875.
LindstrØm, C.D.-V., van Dö, T., Hordvik, I., *et al.* (1996). Cloning of two distinct cDNAs encoding parvalbumin, the major allergen of Atlantic Salmon (Salmo salar). Scandinavian J Immunol, 44, 335-344.
Lopata, A.L., Zinn, C., Potter, P.C. (1997). Characteristics of hypersensitivity reactions and identification of a unique 49 kd IgE-binding protein (Hal-m-1) in abalone (Haliotis midae). J Allergy Clin Immunol, 100, 642-648.
Marsh, G.G., Goodfriend, L., King, T.P., *et al.* (1988). Allergen nomenclature. Int Arch Allergy Appl Immunol, 85, 194-200.
Marianne, T.W. (2007). Quantitative sandwich ELISA for the determination of tropomyosin from crustaceans in foods. J Agric Food Chem, 55, 8025-8032.
Meyer, K., Rapport, M.M. (1951). The inhibition of testicular hyaluronidase by heavy metals. J Biol Chem, 188, 485-490.
Mio, K., Stern, R. (2002). Inhibitors of the hyaluronidases. Matrix Biol, 21, 31-37.
Miyazawa, H., Fukamachi, H., Inagaki, Y., *et al.* (1996). Identification of the first major allergen of a squid (Todarodes pacificus). J Allergy Clin Immunol 98, 5(1), 948-953.
Morgan, J.E., Daul, C.B., Lehrer, S.B. (1990). English M.Characterization of important shrimp allergens by immunoblot analysis. J Allergy Clin Immunol , 85, 170.
Motoyama, K., Ishizaki, S., Nagashima, Y., *et al.* (2006). Cephalopod tropomyosins: Identification as major allergens and molecular cloning. Food and Chemical Toxicology, 44(12), 1997-2002.
Mykles, D.L., Cotton, J.L.S., Taniguchi, H., *et al.* (1998). Cloning of tropomyosins from lobster (Homarus americanus) striated muscles: fast and slow isoforms may be generated from the same transcript. J Muscle Res Cell Motil, 19, 105-115.
Nakamura, A., Sasaki, F., Watanabe, K., *et al.* (2006). Changes in allergenicity and digestibility of squid tropomyosin during theMaillard reaction with ribose. J Agric Food Chem, 54 (25), 9529-9534.
Nakamura, R., Satoh, R., Nakajima, Y., *et al.* (2009). Comparative study of GH-transgenic and non-transgenic amago salmon (Oncorhynchus masou ishikawae) allergenicity and proteomic analysis of amago salmon allergens. Regul Toxicol Pharmacol, 55, 300-308.
Naqpal, S., Rajappa, L., Metcalfe, D.D., *et al.* (1989). Isolation and characterization of heat-stable allergens from shrimp (*Penaeus indicus*). J Allergy Clin Immunol, 83(1), 26-36.
Patwary, M.U., Reith, M., Kenchington, E.L. (1999). Cloning and characterization of tropomyosin cDNAs from the sea scallop Placopecten magellanicus (Gmelin, 1971). J Shellfish Res, 18, 67-70.
Poms, R. E., Klein, C. L., Anklam, E. (2004). Methods for allergen analysis in food: a review. Food Additives & Contaminants, Part A, 21(1), 1-31.
Reese, G., Jeoung, B.J., Daul, C.B., *et al.* (1997). Characterization of recombinant shrimp allergen Pen a 1 (tropomyosin). Int Arch Allergy Immunol, 113(1-3). 240-242.

Reese, G., Ayuso R., Samuel B.L. (1999). Tropomyosin: An invertebrate pan-allergen. Int Arch Allergy Immunol, 119, 247-258.

Salmen, S. (2003). Inhibitors of Bacterial and Mammalian Hyaluronidases: Synthesis and Structure-Activity Relationships. Regensburg: University of Regensburg.

Sanbongi, C., Takano, H., Osakabe, N., *et al.* (2004). Rosmarinic acid in perilla extract inhibits allergic inflammation induced by mite allergen, in a mice Model. Clin Exp Allergy, 34(6), 971-977.

Shibata, T., Fujimoto, K., Kohki, N., *et al.* (2002). Inhibitory activity of brown algal phlorotannin against hyaluronidase. Int J Food Sci Tech, 37(6), 703-709.

Shimakura, K., Tonomura, Y., Hamada, Y., *et al.* (2005). Allergenicity of crustacean extractives and its reduction by protease digestion. Food Chem, 91, 247-253.

Shiomi, K., Sato, Y., Hamamoto, S., *et al.* (2008). Sarcoplasmic calcium-binding protein: Identification as a new allergen of the black tiger shrimp penaeus monodon. Int Arch Allergy Immunol, 146, 91-98.

Shiomi, K., Y. Hamada, K. Sekiguchi, *et al.* (1999). Two classes of allergens, parvalbumins and higher molecular weight substances, in Japanese eel and bigeye tuna. Fisheries Science, 65, 592-595.

Sicherer, S.H., Munoz-Furlong, A., Sampson, H.A. (2004). Prevalence of seafood allergy in the US d etermined by a random telephone survey. J Allergy Clin Immunol, 114, 159-165.

Suzuki, K., Terasaki, Y., Uyeda, M. (2002). Inhibition of hyaluronidases and chondroitinases by fatty acids. J Enzyme Inhib Med Chem 17, 183-186.

Swoboda, I., Bugajska-Schr etter, A., Verdino, P., *et al.* (2002). Recombinant carp parvalbumin, the major cross-reactive fish allergen: A tool for diagnosis and therapy of fish. allergy J Immunol, 168: 4576-4584.

Toida, T., Ogita, Y., Suzuki, *et al.* (1999). Inhibition of hyaluronidase by fully O-sulfonated glycosaminoglycans. Arch Biochem Biophys, 370(2),176-182.

van Do, T., Hordvik, I., Endresen, C., *et al.* (1999). Expression and analysis of recombinant salmon parvalbumin, the major allergen in atlantic salmon (Salmo salar). Scand J Immunol, 50, 619-625.

van Do, T., Hordvik, I., *et al.* (2005). Characterization of parvalbumin, the major allergen in Alaska pollack, and comparison with codfish. Allergen Molecular Immunology, 42, 345-353.

Woods, R.K., Abramson, M., Bailey, M., *et al.* (2001). International prevalences of reported food allergies and intolerances. Comparisons arising from the European Community Respiratory Health Survey (ECRHS) 1991–1994. Eur J Clin Nutr, 55, 298-304.

Wopfner, N., Dissertori, O., Ferreira, F., *et al.* (2007). Calcium-binding proteins and their role in allergic diseases. Immunology and Allergy Clinics of North America, 27(1), 29-44.

Yamamoto, T., Yoshimura, M., Yamaguchi, F., *et al.* (2004). Anti-allergic activity of naringenin chaicone from a tomato skin extract. Biosci Biotechnol

Biochem, 68(8), 1706-1711.
Yoshida, S., Ichimural, A., Shiomi, K. (2008). Elucidation of a major IgE epitope of Pacific mackerel parvalbumin. Food Chem, 111(4), 857-861.
Yu, C.J., Lin, Y.F., Chiang, B.L., *et al.* (2003). Proteomics and immunological analysis of a novel shrimp allergen, Pen m 2. J Immunol, 170(1), 445-453.

7

Food Allergen Epitopes

Hong-Bing Chen[1*], Jin-Yan Gao[2]

[1] State Key Laboratory of Food Science and Technology, Nanchang University, Nanchang, 330047, China
[2] Department of Food Science, Nanchang University, Nanchang, 330047, China
[*] E-mail:chbgjy@hotmail.com

Abstract: Food allergens can trigger the immune system of predisposed individuals towards the development of food allergy symptoms. However, the whole allergen is not involved in the immune response; only its epitopes, which are recognized by a T-cell receptor or an IgE-antibody, contribute to allergic reactions. Thus, identification and characterization of food allergen epitopes can help to better understand food allergies. Historically, linear or continuous and conformational or discontinuous epitopes are defined as two types of epitopes. The terms T-cell epitopes or B-cell epitopes are frequently used. Due to the application of advanced molecular technologies in immunology, food allergen epitopes can be identified and characterized, which can play an important role in the prediction of allergenicity, the definition of cross-reactivity, allergy diagnosis and immuno- therapy. Currently, B-cell epitope mapping is carried out by using enzymatic and chemical cleavage, production of synthetic peptides in various formats, structure resolution via NMR and X-ray crystal diffraction, biological chip technology and by the application of predictive algorithms. Identification of T-cell epitopes and their mapping on the allergen can be achieved by using proliferation assays, flow cytometry, and enzyme-linked immunospot (ELISPOT) assays. Nevertheless, investigations into the role of food allergen epitopes are ongoing, in particular with regard to the conformational epitopes.

7.1 Introduction

An allergen epitope is a specific region on the surface of an antigen. It is the part of a macromolecule (e.g., a protein) that is recognized by the immune system, especially by IgE antibodies (initially on or from B-cells) or the receptors of T-cells, and is termed accordingly as B-cell epitope or T-cell epitope. The recognition specificity of the IgE antibody depends on the uniqueness of the amino acid sequence of the epitopes,which are the interfaces between the immunogenic macromolecule and the immune system.

There are two types of food allergen epitopes, and they are usually categorized as linear or continuous, and conformational or discontinuous epitopes (Barlow *et al.*, 1986). T-cell epitopes are linear amino acid sequences, whereas B-cell epitopes are generally known as conformational. However, some linearly arranged amino acid residues can occur in a conformational epitope. This contributes to the complexity of the structural configuration of epitopes and to the allergenic allergen competence of its harboring protein. Therefore, the proper characterization of epitopes plays a key role in better understanding of food allergy and indeed allergy in general (Lin and Sampson, 2009).

7.1.1 Linear Epitopes of Food Allergens

A linear epitope is an epitope that is recognized by antibodies through its linear sequence of amino acids, i.e., its primary structure. Indeed, the protein primary structure is sometimes mistakenly termed as primary sequence, since it doesn't exist as a biologically functional entity in nature. However, since small peptides, which are the result of protein degradation in antigen presenting cells (APCs) and further exposed on the surface of these cells, can be recognized by the major histocompatibility complex (MHC) proteins on the T-cells according to their linear amino acid sequence, T-cell epitopes are linear by definition (Mohapatra and Lockey, 2001).

It is well known that both T-cell and B-cell epitopes are important for inducing an immune response to a food allergen, and that these epitopes are generally different on a single allergen. B-cell epitopes are the antigenic determinants on the surface of the allergen that are recognized in their natural state by B-cell receptors, which makes these epitopes conformational. However, it is estimated that 10% of the B-cell epitopes can be continuous (Pellequer *et al.*, 1993). This might be the result of several food allergens losing their conformational integrity during cooking and digestion. Consequently, linear B-cell epitopes can be considered to be more relevant to food allergies as thought before. Currently, several linear B-cell epitopes have been identified in various food allergens including allergens from milk, egg, peanut, wheat, soybean and shrimp (Bannon and Ogawa, 2006; Bannon, 2004).

Currently, IgE responses to several major food allergens have been thoroughly investigated, but detailed information about the T-cell responses to food proteins lags behind the current knowledge on B-cell reactivity. Future research should focus more on the T-cell epitope identification and mapping, and on the analysis of T-cell responses to food allergens. This will be helpful for better understanding the cellular and molecular mechanisms underlying the different types of food allergies, as well as the development of novel strategies for treatment and therapy.

7.1.2 Conformational Epitopes of Food Allergen

In contrast to linear epitopes, conformational epitopes are determined by the 3-dimensional folding of the protein and generally involve non-linearly arranged amino acids on the outside of the allergen.

Structurally, conformational epitopes exist in an intact protein that keeps the correct scaffold for a specific antigenitic determinant. Indeed, many allergens are obtained from fresh foods which can induce food allergies on the basis of conformational epitope binding to the antibody. These conformational epitopes are present in pollen-related food allergens through acting as cross-reactive structures for IgE-bindings. For example, Bet v 1, a birch pollen allergen, is a major cause of pollen-related food allergy (Breiteneder and Mills, 2005). Here, the term "conformational epitope" relates to the intact conformation of the native food allergen. Thus, if the protein scaffold is perturbed, the epitope may disappear. This is generally known from the apple allergen, Mal d 1, which loses its allergenicity after a short heating treatment. In some cases, misunderstanding can occur about the linearity and the configuration of the epitopes. Actually, a continuous peptide may form a conformational structure, as in a folded protein, a linear amino acid sequence can be forced into a certain conformation by the specific relationship to the rest of the molecule. Nevertheless, there is a remaining tendency to place epitopes into one or the other categories because food allergens can have both conformational and linear characteristics (Bannon and Ogawa, 2006).

Conformational epitopes are composed of non-adjacent amino acids in the allergen sequence brought into proximity via peptides folding. They cannot be isolated as an independent unit from the whole protein molecule which hinders proper experimental identification. Therefore, systematic characterization of conformational epitopes is much more difficult compared with the identification of linear epitopes (Harrer *et al.*, 2010). In fact, only a few epitopes have been identified by both *in vitro* and *in vivo* approaches including animal testing, X-ray crystallography and panning of random peptide libraries. This epitope data is stored in databases and can be applied to the development of predictive methods in computational immunology.

Historically, researchers have differentiated linear epitopes from conformational epitopes. This traditional classification was argued on the basis of several studies, in particular through the analysis of X-ray crystal structures of allergen-antibody complexes. However, at the present time, the dichotomy between linear and conformational epitopes is becoming less strict. Therefore, the content of the term "conformational epitope" should be interpreted in a broader sense.

7.2 The Role of Epitopes in Food Allergy

Detailed epitope characterization of food allergens is not limited to the mapping of B- and T-cell epitopes, but also includes determination of their role in the etiology of a certain food allergy. In addition, characterization is related to the occurrence of cross-reactivity, and to their application in diagnosis and therapy.

7.2.1 Allergenicity Prediction of Food Allergens

Due to the interest in the production of safe food with regard to food allergies, scientists focus on the prediction and assessment of allergenicity in all kinds of food, including novel food. Fortunately, allergen informatics at the genomics and proteomics level leads to the development of several useful computational tools with which to work. Although the standards for assessing protein allergenicity are still in their formative stage, or even under big discussion, the World Health Organization (WHO) and Food and Agriculture Organization (FAO) have proposed guidelines for evaluating the potential allergenicity of genetically modified foods, based on the screening for the sequence similarity of at least 6 contiguous amino acids shared with allergenic proteins, as potential linear IgE-binding epitopes. According to the Codex Alimentarius, a protein would be potentially allergenic if it has an identity of ≥6 contiguous amino acids or ≥35% sequence similarity over an 80 amino acid window with a known allergen (Taylor, 2006).

For the prediction of allergenicity, knowledge of all clinically important IgE-binding epitopes on food allergens would be ideal. From this information, algorithms could be developed to predict clinically important similarities between a new protein and the epitopes of a known allergen. Unfortunately, only a few food allergen epitopes have been thoroughly identified. Thus, allergenicity prediction of a novel protein on the basis of known linear allergen epitope sequences remains questionable. Therefore, as most of the IgE-binding epitopes are conformational, investigations should focus on structural information to enable incorporatation into the prediction protocol. Accordingly, conformational epitope prediction servers might represent an alternative approach to meet these requirements. For example, the SDAP (Structure Database of Allergenic Proteins, http://fermi.utmb.edu/SDAP) can offer a tool to predict allergenicity by using allergen sequences, known 3D structures of allergens, models of allergens with unknown structure, and IgE epitopes and binding data from literature sources (Pomés, 2009).

7.2.2 Cross-Reactivity of Food Allergen

Cross-reactivity occurs when antibodies, originally created against a given allergen, respond to a similar allergen from a different source. Most of the time, this phenomenon occurs among allergens that are structurally similar and thus are sharing highly similar or identical epitopes. In general, cross-reactivity is quite common among food allergens suggesting the occurrence of homologous molecular regions or conserved protein structures. Pollen-related food allergens are a typical example (Valenta and Kraft, 1996).

Cross-reactivity is particularly important in food allergies, not only because it is often the cause of an allergic reaction and may affect the scope of the disease and the reliability of diagnosis, but also because it has significant impact on current and potential therapies (Bonds *et al.*, 2008). In addition, the possibility that proteins from natural and novel foods may exhibit cross-reactivity with known allergens is of utmost concern to regulatory agencies, food scientists, food industries and physicians.

Previously, the prediction of cross-reactivity was carried out at the level of botanical or animal taxonomy rather than at the molecular classification level (protein family content). However, a high degree of amino acid homology within the profilin family does not at all cause a strong cross-reactivity (Sankian *et al.*, 2005). Another extreme example would be the tropomyosin protein family where chicken and shrimp share about 60% sequence identity but where no IgE cross-reactivity between these foods occurs (Goodman *et al.*, 2002). These and other examples reveal that cross-reactivity should be under consideration with regards to molecular classification. Undoubtedly, the knowledge of epitope cross-reactivity will be helpful to deal with this problem.

Protein evaluation for potential cross-reactivity via comparison of structure similarity to known allergen seems to be a promising approach. This strategy can take into account not only linear, but also conformational epitopes, and thus could overcome the shortcomings of simple comparison of alignments of linear epitopes. Take Bet v 1 for an example, which is the major cause of pollen-related food allergies, the tertiary structure of this allergen shares three surface patches with proteins from a variety of foods including cherry, apple, hazelnut, peach, carrot, celery and soybean (Vieths *et al.*, 2002). To identify clinically relevant IgE cross-reactivity, similarities in conformational epitopes in experimental 3D structures of the allergenic proteins were compared. However, only 5% (45/829) of experimental 3D structures are available in the Structure Database of Allergenic Proteins (SDAP) (Oezguen *et al.*, 2008). Due to this limitation, a new technique was recently developed: A comprehensive 3D-modeling of allergenic proteins and amino acid composition of potential conformational IgE epitopes was reported, which compares these 3D-models with known experimental structures. This finding of the amino acid distribution on epitopes can be used to develop new methods and will increase the predictive power for prediction of cross-reactivity (Oezguen *et al.*, 2008).

7.2.3 Epitopes and Diagnosis of Food Allergy

Double-blind, placebo-controlled food challenges in allergic individuals are still the gold standard in food allergy diagnostics, although such provocation tests with food allergens carry the risk of inducing severe allergic reactions (Sicherer and Sampson, 2010). As an alternative, serum testing, especially for the presence of allergen specific IgE, is much more convenient and also has a high clinical reliability. This made IgE quantification a useful alternative diagnostics approach for the prediction of food allergenicity (Hamilton and Franklin, 2004).

Modern biotechnology has pushed the progress of immunochemistry towards the mapping and synthesis of IgE-binding epitopes for various food allergens. All this progress has enabled component-resolved diagnostics of food allergens by detecting and quantifying IgE antibodies to their linear epitopes. Furthermore, this kind of diagnosis may result in individual sensitization patterns to (1) different proteins of an allergenic food, (2) homologous proteins in different foods, and (3) different epitopes on single allergenic molecules (Steckelbroeck *et al.*, 2008). In addition, it has been suggested that IgE-binding to allergen epitopes is a clinically relevant biomarker, and may be promising in terms of allergenicity prediction purposes. For example, high correlations were found between these IgE-epitope binding intensity and the severity or persistence of food allergy to peanut, shrimp, milk, wheat and egg (Albrecht *et al.*, 2009). Furthermore, taking milk as a typical example, Jaïvinen *et al.* (2001) compared the allergenic epitopes between old and young patients with milk allergies and found that the younger patients were likely to outgrow their allergies, recognized only 3 of these IgE binding epitopes on α-lactalbumin and none on β-lactoglobulin. These results indicated that the presence of IgE antibodies to multiple linear allergenic epitopes may be a marker for persistent cow milk allergies.

Although the IgE-binding peptides may serve as useful tools in diagnosis of food allergies, they are unlikely to act as full epitopes and to be able to trigger real clinical reactions (Albrecht *et al.*, 2009).

7.2.4 Epitopes and Immunotherapy

Although allergen specific immunotherapy has been widely practiced for almost 100 years, this treatment is still under development. Currently, one of the promising approaches is the use of such T-cell epitopes, that only target allergen-specific T-ells without causing adverse IgE mediated effects. Until now, T-cell-epitope-based peptide immunotherapy was successfully documented for both bee venom allergy and cat dander allergy, while similar food allergy strategies have not yet reached the level of clinical trials (Bohle, 2006). However, pepsin-digested peanut allergen-containing T-cell epitopes without IgE-binding have highlighted a possibility of such immunotherapies for food allergies in general (Burks *et al.*, 2008).

7.3 T-Cell Epitope Mapping Approaches

T-cell epitope mapping of food allergens refers to the determination of the specific peptide sequences recognized by CD4+ T-cells responding to an allergen. Because CD4+ T-cells play a central role in the pathophysiology of allergic responses, defining the allergen epitopes that are recognized by CD4+ T-cells are clinically important. There are several kinds of T-cell epitope mapping techniques including (1) through a T-cell proliferation assay, (2) using flow cytometry, (3) by enzyme-linked immunospot (ELISPOT) assays, or (4) otherwise.

7.3.1 T-Cell Epitopes Mapping by a Proliferation Assay

This method is based on the massive *in vitro* proliferation of CD4+ T-cell in response to their cognate antigen. Proliferative responses are quantified according to comparative incorporation of tritiated thymidine in cultures treated with or without peptides. A protocol for this assay applied peripheral blood mononuclear cells (PBMCs) being stimulated with a specific peptide for 4–6 days.

This technique enables an easy and reliable approach to T-cell epitope mapping, but is limited in case of detection of very rare CD4+ T-cells for specific allergens. Generally speaking, this assay can be performed with common laboratory facilities for cell culture, and has been successfully used for mapping of T-cell epitopes in food allergens. For example, the T-cell epitopes on the major peach allergen Pru p 3, the wheat epitopes on α-/β- and γ-gliadins, and the epitopes of bovine serum albumin were mapped using such proliferation assay (Tordesillas *et al.*, 2009; Spaenij-Dekking *et al.*, 2004; Tanabe *et al.*, 2002).

7.3.2 T-Cell Epitope Mapping by Flow Cytometry

Several methods have been developed for T-cell epitope mapping by flow cytometry. The initial protocol is principally based on the fact that T-cells could be identified through detection of intracellular cytokines after *ex vivo* stimulation of PBMCs with a specific peptide (Kern *et al.*, 1998). However, three different methods do not rely on the assay of intracellular cytokines (Malherbe, 2009). One of these methods performs the T-cell proliferation by measuring the decrease of staining of carboxyfluorescein-diacetate-succinimidyl-ester in proliferating cells. The other one is known as tetramer-guided mapping. This assay uses flow cytometry to measure CD8+ or CD4+ T-cells that recognize a specific epitope restricted by single MHC molecules. The third method of epitope mapping uses whole blood in a 6 h intracellular cytokine staining (ICS). This method is simple, requires only a small blood volume without the necessity of cell separation, and permits not only the detection of cytokine-positive cell, but also allows the identification of the responding cells, i.e., the cytokine- producing CD4+ or CD8+ T-cells. By using flow cytometry, T-cell epitopes on some food allergens have been identified, for example, on bovine milk αs1-casein (Elsayed *et al.*, 2004), egg ovomucoid (Kondo *et al.*, 2005), and peach allergen Pru p 3 (Schulten *et al.*, 2009).

7.3.3 T-Cell Epitope Mapping Using the ELISPOT Approach

The ELISPOT assay is widely used to detect antigen-specific immune responses to target antigens. This method is especially useful to measure both clonal size and effector function of low-frequency antigen-specific T-cell populations directly *ex vivo* (Wulf *et al.*, 2009) and is based on the principle that memory CD4+ T-cells secrete effector cytokines upon contact with the antigen. PBMCs are treated with the peptide, followed by cytokine assessment after stimulation during 24 h (Anthony and Lehmann, 2003). This assay is a kind of intracellular cytokine staining assay, which is categorized into a fast and high resolution approach at single cell level. Compared with other assays, it is one or two orders of magnitude more sensitive than the flow cytometry-based techniques, and it is one of the few immune monitoring assays that can be performed with cyropreserved PBMCs samples without significant loss of activity. The disadvantages of this technique are the subjectivity due to manual reading of the plates, and the need for cell separation to discriminate between antigen-specific responses derived from CD4+ T-cells and CD8+ T-cells. Now, several additional reasons also have contributed to the main-stream use of the EL1SPOT assay. The IFN-8 ELISPOT assay and IL-4 ELISPOT have successfully been used to map T-cell epitopes. For example, by using this technique, antagonists and non-toxic variants of wheat gliadin T-cell epitopes were investigated (Anderson *et al.*, 2006).

7.3.4 Other Assays

Due to biotechnological applications in immunology, several new approaches have been developed for T-cell mapping. T-cell epitope mapping using transgenic mice expressing HLA is a good example of such application. In this assay, the draining lymph nodes of immunized HLA transgenic mice provide a more abundant source of allergen-specific CD4+ T-cell to map T-cell epitopes, than the PBMCs of atopic patients (Malherbe, 2009). Other novel techniques involve (1) T-cell mapping by cytokine gene expressing (Provenzano and Spagnoli, 2009), and (2) T-cell mapping by TAD (antigen and epitope discovery), which is on the basis of taking the advantage of paramagnetic beads to augment an antigen presentation (Valentino and Frelinger, 2009). Although these new methods have not been found in food allergen epitope mapping yet, their potential for applications are promising.

7.4 B-Cells Epitope Mapping

Identification of B-cell epitopes on food allergen is of vital importance for better understanding of the complexity of the food allergic reaction, and will then be helpful for designing diagnostic tests, for developing immunotherapy, and for predicting allergenicity and cross-reactivity. Currently, methods for B-cells epitope mapping include enzymatic and chemical cleavage, production of synthetic peptides in various formats, structure resolution via NMR and X-ray crystal diffraction, biological chip technology and the use of predictive algorithms.

7.4.1 B-Cell Linear Epitope Mapping

Although it is believed that a large majority of B-cell epitopes are conformational, experimental epitope identification has focused primarily on linear B-cell epitopes. Several techniques are developed for a reliable approach to localize these linear B-cell epitopes, including single amino acid positions.

7.4.2 Proteolytic and Chemical Fragmentation for Epitope Mapping

Initially, linear B-cell epitope mapping was performed on the basis of fragmenting known allergen by specific proteolytic cleavage of the IgG/IgE-binding peptides via immunoblotting with serum from individual allergic individuals or with pooled sera, followed by the localization of the binding peptides via matching of the identified sequence to the allergen (Mazzoni *et al.*, 2009). Although this approach provides only rough information on the localization of linear epitopes, and cannot define conformational epitopes, this pioneer work led to the successful identification of linear epitopes. For example, two allergenic epitopes of the codfish allergen, Gad c 1, were roughly located onto a dodecapeptide (AA No. 33-44) and a nonapeptide (AA No. 88-96) derived from trypsin digestion (Elsayed *et al.*, 1976; Elsayed and Apold, 1977).

7.4.3 Epitope Mapping Using SPOTTM Peptide Arrays

The SPOT™ peptide array is an effective tool in epitope mapping studies. Here B-cell linear epitope mapping is introduced briefly according to the principle of the SPOT™ peptide approach (Frank, 2002): Short peptides (4–15 amino acids) covering the whole sequence of a known allergen with one or two amino acids offset are synthesized on derivatized membranes by repeated cycles of coupling, blocking and de-protection. After the peptide arrays are synthesized on a membrane, the membrane is incubated with allergic patient serum, followed by the immunological assay of IgE/IgG binding through immunoenzymatic or radioactive detection. This method is a powerful tool to identify and characterize linear epitopes, and many of the available IgE linear epitopes were identified with this method (Bannon and Ogawa, 2006). Even more important, it can be used for critical amino acid determination by amino acid replacement with an alanine, glycine or glutamine (Mine and Rupa, 2003). Both of the epitopes on animal and plant food allergens were identified according to this approach. For example, 4 IgE and 3 IgG-binding epitopes were identified on bovine α-lactogbulin, and 7 IgE and 6 IgG binding epitopes were detected on β-lactogbulin (Järvinen *et al.*, 2001). In the case of peanuts, 4 major IgE-binding epitopes were mapped on Ara h 1, 10 on Ara h 2 and 4 on Ara h 3 (Stanley *et al.*, 1997; Rabjohn *et al.*,1999; Wesley *et al.*, 1997). However, there are some limitations to this technology, including the length of the peptide to be synthesized, the large serum quantity required, and the inability to define conformational epitopes.

7.4.4 Epitope Mapping by the Phage Display Approach

The phage display technique has also been successfully applied in epitope mapping of many allergens. It is based on random peptide display on the phage surface, and the interaction with antibodies (Cwirla *et al.*, 1990). After a specific antibody is immobilized on a solid carrier, such as a microplate, several cycles can be carried out of affinity selection using incubation of the coated antibodies with the phage library, and further identification of the displayed peptides bound to the antibodies. Finally, consensus motives can be deduced from multiple peptide sequences derived from the panning of the phages, followed by position matching on the allergen.

Although many variations in peptide library design and panning strategies have been introduced during the past few years, the basic principle based on the phage display is still maintained (Böttger V. and Böttger A., 2009; Wang and Yu, 2009; Reineke, 2009). Both specific monoclonal and polyclonal antibodies have been used for epitope mapping according to this method. For example, celiac disease toxic gliadin epitopes were identified with monoclonal antibodies (Osman *et al.*, 2001), and several bovine β-lactoglobulin IgG-binding epitopes were identified and mapped with polyclonal antibody (Williams *et al.*, 1998).

7.4.5 Epitope Mapping with Peptides Microarray-Based Immunoassay

Due to some limitations of SPOT™ techniques for epitope mapping, a novel peptide microarray-based immunoassay approach was developed for the growing demand for high-throughput, multiplex allergen analysis in the microliter to nanoliter range, enabling maximization of the biological information from small amounts of precious patient serum material (Lin *et al.*, 2009). Although early studies pioneered on the use of peptide arrays to evaluate the immune response in animals for epitope structural analysis (Geysen *et al.*, 1987), the miniaturized allergen test in microarray format was only developed in 2002 by Hiller (Hiller *et al.*, 2002). For quantitative measurement of serum allergen-specific IgE levels, this method has progressed fast for its application in epitope mapping of food allergens, with further applications in allergy diagnosis and prognosis. Several promising results were documented, which were consistent with other established methods, such as RAST, ELISA, and immnunoblotting tests for milk, peanut and Brazil nut allergens (Cerecedo *et al.*, 2008; Shreffler *et al.*, 2004; Alcocer *et al.*, 2004; Nahtman *et al.*, 2007). All in all, the peptide microarray technology appeared to be much more accurate and sensitive than the SPOT™ immunoassays, and can be used for large scale IgE epitope mapping.

7.4.6 B-Cell Conformational Epitope Mapping

Experimentally, it is very difficult to identify conformational epitopes, and thus studies of conformational epitopes are far behind those of linear epitopes. However, some methods are available to identify conformational epitopes, including phage display, NMR spectrometry and X-ray crystallography.

7.4.7 Conformational Epitope Mapping by Phage Display Technology

One approach to B-cell conformational epitope mapping involves the identification of mimotopes or amino acid residues resembling the epitope of an allergen by the phage display technique. According to the principle of phage display, the phage peptides are panned against purified allergen-specific IgE or IgG. The selected peptides known as mimotopes, which mimic the binding sites of the allergen while not corresponding to its natural (*in situ*) sequence, can be localized on the 3D structure of the allergen with the aid of computational biological techniques. Following this approach, two relevant conformational IgE-binding epitopes of peach Pru p 3 have been identified (Pacios *et al.*, 2008). and 3 IgE-binding epitopes on the parvalbumin surface were defined (Untersmayr *et al.*, 2006). A similar conformational epitope was localized on non-specific lipid transfer proteins (nsLTP) from wheat flour and from peach fruits (Tordesillas *et al.*, 2009).

All this information shows that the mimotope identification strategy can be an alternative way for conformational epitope mapping in different food allergens.

7.5 Conformational Epitope Mapping by Nuclear Magnetic Resonance Spectroscopy Technique

Initially, nuclear magnetic resonance (NMR) spectroscopy was used for the investigation of protein-protein interactions and the dynamics of protein-ligand complexes. Epitope mapping by NMR is developed on the basis of difference in mobility between the amino acid residues of a peptide antigen that bind tightly with the antibody, and the mobility of the residues outside the epitope that do not interact with the antibody. Currently, several NMR techniques have been applied to allergen epitope mapping. These methods can be defined as epitope mapping by dynamic filtering, comparison of 1H-15N HSQC peak intensities, transverse relaxation time, measurements of 1H-15N nuclear overhauser effect (NOE) values, and T1ρmeasurements of relaxation time (Rosen and Anglister, 2009). This method was successfully used for epitope mapping of dust mite allergen Der p 2 (Mueller *et al.*, 2001). Unfortunately, there is no information yet about its use in food allergen epitope mapping, but it represents a very promising tool.

7.6 Epitope Mapping by X-ray Crystallography

Currently, to our knowledge, the ultimate method for the mapping of B-cell epitopes, especially for conformational epitopes, is to define the structures of allergen antibody complexes by X-ray. Initially, in the 1980s, an egg lysozyme allergen was studied using this approach, resulting in defining the conformational epitopes of 3 Fab antibody fragments (Padlan *et al.*, 1989). Since then, progress has been made step-by-step. Three complexes of murine IgG1 Fabs with allergens (the birch pollen allergen Bet v 1, the bee venom allergen Api m 2, and the cockroach allergen Bla g 2) were defined by X-ray crystallography, and their conformational epitopes were reported (Spangfort *et al.*, 2003; Padavattan *et al.*, 2007; Li *et al.*, 2008). However, these identified B-cell epitopes do not necessarily correspond to the real B-cell epitopes in the way they are recognized by serum IgE. This is mainly due to the limitations of this approach for conformational B-cell epitope mapping because of the lack of natural IgE mAbs in milligram amounts that are required for X-ray crystallography studies. Encouragingly, some technological break-throughs have been reported for the bovine β-lactoglobulin allergen and timothy grass pollen allergen (Phl p 2) using human IgE-derived Fab fragments (Niemi *et al.*, 2007; Padavattan *et al.*, 2009). Thus, the mapping of conformational epitopes also on food allergens by X-ray crystallography may successfully be achieved.

7.7 Conclusion

Identification and characterization of food allergen structural features were initiated on the physicochemical definition. Immunological information from the interaction between allergens and the immune system is limited. The identification of epitopes helped to better understand the reactions occurring in food allergies, as well as to deliver improved knowledge regarding the application of the data for diagnostics and therapies. Currently, linear epitopes for many major food allergens have been investigated. However, many or most conformational epitope structural features of food allergens still need to be detected. Meeting this challenge will provide a broader insight into the involvement of protein structures with regard to immunological functions.

Acknowledgements

The work was supported by "National Natural Science Foundation of China" (Nos. 30860220, 31171716 and 21162049), Program for New Century Excellent Talents in University (NCET-08-0704), Research Program of State Key Laboratory of Food Science and Technology (SKLF-MB-201002, SKLF-TS-201109) and National Science and Technology support project, China (No. 2011BAK10B03).

References

Albrecht, M., Kühne, Y., Ballmer-Weber, B.K., *et al.* (2009). Relevance of IgE binding to short peptides for the allergenic activity of food allergens. J Allergy Clin Immunol, 124(2), 328-336.

Alcocer, M.J.C., Murtagh, G.J., Wilson, P.B., *et al.* (2004). The major human structural IgE epitope of the Brazil nut allergen Ber e 1: A chimaeric and protein microarray approach. J Mol Biol, 343(3),759-769.

Anderson, R.P., Van Heel, D.A., Tye-Din, J.A., *et al.* (2006). Antagonists and non-toxic variants of the dominant wheat gliadin T-cell epitope in coeliac disease. Gut, 55(4), 485-491.

Anthony, D.D., Lehmann, P.V. (2003). T-cell epitope mapping using the ELISPOT approach. Methods, 29(3), 260-269.

Bannon, G.A. (2004). What makes a food protein an allergen? Current Allergy and Asthma Reports, 4(1), 43-46.

Bannon, G.A., Ogawa, T. (2006). Evaluation of available IgE-binding epitope data and its utility in bioinformatics. Mol Nutrition Food Res, 50(7), 638-644.

Barlow, D.J., Edwards, M.S., Thornton, J.M. (1986). Continuous and discontinuous protein antigenic determinants. Nature, 322 (6081), 747-748.

Bohle, B. (2006). T-cell epitopes of food allergens. Clinical Reviews Allergy Immunol, 30(2), 97-108.

Böttger, V., Böttger, A. (2009). Epitope mapping using phage display peptide libraries. Methods Mol Biol (Clifton, N.J.), 524, 181-201.

Bonds, R.S., Midoro-Horiuti, T., Goldblum, R. (2008). A structural basis for food allergy: The role of cross-reactivity. Current Opinion Allergy Clin Immunol, 8(1), 82-86.

Breiteneder, H., Mills, E.N.C. (2005). Plant food allergens–Structural and functional aspects of allergenicity. Biotechnology Advances, 23(6), 395-399.

Burks, W., Kulis, M., Pons, L. (2008). Food allergies and hypersensitivity: A review of pharmacotherapy and therapeutic strategies. Expert Opinion on Pharmacotherapy, 9(7), 1145-1152.

Cerecedo, I., Zamora, J., Shreffler, W.G., *et al.* (2008). Mapping of the IgE and IgG4 sequential epitopes of milk allergens with a peptide microarray-based immunoassay. J Allergy Clin Immunol, 122(3), 589-594.

Cwirla, S.E., Peters, E.A., Barrett, R.W., *et al.* (1990). Peptides on phage: A vast library of peptides for identifying ligands. PNAS, 87(16), 6378-6382.

Elsayed, S., Apold, J. (1977). Allergenic structure of allergen M from cod. II. Allergenicity of the limited tryptic hydrolysis peptides of fragment TM2. Int Arch Allergy Appl Immunol, 54(2), 171-175.

Elsayed, S., Apold, J., Aas, K., *et al.* (1976). The allergenic structure of allergen M from cod. I. Tryptic peptides of fragment TM 1. Int Arch Allergy Appl Immunol, 52(1-4), 59-63.

Elsayed, S., Eriksen, J., Øysæd, L.K., *et al.* (2004). T-cell recognition pattern of

bovine milk αS1-casein and its peptides . Mol Immunol, 41(12), 1225-1234.
Frank, R. (2002). The SPOT-synthesis technique: Synthetic peptide arrays on membrane supports-Principles and applications. J Immunol Methods, 267(1), 13-26.
Geysen, H.M., Tainer, J.A., Rodda, S.J. (1987). Chemistry of antibody binding to a protein. Science, 235(4793), 1184-1190.
Goodman, R.E., Silvanovich, A., Hileman, R.E., *et al.* (2002). Bioinformatic methods for identifying known or potential allergens in the safety assessment of genetically modified crops. Comments on Toxicology, 8(3), 251-269.
Hamilton, R.G., Franklin, A. N. Jr. (2004). *In vitro* assays for the diagnosis of IgE-mediated disorders. J Allergy Clin Immunol, 114(2), 213-225.
Harrer, A., Egger, M., Gadermaier, G., *et al.* (2010). Characterization of plant food allergens: An overview on physicochemical and immunological techniques. Molecular Nutrition and Food Research, 54(1), 93-112.
Hiller, R., Laffer, S., Harwanegg, C., *et al.* (2002). Microarrayed allergen molecules: Diagnostic gatekeepers for allergy treatment. The FASEB journal : Official publication of the Federation of American Societies for Experimental Biology, 16(3), 414-416.
Järvinen, K. M., Chatchatee, P., Bardina, L., *et al.* (2001). IgE and IgG binding epitopes on α-lactalbumin and β-lactoglobulin in cow's milk allergy. International Archives of Allergy and Immunology, 126(2), 111-118.
Kern, F., Surel, I.P., Brock, C., *et al.* (1998). T-cell epitope mapping by flow cytometry. Nature Med, 4(8), 975-976.
Kondo, M., Suzuki, K., Inoue, R., *et al.* (2005). Characterization of T-cell clones specific to Ovomucoid from patients with egg-white allergy. Journal of Investigational Allergology and Clin Immunol, 15(2), 107-111.
Li, M., Gustchina, A., Alexandratos, J., *et al.* (2008). Crystal structure of a dimerized cockroach allergen Bla g 2 complexed with a monoclonal antibody. J Bio Chem, 283(33), 22806-22814.
Lin, J., Bardina, L., Shreffler, W.G., *et al.* (2009). Development of a novel peptide microarray for large-scale epitope mapping of food allergens. J Allergy Clin Immunol, 124(2), 315-322.
Lin, J., Sampson, H.A. (2009). The role of immunoglobulin E-binding epitopes in the characterization of food allergy. Current Opinion Allergy Clin Immunol, 9(4), 357-363.
Malherbe, L. (2009). T-cell epitope mapping. Annals of Allergy, Asthma Immunol, 103(1), 76-79.
Mazzoni, M.R., Porchia, F., Hamm, H.E. (2009). Proteolytic fragmentation for epitope mapping. Methods in Molecular Biology (Clifton, N.J.), 524, 77-86.
Mine, Y., Rupa, P. (2003). Fine mapping and structural analysis of immunodominant IgE allergenic epitopes in chicken egg ovalbumin. Protein Engineering, 16(10), 747-752.

Mohapatra, S.S., Lockey, R.F. (2001). Molecular characterization of allergens. Clinical Reviews in Allergy and Immunology, 21(2-3), 203-213.

Mueller, G.A., Smith, A.M., Chapman, M.D., *et al.* (2001). Hydrogen Exchange Nuclear Magnetic Resonance Spectroscopy Mapping of Antibody Epitopes on the House Dust Mite Allergen Der p 2. J Bio Chem, 276(12), 9359-9365.

Nahtman, T., Jernberg, A., Mahdavifar, S., *et al.* (2007). Validation of peptide epitope microarray experiments and extraction of quality data. Journal of Immunological Methods, 328(1-2), 1-3.

Niemi, M., Jylhä, S., Laukkanen, M. L., *et al.* (2007). Molecular Interactions between a Recombinant IgE Antibody and the β-Lactoglobulin Allergen. Structure, 15(11), 1413-1421.

Oezguen, N., Zhou, B., Negi, S.S. (2008). Comprehensive 3D-modeling of allergenic proteins and amino acid composition of potential conformational IgE epitopes. Molecular Immunology, 45(14), 3740-3747.

Osman, A.A., Uhlig, H.H., Valdes, I., *et al.* (2001). A monoclonal antibody that recognizes a potential coeliac-toxic repetitive pentapeptide epitope in gliadins. European Journal of Gastroenterology and Hepatology, 13(10), 1189-1193.

Pacios, L.F., Tordesillas, L., Cuesta-Herranz, J., *et al.* (2008). Mimotope mapping as a complementary strategy to define allergen IgE-epitopes: Peach Pru p 3 allergen as a model. Mol Immunol, 45(8), 2269-2276.

Padavattan, S., Schirmer, T., Schmidt, M., *et al.* (2007). Identification of a B-cell Epitope of Hyaluronidase, a Major Bee Venom Allergen, from its Crystal Structure in Complex with a Specific Fab. J Mol Biol, 368(3), 742-752.

Padavattan, S., Flicker, S., Schirmer, T., *et al.* (2009). High-affinity IgE recognition of a conformational epitope of the major respiratory allergen Phl p 2 as revealed by x-ray crystallography. J Immunol, 182(4), 2141-2151.

Padlan, E.A., Silverton, E.W., Sheriff, S., *et al.* (1989). Structure of an antibody-antigen complex: Crystal structure of the HyHEL-10 Fab-lysozyme complex. PNAS, 86(15), 5938-5942.

Pellequer, J. L., Westhof, E., van Regenmortel, M.H.V. (1993). Correlation between the location of antigenic sites and the prediction of turns in proteins. Immunology Letters, 36(1), 83-99.

Pomés, A. (2009). Relevant B Cell Epitopes in Allergic Disease. Int Arch Allergy Immunol, 152(1), 1-11.

Provenzano, M., Spagnoli, G.C. (2009). T-cell epitope-mapping by cytokine gene expression assay. Methods in Molecular Biology (Clifton, N.J.), 514, 107-118.

Rabjohn, P., Helm, E.M., Stanley, J.S., *et al.* (1999). Molecular cloning and epitope analysis of the peanut allergen Ara h 3. J Clin Investigation, 103(4), 535-542.

Reineke, U. (2009). Antibody epitope mapping using de novo generated synthetic peptide libraries. Methods in Molecular Biology (Clifton, N.J.), 524, 203-211.

Rosen, O., Anglister, J. (2009). Epitope mapping of antibody-antigen complexes by nuclear magnetic resonance spectroscopy. Methods in Molecular Biology (Clifton, N.J.), 524, 37-57.

Sankian, M., Varasteh, A., Pazouki, N., et al. (2005). Sequence homology: A poor predictive value for profilins cross-reactivity. Clin Mol Allergy, 3, 13.

Schulten, V., Radakovics, A., Hartz, C., *et al.* (2009). Characterization of the allergic T-cell response to Pru p 3, the nonspecific lipid transfer protein in peach. J Allergy Clin Immunol, 124(1), 100-107.

Shreffler, W.G., Beyer, K., Chu, T. H.T., *et al.* (2004). Microarray immunoassay: Association of clinical history, in vitro IgE function, and heterogeneity of allergenic peanut epitopes. J Allergy Clin Immunol, 113(4), 776-782.

Sicherer, S.H., Sampson, H.A. (2010). Food allergy. J Allergy Clin Immunol, 125(supple 2), S116-S125.

Spaenij-Dekking, E.H.A., Kooy-Winkelaar, E.M.C., *et al.* (2004). A novel and sensitive method for the detection of T cell stimulatory epitopes of α-/β- and γ-gliadin. Gut, 53(9), 1267-1273.

Spangfort, M.D., Mirza, O., Ipsen, H., *et al.* (2003). Dominating IgE-binding epitope of Bet v 1, the major allergen of birch pollen, characterized by X-ray crystallography and site-directed mutagenesis. J Immunol, 171(6), 3084-3090.

Stanley, J.S., King, N., Burks, A.W., *et al.* (1997). Identification and mutational analysis of the immunodominant IgE binding epitopes of the major peanut allergen Ara h 2. Archives of Biochemistry and Biophysics, 342(2), 244-253.

Steckelbroeck, S., Ballmer-Weber, B.K., Vieths, S. (2008). Potential, pitfalls, and prospects of food allergy diagnostics with recombinant allergens or synthetic sequential epitopes. J Allergy Clin Immunol, 121(6), 1323-1330.

Tanabe, S., Kobayashi, Y., Takahata, Y., *et al.* (2002). Some human B- and T-cell epitopes of bovine serum albumin, the major beef allergen. Biochemical and Biophysical Research Communications, 293(5), 1348-1353.

Taylor, S.L. (2006). Review of the development of methodology for evaluating the human allergenic potential of novel proteins. Mol Nutrition Food Res, 50(7), 604-609.

Tordesillas, L., Cuesta-Herranz, J., Gonzalez-Muñoz, M., *et al.* (2009a). T-cell epitopes of the major peach allergen, Pru p 3: Identification and differential T-cell response of peach-allergic and non-allergic subjects. Mol Immunol, 46(4), 722-728.

Tordesillas, L., Pacios, L.F., Palacin, A., *et al.* (2009b). Molecular basis of allergen cross-reactivity: Non-specific lipid transfer proteins from wheat flour and peach fruit as models. Mol Immunol, 47(2-3), 534-540.

Untersmayr, E., Szalai, K., Riemer, A.B., *et al.* (2006). Mimotopes identify conformational epitopes on parvalbumin, the major fish allergen. Mol Immunol, 43(9), 1454-1461.

Valenta, R., Kraft, D. (1996). Type I allergic reactions to plant-derived food: A consequence of primary sensitization to pollen allergens. J Allergy Clin

Immunol, 97(4), 893-895.

Valentino, M., Frelinger, J. (2009). An approach to the identification of T cell epitopes in the genomic era: Application to Francisella tularensis. Immunologic Research, 45(2-3), 218-228.

Vieths, S., Scheurer, S., Ballmer-Weber, B. (2002). Current understanding of cross-reactivity of food allergens and pollen. Annals of the New York Academy of Sciences, 964, 47-68.

Wang, L.F., Yu, M. (2009). Epitope mapping using phage-display random fragment libraries. Methods Mol Biol (Clifton, N.J.), 524, 315-332.

Wesley, Burks, A., Shin, D., Cockrell, G., *et al.* (1997). Mapping and mutational analysis of the IgE-binding epitopes on Ara h 1, a legume vicilin protein and a major allergen in peanut hypersensitivity. Eur J Biochem, 245(2), 334-339.

Williams, S.C., Badley, R.A., Davis, P.J., *et al.* (1998). Identification of epitopes within beta lactoglobulin recognized by polyclonal antibodies using phage display and PEPSCAN. J Immunolo Methods, 213(1), 1-17.

Wulf, M., Hoehn, P., Trinder, P. (2009). Identification and validation of T-cell epitopes using the IFN-gamma ELISPOT assay. Methods Mol Biol (Clifton, N.J), 524, 439-446.

8

Recombinant Allergens and Applications

Ying-Tao Ma, Zhao-Wei Yang, Zhong-Shan Gao*

Allergy Research Center/Department of Horticulture, Zhejiang University, Hangzhou, 310058, China
* E-mail: gaozhongshan@zju.edu.cn

Abstract: Recombinant DNA technology has great potential in various aspects of allergen-related research and clinical applications. Sufficient amounts of purified wild-type or immunologically modified allergens or fragments have been produced in heterologous expression systems for use in many research fields, such as molecular characterization of the allergen (e.g., three-dimensional structure and epitope mapping), allergen standardization, component-resolved diagnosis (CRD) and patient-tailored specific immunotherapy (SIT). Strategies for obtaining recombinant allergens generally involve three steps, with the choice of heterologous expression system, bacteria, yeast, insect or plant cell, and the purification methods being of major importance. Here we review the major methods used for determining the three-dimensional structure and for epitope mapping, the recent progress in the application of recombinant allergens in clinical research of allergenic disease, such as the recombinant allergen-based microarray diagnosis and therapeutic vaccine.

8.1 Introduction

The first allergen-encoding cDNA was isolated and sequenced a quarter of a century ago (Fang *et al.*, 1988). Since then, the application of recombinant DNA technology to produce sufficient amounts of wild-type or immunologically modified allergens has been a major factor in promoting allergen-related research and clinical practice. Several historical events in research on recombinant allergens illustrate this progress. The first use of recombinant allergens for *in vitro* IgE-based diagnosis of allergies was in 1991 (Valenta *et al.*, 1991), and for T-cell epitope studies in 1992 (Yssel *et al.*, 1992). Then the first recombinant hypoallergens with reduced allergenicity were produced (Breiteneder *et al.*, 1993), followed by recombinant allergen use for *in vivo* skin testing (Moser *et al.*, 1994). In 1996, the three-dimensional structure of the first recombinant allergen was resolved, which can be defined as a milestone in the history of allergy studies (Gajhede *et al.*, 1996). In 1999, *in vitro* studies (Arquint *et al.*, 1999) and allergen-specific immunotherapy (SIT) (Schramm *et al.*, 1999) with recombinant hypoallergenic derivatives began. Since the year 2000, recombinant allergens have become the most commonly used in clinical research, for example the study of the 3D structure of a specific antigen-allergen complex (Mirza *et al.*, 2000), microarray recombinant allergens for allergy diagnosis (Hiller *et al.*, 2002) and recombinant allergen-based vaccination used in high-efficiency SIT (Niederberger *et al.*, 2004; Pauli *et al.*, 2008). In addition, phase III clinical trials using recombinant allergens in SIT were completed (Valenta *et al.*, 2010).

This section summarizes strategies for obtaining recombinant allergens and gives examples of their current applications based on the latest progress made in these fields.

8.2 Advantages of Recombinant Allergens

Historically, crude aqueous allergen extracts were used in serological and provocation tests. These extracts only indicate the allergen source to which a patient is sensitized, while the precise identity of the disease-eliciting molecule(s) remains unknown. Other major limitations of using allergen extracts for diagnosis include contamination from other sources, differential degradation of the allergen by proteolytic enzymes during extraction and low accuracy of the test due to undefined allergenic and non-allergenic components (Bhalla and Singh, 2008). In comparison, recombinant allergens have many advantages, which can be summarized as follows:

(1) They are pure molecules with defined physicochemical and immunological properties which can improve the sensitivity of allergy diagnosis (Crameri and Fluckiger, 2005);

(2) The recombinant wild-type allergens can increase our knowledge of the molecular, immunologic, and biological characteristics of allergens;

(3) Variants with advantageous properties such as reduced allergenic activity or increased immunogenicity can be used to improve the efficiency of allergy therapy and patient safety;

(4) They can be produced as hybrid molecules with the epitopes of several different allergens which can also improve the efficiency of allergy diagnosis and immunotherapy (Valenta and Niederberger, 2007).

There is clear evidence that recombinant forms of allergens are useful tools for basic molecular study or diagnostic and therapeutic purposes. Some extensively studied allergens and their allergen families are shown in Table 8.1.

Table 8.1 Main recombinant allergens of different protein families

Protein family*	Available recombinant allergen	Allergen source	References
Prolamin	Mal d 3	Apple	Zuidmeer *et al.*, 2005
	Pru p 3	Peach	
	Art v 3	Mugwort	Gadermaier *et al.*, 2009
	Cor a 8	Hazelnut	Pokoj *et al.*, 2010
Profilin	Pru p 4	Peach	Willerroider *et al.*, 2003
	Mal d 4	Apple	Ma *et al.*, 2006
	Cap a 2	Bell pepper	Willerroider *et al.*, 2003
	Lyc e 1	Tomato	
Thaumatin-like protein	Pru av 2	Cherry	Fuchs *et al.*, 2006
	Act d 2	Green kiwi	Bublin *et al.*, 2010
Bet v 1-related protein	Api g 1	Celery	Hoffmann-Sommergruber *et al.*, 2000
	Cor a 1	Hazelnut	Luttkopf *et al.*, 2002
	Bet v 1	Birch	Holm *et al.*, 2004
	Bet v 1	Birch	Carnes *et al.*, 2009
	Pru av 1	Cherry	Reuter *et al.*, 2005
	Mal d 1	Apple	Ma *et al.*, 2006
Trypsin-like sertne protease	Try p 3	Dust mite	Liao *et al.*, 2009
Group 5/6 grass pollen allergen	Phl p 5a	Timothy grass	van Ree *et al.*, 2004
Group 2 mite allergen	Der p 2	European house dust mite	Tanyaratsrisakul *et al.*, 2010

*: Allergen families are classified according to AllFam (database freely accessible at http://www.meduniwien.ac.at/allergens/allfam/)

8.3 Strategies for Recombinant Allergen Production

Bacteria, yeast, insect and plant expression systems have been used to produce recombinant allergens. The process has three steps: the first step is to obtain allergen-encoding templates, then to cultivate in a suitable heterologous expression system with appropriate vectors, and the final step is purification according to fusion tags or protein properties. Using this procedure, various kinds of recombinant molecules have been produced with equal or similar properties to their natural counterparts with respect to IgE immunoreactivity, basophil activation and T-cell recognition.

The classical method for obtaining allergen-encoding templates by cDNA cloning was by constructing cDNA expression libraries based on the bacteriophage lambda vector and then screening with antibody (e.g., serum IgE) or DNA probes. By this method, many novel allergenic molecules were successfully isolated and cloned in the last two decades of the 20th century (Lorenz *et al.*, 2001). But the efficiency of this approach was limited because of the need for immobilization to solid phase supports. Screening selectively enriched cDNA clones displayed on filamentous phage coats can be simplified by affinity interaction with serum IgE from allergic individuals, termed phage display technology, linking the phenotype (expressed as fusion with a phase coat protein) to the genotype (the genetic information integrated in the phage genome) (Crameri *et al.*, 1994; Rhyner *et al.*, 2004). In combination with robotic-based screening, this versatile system contributes to a cost-effective, rapid identification of allergenic repertoires from complex allergenic sources (Crameri, 2001).

The strategy using polymerase chain reaction (PCR)-based cloning technology is a more efficient way to amplify allergen-encoding templates with known sequence information. The massive increase in sequence data in various databases, such as the I.U.I.S. database (http://www.allergen.org) and the allergome database (http://www.allergome.org) facilitates allergen cloning. However, the drawback of this approach is the existence of artificial fragments in redundant databases and thus immunological properties of recombinant molecules need to be further evaluated before considering them for diagnostic and therapeutic purposes, and the correlation with allergenicity in natural sources should be clarified. The detailed schemes of these two strategies are illustrated in Fig. 8.1.

8.3.1 Heterologous Expression Systems

The *Escherichia coli* expression system has become the first choice and also one of the most widely used bacterial systems because of its simplicity, inexpensiveness and relatively high yield. Many recombinant forms of clinically important allergens equaling their natural counterparts have been produced by this method. Typically, the biophysical and immunological properties of a recombinant

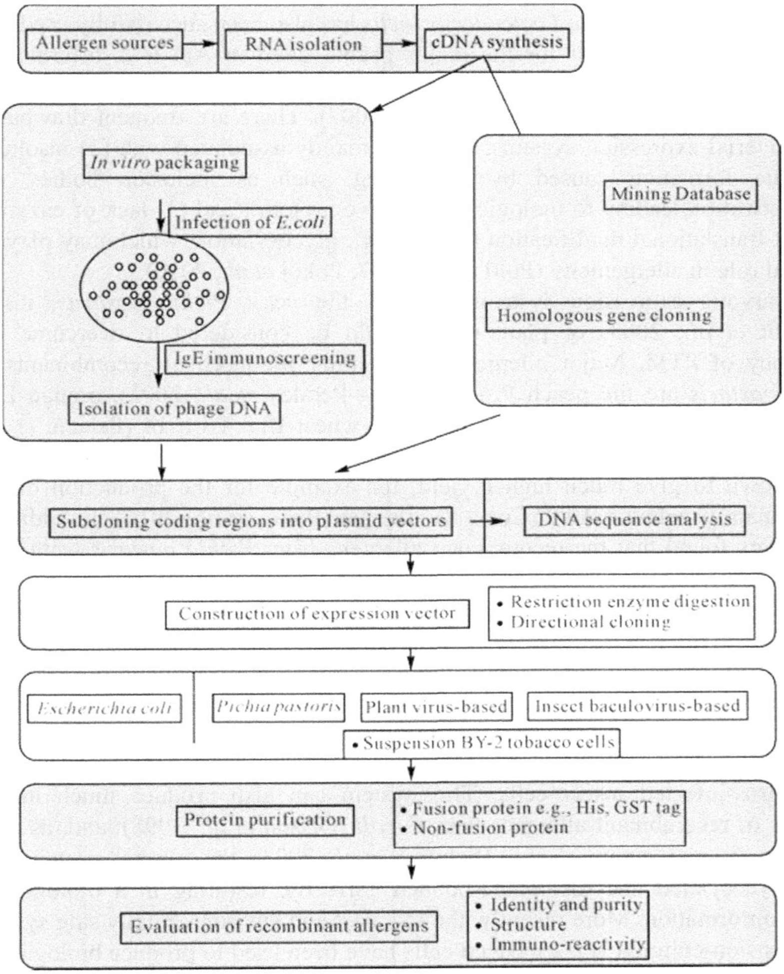

Fig. 8.1. Detailed scheme of the routes for production of recombinant allergens

allergen are investigated, including structure analysis (secondary structure and conformation), stability, homogeneity and aggregation and immunological reactivity (IgE-binding capacity and biological activity). There is abundant evidence suggesting that *E. coli* is a suitable host for production of properly folded (e.g., correct formation of disulfide-bonds) and biologically active, complex eukaryotic proteins (Wallner *et al.*, 2004; de Marco, 2009). Lipid transfer proteins (LTP), important plant panallergens (especially from pollen and food) have been cloned and expressed as soluble proteins in *E. coli* with conserved patterns of eight cysteine residues, from cherry (Scheurer *et al.*, 2001), tomato (Le *et al.*, 2006), grass pollen (Metz-Favre *et al.*, 2007), mugwort pollen (Gadermaier *et al.*, 2009) and apple (Borges *et al.*, 2010).

The lactic acid bacteria *Lactococcus lactis* has also been successfully used as a microbial production host for the major peanut allergen Ara h 2, offering the advantages of being food-grade standard and high protein secretion capacities, while lacking endotoxins (Glenting *et al.*, 2007). There are frequent drawbacks with bacterial expression systems, however, mainly associated with (1) insoluble aggregate formation caused by misfolding, such as inclusion bodies, (2) oligomerization leading to biologically inactive proteins, and (3) lack of enzymes for post-translational modification (PTM), e.g., glycosylation, which may play an essential role in allergenicity (Poltl *et al.*, 2007; Pokoj *et al.*, 2010).

Eukaryotic expression systems such as the yeast *Pichia pastoris*, insect (Schmidt *et al.*, 2003) or plant cells should be considered to overcome the deficiency of PTM. Major allergens successfully produced as recombinants in *Pichia pastoris* are the peach Pru p 3 (Diaz-Perales *et al.*, 2002), orange LTP allergen Cit s 3 (Ahrazem *et al.*, 2005), and wheat LTP Tri a 14 (Palacin *et al.*, 2009). When compared with *E. coli*, the *Pichia pastoris* expression system has been shown to give much higher yield, for example for the production of the recombinant hazelnut nsLTP (Cor a 8) allergen (Pokoj *et al.*, 2010). In addition, the authors found that the recombinant allergens from *Pichia pastoris* were pure IgE-reactive monomer without the impurities detected when using *E. coli*. However, there are also disadvantages of the *Pichia pastoris* eukaryotic expression system, where the recombinant allergens are unable to form correct secondary structure, except disulphide bonds, and have different conformations of hydrophobic cavities from their natural counterpart, as found in the production of the house dust mite allergen Der p 2 (Tanyaratsrisakul *et al.*, 2010).

Another eukaryotic expression system used for the production of allergen is baculovirus-infected insect cells. This system can also produce much higher amounts of recombinant allergens than *E. coli* (Olsson *et al.*, 1998), and, as with the major fire ant venom, rSol i3 (Schmidt *et al.*, 2003), the secreted proteins are often glycosylated and disulphide-bonded correctly, resulting in a biologically active conformation. More recently, the low cost and environmentally safe system of suspension-cultured BY-2 tobacco cells have been used to produce biologically active forms of house dust mite allergens (Lienard *et al.*, 2007).

8.3.2 Isolation and Purification of Recombinant Allergens

For those proteins expressed in *E. coli*, the most widely used isolation procedure involves centrifuging or filtering the insoluble host cells, lysis by sonication (Elmorjani *et al.*, 2004), solubilization of insoluble inclusion bodies using 6 M GuHCl (pH 8.0) (Tanyaratsrisakul *et al.*, 2010), and collecting the supernatant after centrifugation. The difference with isolation of targeted proteins from *Pichia pastoris* is that the secreted proteins are soluble in this medium, so solubilization of inclusion bodies is unnecessary.

Methods for the purification of recombinant allergen include electrophoresis and chromatography. The latter is a widely used technique, including gel-filtration chromatography (GFC), ion exchange chromatography (IEC), affinity chromatography and high-performance liquid chromatography (HPLC). One or more methods are selected for purification, depending on the fusion tags or protein properties. One example is the wheat rLTP fusion protein, purified with a combination of immobilized metal affinity chromatography, ion exchange chromatography and reverse-phase HPLC (Elmorjani *et al.*, 2004). Another example is purification of the yellow mustard allergens, rSin a 3 and rSin a 4, with two chromatographic steps, through fine and superfine Sephadex G-50 (Amersham Biosciences), preceding reverse-phase high-performance liquid chromatography in an m-bondapack C18 column, eluted with an acetonitrile gradient in 0.1% trifluoroacetic acid (Sirvent, 2009).

8.4 Application of Recombinant Allergens

8.4.1 Determination of Three-Dimensional Structures

Advanced DNA technology and the increasing number of available recombinant allergens from different allergenic sources have speeded up the progress of allergen structure determination. The three-dimensional structure of Bet v 1, the major birch pollen allergen, was the first to be determined using X-ray crystallography and NMR spectroscopy (Gajhede *et al.*, 1996). X-ray diffraction, the most common method, has been used to determine the three-dimensional structures of peach Pru p 3 (Pasquato *et al.*, 2006), while the recombinant apple profilin (rMal d 4), the major apple allergen, has been identified as a monomer by small-angle X-ray scattering (Ma *et al.*, 2006). Another common method is NMR spectroscopy, which has been used for modeling the three-dimensional structure of a non-specific wheat LTP (Gincel *et al.*, 1994), and to determine the tertiary structure of the sweet cherry (*Prunus avium*) allergen, Pru av 1, and define cross-reactive IgE epitopes (Neudecker *et al.*, 2003). Other techniques adopted include distance geometry, simulated annealing, energy minimization and molecular dynamics.

8.4.2 B-Cell and T-Cell Epitope Mapping

The availability of recombinant wild-type allergens has also contributed to the immunological characterization of these proteins, such as identifying the relevant T- and B-cell epitopes, which is important for the creation of hypoallergenic variants of allergenic proteins (Mutschlechner *et al.*, 2009). Conformational T-cell epitopes were found throughout the allergen sequence in most cases, and patients with allergies responded to a variety of different epitopes on a given allergen, such as birch Bet v 1 (Beatrice *et al.*, 2005), hazelnut Cor a 1.04 (Bohle *et al.*, 2005) and Pru p 3 (Tordesillas *et al.*, 2008). In the experiments, allergen–specific T-cell lines derived from individuals with a pollen or food allergy were tested with overlapping peptides of the allergen to investigate the major immunogenic T-cell epitope and their roles in cellular cross-reactivity with homologous allergens.

Co-crystallization of an allergen with a bound Fab monoclonal antibody fragment and subsequent resolution of the X-ray structure of the complex is needed for the complete definition of a B-cell epitope (Verdino *et al.*, 2008). Antibodies can selectively recognize continuous epitopes on an antigen or bind to discontinuous, conformational epitopes (Arnon and Vanregenmortel, 1992). Although continuous IgE epitopes of several important allergens were found to bind only relatively low levels of allergen-specific IgE antibodies, they appear to play an important role in food allergies, as food allergens are digested in the gastrointestinal tract before atopic sensitization (Valenta *et al.*, 2010). Individual amino acid residues of the major birch pollen allergen, Bet v 1, have already been shown to be crucial for IgE binding (Ferreira *et al.*, 1998). IgE-binding epitopes of the apple Mal d 1 has also been determined by point mutation and the engineered protein compared to Mal d 1 wild-type to evaluate the allergenic properties, by IgE immunoblotting, ELISA, peripheral blood monocytes proliferation assays, and skin prick tests (Ma *et al.*, 2006). In this way, the key amino acid residues for IgE-binding can be found, raising the possibility of modulating the IgE-binding properties of allergens by single amino acid substitutions at crucial positions, as a future immunotherapy for food allergies.

8.4.3 Cross-Reactive Structures

The extensive occurrence of cross-reactive food allergy suggests that molecules with allergenic potential may share a common structure or immunological properties. Mapping of the T-cell and IgE-binding epitopes has greatly contributed to our understanding of cross-reactivity at the molecular level. The cross-reactive epitopes can be quickly and reliably predicted by determining shared features of cross-reactive allergens at the sequence and structure level (Verdino *et al.*, 2008). Recombinant Mal d 1 has been found to share IgE epitopes with Bet v 1 (Vanekkrebitz *et al.*, 1995), and the hazel allergen Cor a 1.04 shares more IgE-binding epitopes with its homologue in birch than in hazel pollen (Luttkopf *et al.*, 2002). Knowledge of the mechanism underlying the cross allergy will promote diagnosis and immunotherapy of allergic diseases based on recombinant allergens and will help to develop safe and patient-tailored vaccinations by reducing the number of molecular structures needed.

8.4.4 Diagnostic and Therapeutic Applications of Recombinant Allergens

8.4.4.1 Recombinant Allergens for *In Vitro* and *In Vivo* Diagnosis

A detailed case history and clinical examination, including *in vivo* provocation tests and serological tests, are requisites for the diagnosis of IgE-mediated allergies (Bhalla and Singh, 2008). With many conventional allergen extracts, the specificity and reproducibility of *in vitro* diagnosis of a food allergy is low. A multitude of recently reported findings indicate that a recombinant allergen with defined molecules can improve clinical sensitivity and specificity in this diagnosis (Gamboa *et al.*, 2009; Bublin *et al.*, 2010): the use of recombinant allergens has allowed the determination of the patient's sensitization pattern against single components (Valenta and Kraft, 1995; Metz-Favre *et al.*, 2007). A new, more effective method for diagnosis of food allergy, component-resolved diagnostics (CRD), enables assessment of the sensitization profile in individual patients by determining specific IgEs from a panel of allergens derived from one source (Valenta *et al.*, 1999; Pittner *et al.*, 2004; Wohrl *et al.*, 2006).

Progress is being made in IgE-reactivity tests using microarray-chip technology, which enables the simultaneous testing of IgE reactivity against a large number of allergens from various sources spotted onto the surface of a glass slide. A major advantage of this technology is that very small amounts of both allergens and blood samples are required, potentially saving time, money and labor. Another advantage of allergen-based microarray chip technology is that the assay conditions can be controlled to be almost identical to the *in vivo* condition of the patient, where allergens are present in very small amounts as isolated proteins to bind IgE antibodies. Additionally, availability of synthetic peptide-based microarrays has further refined the concept of CRD (Bhalla and Singh, 2008).

8.4.4.2 Recombinant Allergens for Specific Immunotherapy

To date, there are two methods available for food allergy therapy. One is allergen-specific immunotherapy, which gradually desensitizes the allergic individual by administering increasing amounts of natural, crude allergen-containing extracts to induce clinical and immunological tolerance over time (Pons *et al.*, 2005; Durham, 2006; van Overtvelt *et al.*, 2006). The other is pharmacotherapy, with drugs such as antihistamines and topical corticosteroids, used to eliminate symptoms or suppress allergic inflammation without addressing the underlying cause. The gradual desensitization process might involve weekly injections for 8–12 months, and a further period of administration (ranging from 2 to 5 years) to achieve a long-lasting cure (Nelson, 2007).

Both the efficacy and safety of allergen-specific immunotherapy are improved with recombinant allergen-based preparations. The efficacy has been demonstrated in clinical studies with recombinant timothy grass pollen allergens (Jutel *et al.*, 2005) and the birch pollen allergen rBet v 1-FV (Kettner *et al.*, 2007).

For patients' safety and compliance, it is imperative to modify the recombinant allergens into hypo-allergens without changing or increasing immunological properties (T-cell reactivity and the induction of blocking antibodies) or significantly reducing (or removing) IgE reactivity. Approaches modifying allergens to hypoallergenic derivatives which have been used in research and clinical tests are reported to show positive results. These approaches can be summarized as follows: (1) site-directed mutagenesis of individual or groups of amino acids within linear IgE epitopes; (2) disruption of conformational epitopes; (3) generation of oligomeric or hybrid forms; (4) generation of variants by DNA shuffling technology (Bhalla and Singh, 2008). Careful purification of recombinant allergens and elimination of bacterial contaminants, especially endotoxin have also been demonstrated as essential for clinical use (Rolland *et al.*, 2009).

Substituting total protein extracts from allergen sources with defined recombinant proteins as vaccines may improve specific immunotherapy (SIT) of Type I allergies. Some good results have been obtained from initial clinical test on recombinant-allergen based vaccine preparations (Bhalla and Singh, 2008).

8.4.4.3 Recombinant Allergens for Standardization of Allergen

Recombinant allergens which display sufficient structural and immunological similarity and biological potency to their natural counterparts can be established as international reference materials for allergen production and purification. In 2001, in the EU funded project CREATE, recombinant allergens such as rBet v1, rPhl p 5a and rDer p 2 were studied for use as international reference materials for accurate standardization of allergens, which is of great importance for improvement of allergen quality and comparison of allergenic products from different companies (van Ree *et al.*, 2008).

8.5 Conclusion

In recent years, tremendous progress has been made in research on recombinant allergens. Biotechnological approaches offer the possibility of generating large amounts of pure allergens, and further clinical trials could lead to safer and more effective allergen vaccines. Recombinant allergens show promise as a tool for more accurate diagnosis and better treatment of allergenic disorders.

Acknowledgements

This work was supported by the Natural Science Foundation of China (No. 30971970).

References

Ahrazem, O., Ibanez, M.D., Lopez-Torrejon, G., *et al.* (2005). Lipid transfer proteins and allergy to oranges. Int Arch Allergy Immunol, 137(3), 201-210.

Arnon, R., Vanregenmortel, M.H.V. (1992). Structural basis of antigenic specificity and design of new vaccines. FASEB J, 6(14), 3265-3274.

Arquint, O., Helbling, A., Crameri, R., *et al.* (1999). Reduced in vivo allergenicity of Bet v 1d isoform, a natural component of birch pollen. J Allergy Clin Immunol, 104(6), 1239-1243.

Beatrice, J.S., Astrid, R., Dirk, L.T., *et al.* (2005). Bet v 1142-156 is the dominant T-cell epitope of the major birch pollen allergen and important for cross-reactivity with Bet v 1,Äìrelated food allergens. J Allergy Clin Immunol, 116(1), 213-219.

Bhalla, P.L., Singh, M.B. (2008). Biotechnology-based allergy diagnosis and vaccination. Trends Biotechnol, 26(3), 153-161.

Bohle, B., Radakovics, A., Luttkoft., D., *et al.* (2005). Characterization of the T cell response to the major hazelnut allergen, Cor a 1.04: Evidence for a relevant T-cell epitope not cross-reactive with homologous pollen allergens. Clin Exp Allergy, 35(10), 1392-1399.

Borges, J.P., Culerrier, R., Aldon, D., *et al.* (2010). GATEWAY (TM) technology and E. coli recombinant system produce a properly folded and functional recombinant allergen of the lipid transfer protein of apple (Mal d 3). Protein Expression and Purification, 70(2), 277-282.

Breiteneder, H., Ferreira, F., Hoffmannsommergruber, K., *et al.* (1993). Four recombinant isforms of Coralpha-1, the major allergen of Hazel pollen, show different IgE-binding properties. Eur J Biochem, 212(2), 355-362.

Bublin, M., Pfister, M., Radauer, C., *et al.* (2010). Component-resolved diagnosis of kiwifruit allergy with purified natural and recombinant kiwifruit allergens. J Allergy Clin Immunol, 125(3), 687-694.

Carnes, J., Himly, M., Gallego, M., *et al.* (2009). Detection of allergen composition and in vivo immunogenicity of depigmented allergoids of Betula alba. Clin Exp Allergy, 39(3), 426-434.

Crameri, R. (2001). High throughput screening: A rapid way to recombinant allergens. Allergy, 56, 30-34.

Crameri, R., Fluckiger, S., Glaser, A.G., *et al.* (2005). Recombinant allergens in research and clinics. Allergologie, 28(6), 215-218.

Crameri, R., Jaussi, R., Menz, G., *et al.* (1994). Display of expression products of cDNA libraries on phage surfaces-A versatile screening system for selective isolation of genes by specific gene-product ligand interaction. Eur J Biochem, 226(1), 53-58.

de Marco, A. (2009). Strategies for successful recombinant expression of disulfide bond-dependent proteins in Escherichia coli. Microbial Cell Factories, 8, 26.

Diaz-Perales, A., Garcia-Casado, G., Sanchez-Monge, R., *et al.* (2002). cDNA cloning and heterologous expression of the major allergens from peach and apple belonging to the lipid-transfer protein family. Clin Exp Allergy, 32(1), 87-92.

Durham, S.R. (2006). Allergen immunotherapy (desensitisation) for allergic diseases. Clin Med, 6(4), 348-351.

Elmorjani, K., Lurquin, V., Lelion, A., *et al.* (2004). A bacterial expression system revisited for the recombinant production of cystine-rich plant lipid transfer proteins. Biochem Biophys Res Commun, 316(4), 1202-1209.

Fang, K.S.Y., Vitale, M., Fehlner, P., *et al.* (1988). cDNA cloning and primary structure of a white-face hornet venom allergen, antigen-5. PNAS, 85(3), 895-899.

Ferreira, F., Ebner, C., Kramer, B., *et al.* (1998). Modulation of IgE reactivity of allergens by site-directed mutagenesis: potential use of hypoallergenic variants for immunotherapy. FASEB J, 12(2), 231-242.

Fuchs, H.C., Bohle, B., Dall'Antonia, Y., *et al.* (2006). Natural and recombinant molecules of the cherry allergen Pru av 2 show diverse structural and B-cell characteristics but similar T-cell reactivity. Clin Exp Allergy, 36(3), 359-368.

Gadermaier, G., Harrer, A., Girbl, T., *et al.* (2009). Isoform identification and characterization of Art v 3, the lipid-transfer protein of mugwort pollen. Molr Immunol, 46(10), 1919-1924.

Gajhede, M., Osmark, P., Poulsen, F.M., *et al.* (1996). X-ray and NMR structure of Bet v 1, the origin of birch pollen allergy. Nat Structural Biol, 3(12), 1040-1045.

Gamboa, P.M., Sanz, M.L., Lombardero, M., *et al.* (2009). Component-resolved in vitro diagnosis in peach-allergic patients. J Investig Allergol Clin Immunol, 19(1), 13-20.

Gincel, E., Simorre, J.P., Cailleet, A., *et al.* (1994). Three-dimensional structure in solution of a wheat lipid-transfer protein from multidimensional H-1-NMR data-a new folding for lipid carriers. Eur J Biochem, 226(2), 413-422.

Glenting. J., Poulsen, L.K., Kato, K., *et al.* (2007). Production of recombinant peanut allergen Ara h 2 using Lactococcus lactis. Microbial Cell Factories, 6, 28.

Hiller, R., Laffer, S., Harwanegg, C., *et al.* (2002). Microarrayed allergen molecules: diagnostic gatekeepers for allergy treatment. FASEB J, 16(1), 414-416.

Hoffmann-Sommergruber, K., Ferris, R., Pec, M., *et al.* (2000). Characterization of Api g 1.0201, a new member of the Api g 1 family of celery allergens. Int Arch Allergy Immunol, 122(2), 115-123

Holm, J., Gajhede, M., Ferreras, M., *et al.* (2004). Allergy vaccine engineering: Epitope modulation of recombinant Bet v 1 reduces IgE binding but retains protein folding pattern for induction of protective blocking-antibody responses. J Immunol, 173(8), 5258-5267.

Jutel, M., Jaeger, L., Suck, R., *et al.* (2005). Allergen-specific immunotherapy with recombinant grass pollen allergens. J Allergy Clin Immunol, 116(3), 608-613.

Kettner, J., Meyer, H., Cromwell, O., *et al.* (2007). Specific immunotherapy with recombinant birch pollen allergen rBet v 1-FV results of 2 years of treatment (Phase II trial). Allergy, 62 (s83), 262-262.

Le, L. Q., Lorenz, Y., Scheurer, S., *et al.* (2006). Design of tomato fruits with reduced allergenicity by dsRNAi-mediated inhibition of ns-LTP (Lyc e 3) expression. Plant Biotech J, 4(2), 231-242.

Liao, E.C., Hsu, E.L., Tsai, J.J., *et al.* (2009). Immunologic characterization and Allergenicity of Recombinant Tyr p 3 allergen from the storage mite tyrophagus putrescentiae. Int Arch Allergy Immunol, 150(1), 15-24.

Lienard, D., Dinh, O.T., van Oort, E., *et al.* (2007). Suspension-cultured BY-2 tobacco cells produce and mature immunologically active house dust mite allergens. Plant Biotechnol J, 5(1), 93-108.

Lorenz, A.R., Scheurer, S., Haustein, D., *et al.* (2001). Recombinant food allergens. J Chromatogr B-Analytical Technologies in the Biomedical and Life Sciences, 756(1-2), 255-279.

Luttkopf, D., Muller, U., Skov, P.S., *et al.* (2002). Comparison of four variants of a major allergen in hazelnut (Corylus avellana) Cor a 1.04 with the major hazel pollen allergen Cor a 1.01. Mol Immunol, 38(7), 515-525.

Ma, Y., Gadermaier, G., Bohle, B., *et al.* (2006). Mutational analysis of amino acid positions crucial for IgE-binding epitopes of the major apple (Malus domestica) allergen, Mal d 1. Int Arch Allergy Immunol, 139(1), 53-62.

Ma, Y., Zuidmeer, L., Bohle, B., *et al.* (2006). Characterization of recombinant Mal d 4 and its application for component-resolved diagnosis of apple allergy. Clin Exp Allergy, 36(8), 1087-1096.

Metz-Favre, C., Linhart, B., Focke-Tejkl, M., *et al.* (2007). Skin test diagnosis of grass pollen allergy with a recombinant hybrid molecule. J Allergy Clin Immunol, 120(2), 315-321.

Mirza, O., Henriksen, A., Ipsen, H., *et al.* (2000). Dominant epitopes and allergic cross-reactivity: Complex formation between a Fab fragment of a monoclonal murine IgG antibody and the major allergen from birch pollen Bet v 1. J Immunol, 165(1), 331-338.

Moser, M., Crameri, R., Brust, E., *et al.* (1994). Diagnostic value of recombinant aspergillus-fumigatus allergen I/A for skin testing and serology. J Allergy Clin Immunol, 93(1), 1-11.

Mutschlechner, S., Deifl, S., Bohle, B. (2009). Genetic allergen modification in the development of novel approaches to specific immunotherapy. Clin Exp Allergy, 39(11), 1635-1642.

Nelson, H.S. (2007). Allergen immunotherapy: where is it now? J Allergy Clinl

Immunol, 119(4), 769-777.

Neudecker, P., Lehmann, K., Nerkamp, J., *et al.* (2003). Mutational epitope analysis of Pru av 1 and Api g 1, the major allergens of cherry (Prunus avium) and celery (Apium graveolens): Correlating IgE reactivity with three-dimensional structure. Biochem J, 376, 97-107.

Niederberger, V., Horak, F., Vrtala, S., *et al.* (2004). Vaccination with genetically engineered allergens prevents progression of allergic disease. PNAS, 101, 14677-14682.

Olsson, S., Van Hage-Hamsten, M., Whitley, P., *et al.* (1998). Expression of two isoforms of Lep d 2, the major allergen of Lepidoglyphus destructor, in both prokaryotic and eukaryotic systems. Clin Exp Allergy, 28(8), 984-991.

Palacin, A., Varela, J., Quirce, S., *et al.* (2009). Recombinant lipid transfer protein Tri a 14: a novel heat and proteolytic resistant tool for the diagnosis of baker's asthma. Clin Exp Allergy, 39(8), 1267-1276.

Pasquato, N., Berni, R., Folli, C., *et al.* (2006). Crystal structure of peach Pru p 3, the prototypic member of the family of plant non-specific lipid transfer protein pan-allergens. J Mol Biol, 356(3), 684-694.

Pauli, G., Larsen, T.H., Rak, S., *et al.* (2008). Efficacy of recombinant birch pollen vaccine for the treatment of birch-allergic rhinoconjunctivitis. J Allergy Clin Immunol, 122(5), 951-960.

Pittner, G., Vrtala, S., Thomas, W.R., *et al.* (2004). Component-resolved diagnosis of house-dust mite allergy with purified natural and recombinant mite allergens. Clin Exp Allergy, 34(4), 597-603.

Pokoj, S., Lauer, I., Fotisch, K., *et al.* (2010). Pichia pastoris is superior to E-coli for the production of recombinant allergenic non-specific lipid-transfer proteins. Protein Expression and Purification, 69(1), 68-75.

Poltl, G., Ahrazem O., Paschinger, K., *et al.* (2007). Molecular and immunological characterization of the glycosylated orange allergen Cit s 1. Glycobiology, 17(2), 220-230.

Pons, L., Burks, W. (2005). Novel treatments for food allergy. Expert Opin Investig Drugs, 14(7), 829-834.

Reuter, A., Fortunato, D., Garoffo, L.P., *et al.* (2005). Novel isoforms of Pru av 1 with diverging immunoglobulin E binding properties identified by a synergistic combination of molecular biology and proteomics. Proteomics, 5(1), 282-289.

Rhyner, C., Weichel, M., Fluckiger, S., *et al.* (2004). Cloning allergens via phage display. Methods, 32(3), 212-218.

Rolland, J.M., Gardner, L.M., O'Hehir, R.E., *et al.* (2009). Allergen-related approaches to immunotherapy. Pharmacology and Therapeutics, 121(3), 273-284.

Scheurer, S., Pastorello, E.A., Wangorsch, A., *et al.* (2001). Recombinant allergens Pro av 1 and Pru av 4 and a newly identified lipid transfer protein in the in vitro diagnosis of cherry allergy. J Allergy Clin Immunol, 107(4), 724-731.

Schmidt, M., McConnell, T.J., Hoffman, D.R. (2003). Immunologic characterization of the recombinant fire ant venom allergen Sol i 3. Allergy,

58(4), 342-349.

Schramm, G., Kahlert, H., Suck, R., *et al.* (1999). "Allergen engineering": Variants of the timothy grass pollen allergen Phl p 5b with reduced IgE-binding capacity but conserved T-Cell reactivity. J Immunol, 162(4), 2406-2414.

Sirvent, S., Palomares, O., Vereda, A., *et al.* (2009). nsLTP and profilin are allergens in mustard seeds: cloning, sequencing and recombinant production of Sin a 3 and Sin a 4. Clinical and Experimental Allergy, 39(12), 1929-1936.

Tanyaratsrisakul, S., Malainual, N., Jirapongsananuruk, O., *et al.* (2010). Structural and IgE Binding Analyses of Recombinant Der p 2 Expressed from the Hosts Escherichia coli and Pichia pastoris. Int Arch Allergy Immunol, 151(3), 190-198.

Tordesillas, L., Cuesta-Herranz, J., Gonzalez-Munoz, M., *et al.* (2008). T-cell epitopes of the major peach allergen, Pru p 3: Identification and differential T-cell response of peach-allergic and non-allergic subjects. Mol Immunol, 46(4), 722-728.

Valenta, R., Duchene, M., Vrtala, S., *et al.* (1991). Recombinant allergens for immunoblot diagnosis of tree pollen allergy. J Allergy Clin Immunol, 88(6), 889-894.

Valenta, R., Kraft, D. (1995). Recombinant allergens for diagnosis and therapy of allergic diseases. Current Opinion in Immunology, 7(6), 751-756.

Valenta, R., Lidholm, J., Niederberger, V., *et al.* (1999). "The recombinant allergen-based concept of component-resolved diagnostics and immunotherapy (CRD and CRIT)". Clin Exp Allergy, 29(7), 896-904.

Valenta, R., Ferreira, F., Focke-Tejkl, M., *et al.* (2010). "From Allergen Genes to Allergy Vaccines". Annual Review of Immunol, 28(28), 211-241.

Valenta, R., Niederberger, V. (2007). Recombinant allergens for immunotherapy. J Allergy Clin Immunol, 119(4), 826-830.

van Overtvelt, L., Batard, T., Fadel, R., *et al.* (2006). Immune mechanisms of allergen-specific sublingual immunotherapy. Revue Francaise D Allergologie Et D Immunologie Clinique, 46(8), 713-720.

van Ree R. (2004). The CREATE project: EU support for the improvement of allergen standardization in Europe. Allergy, 59(6), 571-574.

van Ree, R., Chapman, M.D., Ferreira, F., *et al.* (2008). The CREATE Project: development of certified reference materials for allergenic products and validation of methods for their quantification. Allergy, 63(3), 310-326.

Vanekkrebitz, M., Hoffmannsommergruber, K., Machado, M.L.D., *et al.* (1995). Cloning and sequencing of Mal d 1, the major allergen from apple (Malus-domestica), and its immunological relationship to Bet v 1, the major birch pollen allergen, Biochemical and Biophysical Research Communications, 214(2), 538-551.

Verdino, P., Barderas, R., Villalba, M., *et al.* (2008). Three-dimensional structure of the cross-reactive pollen allergen Che a 3: Visualizing cross-reactivity on the molecular surfaces of weed, grass, and tree pollen allergens. J Immunol, 180(4), 2313-2321.

Wallner, M., Gruber, P., Radauer, C., *et al.* (2004). Lab scale and medium scale production of recombinant allergens in Escherichia coli. Methods, 32(3), 219-226

Willerroider, M., Fuchs, H., Ballmer-Webe, B.K., *et al.* (2003). Cloning and molecular and immunological characterisation of two new food allergens, Cap a 2 and Lyc e 1, profilins from bell pepper (Capsicum annuum) and tomato (Lycopersicon esculentum). Int Arch Allergy Immunol, 131(4), 245-255

Wohrl, S., Vigl, K., Zehetmayer, S., *et al.* (2006). The performance of a component-based allergen-microarray in clinical practice. Allergy, 61(5), 633-639.

Yssel, H., Johnson, K. E., Schneider, P.V., *et al.* (1992). T-cell activation inducing epitopes of the house dust mite allergen Der p 1. Proliferation and lymphokine production patterns by Der p 1-specific CD4+ T-cell clones. J Immunol, 148(3), 738-745.

Zuidmeer, L., van Leeuwen, W.A., Budde, I.K., *et al.* (2005). Lipid transfer proteins from fruit: Cloning, expression and quantification. Int Arch Allergy Immunol, 137(4), 273-281.

Part III

Diagnosis

9

The CREATE Project: Development of Certified Reference Materials for Allergenic Products and Validation of Methods for Their Quantification

R. van Ree[1,#,*], M. D. Chapman[2,*], F. Ferreira[3,*], S. Vieths[4,*], D. Bryan[5,*], O. Cromwell[6,*], M. Villalba[7,*], S.R. Durham[8,*], W.M. Becker[9,*], M. Aalbers[10], C. André[11], D. Barber[12], A. Cistero Bahima[13], A. Custovic[14], A. Didierlaurent[15], C. Dolman[5], J.W. Dorpema[15], G. Di Felice[16], F. Eberhardt[9], E. Fernandez Caldas1[7], M. Fernandez Rivas1[8], H. Fiebig[6], M. Focke[19], K. Fötisch[4], G. Gadermaier[3], R. Gaines Das[5], E. Gonzalez Mancebo[20], M. Himly[3], T. Kinaciyan[19], A.C. Knulst[21], A.M. Kroon[22], U. Lepp[9], F.M. Marco[23], A. Mari[24], P. Moingeon[11], R. Monsalve[12], A. Neubauer[25], S. Notten[10], P. Ooievaar-de Heer[10], G. Pauli[26], C. Pini[16], A. Purohit[26], J. Quiralte[27], S. Rak[28], M. Raulf-Heimsoth[29], M. San Miguel Moncin[13], B. Simpson[14], A. Tsay[2], L. Vailes[2], M. Wallner[3], Bernhard Weber[6]

[1] Academic Medical Center, Amsterdam, the Netherlands; [2] Indoor Biotechnologies, Charlottesville, VA, USA; [3] University of Salzburg, Salzburg, Austria; [4] Paul-Ehrlich Institute, Langen, Germany; [5] NIBSC, Potters Bar, United Kingdom; [6] Allergopharma, Reinbek, Germany; [7] Complutense University, Madrid, Spain; [8] Imperial College, London, United Kingdom; [9] Research Center Borstel, Borstel, Germany; [10] Sanquin, Amsterdam, the Netherlands; [11] Stallergènes SA, Antony, France; [12] ALK-Abello, Madrid, Spain; [13] University Institute Dexeus, Barcelona, Spain; [14] University of Manchester, Manchester, United Kingdom; [15] HAL Allergy BV, Haarlem, the Netherlands; [16] Istituto Superiore di Sanita, Rome, Italy; [17] CBF Leti SA, Madrid, Spain; [18] Hospital Clinico San Carlos, Madrid, Spain; [19] Medical University Vienna, Vienna, Austria; [20] Ramon y Cajal Hospital, Madrid, Spain; [21] University Medical Center Utrecht, Utrecht, the Netherlands; [22] European Allergen Manufacturers Group, Haarlem, the Netherlands; [23] ASAC Pharmaceutical International SA, Alicante, Spain; [24] CACeS, IDI-IRCCS, Rome, Italy; [25] Biomay AG, Vienna, Austria; [26] University Hospital Strasbourg, Strasbourg, France; [27] Complejo Hospitalario de Jaen, Jaen, Spain; [28] Sahlgrenska

University Hospital, Gothenborg, Sweden; [29] BGFA, Ruhr-University Bochum, Bochum, Germany
[#] Members of the Steering Committee of CREATE
[*] E-mail: r.vanree@amc.uva.nl

(Reprint from Allergy 2008(63): 310-326, with permission from John Wiley and Sons, License Number 2485341387681)

Abstract: Standardization of allergen extracts for diagnosis and immunotherapy of allergies is based on biological standardization, i.e., IgE-binding potencies. Skin tests and competitive IgE-binding assays are important elements of allergen standardization, especially from a safety perspective. However, biological standardization does not provide specific information about the major allergen content of allergen vaccines, i.e., the content of the active ingredients needed for attaining efficacy of immunotherapy. Another disadvantage of the current system is that allergen manufacturers express potencies of their products in company-specific units that do not allow product comparison and this situation is not tenable as an international system of standardization. In the 1980s, the WHO/IUIS Allergen Standardization Subcommittee developed International Reference Preparations (IRP) of several extracts to facilitate product comparison. Unfortunately, these IRP standards were not adopted by the industrial or regulatory authorities. During the 1990s, most major respiratory allergens became available as recombinant molecules, and the dependence of effective immunotherapy on administration of defined quantities of major allergen had become well-accepted. That was the moment for the IUIS-WHO committee to revive the endeavor for setting up a system of allergen standardization that would allow comparison of products and at the same time give accurate information on the content of active ingredients of the major allergens. The initiative was supported by a grant from the European Union, and the CREATE project was born. From November 2001 to April 2005, a consortium of basic and clinical researchers, regulators, allergen manufacturers and biotech companies carried out the project that aimed at evaluating the potential of recombinant allergens to serve as future major allergen reference materials and the potential of available sandwich ELISAs for their accurate measurement. The approach was to produce purified recombinant allergens and compare these to their natural counterparts serving as a gold standard. Nine recombinant molecules representing 8 major allergens were produced: rBet v 1, rPhl p 1, rPhl p 5a and rPhl p 5b, rOle e 1, rDer p 1, rDer p 2, rDer f 1 and rDer f 2. They were compared with purified natural allergens for physico-chemical (identity, purity, folding, aggregation state, solubility and stability) and immunological (IgE-binding potency, biological activity and dose-response behavior in ELISA) characteristics. As part of these studies, panels of sera from allergic patients with seasonal or perennial rhinitis and/or asthma were collected from clinical centers in 8 countries for use in comparing IgE antibody responses and in assessments of biological potency. Overall, 150–350 sera with IgE antibodies to each of the allergens were obtained, comprising a bank of 961 sera. Three recombinant allergens displayed sufficient structural and immunological similarity, and biological potency, to their natural counterparts to be selected for a follow-up project that should ultimately lead to their establishment as international reference materials: rBet v 1, rPhl p 5a and rDer p 2. In parallel with these studies, several ELISAs were evaluated for their measurement of major allergens. For most allergens, one or two ELISAs were identified, which showed comparable dose response curves for the recombinant and natural allergens (a requirement for measurement of major allergens in allergenic products, using a recombinant reference). This fulfilled an important aim of the CREATE project which was to develop purified allergen references together with complementary ELISA systems. A follow-up project has recently been initiated by the CREATE partnership and the European Directorate for the Quality of Medicines (EDQM). Together, three allergen manufacturers from CREATE provided EDQM with sufficient GMP-grade rBet v 1 and rPhl p 5a and with two suitable ELISA kits per allergen. After physico-chemical characterization of both allergens, these reagents will be used in a ring trial together with the selected ELISAs to validate their applicability in allergen standardization based on mass units of major allergen. One of the conclusions from CREATE is that certified recombinant references will have to be linked to certified assays so that products can be directly compared on the basis of mass units of major allergens. The rBet v 1 and rPhl p 5a allergens developed through CREATE will soon be established as international references and will become part of the European Pharmacopeia. The task for the future is to apply the approach used in CREATE to produce a repertoire of purified allergens that can be used as international references for standardization purposes and to harmonize allergen measurements worldwide.

9.1 WHO/IUIS Allergen Standardization Initiatives

The need for a coherent strategy to develop standards for allergenic products was recognized by the WHO/IUIS Allergen Standardization Subcommittee in the 1970s. In 1981, the committee established a program, funded by academic organizations and allergen manufacturers, to produce allergen extracts that would meet WHO specifications as International Standards (IS). The potency of candidate reference preparations was compared using *in vivo* and *in vitro* testing, including skin tests, RAST inhibition, crossed radio-immuno electrophoresis and measurements of specific major allergens. Following a series of international collaborative studies, WHO-approved IS were established for birch, timothy and short ragweed pollens, dust mite (*D. pteronyssinus*) and dog hair (Table 9.1) (Helm *et al.*, 1984; Ford A *et al.*, 1985; Ford AW *et al.*, 1985; Platts-Mills *et al.*, 1985; Gjesing *et al.*, 1985; Baer *et al.*, 1986; Larsen *et al.*, 1988; Arntzen *et al.*,

1989; Platts-Mills *et al.*, 1991). The IS were freeze-dried and stored in ampoules at −70C at the National Institute of Biological Standards and Controls (Potters Bar, UK), which is a WHO-approved repository. Approximately 4,000 ampoules of the IS were prepared and each IS was allotted a potency of 100,000 International Units.

Table 9.1 International Standards and Reference Reagents produced by the WHO/IUIS Allergen Standardization Subcommittee (1981−1990). WHO IS are available from NIBSC (www.nibsc.ac.uk). A reference serum pool from mite allergic individuals was also produced as part of these studies (NIBSC 82/528) (Ford A *et al.*, 1985; Platts-Mills *et al.*, 1985) (with Courtesy of Ford *et al.* and Platts-Mills *et al.*)

WHO International Standards	Code No.	International Units
Timothy pollen extract	NIBSC 82/520	100,000 (Larsen *et al.*, 1989)
Short ragweed pollen extract	NIBSC 84/581	100,000 (Helm *et al.*, 1984)
D. pteronyssinus extract	NIBSC 82/518	100,000 (Ford *et al.*, 1985b; Platts-Mills *et al.*, 1985)
Birch pollen extract	NIBSC 82/522	100,000 (Platts-Mills *et al.*, 1991)
Dog-hair/Dander extract	NIBSC 84/685	100,000 (Larsen *et al.*, 1988)
International Reference Reagent Bermuda grass pollen extract	ATCC	(Ford A *et al.*, 1985)

The WHO/IUIS program to establish allergen standards was led by Drs. Henning Lowenstein, Philip Norman, Thomas Platts-Mills and Alain de Weck. The aim of the program was to develop IS that could be used as "yardsticks" to which other allergen extracts could be compared for both total potency and specific allergen content (Norman, 1986; 1994). The program did much to establish the importance of allergen standardization within the academic allergy community and encouraged allergen manufacturers to increase their in-house standardization efforts. The intent was to establish standards for a large panel of allergenic products. However, this ambitious goal was compromised because regulatory agencies in Europe and in the US did not adopt the WHO IS as national standards. There was no statutory requirement that allergen manufacturers use the IS as standards in the licensing of allergenic products and the standards were not adopted.

Nonetheless, the WHO/IUIS standards were important references for research purposes. In most cases, estimates were made of the major allergen content of the IS in absolute units (μg/ampoule). At the time, monoclonal and polyclonal immunoassays were being developed for important allergens (e.g., Der p 1, Der f 1, Can f 1 etc.) and the WHO/IUIS standards provided an important resource of standards with known allergen content. The Der p 1 content of the *D. pteronyssinus* IS (NIBSC 82/518) was estimated at 12.5 μg/ampoule. Measurements of Der p 1 over the past 25 years have been calibrated based on this standard (Ford A *et al.*, 1985; Ford AW *et al.*, 1985). The IS were also used for measuring ragweed allergen Amb a 1 and dog allergen, Can f 1(Helm *et al.*, 1984; Larsen *et al.*, 1988). The use of the IS standards established the principle of using major allergen measurements as a key element of allergen standardization, a position which was presented at international seminars on the "Regulatory Control and Standardization of Allergenic Extracts" held at the Paul-Ehrlich-Institut and by a Position Statement of the American Academy of Allergy, Asthma and Immunology (Norman, 1994; Position Statement, 2007; van Ree, 1999).

Measurement of specific allergens became increasingly important in several areas of allergy and immunology studies. The most widespread use of allergen measurements was in environmental studies to assess sensitization and exposure to indoor allergens (dust mite, cat, dog, cockroach, rodents) in relation to chronic allergic diseases (rhinitis, asthma and atopic dermatitis) (Platts-Mills *et al.*, 1997). In addition to epidemiologic studies, allergen measurements were also used to assess the aerodynamic properties of allergens, their environmental distribution and the effects of allergen control procedures, products and devices. Together, assays in these research areas accounted for the largest use of major allergen measurements. The widespread use of molecular cloning in the 1990's facilitated the identification of the most important major and minor allergens and the use of purified allergens for diagnostic and therapeutic purposes. It was envisaged that measuring IgE to a panel of 2–4 major allergens from a given source could be used for allergy diagnosis and that standardization of purified allergens would be much less complex than for heterogeneous allergenic products (Scheiner *et al.*, 1994; Chapman *et al.*, 2000; Valenta and Kraft, 2004). Moreover, measurements of major allergens could be used to more closely monitor the doses of allergens that were used in allergen vaccines for conventional immunotherapy. In 1997, the WHO/IUIS Allergen Standardization Subcommittee participated in a conference in Geneva, Switzerland, to discuss the use of allergen vaccines in immunotherapy. This conference was organized by the Chair of the WHO/IUIS committee Dr. Jean Bousquet, together with Drs. Richard Lockey and Hans-Jorgen Malling. The WHO Position Paper that was produced from the conference recommended the use of standardized allergy vaccines of defined allergen content for immunotherapy to achieve maximal clinical efficacy. The Position Paper reviewed studies comparing the doses of major allergens that had been used in successful immunotherapy studies and concluded that "there is good evidence from immunotherapy studies with ragweed, grass, mite, cat and venom allergens that a maintenance dose of 5–20 μg of major allergen per injection is associated with significant improvement in patient symptom scores" (Bousquet *et al.*, 1998). Recent studies of the efficacy of immunotherapy for cat and dog allergies have confirmed these maintenance doses (Nanda *et al.*, 2004; Lent *et al.*, 2006; Nelson, 2007).

Over the past 20 years, the WHO/IUIS Allergen Standardization Subcommittee has played an influential role in promoting and coordinating international standardization efforts and in producing standards that are still being used today.

In 1999, the committee, under the leadership of Drs. Martin Chapman and Ronald van Ree, embarked on a new initiative to develop purified allergen standards that could be used for the standardization of *in vitro* assays. The use of specific allergen measurements for environmental studies and for assessing the allergen content of diagnostic and therapeutic products was widespread, and yet there were no approved purified allergen standards that could be used for calibration purposes. Recombinant allergens were being widely used for research and in clinical trials to produce a new generation of allergy vaccines, yet the immunologic reactivity of the recombinant allergens and their natural counterparts had not been systematically compared in international collaborative studies. This background provided the genesis for the European Union CREATE project (van Ree, 2004). The aim was to produce international standards of purified natural or recombinant allergens with verifiable allergen content. This would enable allergen manufacturers, academic organizations, government and regulatory agencies to use a common international standard as a reference for specific allergen measurements. A second aim was to compare the specificity, sensitivity and reproducibility of ELISA assays that were being used for allergen analysis.

Allergens were selected for the project based on the following criteria:

(1) The allergen was a major allergen of well-documented clinical importance;
(2) Purified natural and recombinant forms of the allergen were available in >20 mg amounts from academic or commercial laboratories;
(3) Strong evidence that the recombinant allergen had equivalent IgE-binding to its natural counterpart and extensive structural data on the allergen;
(4) ELISA kits to measure the allergen were available from one or more laboratories.

A consensus was reached and the following allergens were included in the study: pollens – Bet v 1, Phl p 1, Phl p 5, Ole e 1; mites – Der p 1, Der f 1, Der p 2, Der f 2.

It was recognized that this list excluded other important allergens, such as ragweed Amb a 1 and cat allergen, Fel d 1. However, these allergens did not fully satisfy the selection criteria and it was considered that if the study was successful such allergens could be part of a second phase. Dr. Ronald van Ree was the project co-ordinator for a grant application that was funded by the European Union under the 5th Framework Program in 2000.

9.2 Partnership and Aims of CREATE

The full title of the CREATE project was "Development of Certified Reference Materials for Allergenic Products and Validation of Methods for Their Quantification" (van Ree, 2004). The project partnership consisted of 28 organizations, i.e., 6 allergen manufacturers, 2 biotech companies, 11 clinical researchers and 9 research laboratories (Table 9.2). The research laboratories

Table 9.2 Partnership of CREATE

Research Laboratories		
Sanquin Research (co-ordinator)	R. van Ree /M. Aalbers/ S. Notten/P. Ooievaar-de Heer	NL
Universität Salzburg	F. Ferreira/M. Himly/ M. Wallner / G. Gadermaier	AT
Universidad Complutense Madrid	M. Villalba / R. Rodriguez	ES
Forschungszentrum Borstel	W-M. Becker	DE
Berufsgenossensch. Forschungsinstitut für Arbeitsmedizin	M. Raulf-Heimsoth	DE
Universität Wien	M. Focke / R. Valenta	AT
Research Laboratories with Regulatory Role		
National Institute for Biological Standards and Control	D. Bryan / C. Dolman	UK
Paul-Ehrlich-Institut	S. Vieths / K. Fötisch	DE
Istituto Superiore di Sanità	G. di Felice / C. Pini	IT
Allergen Manufacturers		
Allergopharma Joachim Ganzer	O. Cromwell / H. Fiebig / B.Weber	DE
HAL Allergy	H. van Schijndel / J.W. Dorpema	NL
ASAC Pharmaceutical Int.	F.M. Marco	ES
ALK-Abelló	R. Monsalve / D. Barber	DK
CBF Leti	E. Fernandez Caldas	ES
Stallergènes	P. Moingeon / A. Didierlaurent	FR
European Allergen Manufacturers Group	A. Kroon	EU
Biotech Companies		
Biomay	A. Neubauer	AT
Indoor Biotechnologies	M. Chapman / L. Vailes / A. Tsay	UK
Clinical Researchers		
Imperial College (London)	S. Durham	UK
North West Lung Centre (Manchester)	A. Custovic / B. Simpson	UK
University Medical Centre (Utrecht)	A. Knulst	NL
Fundación Hospital Alcorcón (Madrid)	M. Fernandez Rivas	ES
Forschungszentrum Borstel	U.Lepp / F. Eberhardt	DE
Institut Universitari Dexeus (Barcelona)	A. Cistero Bahima / M. San Miguel Moncin	ES
Adriano Mari (as private clinician / Rome)	A. Mari	IT
University Medical School of Vienna	T. Kinaciyan	AT
Hospital Ciudad de Jaén	J. Quiralte	ES
Hôpiteaux Universitaires de Strasbourg	G. Pauli / A. Purohit	FR
Sahlgrenska University Hospital (Götenborg)	S. Rak	SE

included 3 regulatory agencies from Germany (Paul-Ehrlich-Institute), from Italy (Istituto Superiore di Sanità) and from the United Kingdom (National Institute for Biological Standards and Control). The latter organization is also a WHO-approved repository. The partnership originated from 9 EU member states: the Netherlands, Spain, United Kingdom, Germany, Austria, Italy, Sweden, Denmark and France.

Allergen standardization has previously been dominated by IgE-based biological standardization (van Ree, 2007). Overall IgE-binding potencies of extracts are monitored by skin reactivity and by competitive IgE tests like RAST- or ImmunoCAP-inhibition. Potencies are expressed in company-specific units, relative to in-house reference preparations. By focusing on overall IgE-binding potencies, current allergen standardization requirements concentrate on the safety aspect of allergen products. To some extent this is understandable because IgE mediated reactions are responsible for the immunological events that result in adverse reactions to immunotherapy. The problem with total potency measurements is that they do not include the content of major allergens. It is a well-accepted fact that the presence of a sufficient major allergen is crucial for the outcome of treatment (Bousquet *et al.*, 1998; Nelson, 2007). Major allergens are the active ingredients of allergen-specific immunotherapy and their presence should therefore be monitored. Quantification of major allergens as active ingredients of immunotherapy will not only help to further establish the dose-response relation between allergen and treatment efficacy, but also it will allow comparison of allergenic products from different companies. A system of allergen standardization based on micrograms of major allergen requires internationally recognized reference materials, as well as validated assays for their measurement. The aims of the CREATE projects were formulated based on these needs: (1) Evaluation of the potential of purified recombinant allergens as certified reference materials (CRM); (2) Evaluation of available ELISAs for measurement of major allergens using the candidate CRM as standard.

The logo of the CREATE project (Fig.9.1) symbolizes the transition from extract-based standardization to molecular standardization, in this case represented by the transformation of a mite body into the Der p 2 major allergen molecule (Mueller *et al.*, 1998). The advancement of allergen standardization based on mass units of major allergens is furthermore a logical counterpart of the development of recombinant allergen-based immunotherapy to replace extract-based approaches.

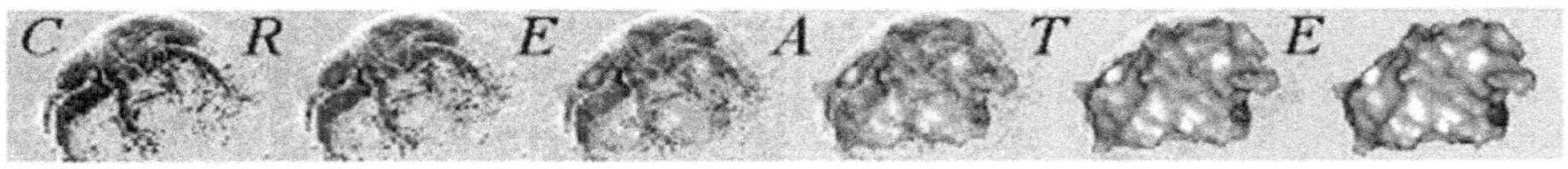

Fig. 9.1. CREATE logo. The CREATE logo symbolizes the transition from extract-based (mite body) to molecule-based (3D-structure of Der p 2) standardization

The CREATE project included the production of purified natural and recombinant allergens (20 mg of each), detailed physico-chemical characterization and stability testing, characterization of IgE-binding potencies and biological activity, and performance evaluation as standards in ELISA. An overview will be given of the results obtained for each of these areas of investigation.

9.3 Allergen Production and Purification

In total, 18 purified allergen preparations (8 natural allergens and 10 recombinant versions) were produced by the CREATE consortium. Table 9.3 lists these 18 allergen preparations, their extraction sources or host organisms used for the recombinant production, and their purification strategies. SDS-PAGE and Coomassie staining showed that all proteins were >98% pure (not shown). PAS staining after SDS-PAGE and blotting onto nitrocellulose membranes revealed the presence of sugars on group 1-mite allergens (nDer p 1, rDer p 1, nDer f 1, rDer f 1), nPhl p 1, and on natural and recombinant Ole e 1 (data not shown).

Table 9.3 Purification of recombinant allergens and their natural counterparts

Allergen	Source	Purification Method	Remarks
Bet v 1	Natural (birch pollen)	Gel filtration, anion exchange, IMAC	Method developed by ALK; Cu^{2+} used as immobilized metal ion
	Recombinant (*E. coli*)	Phosphate precipitation, hydrophobic interaction, anion exchange, RP-HPLC	Method developed by Biomay AG; soluble and inclusion bodies (solubilized in urea) fractions combined for purification
Phl p 1	Natural (timothy grass pollen)	Hydrophobic interaction, gel filtration	Method developed by Allergopharma
	Recombinant (*E. coli*)	Ammonium sulfate precipitation, anion exchange, hydroxyapatite chromatography	Method developed by Biomay AG; protein purified from inclusion bodies
Phl p 5	Natural (timothy grass pollen)	(1) Hydrophobic interaction, gel filtration; (2) Immunoaffinity chromatography	Method (1) developed by Forschungsinstitut Borstel; Method (2) developed by Allergopharma
	Recombinant Phl p 5a (*E. coli*)	IMAC, TEV cleavage, IMAC	Produced by Allergopharma (GMP); His-tag cleaved by His-tagged TEV; purified from soluble fraction
	Recombinant Phl p 5a (*E. coli*)	Ammonium sulfate precipitation, cation exchange, hydrophobic interaction	Method developed by Biomay AG

(To be continued)

(Table 9.3)

Allergen	Source	Purification Method	Remarks
	Recombinant Phl p 5b (*E. coli*)	IMAC, TEV cleavage, IMAC	Produced by Allergopharma under GMP conditions; His-tag cleaved by His-tagged TEV; purified from soluble fraction
Ole e 1	Natural (olive pollen)	Two gel filtration steps	Method developed by Universidad Complutense Madrid
	Recombinant (*Pichia pastoris*)	Anion exchange, gel filtration, RP-HPLC	Method developed by Universidad Complutense Madrid; purified from culture medium
Der p 1	Natural (*D. pteronyssinus*)	Immunoaffinity chromatography, SEC-HPLC	Method developed by Indoor Biotechnologies; purified from mite bodies-free culture
	Recombinant (*Pichia pastoris*)	SEC-HPLC	Method developed by Indoor Biotechnologies; purified from culture medium
Der f 1	Natural (*D. farinae*)	Immunoaffinity chromatography, SEC-HPLC	Method developed by Indoor Biotechnologies; purified from mite bodies-free culture
	Recombinant (*Pichia pastoris*)	SEC-HPLC	Method developed by Indoor Biotechnologies; purified from culture medium
Der p 2	Natural (*D. pteronyssinus*)	Immunoaffinity chromatography, SEC-HPLC	Method developed by Indoor Biotechnologies; purified from mite bodies-free culture
	Recombinant (*E. coli*)	Immunoaffinity chromatography, SEC-HPLC	Method developed by Indoor Biotechnologies; purified from inclusion bodies
Der f 2	Natural (*D. farinae*)	Immunoaffinity chromatography, SEC-HPLC	Method developed by Indoor Biotechnologies; purified from mite bodies-free culture
	Recombinant (*E. coli*)	Immunoaffinity chromatography, SEC-HPLC	Method developed by Indoor Biotechnologies; purified from inclusion bodies

Abbreviations: TEV, Tobacco Etch Virus protease; IMAC, Immobilized Metal Ion Affinity Chromatography; SEC, Size Exclusion Chromatography; HPLC, High Performance Liquid Chromatography; RP, reversed phase; GMP, Good Manufacturing Practice. *D. pteronyssinus* and *D. farinae* culture for mite allergen purification was provided by by Laboratorios Leti S.L., Madrid, Spain

9.3.1 Physico-chemical Characterization

Physico-chemical parameters selected for evaluating the quality of purified natural and recombinant allergens (Chirino and Mire-Sluis, 2004; Ferreira *et al.*, 2006) were identity, purity, homogeneity and (secondary) structure (folding):

- Identity. The allergens must display an amino acid composition and sequence that is in agreement with the known (published) primary sequence(s). Amino acid analysis and mass spectrometry (MS) were the methods selected for this task.
- Purity. The allergen preparations must be at least 95% pure with respect to protein content. The method of choice was SDS polyacrylamide gel electrophoresis (SDS-PAGE) in conjunction with silver staining. Amino acid analysis provided additional information on the purity.
- Homogeneity. Ideally, recombinant allergen preparations are homogeneous with respect to molecular size. One method to assess this is analytical size-exclusion chromatography (SEC) or gel filtration. The sample should then contain a single peak (>95% of the area under the curve) of the expected molecular size, or in some cases of a molecular size representing, e.g., a dimer or trimer, because some proteins behave as oligomers under "native" conditions (in solution). Multiple peaks can point towards impurities, but could also be explained by partial oligomerization/aggregation. Small angle X-ray scattering (SAXS) was used as an additional technique to monitor molecular size and aggregation. Expression of recombinant proteins in eukaryotic expression systems like *Pichia pastoris* can result in variable post-translational modifications (e.g., glycosylation). If capable of delivering well-folded protein, production of recombinant allergens in prokaryotic systems like *E. coli* with limited machinery for post-translational modifications therefore has preference. For purified natural allergens, deviations from homogeneity can be accepted if explained by the presence of multiple isoforms with significantly different molecular size, or by the existence of non-glycosylated and glycosylated versions.
- Structure. Purified recombinant allergen preparations must be correctly folded proteins. Circular dichroism (CD) spectroscopy was used to evaluate folding, i.e., secondary structure (α-helix, β-sheet and random coil; Fig. 9.2). Several candidate molecules were correctly folded and showed typical spectra of folded proteins with peak amplitudes similar to reference spectra of natural allergen preparations (Fig. 9.2). Table 9.4 gives an overview of the methods used for physico-chemical characterization of the candidate allergens and the information that can be derived concerning identity, purity, homogeneity and secondary structure.

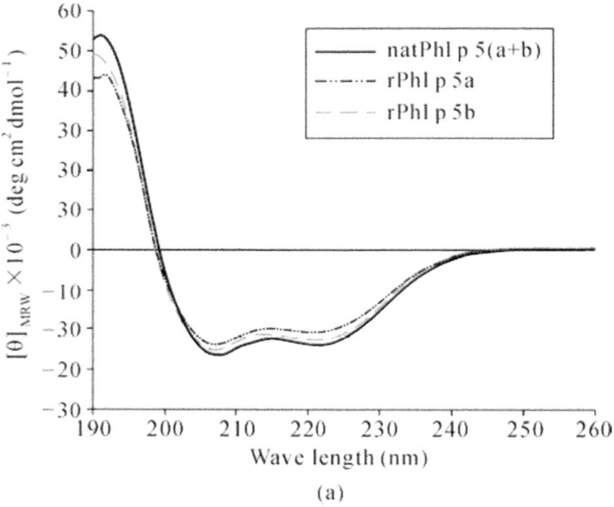

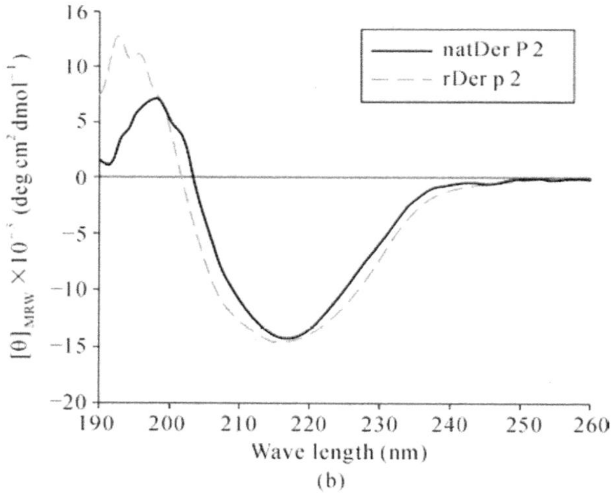

Fig. 9.2. CD spectroscopy. (a) CD spectra of the Phl p 5 preparations evaluated in CREATE, representing an allergen dominated by α-helical structures; (b) CD spectra of the Der p 2 preparations evaluated in CREATE, representing an allergen dominated by β-sheet structures

9.3.1.1 Amino Acid Analysis

Amino acid analysis is an indispensable tool in the analysis of protein pharmaceuticals. When performed on purified proteins, the technique allows the identification and quantification of the protein, providing more accurate results than colorimetric methods. The accuracy of amino acid analysis depends on the

Table 9.4 Parameters and selected analytical methods for the physico-chemical characterization of Candidate Reference Materials in the CREATE project

	Identity	Purity	Homogeneity	Secondary Structure (Folding)
Amino acid analysis	Amino acid content according to primary structure	Contamination by other proteins might change the content of individual amino acids		
Mass spectrometry/ peptide mapping	Molecular mass according to primary structure. MS/MS allows confirmation of primary structure	Contamination by other proteins, glycosylation heterogeneity or non-protein compounds gives rise to additional peaks	Post-translation modifications and chemical modifications resulting from the recombinant production can be detected and accounted for heterogeneity	
HPLC size exclusion chromatography (HPLC-SEC)		Contamination by proteins with different molecular weight might give rise to additional peaks	Homogeneity in solution: oligomeric/ monomeric forms, aggregates, and degradation products can be detected	Changes in retention times might also be attributable to denatured protein
Circular Dichroism (CD)				Folded status of proteins. Quantification of secondary structure elements is possible, but usually not reliable. Thermo-stability
Small angle X-ray scattering (SAXS)			Molecular size and aggregation status of proteins in solution	Denatured proteins might show altered PDDF
SDS-PAGE/ PAS staining		Contaminating proteins with different molecular weight can be detected	Post-translation modifications, aggregation, or degradation might give rise to additional bands. Sugars can be detected	

Abbreviation: PDDF, pair distance distribution function

integrity of the sample and other factors including purity of the reagents used, presence of salts, metals or detergents, and sample handling (Tyler, 2000). All allergen preparations in CREATE were quantified using amino acid analysis to assure standardization of protein measurements in mass units. Identity of all allergen preparations was confirmed by close correlations found between the theoretical and experimental amino acid composition, e.g., for Bet v 1 (Fig. 9.3(a)). An example revealing likely contamination was the analysis of nPhl p 1 (Fig. 9.3(b)). An increase in alanine content pointed towards contamination with Phl p 5, which has been shown to be rich in alanine (~30%).

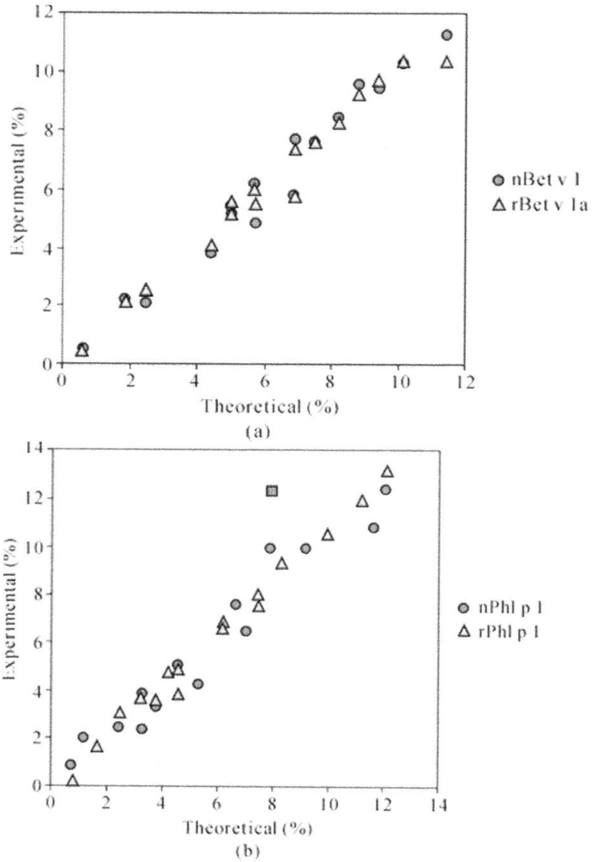

Fig. 9.3. Amino acid composition analysis: Identity and purity. Each data point represents an amino acid. There is an excellent correlation between the expected (theoretical) and observed (experimental) amino acid composition (identity). (a) Shows natural and recombinant Bet v 1; (b) Shows natural and recombinant Phl p 1. For the natural allergens, the average of known isoforms was calculated for the theoretical value. In case of natural Phl p 1, a single amino acid (alanine) deviates from theoretical (indicated by a square) suggesting contamination with alanine rich (30%) Phl p 5

9.3.1.2 Mass Spectrometry

Mass spectrometry of the allergen preparations reveals whether the expected molecular mass is the dominant component, but more reliable confirmation of identity requires digestion. A given protein with a known primary sequence can be digested with a specific protease like trypsin to give rise to a predictable set of peptides. Accurate determination of the masses of even a few peptides generated through trypsin digestion (to an accuracy of 1 Dalton) can be compared to the masses of peptides obtained through *in silico* digestion of the known protein sequence, upon which matching masses are assigned to predicted peptides. However, the unambiguous confirmation of identity is only possible after sequencing the peptides. This requires tandem mass spectrometry, which is usually denoted MS/MS (Lill, 2003; Reinders *et al.*, 2004). During this process, peptides identified during the first MS are dissociated, amino acid by amino acid, by collision. The mass spectrometer determines the mass after each subsequent collision-induced removal of an amino acid. The mass differences will provide information about the nature of the amino acid cleaved off. In the CREATE project, peptides were sequenced using electrospray-ionization (ESI) and collision-induced dissociation (CID) in a quadrupole time-of-flight (Q-TOF) instrument. This technique also allows the isoform composition of natural purified allergen preparations to be evaluated. Full sequencing of protein molecules using MS/MS is rarely achieved because not all peptides can effectively be ionized, which is necessary to "fly" to the detector. For some allergens, >95% of the sequence was confirmed, e.g., for rBet v 1a, rPhl p 5a, rPhl p 5b, and rDer f 2. For all allergens (except rPhl p 1), MS/MS provided convincing evidence of identity, including information about isoform-composition of natural allergens, and the presence of post-translational modifications of amino acids by glycosylation, carbamylation, deamidation and oxidation etc.

9.3.1.3 High-Performance Liquid Chromatography-Size-Exclusion Chromatography (HPLC-SEC)

In the CREATE project, purified natural and recombinant allergen preparations were analyzed by SEC using a HP 1100 HPLC system (Agilent Technologies, Palo Alto, CA, USA) with a TSK-Gel G2000SWXL column (Tosoh Biosep, 7.8 mm ID × 30.0 cm L). All but two of the natural allergens presented as essentially homogeneous preparations, eluting at a volume indicative for a monomeric appearance. Natural Ole e 1 and nDer p 1 were heterogeneous in solution. The sensitivity of HPLC-SEC did not allow firm conclusions to be drawn about the nature of the observed heterogeneity, but variable degrees of glycosylation and degradation are possible explanations. Heterogeneity was more common among the recombinant allergens. The elution profiles of rBet v 1, rPhl p 5a, rPhl p 5b and rDer f 2 showed a single homogeneous peak suggesting that these allergens were monomeric.

9.3.1.4 Circular Dichroism (CD) Spectroscopy

CD spectroscopy is widely applied to evaluate folding of allergen molecules. CD analysis of proteins uses the differential absorption of left and right circularly polarized light in the far UV region by chiral chromophores. In proteins, peptide bonds serve as chromophores. Absorption characteristics of peptide bonds are measurable depending on the environment and therefore on the secondary structure surrounding the chromophore. CD spectra give valuable information on the folding of proteins (Verdino and Keller, 2004; Kelly *et al.*, 2005). In the CREATE project, a protocol was set up to compare a series of protein pairs (natural and recombinant) in terms of their secondary structures and thermo-stability. Measurements were performed with a JASCO J-810 spectro-polarimeter fitted with a Neslab RTE-111M temperature control system. For the major allergens under investigation in CREATE, 3D structures were either reported based on X-ray crystallography and/or NMR, or modeled on the basis of homology with related and structurally defined proteins. On the basis of this information, secondary structure elements of correctly folded molecules could be predicted. Bet v 1 is dominated by β-sheets with a single α-helix. The structure of Phl p 1 has not been determined, but that of Zea m 1, the homologous allergen from corn pollen, displays a two-domain structure dominated by β-sheets. Phl p 5 is a typical α-helical protein. No structure is available for Ole e 1. Group 1 allergens from house dust mites belong to the papain-like cysteine proteases and display a typical mixed α-helical and β-stranded structure, whereas the group 2 allergens have a typical immunoglobulin-like tertiary fold with two anti-parallel β-pleated sheets.

Natural and recombinant versions of Bet v 1, Phl p 5 and Der f 1 gave CD spectra in close agreement with each other, respectively, and with their known tertiary structures. Ole e 1, Der p 1 and the group 2 house dust mite allergens showed the expected CD spectra based on tertiary structures, though some differences were observed in peak amplitudes. The secondary structure of Phl p 1 did not agree with its proposed structure based on the crystal structure of the homologous allergen from corn pollen (Zea m 1). The natural protein displayed some random coil structure, but no clear evidence of the expected β-sheet structure. The recombinant protein completely lacked organized secondary structure.

Upon heating to 95℃, the CD spectra of all allergen preparations with the exception of Phl p 1 changed to that of a random coil. This is a normal behavior of polypeptides undergoing thermal unfolding. In general, the calculated values of melting temperature (Tm) were very similar for natural and recombinant preparations of the same allergen. Upon cooling to 25℃, Bet v 1, Phl p 5, and rDer f 1 showed CD spectra similar in shape and peak amplitudes to those recorded before heating, suggesting that these proteins are able to correctly refold after thermal denaturation. Both natural and recombinant Ole e 1 were partially refolded after thermal denaturation. In contrast, thermal denaturation irreversibly denatured all the other natural and recombinant house dust mite group 1 and 2 allergens.

9.3.1.5 Small-Angle X-ray Scattering (SAXS)

Small-angle X-ray scattering (SAXS) resembles the familiar situation when a beam of visible light is scattered by a colloidal suspension like milk or when the front lights of a car are scattered in the mist (Pilz *et al.*, 1979). Clear solutions of non-aggregated proteins, in contrast, require an electromagnetic beam of much smaller wavelength to interact with. Monochromatic X-rays are guided through the protein solution and the scattering curve [intensity×scattering vector] is recorded. Using indirect Fourier transformation, the pair distance distribution function [distance frequency×dimension] can be calculated from the scattering curve (Glatter, 1977). The pair distance distribution function (PDDF) expresses the frequency of intra-molecular distances between electrons. Thus, conclusions on the molecular size, shape, and aggregation behaviour of proteins in solution can be drawn (Glatter, 1997; Svergum and Koch, 2002).

For the CREATE samples, measurements were performed with the SAXSess camera (Anton Paar KG, Graz, Austria) attached to a conventional copper K X-ray generator (Philips, Eindhoven, the Netherlands). Data evaluation was done by indirect Fourier transformation (Glatter, 1977) using the PCG software package (Karl-Franzens, University of Graz). Briefly, SAXS measurements showed that 5 of the allergen preparations (nBet v 1, nPhl p 1, nDer p 1, nDer f 1, and nDer p 2) could be assigned to monomeric or dimeric molecules. Six allergens (rBetr v 1a, rPhl p 5a, rPhl p 5b, nOle e 1, rDer p 2, and nDer f 2) displayed some tendency to oligomerize at the rather high concentrations (3.3 mg/mL) used in the SAXS measurements. A non-aggregated state (i.e., not multimerized) could be assigned to 11 of the 17 samples tested. Three preparations (rDer p 1, rDer f 1, and rDer f 2) showed moderate aggregation and rOle e 1 showed extensive aggregation. Two allergen preparations (rPhl p 1 and nPhl p 5) did not give conclusive results, probably due to low protein concentration in the case of rPhl p 1 or a too high scattering contribution of the buffer in the case of nPhl p 5.

9.3.2 Immune Reactivity

Another criterion to judge the quality of candidate reference molecules was their immune reactivity relative to the reactivity of their purified natural counterparts. Within CREATE, immune reactivity was essentially analyzed in two ways: IgE reactivity and monoclonal antibody reactivity in ELISA. IgE reactivity was analyzed in four ways:

Direct IgE-binding in RAST
Direct binding in dot-blot
RAST-inhibition
Biological activity in basophil histamine release

9.3.2.1 RAST, RAST Inhibition and Dot-Blot Analysis

For RAST analysis, natural and recombinant versions of the eight major allergens evaluated in CREATE were coupled to Sepharose (Aalberse *et al.*, 1981). For each of the four allergen sources (birch, grass, olive pollen and house dust mite), the clinical partners of CREATE collected sera of patients with a positive SPT and a convincing history of symptoms upon exposure. These sera were used in RAST analyses to compare the IgE-binding potencies of natural and recombinant allergens. Linear regression analysis showed a very good quantitative relationship between IgE antibody levels to natural and recombinant allergens for each of the allergens tested ($R_{Spearman}$ >0.9; P<0.01). The tightest correlation was seen for Bet v 1, with other allergens showing broader distributions of data points, especially at low IgE levels (<1IU/mL). Although the levels of allergen-specific IgE were highly correlated, with the exception of Bet v 1, recombinant allergens showing a mean of 50%–80% of the IgE antibody binding of their natural counterparts (Table 9. 5).

Table 9.5 IgE-binding potencies of recombinant allergens

Allergen	# of Patients	Mean RAST (± SE) (IU/mL) Natural Allergen	Mean Ratio Rec/Nat (Sera >0.3 IU/mL)	*P*-Value (paired *t*-Test)
Bet v 1	187	17.5 (± 2.6)	1.0	> 0.2
Phl p 1	345	40.2 (± 2.2)	0.6	< 0.001
Phl p 5		24.5 (± 1.3)	0.7 / 0.5*	< 0.001
Ole e 1	166	10.9 (± 1.9)	0.6	< 0.001
Der p 1	280	6.4 (± 0.6)	0.5	< 0.001
Der p 2		10.9 (± 1.3)	0.7	< 0.001
Der f 1		5.1 (± 0.4)	0.5	< 0.001
Der f 2		10.6 (± 1.4)	0.8	< 0.001

* rPhl p 5a/rPhl p 5b

Recombinant Bet v 1 showed almost identical IgE binding as the natural allergen. These results suggested that in many cases the recombinant allergens could be used for *in vitro* diagnostic purposes, but that in some cases improvements to the recombinant allergens would be needed to achieve equivalence with the natural allergens. Dot-blot analysis (Fig. 9.4) confirmed the similarity in IgE-binding potency of some natural and recombinant allergens (e.g., Bet v 1) but demonstrated that differences were more significant for the others (e.g., Phl p 1). RAST-inhibition analysis was used to quantify the differences in IgE-binding potencies. Only rBet v 1 and rDer f 2 demonstrated similar inhibitory potency as their natural counterpart. For all the other allergens, recombinants were at least 10-fold weaker in RAST-inhibition. For recombinant versions of Phl p 1 and Der p 1, the bserved difference was >100-fold. This implies that the minor decrease in direct binding (≤2-fold) is in fact an overestimation of IgE-binding potencies. This can most likely be explained by the high allergen concentrations used in RAST (or CAP) resulting in saturating assay conditions, even for correctly folded allergen being under-represented in some of the recombinant allergen preparations.

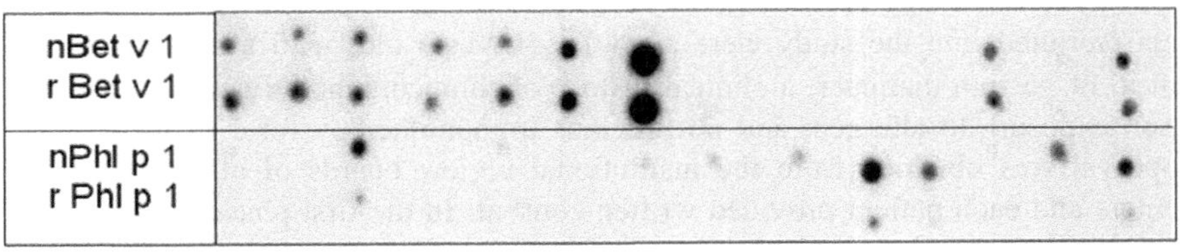

Fig. 9.4. Dot blot analysis: Natural versus recombinant

9.3.2.2 Biological Activity

Another way to evaluate the immune-reactivity of recombinant allergens is by comparing their potency to induce mediator release from basophils. In CREATE, we used the so-called stripped basophil protocol (Kleine-Budde *et al.*, 2001) in which basophils from a non-allergic donor are treated with lactic acid to remove IgE. These cells were subsequently re-sensitized with IgE from the patients' sera. Using 25 different sera of allergic patients per allergen, each of the recombinant allergens of CREATE was compared to its natural counterpart with respect to their potency to induce histamine release (biological activity). In some cases, the biological activity was in agreement with the potency detected by RAST-inhibition (Bet v 1, Phl p 1, Der f 1 and Der f 2). For Phl p 5a and Phl p 5b, Ole e 1, Der p 1 and Der p 2, the biological activity was higher than what was expected on the basis of RAST-inhibition. In two cases (Phl p 5 and Ole e 1), recombinant allergens were significantly more active than their natural counterparts (10-fold). A possible explanation for these observations could be that oligomerization and aggregation favor efficient histamine release by increasing the epitope valency (repetitive epitopes). Alternatively, the chosen isoforms for recombinant production could exhibit higher allergenicity.

Overall, the IgE-based analysis demonstrated that simple direct binding assays were not sufficient to get a good impression of the IgE-binding characteristics and allergenic potency of recombinant allergens. Competitive and functional assays are needed to provide a balanced evaluation.

9.3.3 The CREATE Serum Bank

The development of a serum bank from a broad spectrum of allergic patients was an essential part of the CREATE project and was co-ordinated by Drs. Stephen Durham (London, UK) and Montserrat Fernández-Rivas (Madrid, Spain). The serum bank was established by recruiting allergic patients in 11 clinical centers across Europe (London, Manchester, Madrid, Jaén, Barcelona, Rome, Borstel, Utrecht, Gothenburg, Strasbourg, Vienna) by researchers listed in Table 9. 2. Patients recruited into the study were aged 16–50 years old, with a skin prick test wheal of ≥5 mm diameter, a clinical history of rhinoconjunctivitis and/or asthma upon exposure to allergen; and no previous immunotherapy for 5 years. Ethical approval was obtained from the institutional review boards of all participating centers and each patient provided written consent. In the first phase of the study, each patient who was enrolled donated ~20 mL blood to the project and ~10 mL serum was obtained for IgE antibody measurements. Serum was obtained from a total of 961 patients, as follows: birch, n=186; timothy pollen, n=342; olive pollen, n=155; mite, n=278. In a second phase of the study, candidates with a range of IgE antibody levels by RAST were selected to donate 100 mL blood to the project and the mean serum volume collected was ~35 mL. The "large" serum bank contains sera from 155 patients, 34 for birch pollen, 60 for grass pollen, 20 for olive pollen and 41 for house dust mite. The CREATE serum bank is maintained by the project co-ordinator, Dr. Ronald van Ree at the Academic Medical Center, Amsterdam, The Netherlands, and is available to researchers upon agreement of the CREATE partners.

9.3.4 Stability

Stability of allergen preparations that would in the future serve as international reference preparations is of the utmost importance. References that rapidly decrease in quality could not be used as certified standards. CREATE evaluated the process of freeze-drying and the stability of freeze-dried allergens using accelerated degradation studies which were performed at the National Institute of Biological Standards and Control, a WHO-approved repository based in St. Albans, UK.

9.3.4.1 Freeze Drying Cycle Conditions

Seventeen purified allergen preparations and a timothy (*Phleum pratense*) grass pollen extract (NIBSC code 02/322) were formulated in a sterile normal saline solution containing 0.1% D-(+) trehalose dehydrate and 0.2% human serum albumin. One milliliter aliquots of the formulation containing 5 μg allergen product were dispensed into sterile, acid washed glass ampoules baked at 250℃ for 1 h prior to filling. A sub-sample of ampoules was heat-sealed and stored at −150℃ serving as frozen baseline for all future measurements. rPhl p 1 was not included in the study because of its poor solubility. The timothy grass pollen extract was used to optimize freeze drying conditions. A freeze drying cycle was carried out to ensure that the freezing temperature, the freeze-drying (shelf temperature), the time profile of the vacuum, the properties of the fill solution and the glass transition temperature were all optimized.

The material was freeze-dried for over 100 h. During the freeze-drying process the shelf temperature of the freeze drier was maintained at −40℃ for at least 75 h and then raised over a 20 h period to +20℃ and then maintained at +20℃ for 10 h.

The cycle condenser temperature was maintained at between −60℃ and −70℃ and the vacuum between 4×10^{-2} − 5×10^{-2} mbar. The ampoules were fitted with a capillary plug and backfilled with dry nitrogen (O_2 level 10 ppm, H_2O 5 ppm) and heat sealed by fusion of the glass neck. The residual moisture in each case should be less than 1%.

9.3.4.2 Lyophilization Has No Adverse Effects on Allergen Activity

The activity of the original allergen product (by definition being 1), the formulated allergen product (frozen base line) and the formulated lyophilized allergen product was assessed. Lyophilization of the various allergen products did not significantly affect the activity of the original product as illustrated in Table 9.6, for one of the allergens.

Table 9.6 Potency estimates for frozen baseline and freeze-dried Phl p 5 preparations

Allergen	Frozen Baseline	Freeze-Dried	Original Allergen
rPhl p 5a (03/102)	0.98	1.08	1
rPhl p 5a (03/106)	1.15	1.1	1
rPhl p 5b (03/108)	1.08	1.0	1
nPhl p 5 (02/322)	0.9	1.09	1

Of all allergen vials filled, 1%–2% were selected at random and tested for the control of variation of the fill, the oxygen level, and moisture content. All met the WHO criteria for these parameters: CV of 1 mL/1 g fill <0.25%, O_2-content < 40 mM/L and the moisture content <1%).

9.3.4.3 Study Design: Accelerated Degradation Studies

Twenty ampoules of each allergen product filled were placed at each of the different storage temperatures, −150, −70, −20, +4, +20, +37, +45 and +56 ℃ in order to assess the effect of temperature over a period of time on the stability of the product. Two ampoules per temperature point were removed and tested post-lyophilization during the 1st, 3rd, 6th, 9th, 12th and 24th month using an appropriate allergen-specific sandwich ELISA.

Raw data were assessed graphically to check for any anomalies and to ensure that the dose-response relation was monotonic. Additional tests for statistically significant outliers and for homogeneity have been carried out using an in-house program, SCAN (Gaines and Rice, 1985). Row and column effects were assessed using analysis of variance; effects within-plates as well as between plates were investigated. In most cases, the dose-response curves were sigmoid and could be satisfactorily described using a four parameter logistic function. For each plate in each assay, asymptotic limits for the logistic function were determined and logit transformed responses were analyzed using the methods of parallel line bioassay and an in-house program (Gaines and Tydeman, 1982) to estimate potency at each temperature for each time point. The potency estimates used for calculation of degradation rates are based on geometric means of the estimates from the individual plates. Potency estimates calculated for each temperature for each time point relative to −20℃ were used to predict annual loss of potency (Kirkwood and Tydeman, 1984), assuming that the relationship between the activity and temperature is described by the Arrhenius equation. At the earlier time points, samples stored at the higher temperatures showed little or no loss of activity and reliable predictions based on this data could not be obtained. Estimates for preparations stored at 12 and 24 months post-lyophilization were more likely to show a significant loss of activity and thus gave a more reliable estimate for degradation. For 88% of the allergen products tested, rates for degradation could be determined using data at these two time points. In the remaining cases, the maximum likelihood procedure used for estimation failed to converge, and no estimate of degradation rate was obtained.

9.3.4.4 Stability of Lyophilized Allergens during 12 and 24 Months

Most allergens (9/17) remained stable at −20℃ for 2 years, with the annual predicted potency loss being < 0.5% (Table 9.7). For an additional 6/17 the loss was < 0.5% as well at year 2, but a firm conclusion can not be drawn because loss had been > 0.5% at year 1. Additional studies are therefore warranted. At +4℃,

6/17 allergens remained stable for 2 years (consistent <0.5% loss), and 6/17 again had discrepant results between year 1 (>0.5%) and year 2 (<0.5%). Only for natural and recombinant Der p 1, potency loss at one year was 1.6% and 4.2%, respectively, and 6.7% and 9.7%, respectively, at two years. In summary, when kept at −20℃ stability was good (year 1 and year 2 in agreement) or promising (year 2 <0.5% but conflicting with year 1) for all other allergens when evaluated at year 2.

Table 9.7 Stability at 24 months post-lyophilization. Question marks indicate that data for the 2nd year were <0.5% but data for the 1st year >0.5%

Allergen	NIBSC Codes	<0.5% loss (−20℃)	<0.5% loss (+4℃)
nBet v 1	03/180	?	N
rBet v 1	03/184	Y	N
nPhl p 1	03/234	Y	Y
rPhl p 1		no data available	
nPhl p 5	02/322	Y	Y
rPhl p 5a (1)	03/106	Y	?
rPhl p5a (2)	03/108	Y	Y
rPhl p 5b	03/102	?	?
nOle e 1	03/230	?	N
rOle e 1	03/232	Y	?
nDer p 1	03/120	N	N
rDer p 1	03/122	N	N
nDer f 1	03/156	Y	Y
rDer f 1	03/154	?	?
nDer p 2	03/174	?	?
rDer p 2	03/176	?	?
nDer f 2	03/160	Y	Y
rDer f 2	03/162	Y	Y

9.3.5 ELISA Evaluation

Incorporating quantification of major allergens in standardization protocols not only requires availability of adequate allergen references, but is also dependent on well-validated immuno-assays for their measurement. Over the past decades numerous assays have been developed, covering most major inhalant allergens (Platts-Mills *et al.*, 1997; van Ree, 1997). CREATE selected multiple sandwich ELISAs for each allergen, i.e., 4 different assays for Bet v 1, 3 for Phl p 1, 4 for Phl p 5, 3 for Ole e 1, 2 for Der p 1 and 2 for Der f 1 and 3 for Der p 2/Der f 2. The protocol to evaluate the performance of these assays was not set up as a true ring trial, necessary for validation according to international regulations. For practical reasons the number of participating laboratories per ELISA was too small (n=5) for a true ring trial. In addition, ELISAs were not delivered as complete kits, but as sets of antibodies in some cases pre-coated to microtiter plates. Although uniformity of protocols was the original intention, in the end the absence of complete ready-to-use kits resulted in minor deviations between protocols used at various locations. Differences were reported in the types of microtiter plates, enzyme-conjugates, and substrates used. Despite these shortcomings, the evaluation has resulted in very valuable information about the performance of the various assays, in particular with respect to differences in specificity and the consequences of those differences. A decisive criterion for the evaluation was the comparison between results obtained when using the natural allergen as standard or the recombinant. Two factors play a role here, isoform specificity and sensitivity to folding differences including aggregation state. Obtaining similar results when using either a natural or a recombinant standard would point towards a broad spectrum of reactivity with regard to isoforms and folding and aggregation status. Such a "lack of specificity" should be seen as an advantage for the measurement of allergens that appear in several isoforms and are extracted from biological source materials using different protocols, or in the future produced in different expression systems. Five laboratories carried out 2–4 ELISAs for the 8 allergens of CREATE, using natural and recombinant allergens as standards and freeze-dried intermediate bulk extracts from the 6 allergen manufacturers in the project as samples.

9.3.5.1 Recombinant Versus Natural: Isoforms Specificity

It has been reported elsewhere that monoclonal antibody-based assays can be too specific, i.e., have preference for specific isoforms (Hakkaart *et al.*, 1998; Smith *et al.*, 2001). Such assays are less likely to be very suitable for standardization purposes, because they will not pick up all isoforms present in an extract with similar efficacy. In extreme cases, they do not react with a specific isoform at all. Within the CREATE project, two examples of assays were found where a specific (recombinant) isoform was not picked up. This was observed for an assay for measurement of Phl p 5 when using rPhl p 5b as a standard, and for an assay for detection of house dust mite group 2 allergen using rDer p 2.0101 as a standard. Assays based on such combinations of standard and antibodies will of course not be implemented because a standard curve cannot be produced. However, an assay with decreased but significant recognition of a relevant allergen isoform will not be that easily unmasked as inappropriate, if no thorough comparison is made between purified natural and recombinant allergen as standard. A good example is one of the three assays used in CREATE for the measurement of Bet v 1. This assay was shown to be approximately 5-fold less reactive with rBet v 1.0101 (Bet v 1a) than with natural Bet v 1 containing around 50% of this isoform. Theoretically, the difference could also be explained by incorrect folding of the recombinant allergen but physico-chemical analysis (e.g., CD-spectroscopy) did not provide support for this (see above). Application of this rBet v 1.0101-insensitive assay, with that particular isoform as standard, resulted in overestimation of Bet v 1 in extracts, because the absorbance per ng Bet v 1 was lower for pure rBet v 1.0101 reference than for a mix of isoforms in the sample extract. The other 2 assays did not distinguish significantly between rBet v 1.0101 and nBet v 1. The results obtained with these 2 assays closely correlated and were approximately 5 times lower than that with the isoform-specific assay (Fig. 9.5). In Fig. 9.5 the dot blot illustrates that sera of birch pollen-allergic patients see no significant difference between nBet v 1 and rBet v 1a. On the other hand, grass pollen-allergic patients clearly react weaker to rPhl p 1 than to nPhl p 1. These observations clearly illustrate the importance of detailed evaluation of the fine-specificity of major allergen immuno-assays.

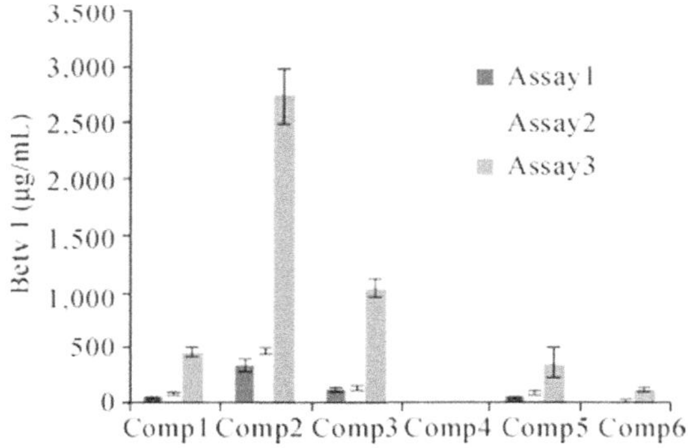

Fig. 9.5. Role of isoform-sensitivity of Bet v 1 ELISAs

9.3.5.2 Folding or Aggregation Sensitive Assays

Folding or aggregation of recombinant allergens can also influence their detection in major allergen ELISAs. Monoclonal antibodies can be directed to conformational or linear epitopes. The latter will most likely be less sensitive to incorrect folding. Oligomerization can mask one epitope and leave another available for antibody binding. A likely example of a folding/aggregation-sensitive and -insensitive (or isoform-specific) assay was observed amongst the assays used to detect Ole e 1.

Using nOle e 1, both ELISAs gave very similar results (Fig. 9.6). In contrast, rOle e 1 resulted in close to identical results for one assay (compared to using nOle e 1 as standard) and around 10 times lower for the other assay (Fig. 9.6). Three ELISAs were used to measure Bet v 1 in 6 extracts. Assay 3 was relatively insensitive for detecting Bet v 1a. Using a recombinant version of the isoform as standard results in clear over-estimation of Bet v 1 titres compared to both other assays that see no significant difference between nBet v 1 and rBet v 1a. Differences in sensitivity of antibodies for natural and recombinant allergens find their origin in the epitope recognized but also in the preparation used for immunization of mice for the production of monoclonal antibodies. A specific isoform or an incorrectly folded or aggregated allergen used as immunogen is likely to induce different antibodies from those a native mixture of well-folded isoforms does. These factors should be taken into consideration when selecting immunoassays and references for standardization purposes.

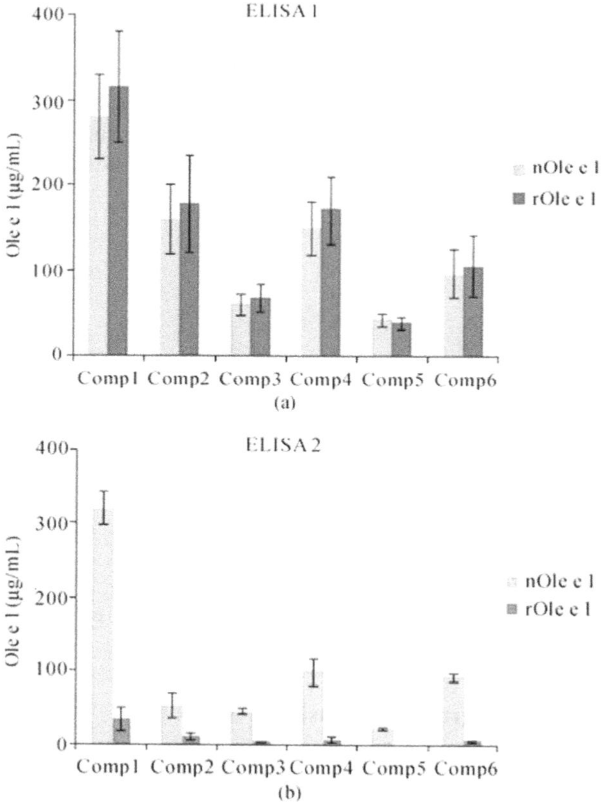

Fig. 9.6. Effect of structural differences between n and rOle e 1 on ELISA performance. (a) shows an ELISA that sees no significant difference between nOle e 1 and rOle e 1, resulting in similar Ole e 1 titres of six extracts using either of these molecules as standard; (b) illustrates that an ELISA that clearly reacts differently to nOle e 1 and rOle e 1 delivers Ole e 1 titres that differ up to 10-fold

9.4 Future Perspectives of CREATE

9.4.1 Summary of Achievements of CREATE

The CREATE project has delivered a wealth of data about a panel of 9 candidate reference molecules representing 8 of the most important major respiratory allergens. The starting point of the project was to evaluate whether recombinant molecules representing these allergens are suitable candidates to serve as certified reference materials for standardization purposes in the future. Moreover, CREATE aimed at evaluating whether available sandwich ELISAs are adequate tools for measuring these allergens in allergen extracts.

Recombinant allergens were evaluated using their natural counterparts as the gold standard. The comparison consisted of a detailed physico-chemical (identity, purity, folding, aggregation state, solubility and stability) and immunological characterization (IgE-binding, biological activity and dose-response behaviour in ELISA). Clearly, not all recombinant allergens tested fulfilled the requirements for serving as a future reference material. Shortcomings included incorrect folding, significant aggregation, poor solubility, decreased IgE-binding and biological activity, and insufficient stability. Obviously, these parameters are closely related and consequently some allergens scored poorly on a combination of them. Three allergens were characterized as well-folded, soluble and stable proteins with comparable IgE reactivity and biological activity as their natural counterparts: rBet v 1, rPhl p 5a and rDer p 2. However, this did not guarantee similar dose-response curves in all monoclonal antibody-based sandwich ELISAs. For both pollen allergens, two ELISA fulfilled this requirement, for Der p 2 only one assay. In other words, the application of a recombinant reference molecule in allergen standardization protocols is not enough to facilitate comparability of products on the basis of mass units of major allergens. To achieve that goal, recombinant references need to be linked to one or two validated assays that have been demonstrated to measure natural and recombinant allergen with similar performance characteristics, thus leading to the same designation of potencies in mass units of major allergens.

9.4.2 How to Proceed

One of the major achievements of CREATE has been the willingness of the commercial partners to collaborate and share reagents, in a joint effort to achieve the common goal of a validated system of allergen standardization in mass units of major allergens. Two of the recombinant allergens qualified as good reference candidates have recently been used in immunotherapy clinical trials, rBet v 1

(Batard *et al.*, 2005) and rPhl p 5a (Jutel *et al.*, 2005). These reagents were produced under GMP conditions required for *in vivo* application. Illustrating the spirit for cooperation and progress, both companies have made available 200 mg of each of these GMP-produced allergens to facilitate a follow-up program of CREATE aiming at establishing the first two internationally recognized certified recombinant reference materials. In addition, these companies and a third commercial partner of CREATE have agreed to provide the follow-up project with ELISA kits for both allergens that had demonstrated similar dose-response curves with natural and recombinant allergens. The follow-up project will be coordinated by the European Pharmacopoeia Commission of the European Directorate for the Quality of Medicines (EDQM) under their Biological Standardization Programme. The two allergens will be freeze-dried and formulated in at least 10,000 vials. The freeze-dried product will undergo the same physico-chemical characterization as was carried out in the CREATE programme. In addition, the influence of formulation (e.g., freeze-drying, small molecule additives) on protein stability will be investigated by Fourier Transform Infrared (FTIR) spectroscopy. Only SAXS will be left out because the high protein concentrations needed for this technique were thought to lead to over-interpretation of the problem of aggregation. Following that analysis, a true ring trial will be set up for each of the allergens using two ELISAs per allergen. The outcome of this project is expected to be the establishment of both recombinant allergens as certified reference materials in conjunction with at least one but preferably two certified ELISAs. The companies have agreed to allow production of allergens and mAbs also in the future for unrestricted use for standardization purposes.

The ring trial will be carried out by at least eight Official Medicine Control Laboratories (OMCL). In addition, non-European laboratories like the FDA will be invited to join. This is of great importance to ensure acceptance of the new reference materials beyond the borders of the European Union. Upon successful completion of this program, other allergens of CREATE, and also other allergens, including Amb a 1 from ragweed and Fel d 1 from cat, will need to follow.

Acknowledgements

The CREATE project was financially supported by the 5th Framework Programme of the European Union (Contract No. G6RD-CT-2001-00582), and by ALK-Abelló, Allergopharma, ASAC Pharmaceutical Int., Biomay, CBF Leti, HAL Allergy, Indoor Biotechnologies and Stallergènes. All CREATE partners are acknowledged for their contributions way beyond this financial support, without which the project could never have succeeded.

References

Aalberse, R.C., Koshte, V., Clemens, J.G. (1981). Immunoglobulin E antibodies that crossreact with vegetable foods, pollen, and Hymenoptera venom. J Allergy Clin Immunol, 68, 356-364.

Arntzen, F.C., Wilhelmsen, T.W., Lowenstein, H., *et al.* (1989). The international collaborative study on the first international standard of birch (Betula verrucosa)-pollen extract. J Allergy Clin Immunol, 83(1), 66-82.

Baer, H., Anderson, M.C., Helm, R.M., *et al.* (1986). The preparation and testing of the proposed International Reference (IRP) Bermuda grass (Cynodon dactylon)-pollen extract. J Allergy Clin Immunol, 78(4 Pt 1), 624-631.

Batard, T., Didierlaurent, A., Chabre, H., *et al.* (2005). Characterization of wild-type recombinant Bet v 1a as a candidate vaccine against birch pollen allergy. Int Arch Allergy Immunol, 136(3), 239-249.

Bousquet, J., Lockey, R., Malling, H.J. (1998). Allergen immunotherapy: therapeutic vaccines for allergic diseases. A WHO position paper. J Allergy Clin Immunol, 102, 558-562.

Chapman, M.D., Smith, A.M., Vailes, L.D. (2000). Recombinant allergens for diagnosis and therapy of allergic disease. J Allergy Clin Immunol, 106 (3), 409-418.

Chirino, A.J., Mire-Sluis, A. (2004). Characterizing biological products and assessing comparability following manufacturing changes. Nat Biotechnol, 22(11), 1383-1391.

Ferreira, F., Wallner, M., Gadermaier, G., *et al.* (2006). Physico-chemical characterization of candidate reference materials. Arb Paul Ehrlich Inst Bundesamt Sera Impfstoffe Frankf A M, 95, 75-82.

Ford, A., Seagroatt, V., Platts-Mills, T.A., Lowenstein, H. (1985). A collaborative study on the first international standard of Dermatophagoides pteronyssinus (house dust mite) extract. J Allergy Clin Immunol, 75(6), 676-686.

Ford, A.W., Rawle, F.C., Lind, P., *et al.* (1985). Standardization of Dermatophagoides pteronyssinus: assessment of potency and allergen content in ten coded extracts. Int Arch Allergy Appl Immunol, 76(1), 58-67.

Gaines Das, R.E., Rice, L.R. (1985). SCAN, an exploratory program for preliminary analysis of bioassay and immunoassay data. Comput Methods Programs Biomed, 21(1), 25-33.

Gaines Das, R.E., Tydeman, M.S. (1982). Iterative weighted regression analysis of logit responses: A computer program for analysis of bioassays and immunoassays. Computer Programs in Biomedicine, 15, 13-22.

Gjesing, B., Jager, L., Marsh, D.G., *et al.* (1985). The international collaborative study establishing the first international standard for timothy (*Phleum pratense*) grass pollen allergenic extract. J Allergy Clin Immunol, 75(2), 258-267.

Glatter, O. (1977). A new method for the evaluation of small angle scattering data. J Appl Cryst, 10, 415-421.

Hakkaart, G.A., Chapman, M.D., Aalberse, R.C., *et al.* (1998). Immune-reactivity of recombinant isoforms of the major house dust mite allergen Der p 2. Clin Exp Allergy, 28(2),169-174.

Helm, R.M., Gauerke, M.B., Baer, H., *et al.* (1984). Production and testing of an international reference standard of short ragweed pollen extract. J Allergy Clin Immunol, 73(6), 790-800.

Jutel, M., Jaeger, L., Suck, R., Meyer, H., *et al.* (2005). Allergen-specific immunotherapy with recombinant grass pollen allergens. J Allergy Clin Immunol, 116(3), 608-613.

Kelly, S.M., Jess, T.J., Price, N.C. (2005). How to study proteins by circular dichroism. Biochim Biophys Acta, 1751(2), 119-139.

Kirkwood, T.B.L., Tydeman, M.S. (1984). Design and analysis of accelerated degradation tests for the stability of biological standards II. A felxible computer program for data analysis. J Biol Standardization, 12, 207-214.

Kleine-Budde, I., de Heer, P.G., van der Zee, J.S., *et al.* (2001). The stripped basophil histamine release bioassay as a tool for the detection of allergen-specific IgE in serum. Int Arch Allergy Immunol, 126(4), 277-285.

Larsen, J.N., Ford, A., Gjesing, B., Levy, D., *et al.* (1988). The collaborative study of the international standard of dog, Canis domesticus, hair/dander extract. J Allergy Clin Immunol, 82(3 Pt 1), 318-330.

Lent, A.M., Harbeck, R., Strand, M., *et al.* (2006). Immunological response to administration of standardized dog allergen extract at differing doses. J Allergy Clin Immunol, 118(6), 1249-1256.

Lill, J. (2003). Proteomic tools for quantitation by mass spectrometry. Mass Spectrom Rev, 22(3), 182-194.

Mueller, G.A., Benjamin, D.C., Rule, G.S. (1998). Tertiary structure of the major house dust mite allergen Der p 2: Sequential and structural homologies. Biochemistry, 37(37), 12707-12714.

Nanda, A., O'connor, M., Anand, M., *et al.* (2004). Dose dependence and time course of the immunologic response to administration of standardized cat allergen extract. J Allergy Clin Immunol, 114(6), 1339-1344.

Nelson, H.S. (2007). Allergen immunotherapy: Where is it now? J Allergy Clin Immunol, 119(4), 769-779.

Norman, P.S. (1986). Why standardized extracts? J Allergy Clin Immunol, 77(3), 405-406.

Norman, P.S. (1994). WHO-IUIS International Standards: Advantages of these extracts. Arb Paul Ehrlich Inst Bundesamt Sera Impfstoffe Frankf A M, (87), 59-64.

Pilz, I., Glatter, O., Kratky, O. (1979). Small-angle X-ray scattering. Methods Enzymol, 61, 148-249.

Platts-Mills, T.A., Chapman, M.D. (1991). Allergen standardization. J Allergy Clin Immunol, 87(3), 621-625.

Platts-Mills, T.A., Rawle, F., Chapman, M.D. (1985). Problems in allergen standardization. Clin Rev Allergy, 3(3), 271-290.

Platts-Mills, T.A., Vervloet, D., Thomas, W.R., *et al.* (1997). Indoor allergens and asthma: report of the Third International Workshop. J Allergy Clin Immunol, 100(6 Pt 1), S2-24.

Position, S. (2007). The use of standardized allergen extracts. J Allergy Clin Immunol, 99(5), 583-586.

Reinders, J., Lewandrowski, U., Moebius, J., *et al.* (2004). Challenges in mass spectrometry-based proteomics. Proteomics, 4(12), 3686-3703.

Scheiner, O., Breiteneder, H., Dolocek, C., *et al.* (1994). Molecular and functional characterization of allergens: basic and practical aspects. Arb Paul Ehrlich Inst Bundesamt Sera Impfstoffe Frankf A M, (87), 221-232.

Smith, A.M., Benjamin, D.C., Derewenda, U., *et al.* (2001). Sequence polymorphisms and antibody binding to the group 2 dust mite allergens. Int Arch Allergy Immunol, 124, 61-63.

Svergun, D.I., Koch, M.H. (2002). Advances in structure analysis using small-angle scattering in solution. Curr Opin Struct Biol, 12(5), 654-660.

Tyler, M.I. (2000). Amino acid analysis. Methods Mol Biol, 159, 1-7.

Valenta, R., Kraft, D. (2004). Recombinant allergens: From production and characterization to diagnosis, treatment, and prevention of allergy. Methods, 32(3), 207-208.

van Ree, R. (1997). Analytic aspects of the standardization of allergenic extracts. Allergy, 52(8), 795-805.

van Ree, R. (1999). A new start for allergen references and standardization based on purified (natural or recombinant) allergens and monoclonal and monospecific polyclonal antibodies. Arb Paul Ehrlich Inst Bundesamt Sera Impfstoffe Frankf A M, (93), 87-90.

van Ree, R. (2004). The CREATE project: EU support for the improvement of allergen standardization in Europe. Allergy, 59(6), 571-574.

van Ree, R. (2007). Indoor allergens: Relevance of major allergen measurements and standardization. J Allergy Clin Immunol, 119, 270-277.

Verdino, P., Keller, W. (2004). Circular dichroism analysis of allergens. Methods, 32(3), 241-248.

10

Diagnosis of Skin Allergy Diseases

Huan Qian and Min Zheng*

Department of Dermatology, the Second Affiliated Hospital, School of Medicine, Zhejiang University, Hangzhou, 310009, China
* E-mail: minz@zju.edu.cn

Abstract: The diagnosis of skin allergy diseases is crucial to treatment and prevention. A complete diagnosis of allergic skin diseases includes the clinical history, skin tests or laboratory markers for specific allergens. Clinical *in vitro* and *in vivo* tests are often preformed, the cutaneous test *in vivo* are especially designed for determining pathogenic allergens leading to immediate and delayed allergic reaction. All the laboratory methods are based on various stages and types of human immune response. Consequently, prior familiarity with their indications, contraindications, precautions and result interpretation is advocated. Advances in awareness and diagnostic procedures may aggravate in the recognition of allergic disease, since detecting some "inappropriate allergens" may define one as a patient. However, as the causes of allergic skin diseases are complicated, sometimes it is difficult to identify the original allergens. Further investigation of allergy triggers and risk factors, such as more specific serologic tests, molecular biologic tests, are under development.

10.1 Introduction

Allergic skin diseases are generally referred to as urticaria, atopic dermatitis, atopic eczema and angioedema. To identify allergens is the key point to the diagnosis, treatment and prevention of allergic skin diseases. Inhalation allergens can include house dust, dust mites, animal fur, elm, mulberry, the Indus, multiple mixture of molds, and Artemisia pollens; while the most common food allergens are fish, milk, crab, shrimp, meat, eggs, wheat, soy, and peanuts (Zhou, 2003; Liu Y.M. *et al.*, 2001; Yu *et al.*, 2009; Feng *et al.*, 2008; Zhang *et al.*, 2006; Zeng *et al.*, 2008; Wang *et al.*, 2002; Li *et al.*, 2007; Gao *et al.*, 2006). Allergens are closely related to the incidence, and recurrence as well as the duration of allergic skin diseases. Together with changes or variations in environmental and genetic statutes, allergens can be altered over time and in different areas. On the other hand, the uniqueness of skin allergic diseases lies in their "superficial" manifestation and the usage of cutaneous tests.

10.2 Clinical Evaluation

A thorough disease history is important for a reliable diagnosis. Patients with water-running symptoms like tears, nasal itching and eye symptoms are highly suspicious of allergies. The exact symptoms and its occurrence frequencies are important. Skin allergy diseases mainly manifest as cutaneous flushing, urticaria, angioedema, eczema, many of which include an itching sensation. Based on clinical diagnosis, urticarial lesions generally appear as pruritic, erythematous cutaneous lesions with regular or irregular borders that may vary in size and location, and persist for a few hours and rarely for more than 24 h. Angioedema occurs most frequently in the face, hands, and other soft tissues and is generally accompanied by symptoms of stretching, tingling, and tightness of the skin that typically last 24–36 h.

Because of a wide range of allergens, a detailed disease history inquiry can help to determine the possible allergens. Different types of clinical histories indicate different kinds of allergens. Pollen allergies generally have a seasonal onset. Allergies caused by fungi will increase in rainy and wet seasons. Symptoms attacking children throughout the year are often associated with mites, cockroaches, animal fur, food and fungi. Mite allergies often present symptoms which cause aggravation at night or early morning, in wet rooms or during bed making, and relieve or disappear when going outdoors. Patients get allergic to animal furs and dander, which may be as a result of having pets and having a history of animal contacts (Wang, 2007).

Moreover, it is important to identify foods and medications used by patients, particularly those substances ingested within 2–4 h before the development of lesions. For skin lesions induced by food allergies, the inquiry should include: (1) suspected allergy-induced food; (2) the amount of intake; (3) the duration between onset of the symptom and time of food intake; (4) having the same food at other times if there is the possiblity of the same symptoms; (5) the time of the latest incidence; (6) other interfering factors such as exercise (Ding and Wang, 2007).

Since atopic dermatitis has a high degree of heritability, family history should always be checked. Other potential clues are occupational history, recent infection, insect stings, and etc. Combined with physical examinations, further supplementary examination could be conducted to disclose possible causes.

10.3 *In Vitro* Tests

10.3.1 Eosinophil Count and Examination of Secretions

Complete blood count (CBC) and examination of secretions can suggest but not confirm an allergic origin of symptoms. Although total white blood cell (WBC) of patients with allergic diseases is usually normal, CBC should be applied in all allergy-suspected patients to detect eosinophilia. The identification of eosinophil differential depends on its percentage and the patient's condition. An eosinophil differential of 5%–15% of total WBC is considered nonspecific atopy; 16%–40% may suggest atopy, while other possibilities including drug hypersensitivity, cancer, autoimmune disorders, parasitic infection or other conditions; a differential of 50%–90% almost never occurs in atopic disorders but rather represents hypereosinophilic syndrome or visceral larva migrans. Various secretions, conjunctive or nasal secretions or sputum for instance, can be examined for leukocytes, especially for eosinophils, which indicates TH2-mediated allergic inflammation. The quantification of blood, sputum, nasal mucus, or tissue eosinophilia are used in detection, differentiation and management, such as defining disease severity, and monitoring response to corticosteroid therapy of allergic diseases (Goldman and Ausiello, 2007).

10.3.2 Total Immunoglobulin E (IgE)

Serum IgE levels are commonly elevated in atopic disorders, parasitic infections, infectious mononucleosis, autoimmune disorders, drug reactions, immunodeficiency disorders, and in some forms of multiple myeloma. The measurement of total serum IgE is rarely useful in diagnosis, since a large proportion of IgE in a given individual may be directed toward a single antigen, while total IgE levels may be normal in the presence of allergic disease. However, to some extend, total serum IgE levels can indicate disease severity or the risk of exacerbation and are probably helpful for following responses to therapy (Nichols and Cook-Bolden, 2009). Therefore, the identification and quantification by *in vitro* or *in vivo* tests (skin prick tests for instance) of allergen-specific IgE in suspected patients is of great importance.

10.3.3 Allergen-Specific Immunoglobulin E (IgE)

In vitro assessment performed for specific IgE tests improves diagnostic accuracy because it is highly specific and sensitive (Gao *et al.*, 2006). Yet, as origins of allergic skin diseases are so diversified and complicated, limitations of the specific antigen test are inevitable. On the one hand, the magnitude of the reaction is weakly correlated with the degree of sensitization and expression of the allergy, even if there are exceptions for certain foods. We can not conclude final clinical diagnosis based only on the test results. On the other hand, only a limited number of allergen extracts are available for testing; especially allergens from trees, whose cross-reactivity is very limited. Therefore, kits for IgE testing is a preliminary estimation of serum total IgE level and appearance of specific IgE antibodies (Zeng *et al.*, 2008).

The current status of allergic disease is a determining factor for fluctuation of serum IgE levels. A large sample screen of specific IgE level studies have shown that for people suffering from asthma, the IgE allergen-positive rate was significantly higher in acute exacerbation than that in remission, and for patients associated with allergic rhinitis the IgE levels are higher than those who are not. Anti-allergy agent consumers had higher positive rates than people in the control groups (Liu *et al.*, 2006). It is also influenced by the degree of sensitization to allergens, drugs like prednisone, other allergic diseases, atopic disorders, parasitic infections and autoimmune disorders. The most common ways to detect specific IgE are radioallergosorbent testing (RAST) and enzyme-linked immunosorbent assay (ELISA).

10.3.3.1 Radioallergosorbent Testing (RAST)

RAST is a sensitive and specific method forto detecting specific IgE. It is preferred in clinical practice for its specificity and sensibility. The only difference between RAST and ELISA is that the second antibody is labelled with ^{125}I and tests the radioactivity of samples. The more radioactive it is, the more specific IgE there is. Although it is more sensitive than ELISA, its application in China is very limited because of the high expenses of time and money, radioactive environmental problems, lacking qualified technicians and limitations specific antibodies.

10.3.3.2 Enzyme-Linked Immunosorbent Assay (ELISA)

No isotopes are utilized in ELISA and the enzyme-linked antibody is stable enough to be stored for a long time. Consequently, ELISA is generally used in detecting specific anti-IgE in serum in China. Basically, purified allergens are adsorbed on a solid phase carrier, and then the tested serum is washed over the surface. The antibodies in the serum are linked to a kind of enzyme, which could make the substance added afterwards detected by fluorescence (Xu *et al.*, 2006; Liu and Lai *et al.*, 2001).

10.4 *In Vivo* Tests

In vivo test involves introducing small amounts on or into the skin or by ingesting potentially allergic food for provocative testing. These methods are simple, direct and inexpensive, and provide immediate results and allow several allergens to be tested simultaneously.

Choice of allergen in question is based on the patient's past medical history and local geographic prevalence. Skin tests have a specific diagnostic value, especially when a detailed history and physical examination fails to reveal possible triggering allergens. The most commonly used allergens are from pollens (tree, grass, and weed), molds, house dust mites, animal dander and sera, insect venom, foods, and β-lactam antibiotics (Nichols and Cook-Bolden, 2009).

In vivo test could induce systemic allergic reactions, exacerbation, and sometimes even death. If a patient has a serious allergy history or is in the acute phase, *in vivo* tests are not suggested. Moreover, patients under the following conditions are not suitable for skin tests: (1) with severe skin lesions (such as large area of atopic dermatitis) affecting the results of skin tests; (2) taking anti-histamine drugs which could not be interrupted; (3) a high risk of lethal complications of skin test.

For all *in vivo* tests, certain drugs, including OTC and prescription antihistamines, prednisone, tricyclic antidepressants, and monoamine oxidase inhibitors can interfere with results and should be stopped at least 3 days before testing. Long-acting antihistamines may have to be avoided for 2–3 weeks. Patients taking β-blockers should not be tested (Wu *et al.*, 2003).

In vivo test results should be combined with clinical history and *in vitro* tests in order to identify the causes of the symptoms. If there is an absence of symptoms, the patient with allergen-specific IgE is identified as sensitization but not as allergic. Allergens can be identified only when the results of clinical manifestation, *in vitro* tests and skin tests are consistent. Even for those who can not afford *in vitro* tests in some areas of China, administrating skin tests more than once and along with a careful medical history are much more helpful in making the final diagnosis (Zhang, 2008). More elaborate tests may be required for a definitive diagnosis or to prescribe immunotherapy.

10.4.1 Cutaneous Test

10.4.1.1 Skin Prick Test (SPT)

Skin prick tests (Fig. 10.1) are widely used for diagnosis of allergic diseases. This kind of *in vivo* assessment of allergen-specific IgE is undertaken by introducing into the skin a small amount of highly purified allergen or allergic food by a prick puncture technique and assessing the cutaneous response 15–20 min thereafter. This method can detect multiple allergens with high sensitivity at one time and is quick and easy to use.

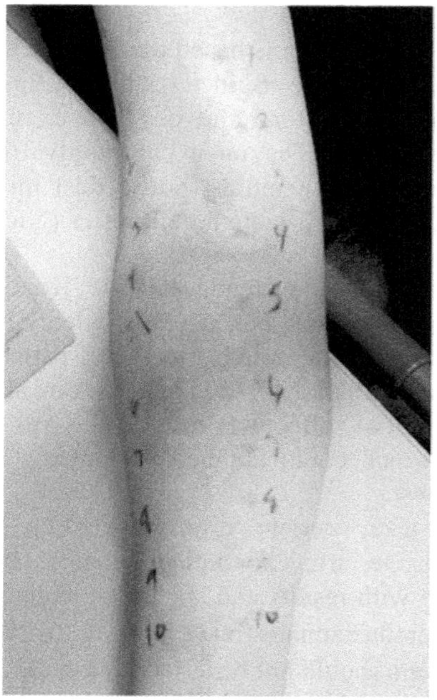

Fig. 10.1. Skin prick test

In both SPT and traditional intradermal tests, it is necessary to set up control groups. Normal saline is set as a negative control, while patients display a positive response to a control employing histamine. For patients who have had a recent (<1 year) generalized reaction to the test allergen, testing begins with the standard reagent diluted 100-fold, then 10-fold, and then the standard concentration.

Because there are no uniform agents and allergens for SPT in China, there is so far no standard criteria for wheal-and-flare responses. Different agents will have their own measurements. Generally speaking, a positive response is judged as a wheal and flare when it is at least 2 mm larger than those caused by a saline control. The larger the red wheal area is, the greater the allergy severity is. False positives occur in dermatographism, which is a wheal and flare reaction provoked by stroking or scraping the skin, and when allergen extracts have been stored incorrectly or are outdated (Rong *et al.*, 2007).

SPT results are subject to a number of factors. As mentioned previously, medicines such as antihistamines and prednisone may result in false negatives. Secondly, the operating skills also affect the test results, such as acupuncture depth in the human skin, force, speed, and other factors which possibly lead to variations of allergen quantities, since different amounts of allergens can give rise to different responses. Furthermore, age is a criterion for the results of SPT. Children under 1 year old are more likely to manifest false negative results than older children, because of low levels of *in vivo* allergen-specific IgE and skin reactivity under an immature immune system (Nichols and Cook-Bolden, 2009).

Compared with the traditional intradermal skin test, SPT gains its prevalence for fewer adverse reactions, better cost-effectiveness, higher specificity and less pain and trauma, which makes it more suitable for children. Subsequently, SPT has been gradually applied as a clinical routine skin test, especially for children (Liu and Pei, 2005).

10.4.1.2 Intradermal Test (Intracutaneous)

The intradermal test is another allergen-specific IgE related skin test in which a small amount of the suspected allergen is injected intracutaneously. After about 20 min the area is examined for a reaction at the site. A typical reaction looks like a small hive with swelling and redness. It's generally safe, although the intradermal test has a slightly higher risk of a severe reaction than others. Highly allergic individuals may experience arm swelling, particularly after intradermal testing. High sensitivity to suspected allergens is the contradiction for intradermal tests in case of accidents.

In comparison with SPT, the intradermal test is more sensitive and reproducible than SPT, but less specific. When the SPT or RAST test is negative or equivocal but allergy is highly suspected, an intradermal test may be recommended or a skin test may be repeated at a later date. If there is doubt, both tests should be performed. In addition, the intradermal test may be much easier for infrequent users.

10.4.1.3 Patch Test (Percutaneous)

The patch test relies on the principle of a type Ⅳ hypersensitivity reaction, which is directly related with the immune response in contact dermatitis. Currently, the patch test is the most widely applied skin test to detect allergic hypersensitivities. Test indications include persistent or frequent dermatitis; hand, foot, leg or facial dermatitis; discoid dermatitis; atypical allergic symptoms; perianal or perineal dermatitis; dermatitis or urticaria reactions after ingestion of suspected allergens; no improvement after treatment; sudden appearance without past history; unusual distribution or pattern and other undetermined causes. It is a golden standard forto diagnosing allergic contact dermatitis (Davis, 2009).

Allergen is extracted to soak a small piece of gauze and placed on the upper back or arm for 48 h, after which the skin reaction is examined and scored from no reaction to severe blistering and redness in the area of exposed skin. The test sites are marked, since positive reactions at day 2 will not be considered positive unless the reaction persists for 3 days or more. Reactions occurring at day 7 or later are considered "late reactions" and delayed positive reactions may appear after day 5.

The interpretation of the results requires considerable experience and training. Other skin diseases may complicate interpretations and it may be difficult to distinguish between irritant contact dermatitis and allergy dermatitis. Irritant reactions including sweat rash, follicular pustules and burn-like reactions are often sharply delineated and tend to be maximal on day 2 and fade on removal of patchs. If the results are considered positive, the person is probably allergic to that substances (Kurowski and Boxer, 2008).

Tests substances may be ready-to-use (e.g., TRUE TEST) or prepared using patient-provided products or commercially available allergens (e.g., Trolab, Chemotechnique, allergEAZE). The most frequently used allergen in patch testing around the world is nickel (Douglas, 2009).

The following reasons should be considered for a false negative: Remove panels too early or poor contact with skin, read too early, missed second reading or late reactions, topical or systemic immunosuppressant used, results affected by UV light or tanning, possible photoactive allergens, poor skin condition at the test site and low allergen concentration. False positive reactions may also be caused by high test concentration, technical errors, impure or contaminated test substances, irritant vehicles especially solvents, excessive test preparation, adjacent allergen's reaction, current or recent dermatitis at the test site or distant skin sites away from tested locations, recent (<3 weeks) contact allergen testing, generally irritable skin, tape or patch reaction, pressure effect of tape, or allergen degradation.

Unexplained positive reactions are related to incomplete medical or occupational history, exposure to unsuspected, undetected, unrecognized allergens, cross-reactivity to a related allergen, or low environmental exposure to an allergen prior to testing. If a patient with a strong history of allergies shows a doubtful reaction, it is highly recommended to have the test read again at a later time by other dermatologists and retest the patient, if necessary.

10.4.2 Provocative Testing

Provocative testing involves direct exposure of the mucosa to allergens. While some allergens are highly suspected but can not be determined by the approaches mentioned above, provocative testing can be preformed. Patients are commonly told to document their reactions (e.g., occupational or disability claims, and sometimes for diagnosis of food allergies). Ophthalmic testing and nasal and bronchial challenge have no advantage over skin testing in the diagnosis of allergic skin disease. Thus, they are rarely used (Kurowski and Boxer, 2008).

For skin lesions caused by food, double-blind and placebo-controlled food challenge (DBPCFC) is the golden standard for diagnosis. And it may be useful in separating allergies from sensitization or in eliminating a suspect food from consideration. Subjects and physicians do not know what things can be detected with food. However, DBPCFC, including placebo groups as control subjects to rule out the impact of psychological factors of patients and doctors, are much more objective and reliable. A DBPCFC exclusion diet should be conducted after 7–14 d and after overnight fasting. Food intake usually starts from 10 mg, and the amount doubles every 15–60 min. We stop if symptoms emerge, if not, the maximum dosage is 8–10 g. A negative conclusion could be made if it is still asymptomatic.

As a provocative test can cause severe allergic reactions, patients who have a clear history of severe allergic reactions to specific allergens or where a rescue facility is not readily available, are not suitable for this type of test.

10.5 Conclusion

The diagnosis of skin allergy diseases demands a complete disease history, including present symptoms and physical signs, environmental exposures, past medical conditions, family history, occupation factors and as much other relevant information as the patient can provide. Common laboratory tests consist of blood routine, examination of secretions, total and allergen-specific serum IgE levels by, RAST or ELISA. For more visualized detecting methods, cutaneous tests such as the skin prick test, the intradermal test and the patch test can introduce certain allergens on the skin to induce specific allergic reactions, while provocative tests are used for suspect allergic food analysis. Allergists or dermatologists should choose appropriate tests according to the patients' indications and assess the results in accordance with their comprehension of the merits and limitations from the findings. Therefore, evaluation of the disease severity, the risk of exacerbation and suitable treatment will be determined afterwards. However, sometimes determination of the precise allergens can be beyond our existing technology. In order to keep track of allergens, it is highly recommended to keep a diary of disease fluctuation and its possible triggers. Physicians and patients should cooperate closely for identification and treatment with many follow-ups as necessary.

References

Davis, M.D. (2009). Unusual patterns in contact dermatitis: Medicaments. Dermatol Clin, 27(3), 289-297.

Ding, H.M., Wang, X.Q. (2007). Current study and diagnosis of allergic diseases. Clinical Medicine of China, 23(7), 664-666 (in Chinese).

Douglas, J.D. (2009). Allergic contact dermatitis. Dermatol Nurs, 21(5), 287-291.

Feng, L., Geng, C.F., Xue, W., *et al.* (2008). Analysis of allergen-specific IgE test in 442 patients with allergic skin diseases. Jiang Xi Medicine, 43(1), 34-36 (in Chinese).

Gao, Z.Q., Wu, L., Gu, W.M. (2006). *In vitro* assay of serum total and specific IgE tests in 288 patients with allergic diseases. China Journal of Leprosy and Skin Disease, 22(10), 869 (in Chinese).

Gao, Z.Q., Wu, L., Gu, W.M. (2006). *In vitro* Assay of Serum Total and Specific IgE Tests in 288 Patients with Allergic Diseases. China Journal of Leprosy and Skin Diseases, 22(10), 869 (in Chinese).

Goldman, L., Ausiello, D.A. (2007). Goldman's Cecil Medicine (23rd Edition). Amsterdam: Elsevier.

Kurowski, K., Boxer, R.W. (2008). Food allergies: detection and management. Am Fam Physician, 15, 77(12), 1678-1686.

Li, M.Z., Lin, A.Z., Feng, Z.H., *et al.* (2007). Analysis about allergen detection of serum in allergic dermatitis patients. Laboratory Medicine, 22(2), 122-124 (in Chinese).

Liu, M.L., Pei, W.X. (2005). A study on the allergy test method compared the skin prick test with the intradermal skin test. Chinese Journal of Clinical Medicine, 4(7), 25-26 (in Chinese).

Liu, Y.M., Ye, S.Y., Li, P., *et al.* (2001). Analysis of allergen-specific IgE test in urticaria patients. Journal of Norman Bethune Medical University, 27, 337-338 (in Chinese).

Liu, Y.P., Lai, N.K. (2001). Study on the relationship between serum sIgE and allergic skin test among 100 cases allergic disease. Chin J Microbiol Immunol, 21, 91-92 (in Chinese).

Nichols, K.M., Cook-Bolden, F.E. (2009). Allergic skin disease: major highlights and recent advances. Med Clin North Am, 93(6), 1211-1224.

Rong, G.S., Liu, S.W., Gao, X. (2007). The applied evaluation of the skin prick test in the diagnosis of the allergies. Chinese J Current Practice Med, 6(7), 1-3 (in Chinese).

Wang, X.C. (2007). Several Issues of Diagnosis of Pediatric Allergic Diseases. J Appl Clin Pediatr, 22(9), 647-649 (in Chinese).

Wang, Y.X., Zhang, H., Li, H., *et al.* (2002). Analysis of Allergen Assay of 121 Patients with Allergic Dermal Diseases. Chin J Derm Veaneol, 16(2), 96-98 (in Chinese).

Wu, Q.H., Li, F., Wang, J.Y. (2003). Clinical application and analysis of allergen intracutaneous test. J dermatology and Venereology, 25(4), 4-6 (in Chinese).

Xu, L.L., Lin, X.Y., Liu, M.Z., *et al.* (2006). Comparison of specific IgE test and SPT in allergic diseases. J Fujian Med Univ, 40(5), 531-533 (in Chinese).

Yu, W., Chen, F., Luo, B., *et al.* (2009). Investigation of specific IgE against allergens detection in patients with allergic disease in Nanjing area. Chin J Lab Diagn, 13(6), 784-786 (in Chinese).

Zeng, B.B., Li, J.J., Zhang, J., *et al.* (2008). Analysis of serum specific IgE in 1120 patients with allergic dermatosis. Chinese Journal of General Practice, 6(12), 1301-1302. (In Chinese)

Zeng BB, Li JJ, Zhang J, *et al.* (2008) Analysis of serum specific IgE in 1120 patients with allergic dermatosis. Chinese Journal of General Practice, 6(12), 1301-1302 (in Chinese).

Zhang, H.Y. (2008). Allergy diseases in China. Chinese Journal of Allergy and Clinical Immunology, 2(4), 255-260 (in Chinese).

Zhang, Y.J., An, R.Z., Zhang, F., *et al.* (2006). Examination of specific serum IgE antibodies in 739 patients with allergic cutaneous disorder. Chin J Dem Venereol, 20(3), 153-154 (in Chinese).

Zhang, Z.F., Zhu, P.F. (2005). The clinical factors of serum specific IgE assay. Practical J Clin Med, 2(3), 49-50 (in Chinese).

Zhou, X.M. (2003). Reports of allergens tests results in 1748 allergic cases. China J Leprosy and Skin Dis, 19(5), 453-454 (in Chinese).

11

Novel Approaches for the *In-Vitro* Diagnosis of Type Ⅰ Allergies

Margrit Fooke

Dr. Fooke Laboratorien GmbH, Neuss, Germany
E-mail: mfooke@fooke-labs.de

Abstract: Diagnosis of type Ⅰ allergic reactions are based on the combination of a patient's history, skin testing and laboratory tests to detect specific IgE (sIgE). The methodology for measuring sIgE has evolved during recent years. Furthermore stimulation tests on the cellular level measure mediators such as histamine and leukotriens. Cellular stimulation tests and histamine measurement can be performed only in specialized laboratories, require fresh blood samples and are not applicable for routine diagnosis. The following chapter summarizes recent advances in the detection of sIgE.

11.1 Introduction

Type Ⅰ hypersensitivity reactions (type Ⅰ allergies) are characterized by the involvement of class E (specific immunglobulin E, sIgE) allergen-specific immunoglobulins (antibodies), so the detection of sIgE is an important tool for modern allergy diagnostics (Hamilton and Adkinson, 2003; Johansson *et al.*, 2004; Hamilton and Franklin, 2004). State-of-the-art allergy diagnosis includes a detailed case history of the patient, physical examination, skin prick testing (SPT) and *in-vitro* tests to detect sIgE. Furthermore, provocation challenges and/or cellular tests such as the basophile degranulation test are needed in case of food allergies (Hamilton and Adkinson, 2003; Hamilton and Franklin, 2004).

11.2 Skin Prick Test

Although the SPT has been proven to be a reliable and widespread method for the diagnosis of type I allergies, the method has several drawbacks. Since SPT is based on subcutaneous provocation *in-vivo*, it is a very unpleasant experience for most patients, on occasion producing severe local reactions. Moreover, rare cases of fatal or near-fatal systemic reactions have been reported. Patients with a history of anaphylactic reactions, young children, pregnant women and uncontrolled asthma patients should be considered at higher risk for systemic reactions (Liccardi *et al.*, 2003; Liccardi *et al.*, 2006). SPT can also not be performed on patients taking anti-histamines and several other types of anti-allergy medication.

Noteworthy, from the regulatory point of view, SPT products require a high level of quality control and in many countries a batch control process by an independent regulatory institution is mandatory. In Germany for instance, each lot of SPT solution must be quality controlled and released by the Paul-Ehrlich Institute. As this leads to a significant reduction of available SPT solutions, novel approaches, for example using rapid assays, may be the future for first line screening of type I allergies.

11.3 Rapid Assays for the Detection of Specific and Total IgE

In 2007, the first study on a lateral flow assay for the detection of allergen-specific IgE was published. The Allergy Lateral Flow Assay (ALFA) developed by Lucassen *et al.* (2007) is based on a universal device that can be combined with a variety of different allergens offering a flexible assay system for the rapid detection of allergen-specific IgE, with an assay time of 20–25 min. The principle of ALFA is illustrated in Fig.11.1. The agreement found by these authors between ALFA and SPT ranged from 87.7% (for d1, house dust mite) to 95.5% (for m6? *Alternaria alternata*), and that between ALFA and ImmunoCAP® ranged from 90.8% (for d1) to 96.7% (for g6, timothy grass pollen) (Lucassen *et al.*, 2010).

The ImmunoCAP® Rapid assay is also based on lateral flow technology. The major difference between ImmunCAP® Rapid and ALFA is the format of the allergens: ImmunoCAP® Rapid utilises immobilised allergens on a solid support while ALFA employs liquid phase allergens. Both systems have advantages and disadvantages. On the one hand, using ImmunoCAP® Rapid, simultaneous detection of specific IgE to ten allergens in a fixed panel is possible. On the other hand, ALFA can be used to test various allergens or allergen mixtures.

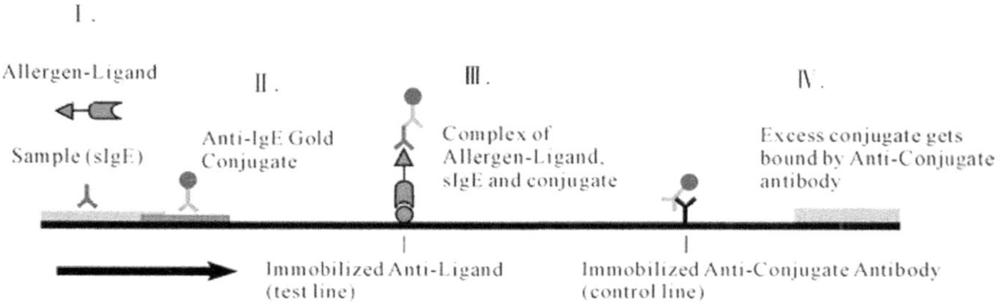

Fig. 11.1. Principle of the ALFA (Allergy Lateral Flow Assay)

In a study using ImmunoCAP®Rapid in 175 undiagnosed children, Donnano *et al.* (2007) found an overall sensitivity of 78% with a specificity of 96%. When a cut-off value of 1.0 kUA/L was applied, the sensitivity and specificity of ImmunoCAP®Rapid compared to the standard ImmunoCAP® Specific IgE blood test were 79% and 96%, respectively. Comparison with SPT revealed a sensitivity and specificity of 75% and 90%, respectively (Donnanno *et al.*, 2007). Similar findings have been reported by Hedlin and co-workers, in a multi-centre study at three sites in Sweden and two in Spain (Hedlin *et al.*, 2006), with overall agreement between the doctor's diagnosis and ImmunoCAP®Rapid of 91%. However, further studies are required to confirm the results of these studies.

11.4 Laboratory Methods for the Detection of Specific IgE

Historically, sIgE to various allergens was determined by the radioallergosorbent test (RAST) using allergen-coupled Sephadex and/or cellulose paper discs, first described by Wide (Wide *et al.*, 1967). Later on, the enzyme allergosorbent test (EAST) and more recently the reverse allergosorbent test (REAST) were developed and used for the detection of sIgE (Hamilton and Adkinson, 2003; Hamilton and Franklin, 2004). With the introduction of these second generation methods, significant improvements in speed and accuracy, use of nonisotopic labels, as well as reporting of IgE concentrations in a continuous range (IU/mL) according to the WHO reference preparation for IgE have been achieved (WHO 75/702). Although a variety of methods are available for detecting specific IgE in human serum or plasma, the assay generally follows the historical RAST protocol, using allergens immobilised on a solid phase support.

11.5 Reversed Enzyme Allergosorbent Test (REAST)

In the vast majority of test systems, allergens are immobilised on a solid support such as cellulose discs, cellulose membranes or so-called carrier polymers (CAP). In contrast, the ALLERG-O-LIQ System (Dr. Fooke Laboratorien GmbH, Neuss, Germany) follows the REAST protocol using anti-IgE coated microtitre plates and biotinylated allergens combined with streptavidin horseradish peroxidase (HRP) conjugate (Fig. 11.2).

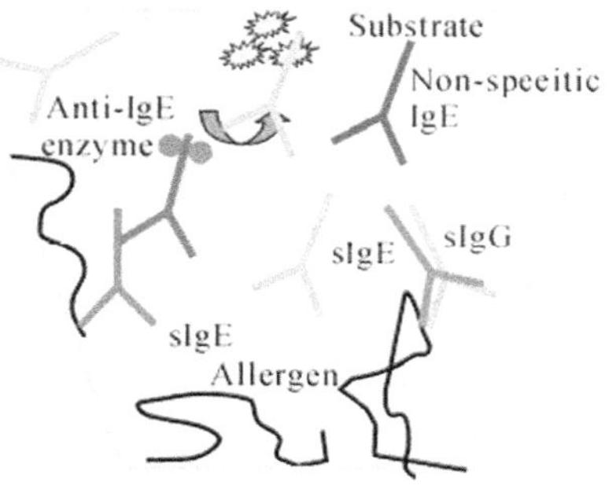

Fig. 11.2. Principle of the ALLERG-O-LIQ System

Patient samples, calibrators and controls (50 L each) are first applied to wells of an anti-IgE coated microtiter plate, then the IgE fraction of the samples is immobilised to the surface by incubating for 1 h. The subsequent washing step removes all other serum components, including IgG which may interfere with the specific IgE-allergen interaction. Biotinylated allergens are then added and incubated for 1 h. IgE binding proteins are bound in the presence of allergen specific IgE. Unbound allergens and proteins are removed by washing followed by an addition of streptavidin HRP conjugate and a further washing step. Finally, substrate is added and, after 30 min incubation, the reaction is stopped. Optical densities are photometrically determined at 450 nm wavelength (reference filter 620 nm).

A major challenge for developing reliable sIgE assays is allergen-specific IgG (asIgG) which can compete with the binding to the allergens (Steckelbroeck *et al.*, 2008). In addition, the concentration of IgG in normal conditions is significantly higher (up to 1000 folds) than that of IgE. With the development of the ImmunoCAP® System the blocking effect of asIgG was reduced by the increased allergen binding capacity of the CAP solid phase (Dolen, 2003), although, high titers of asIgG may still have a blocking effect. Following the protocol of the ALLERG-O-LIQ System, the IgE fraction is purified from the patient sample during the first incubation step. All other serum components which can compete with allergen binding, including asIgG, are removed. During the next incubation step, biotinylated allergens in the allergen extract are purified and separated from non-IgE binding molecules. These two purification steps contribute to the good performance of the ALLERG-O-LIQ as well as the ADVIA Centaur® system, which also follows the REAST protocol (Ollert *et al.*, 2005; Contin-Bordes *et al.*, 2007).

Alternatives to the REAST procedure have been developed. In the CARLA®

System (Radim Diagnostics, Italy), the serum samples are co-incubated with the biotinylated allergens, which saves on assay time but allows the allergens to interact with specific IgG which then can interfere with the binding of specific IgE. This may lead to a significant loss of sensitivity in samples containing allergen-specific IgG.

In 2004, a comparison of the ALLERG-O-LIQ and the ImmunoCAP® systems for the detection of specific IgE to common food and inhalant allergens showed good agreement for inhalant allergens but less for food allergens (Kleine-Tebbe *et al.*, 2004). More recently, the sIgE results of the ALLERG-O-LIQ System and ImmunoCAP® System for inhalant (d1; d5; e2; m3; *blomia tropicalis*) and food allergens (egg white, f1; cow's milk, f2; crab, f23 and shrimp, f24) have been compared in a Chinese patient cohort. The qualitative kappa agreement between both methods ranged from 0.6 to 1.0. When the sIgE results of all allergens obtained with the two systems were compared, an AUC (area under the curve) of 0.92 (confidence interval 0.881 to 0.963) was found (Sun *et al.*, 2008a). In earlier studies comparing the ImmunoCAP® and the ImmuLite® System (formerly Diagnostic products cooperation, DPC, Los Angeles, CA; now Siemens Healthcare Diagnostics) to SPT the ImmuLite® System gave better agreement to SPT for most allergens (Ollert *et al.*, 2005; Contin-Bordes *et al.*, 2007). When ImmunoCAP® was compared to ADVIA Centaur® (formerly Bayer Diagnostics, now Siemens) in the diagnosis of food allergies in children with atopic dermatitis, in 34/40 discrepant samples the results of the ADVIA Centaur® were consistent with those of SPT, compared to only 6/40 cases using ImmunoCAP® (Contin-Bordes *et al.*, 2007). The discrepancies between SPT and the ImmunoCAP® system were mainly due to false positives for sIgE against milk, wheat and soy. The authors concluded that the ImmunoCAP® System detects low affinity polyreactive IgE due to high allergen concentration on the CAP surface and thus recommended the ADVIA Centaur® System for these allergens (Contin-Bordes *et al.*, 2007). In other comparative studies, performance of the ImmunoCAP® and ADVIA Centaur® system was similar for all allergens tested, including d1, d2, e1, f1, f2, g6 and *Cynodon dactylon* (g2) (Ricci *et al.*, 2003; Petersen *et al.*, 2004, Li *et al.*, 2005).

Recent studies have provided evidence that the number of positive sIgE results and the total amount of sIgE correlate with disease severity and the number of clinical symptoms Wickman *et al.*, 2005; Nopp *et al.*, 2006). However, the importance of sIgE density rather than the amount of sIgE with respect to the clinical response has been emphasised (Crimi *et al.*, 1999), and it has been shown that low sIgE titers, especially against allergens of grass pollen and house dust mites, do not necessarily have clinical impact (Sun *et al.*, 2008b).

11.6 Standardization of Allergen Extracts

The major problem with comparing results of different test systems is the diversity of the allergen extracts. Currently available standard products are based on allergen extracts prepared from biological raw materials. The content of major and/or minor allergen epitopes of the natural mixtures of allergenic and non-allergenic molecules are generally not fully standardized. Many biological sources contain highly cross-reactive allergens, for example profilin, which is present in a broad variety of plant pollen and plant-derived food. Sensitization towards these allergens gives positive test results against numerous allergen extracts. Consequently, using extract based specific IgE testing, it is sometimes difficult to identify the correct allergen source when only cross-reactive allergen components are involved. There is also a lack of standardized and reproducible positive controls. Currently there is a very promising approach to develop these positive controls, based on chimeric avian antibody constructs (Braren *et al.*, 2010).

11.7 Molecular Allergy Diagnostics

Because of the diversity in allergen extracts and natural mixes of allergenic and non-allergenic molecules, and in the content of the biological material, it is sometimes difficult to correctly identify the allergen which causes the clinical symptoms. Cross-reactive carbohydrate determinants show structural similarities with allergens and can bind IgE antibodies but seem not to have clinical significance. The current diagnosis of Type I allergy therefore only permits the identification of a given allergen source, and not the molecular entities involved in the pathogenesis of the disease.

The latest approach is the use of highly purified native (n) or recombinant (r, produced by biotechnological methods) allergens for *in-vitro* diagnosis and for future immunotherapy. The nomenclature for these identified single allergens is internationally fixed (www.allergen.org). The growing number of known allergen molecules offers new diagnostic opportunities which can be increasingly used to determine sIgE antibodies. Highly purified and recombinant allergens can be used as single allergens in microarrays for component-resolved, molecule-specific diagnosis (Phadia), or for spiking of extracts or mixed as substitutes for natural extracts to increase the sensitivity and liability of *in-vitro* tests.

Many allergens have a complex structure of disulfide bridges and carbohydrate side chains and can therefore not be produced as recombinants. In these cases, native, highly-purified allergens have to be employed. Also the presence of iso-allergens has to be taken into account, for instance 20 isoforms of Bet v 1 can be isolated, differing in their IgE-reactivity and immunogenicity. The significance of the test results still need to be verified clinically, making allergy diagnosis more complex and expensive.

11.8 Conclusion

Modern allergy *in-vitro* diagnostic systems represent reliable tools for the detection of specific IgE, especially in combination with native, highly purified allergens or recombinant functional allergens. In addition, rapid assays such as ALFA have high potential to improve diagnosis in the future, also in combination with native, highly purified or recombinant allergens, and may lead to changes in the common diagnostic workflow of type I allergies. Whether the results produced by these tests are of clinical relevance has to be decided by the clinician. Rapid tests for the detection of allergen-specific IgE are appropriate tools for first line screening and diagnosis of type I allergy and offer new avenues for early diagnosis of this common disease. In combination with component-resolved diagnosis, rapid tests might have a future when deciding on the need for immunotherapy.

References

Braren, I., Heinrich, D., Ruehl, D., Bredehorst, R., Grunwald, T., Ollert, M., Spillner, E. (2010). Establishment of artificial control sera for diagnostic applications. 29th Congress of the European Academy of Allergology and Clinical Immunology, Oral presentation.

Contin-Bordes, C., Petersen, A., Chahine, I., Boralevi, F., Chahine, H., Taieb, A., Sarrat, A., Moreau, J.F., Taupin, J.L. (2007). Comparison of ADVIA Centaur and Pharmacia UniCAP tests in the diagnosis of food allergy in children with atopic dermatitis. Pediatr Allergy Immunol, 18, 614-620.

Crimi, E., Voltolini, S., Minale, P., Falagiani, P. (1999). Value of immunoglo-bulin E density in predicting nasal and bronchial response to inhaled allergens in rhinitic and asthmatic subjects with multiple sensitizations. Clin Exp Allergy, 29, 1663-1670.

Dolen, W.K. (2003). IgE antibody in the serum–detection and diagnostic significance. Allergy, 58, 717-723.

Donnanno, S., Sarratud, T., Trimarco, G., Terracciano, L., Nilsson, M., Cavagni, G., Fiocchi, A. (2007). Performance of a new rapid allergy point of care test among undiagnosed children with wheeze/asthma and rhinitis. Allergy, 62 (Suppl 83), 148.

Hamilton, R.G., Adkinson, N.F. (2003). Clinical laboratory assessment of IgE-dependent hypersensitivity. J. Allergy Clin. Immunol., 111, S687-S701.

Hamilton, R.G., Franklin, A.N. (2004). *In vitro* assays for the diagnosis of IgE-mediated disorders. J Allergy Clin. Immunol., 114, 213-225.

Hedlin, G., Nieto, G., Lilja, G., Nordvall, L., Lorente Toledano, F., Carlsson, L., Borres, M. (2006). A new point-of-care test for first line diagnosis of allergy in children. 25th Congress of the European Academy of Allergology and clinical Immunology Poster Presentation.

Johansson, S.G., Bieber, T., Dahl, R., Friedmann, P.S., Lanier, B.Q., Lockey, R.F., Motala, C., Ortega Martell, J.A., Platts-Mills, T.A., Ring, J., Thien, F., van Cauwenberge, P., Williams,H.C. (2004). Revised nomenclature for allergy for global use: Report of the Nomenclature Review Committee of the World Allergy Organization, October 2003. J. Allergy Clin. Immunol., 113, 832-836.

Kleine-Tebbe, J., Lepp, U., Vieths, S., *et al.* (2004). Spezifische IgE-Werte gegen Aero und Nahrungsmittelallergene im Vergleich: Dr. Fooke ALLERG-O-LIQ und Pharmacia CAP-System. Allergologie, 27, 129-135.

Li, T.M., Fu,P., Zic, V. (2005). Performance validation of a third-generation allergen-specific IgE assay in the clinical laboratory: interlaboratory and intermethod comparison. Clin Chim Acta, 361, 199-205.

Liccardi, G., D'amato, G., Canonica, G.W., *et al.* (2006). Systemic reactions from skin testing: literature review. J Investig Allergol Clin Immunol, 16, 75-78.

Liccardi, G., Salzillo, A., Spadaro, G., *et al.* (2003). Anaphylaxis caused by skin prick testing with aeroallergens: Case report and evaluation of the risk in Italian allergy services. J Allergy Clin Immunol, 111, 1410-1412.

Lucassen, R., Schulte-Pelkum, J., Csuvarszki, C., *et al.* (2010). Evaluation of a novel rapid test system for the detection of allergic sensitization to timothy grass pollen against established laboratory methods and skin prick test. Journal of Allergy, Volume 2010, Article ID 524084, doi:10.1155/2010/524084.

Nopp, A., Johansson, S.G., Lundberg, M., *et al.* (2006). Simultaneous exposure of several allergens has an additive effect on multisensitized basophils. Allergy, 61, 1366-1368.

Ollert, M., Weissenbacher, S., Rakoski, J., *et al.* (2005). Allergen-specific IgE measured by a continuous random-access immunoanalyzer: interassay comparison and agreement with skin testing. Clin Chem, 51, 1241-1249.

Petersen, A.B., Gudmann, P., Milvang-Gronager, P., *et al.* (2004). Performance evaluation of a specific IgE assay developed for the ADVIA centaur immunoassay system. Clin Biochem., 37, 882-892.

Ricci, G., Capelli, M., Miniero, R., *et al.* (2003). A comparison of different allergometric tests, skin prick test, Pharmacia UniCAP and ADVIA Centaur, for diagnosis of allergic diseases in children. Allergy, 58, 38-45.

Steckelbroeck, S., Ballmer-Weber, B.K., Vieths, S. (2008). Potential, pitfalls, and prospects of food allergy diagnostics with recombinant allergens or synthetic sequential epitopes. J Allergy Clin Immunol, 121, 1323-1330.

Sun, B.Q., Mahler, M., Jiang, M., *et al.* (2008a). Allergen-Specifi c IgE to inhalant and food allergens and total IgE values in China: Comparison of 2 commercial immunoassays. J Investig Allergol Clin Immunol, 18, 319-321.

Sun, B.Q., Mahler, M., Jiang, M., *et al.* (2008b) Comparison between two commercial immunoassays: Dr. Fooke ALLERG-O-LIQ versus Phadia

ImmunoCAPR System in detecting allergen-specific IgE and total IgE values. J Nanjing Med University, 22(5), 273-278.
Wickman, M., Lilja, G., Soderstrom, L., *et al.* (2005). Quantitative analysis of IgE antibodies to food and inhalant allergens in 4-year-old children reflects their likelihood of allergic disease. Allergy, 60, 650-657.
Wide, L., Bennich, H., Johansson, S.G. (1967). Diagnosis of allergy by an in-vitro test for allergen antibodies. Lancet, 2, 1105-1107.

12

Influence of Food Processing, Digestion and the Food Matrix on Allergenicity & Cellular Measures of Allergenicity

Y.M. Vissers[1,2], H.J. Wichers[1,2,3], H.F.J. Savelkoull[1,*]

[1] Cell Biology and Immunology Group, Wageningen University, P.O. box 338, 6708 WD Wageningen, the Netherlands
[2] Food Chemistry Group, Wageningen University, P.O. box 8129, 6700 EV, Wageningen, the Netherlands
[3] Food and Biobased Research, Wageningen University, P.O. Box 17, 6700 AA, Wageningen, the Netherlands
[*] E-mail: huub.savelkoul@wur.nl

Abstract: The degree of processing and digestion and the impact of embedding in food matrix components could influence the way in which allergens are presented to the immune system, and thus contribute to the severity of an allergic reaction. In addition, processing, digestion and matrix effects are also of importance in diagnostic procedures, in particular when food challenges are part of the diagnostic process. Food processing, for example, thermal processing, could alter the nature of the epitopes of a protein; however, there is no general rule on how proteins respond and the type of allergen, type of food structure, and type of thermal processing may all contribute to the impact of processing. As most proteins are digested in the stomach and the intestine, the immune system is mostly exposed to digested protein. However, digestion, even with a variety of gastric enzymes, is not always sufficient to abolish the allergo-reactivity of allergens. Recently developed *in vitro* models of digestion (e.g., including surfactants) are improving our knowledge on digestion. Furthermore, matrix compounds such as the fat content and proteolytic and oxidative enzymes could have an effect on the allergenicity. Enzyme immunoassays and immunoblotting are often used to assess IgE binding. However, these methods do not determine whether the allergen can cause cross-linking of FcεR1 bound IgE antibodies; for this a cellular assay is needed. Histamine release tests, CAST, BAT and mediator release from RBL cells are examples of such cellular measurements looking at cross linking of allergens while T-cell polarization assays can be used to assess the potency of an allergen to activate CD4+ helper T-cells and skew them towards a Th2 like profile. A few cellular assays are discussed more extensively and are considered crucial to providing a comprehensive analysis of the effects of food matrix and processing methods on individual allergenic proteins.

12.1 Introduction

Influence of food processing, digestion and the food matrix on allergenicity is a part of an EU-funded project on food allergies (EuroPrevall, FOOD-CT-2005-514000; www.europrevall.org). The primary objective of EuroPrevall is to gain insight into the prevalence of food allergies across Europe. To improve the quality of life for all food allergic consumers, EuroPrevall will also develop the tools that are required to manage food allergies more effectively (Mills *et al.*, 2007). One of the research questions within the themes of EuroPrevall is to assess the influence of food processing, digestion and the food matrix on allergenicity of individual allergens.

12.1.1 Prevalence Data of Food Allergy

IgE mediated food allergy is a disease affecting all age groups. At the present time, the only treatment is avoidance of problematic allergens, and as a consequence having a food allergy changes the quality of life experienced by a food allergy sufferer, or their carers, in a profoundly negative way. As a minimum precaution, in addition to medication against some clinical symptoms, those affected need to be very vigilant regarding their diet, and are very restricted in their possibilities to participate in many aspects of social life, such as attending meetings, going on vacations, travelling to countries where different languages are spoken, visiting restaurants, etc., because they might unexpectedly be exposed to the problematic food allergens.

The true prevalence of adverse reactions to food and food allergies is still being assessed, but is estimated to be around 3%–4% for adverse food reactions, and 1%–2% for food allergies, in the adult population. In young children below 3 years of age, the prevalence of adverse food reactions is estimated to be 6%–8% (Burks and Ballmer-Weber, 2006; Sicherer and Sampson, 2010). The determination of accurate food allergy prevalence rates is hampered by the lack of studies applying reliable diagnostic methodologies, such as double blind placebo controlled food challenges, to large unselected populations. New prevalence data, obtained from the multi-centre birth cohort study within the EuroPrevall project, is currently being collected which may provide new insights regarding the prevalence of food allergies (Keil *et al.*, 2010; Kummeling *et al.*, 2009).

12.1.2 Exposure to Food Allergens

The intensity of the allergic reactions to food is not only determined by individual-related parameters, such as genetic background, but also the time of the day or the stage of the year of exposure to allergens, medication history, disease history, etc. Furthermore, geographical parameters could be of importance. For example, Central and Northern Europe have a high exposure to birch pollen, resulting in many people suffering from related food allergies compared to Mediterranean areas where birch pollen exposure is less prevelent. Variation in diet in different parts of the world partly explains the reported differences in the most important allergens from different countries. Another parameter which could influence the severity of an allergic reaction is the way in which the allergens are presented to the gut-based immune system. Many allergens are present in the food product in high protein concentration. For example, the major allergens in milk, egg white, fish and peanuts are often abundantly present in the food product and consist of 25%–50% of the total protein content in the normal diet. Another condition which can have influence on the immunological reactivity of allergenic proteins is the degree of digestion by the gastro-intestinal tract, the degree of processing, and the impact of embedding in food matrix components.

12.1.3 Diagnostics of Food Allergy

In addition to its clinical impact in terms of severity of reactions, processing, digestion and matrix effects are also of importance in diagnostic procedures, in particular when food challenges are part of the diagnostic process. Also, when comparison of challenge results between various test centres is required, it is important to perform challenges under optimally defined and comparable conditions. The selection of the tested challenge material is of importance in the standardization of the challenges (Vlieg-Boerstra *et al.*, 2004). Furthermore, such existing information could help in modulating the allergenic properties of foods. In addition, this information can be of help to design foods, to control and manage food production chains and to develop desensitisation strategies, all of which may potentially reduce the disease-burden for those affected. However, because of its complex nature, the impact of food processing and the food matrix on allergenicity of proteins has only recently become a subject of research. Such investigations are fraught with difficulties, not least the fact that food processing often renders food proteins insoluble in the simple salt solutions frequently employed in serological or clinical studies. As a consequence, our understanding of the impact of food processing on allergenicity is limited to the more soluble and extractable residues in foods, and the allergenic potential of insoluble protein complexes is virtually unstudied despite representing the vast bulk of food proteins consumed (Mills *et al.*, 2009). In this chapter, several cellular assays are used as a read-out parameter allowing the analysis of the immune reactivity in white blood cells of allergic individuals and healthy controls.

12.2 Influence of Food Processing, Digestion and Food Matrix on the Allergenicity of Food Proteins

Food protein allergenicity is largely determined by the nature of protein structure and property in the primitive state. Here we will discuss the food processing, human digestion and food matrix issues.

12.2.1 Food Processing

Nowadays, foods are increasingly processed, which may serve a variety of purposes. These could include improving general food qualities such as flavour, texture, taste and colour; improving food preservation and safety; enhancing suitability of food components for specific product applications; consumers' convenience, pleasure and variety; and finally, obtaining or generating useful by-products and increasing marketability of foods (Sathe and Sharma, 2009). Frequently applied methods of food processing performed at home or in a restaurant are, for example, peeling, soaking, blending, grinding, cutting, cooking, pressure cooking, frying, roasting, toasting, baking, smoking, grilling, microwaving, cooling, freezing and canning. Industrial processing also involves other techniques such as mechanical harvesting, various specialized drying methods, ultrafiltration, freeze concentration, high pressure processing, sonication, several sterilization methods, various chilling, and freezing methods and many other processing methods. Food processing offers opportunities to alter the nature of epitopes of a protein, which is important for immune recognition by food component-specific antibodies. For example, three-dimensional epitope conformation may be modified as a result of protein denaturation treatments (e.g., various thermal processing treatments) leading to destruction or break-up of epitopes and therefore leading to reduction in IgE binding capacity. However, denaturation treatments can also lead to generation of new epitopes, or to exposure of formerly hidden antigenic sites, which results in an increase in IgE binding. Acid or enzyme hydrolysis of an allergenic protein may help to delete critical amino acids of an epitope. Whether caused by protein denaturation or hydrolysis, loss of epitope and ensuing loss of IgE binding may help to reduce the bioactivity of an allergen (Sathe and Sharma, 2009). Different food processing methods may impact the allergenic potential of foods or proteins, but there is no general rule on how different allergenic foods or proteins respond to physical, chemical, or biochemical exposures during processing (Ladics, 2009). The effects of food processing may be governed by the molecular properties of an allergen and its interactions with food components.

Thermal processing is one of the most commonly used methods in food processing and effects of thermal processing on food protein allergenicity have therefore been reviewed in several recent articles (Wal, 2003; Maleki, 2007; Mills and Mackie, 2008; Mills *et al.*, 2009; Nowak-Wegrzyn and Fiocchi, 2009; Paschke, 2009; Sathe and Sharma, 2009). Typically, loss of tertiary structure is followed by (reversible) unfolding, loss of secondary structure (55–70℃), cleavage of disulphide bonds (70–80℃), formation of new intra-/inter-molecular interactions, rearrangements of disulphide bonds (80–90℃) and the formation of aggregates (90–100℃) (Davis and Williams, 1998). These modifications reflect a progressive passage to a disorganized structure with denaturation of the proteins that adopt an unfolded, random-coil conformation. The denatured molecules associate to form aggregates and then gels resulting in a modification of the surface properties and an increase in size (Wal, 2003).

One of the most important chemical modifications occurring in foods during thermal processing is the Maillard reaction, which involves the reaction of free amino groups on proteins (generally lysine residues) with reducing sugars. The extent of glycation depends on different environmental conditions such as temperature, pH, water activity, duration of the heating and the concentration of the reducing sugars present (Renn and Sathe, 1997). The Maillard reaction may play a possible role in the allergenicity of foods as shown in several studies (Chung and Champagne, 1999; Maleki *et al.*, 2000; Gruber *et al.*, 2004; Gruber *et al.*, 2005; Mondoulet *et al.*, 2005; Sancho *et al.*, 2005; Nakamura *et al.*, 2008; Taheri-Kafrani *et al.*, 2009; Ilchmann *et al.*, 2010). As mentioned before, the impact of processing methods may be different from food to food or protein to protein. Depending on the system, heating may have no effect or it may decrease or increase allergenicity. The existence of sequential and conformational epitopes, the demasking of new epitopes or the modification of epitopes through Maillard reactions can explain some of the results reported in the literature.

Some plant proteins such as the Bet v 1 homologues from apple and cherry are considered to (partially) lose their allergenic activity upon food processing. This is well reflected by the fact that thermally processed fruit and vegetable products often show non- or less severe effects towards patients than the corresponding fresh plant material (Gruber *et al.*, 2005). Also Bohle *et al.* (2006) showed that *in vivo* ingestion of cooked birch pollen-related foods did not induce OAS and that, in general, cooked birch pollen-related foods can be consumed without difficulty because PR-10-like food proteins are easily denatured by thermal processing. However, *in vivo* ingestion of cooked birch pollen-related foods did cause atopic eczema to worsen. In a study it was shown that 60 min cooking of the different recombinant Bet v 1-related allergens completely abolished IgE binding, but no reduction of the capacity to activate allergen-specific T-cells was observed (Bohle *et al.*, 2006). Other commonly plant-derived ingredients, such as soya isolates, often comprise 11S and 7S seed storage globulins, which are relatively thermostable. 7S globulins have their major thermal transition at around 70–75℃, whilst 11S globulins unfold at temperatures above 94℃, as determined by differential scanning calorimetry. However, even upon heating to such temperatures these proteins only appear to partially unfold, undergoing only minor conformational changes suggesting that the β-barrel motif, characteristic of these proteins, is a highly stable structure (Mills *et al.*, 2003; Mills *et al.*, 2009).

The type of allergen, type of food structure, and type of thermal processing may be of great importance when studying the impact of processing on food allergenicity. However, lack of knowledge on the influence of processing of different allergens and in different matrices makes it still difficult to predict the allergenicity of foods and provide allergic patients with appropriate advice over what is safe to eat (Mills and Mackie, 2008).

12.2.2 Digestion

Due to the very acidic conditions in the stomach and the intense proteolysis occurring in the stomach and the intestine, only small amounts of intact or immunologically active proteins are taken up by the gut mucosa. This suggests that food allergens are, at least partially, resistant to gastro-duodenal digestion in order to be able to sensitize the mucosal immune system. Susceptibility to digestion has therefore been considered an important biomarker for food allergies, the idea being that proteins or at least peptides of a few kDa that survive digestion are more likely to be allergens. While this idea has some appeal, the evidence is rather equivocal; showing that peptide fragments of various sizes, produced during the digestion of a protein can still be immunologically active (Mackie and Macierzanka, 2010). Furthermore, it is shown that some food allergens are rapidly and extensively degraded during digestion, whereas some other food proteins that are resistant to digestion are not allergenic. It is also noteworthy that a pre-gastric absorption also occurs, i.e., in the oral cavity, which explains the occurrence of symptoms a few minutes after ingestion of food allergens (Adel-Patient and Wal, 2008).

Furthermore, even though the immune system is mostly exposed to digested proteins, it is, in addition, exposed to intact allergens to a lesser extent. It is assumed that a small fraction of intact dietary proteins are absorbed from the mature gut (Warshaw *et al.*, 1977; Gardner, 1988; Moreno *et al.*, 2006; Yu, 2009), thus leaving the option open that undigested proteins play a role in the allergic sensitisation process. A recent review on intestinal barrier function also summarized evidence from studies demonstrating that intestinal barrier dysfunction leading to increased intestinal permeability and exposure to intact proteins which in turn might promote sensitization and enhance the severity of food induced allergic reactions (Groschwitz and Hogan, 2009). Additionally, it has been widely documented both in humans and in animal models that an increase in gastric pH (e.g., caused by antacid therapies) impedes the gastric protein digestion (Untersmayr and Jensen-Jarolim, 2008) and presumably facilitates the presentation of food peptides to intestinal T-cells.

There are many processes that turn the protein in a structured food into a nutrient absorbed by the gastrointestinal mucosa. After processing mastication and bolus formation etc. in the mouth, food passes into the stomach where the pH can vary widely but tends to be in the range of 1.5 to 3. This may have the effect of aggregating any protein consumed. Gastric pepsin may hydrolyse the protein into peptides ranging widely in size depending on the protein. Biosurfactants such as phosphatidylcholine (PC) may interact with the protein changing its properties. Once in the small intestine, protein is exposed to a wide range of enzymes including trypsin, chymotrypsin, elastase, etc. and a more extensive range of biosurfactants including bile acids and phospholipids and the products of lipolysis. The complex nature of the gastrointestinal tract means that we are still a long way from fully understanding the processes that govern digestion. In the last few years there has been an increase in research on protein digestion using increasingly sophisticated models of digestion that include oral, gastric and duodenal phases etc., and also include physiological surfactants such as phosphatidylcholines, which are the most abundant among the phospholipids and are present in the proximal small intestine and in the stomach. Interaction with a range of physiological surfactants has been shown to be extremely important in protein digestion. The development of more physiologically relevant *in vitro* models of digestion has significantly improved our understanding of the importance of food structures in digestion (Mackie and Macierzanka, 2010).

For the major peanut allergen Ara h 1, gastric digestion is primarily the result of the action of pepsin. After gastric passage, Ara h 1 is almost completely degraded into much smaller fragments. Digestion with either porcine gastric fluid or pepsin produced very similar results (Kopper *et al.*, 2004). However, even after simulated gastroduodenal digestion of Ara h 1, the remaining fragments retained the ability to stimulate T cells and IgE binding and cross-linking properties of the intact protein (Eiwegger *et al.*, 2006). Furthermore, part of Ara h 1 can reach the intestinal lumen in an undigested form, due to the aggregated nature of the oligomeric allergen, indicating that the quaternary structure of Ara h 1 may play an important role in protecting epitopes from digestive enzymes (Maleki *et al.*, 2000). However, digestion, even with a variety of gastric enzymes, is not always sufficient, to abolish the allergo-reactivity of allergens, as evidenced, for instance, by another major peanut allergen, Ara h 2 (Sen *et al.*, 2002). In contrast, the PR-10 family allergens of apple, hazelnut, and celery were shown to undergo rapid gastric digestion leading to total loss of IgE-binding activity. Intriguingly, this digestive treatment did not destroy the T-cell-activating capacity of these PR-10 family allergens, suggesting that small peptides representing T-cell epitopes may survive gastrointestinal degradation (Schimek *et al.*, 2005).

Changes in protein structure by processing methods might enhance or decrease IgE binding, but the changes by processing might also enhance or hinder the digestibility of a particular protein. Several studies have shown altered rate and pattern of proteolysis by processing methods such as heating and high pressure treatment (Chicon *et al.*, 2008; Stanciuc *et al.*, 2008; Zeece *et al.*, 2008). An explanation could, for example, lie in the fact that thermal treatment causes unfolding of a protein leading to an increased accessibility of the specific peptide bonds to digestive enzymes like trypsin and chymotrypsin, which increase the rate of hydrolysis. Against this, prolonged heating time causing aggregation of the protein could cause a lower accessibility of the specific bonds and thereby decrease the rate of hydrolysis. This altered digestibility could modify the form in which allergens are taken up across the gut mucosal barrier and presented to the immune system with regards to both sensitisation and elicitation.

12.2.3 Food Matrix

Few detailed reports have addressed the effects of matrices on allergenicity. Grimshaw *et al.* (2003) demonstrated a profound impact of the fat content of chocolate, in which peanut allergen was embedded, on both the clinical reactions and the recognition of peanut allergen in ELISA-testing. Higher fat contents resulted in a lower ELISA-detection of the peanut and a larger dose of peanut protein was needed in the food challenge to cause a clinical reaction, but the reaction was more severe. It was suggested that the allergenic epitopes are concealed by the relatively high-fat food matrix, and are detected only after digestion of the fat. The allergens are thus released and absorbed more slowly than they were in a lower-fat matrix. This slower absorption may explain why the subjects did not experience the oral warning symptoms preceding more severe symptoms. This research indicated the importance of the fat content of provocation recipes, because it can influence the clinical reaction of the provoked person (Grimshaw *et al.*, 2003). The influence of the matrix in a peanut extract was also tested by using a mouse model. The authors observed that purified peanut allergens possessed little intrinsic immune-modulating capacity in contrast to the whole peanut extract. The matrix seemed to serve as an adjuvant probably by activating the dendritic cells. In this model peanut allergens needed an accompanying adjuvant to be able to activate antigen presenting cells (APCs) and to induce subsequent immune stimulation (van Wijk *et al.*, 2005).

Non-enzymatically active food matrix components, such as fats, may affect allergo-reactivity of proteins. Effects of other matrix components are also obvious, such as raw-material borne proteolytic (Wichers *et al.*, 2004) and oxidative enzymes (Chung *et al.*, 2004; Gruber *et al.*, 2004; Garcia *et al.*, 2007). During in-plant maturation of the peanut allergen Ara h 1, for instance, an immunologically active N-terminal peptide is cleaved off (Wichers *et al.*, 2004). For allergens belonging to the PR-10 family and the 2S-albumin family, treatment with polyphenol oxidases and/or with peroxidases decreases their IgE-reactivity, and both enzymes are ubiquitously present in plant material (Chung *et al.*, 2004; Gruber *et al.*, 2004; Garcia *et al.*, 2007). The mechanism that underlies these activities is as yet not fully clear; but both allergen cross-linking, or binding of phenolics to the allergens might be involved.

12.3 Determination of Allergenicity of Food Proteins

It can be concluded that food processing, digestion and the food matrix affect food allergenicity, and the magnitude of these effects can be examined by combining various assays. To assess the potential of allergens to elicit clinical manifestations in sensitised individuals (elicitation potential), IgE binding studies are usually performed using methods of enzyme immunoassays and immunoblotting. In those studies, sera from individuals with a known allergy that contain specific IgE antibodies are used to bind a solid phase-bound allergen. A drawback of these assays is that they detect only the binding capacity of the allergen to IgE and do not determine whether the allergen can cause cross-linking of FcεR1 bound IgE antibodies on a basophil. This is crucial since such interactions modulate the ability of proteins to trigger degranulation of effector cells, resulting in mediator release and finally in clinical manifestations of the allergic reaction. More than one epitope or IgE binding site is required to cause IgE cross-linking and a molecule with a single IgE binding site must be bound or cross-linked to another molecule with an IgE binding site in order to cause histamine release. Therefore, the biological activity of proteins can not be assessed by IgE binding tests such as an ELISA, while cellular assays, in which cross-linking is required to provoke a response, are necessary.

In vitro cellular tests such as histamine release tests using autologous or passively sensitized heterologous human basophiles, or the cellular allergen stimulation tests (CAST) measuring sulfidoleukotriene release, can assess the ability of proteins (allergens) to degranulate basophils from human blood. An alternative readout of allergenic activity is to analyse upregulation of basophil-specific activation markers CD63 or CD203c by flow cytometry, which is called basophil activation test (BAT). BAT and CAST have been mainly used to address clinically relevant research issues and to assess the biological function of observed IgE reactivity. This is in contrast to skin and serological assays which solely indicate sensitization rather than clinically relevant IgE reactivity resulting in food allergies. Examples of using BAT and CAST in the diagnosis of several IgE-mediated allergies are those classical inhalant allergens (house dust mite, cat epithelium, and pollen), hymenoptera venom allergy, natural rubber latex allergy, primary and secondary food allergies, drugs, NSAID hypersensitivity and component-resolved diagnosis which is based on the use of single purified or recombinant components instead of often ill-defined whole extract preparations from native allergens (Ebo *et al.*, 2008).

In vitro functional tests with a humanized stable basophil cell line could be used instead of primary basophils from a donor to study the biological activity of purified proteins or to study the effect of processing or digestion on their allergenic potential. Elicitation of an allergic reaction upon allergen exposure in recently developed animal models can be used to assess the potential of such an allergen to elicit clinical manifestations in already sensitised individuals (elicitation potential). A comprehensive and complete overview on *in vivo* and *in vitro* techniques to determine the biological activity of food allergens can be found in the review by Poulsen (2001). A recent review extensively discussed physicochemical and immunological techniques to characterize plant food allergens (Harrer *et al.*, 2010).

Besides assessing IgE binding and cross linking of allergens, another informative aspect is to determine T-cell mediated immune reactions. To assess whether the processed or digested food protein is still able to activate $CD4^+$ helper T-cells and skew them towards a Th2 like profile, T-cell polarisation assays can be used. Some treatments (such as cooking or *in vitro* digestion) could have a different effect on IgE binding and cross linking compared to T-cell mediated immune responses (Schimek *et al.*, 2005; Bohle *et al.*, 2006), which makes it interesting to perform both types of assay. Below, basophil assays (histamine release tests and the mediator release assay using RBL cells) and T-cell assays will be discussed more in detail.

12.3.1 Histamine Release Tests

The principle of the histamine release test is to challenge sensitized basophils with an allergen which will cross-link surface-bound specific IgE antibodies causing histamine to be released from the cells. Histamine can be measured fluorometrically after coupling to a fluorophore (*O*-phthaldialdehyde), immunochemically (by the Immunotech Radioimmunoassay) or by performing an automated fluorometric histamine assay. In one application of the histamine release method, glass-fibre coated microtiter plates are used for separation of histamine from other constituents of the assay. Histamine content is determined and a dose-response curve can be constructed and be compared with an appropriate standard. The amount of released histamine can also be expressed as a percentage of total histamine content of non-challenged cells. Furthermore, besides using blood from sensitized patients, a passive sensitization method can be applied. Basophils from a non-sensitized person are used, the receptor-bound IgE which is natively present is stripped from the surface of the donor basophils, and the cells are subsequently passively sensitized with human serum containing specific and relevant IgE-antibodies (Stahl Skov *et al.*, 1984; Skov *et al.*, 1997; Poulsen, 2001).

Basophils are present in blood in low numbers (less than 0.5% of leukocytes), and are very hard to purify (Gibbs *et al.*, 2007). Furthermore, fresh blood cells from specific allergic donors or non-allergic donors in case of passive sensitization are needed for each experiment and it requires processing of blood samples immediately after collection, which may present logistical obstacles. Another drawback is that around 10% of basophil donors have non-responsive basophils (Kepley *et al.*, 1999; Palmer *et al.*, 2005) which makes the test less reliable in assessing the level of individual responsiveness. An option to overcome these problems is by using a stable cell line which can be passively sensitised with serum IgE from allergic individuals.

12.3.2 Mediator Release Assay

The mediator release assay which mimics the mechanism of a type 1 allergic reaction (Hoffmann-Sommergruber *et al.*, 1999) uses transfected rat basophilic leukaemia (RBL) cells that recombinantly express the human FcεR1 receptor (Gilfillan *et al.*, 1992; Dibbern *et al.*, 2003; Takagi *et al.*, 2003; Vogel *et al.*, 2005; Kaul *et al.*, 2007; Ladics *et al.*, 2008). These cells bind human IgE antibodies and have all functional properties of mast cells and basophils. β-hexosaminidase, which is present in the granules and released together with histamine, has been chosen as a surrogate marker for histamine release. The added substrate is hydrolysed by this active enzyme, resulting in a coloured product that can be measured in a spectrophotometer. A microscopic picture of RBL cells in culture and a simplified graph of the histamine release test are depicted in Fig.12.1.

The mediator release assay allows, among others, the measurement of the biological activity (or potency) of allergens, the detection of allergens in various samples and the analysis of cross-reactivities between allergens. A good indicator of the allergenic potency of allergens is the dose needed to efficiently induce cell degranulation and values corresponding to the dose of allergen that induced 50% of the maximum release (EC50 values). A limitation of this cell-based test is that in general only sera with higher percentages of allergen-specific versus total IgE can be used. For example, Dibbern *et al.* (2003) found that, for sera from peanut allergic patients, the most effective sera had at least 50 kU/L of total IgE and 15 kU/L of peanut-specific IgE and contained >10% peanut-specific IgE. Furthermore, the number of FcεRI on the transfected RBL cell-lines may be limiting. For example, RBL SX-38 cells have approximately 100,000 receptors/cell (Wiegand *et al.*, 1996) versus 500,000 receptors/cell on basophils from atopic individuals (MacGlashan, 2007). Together, the low levels of specific IgE in the serum and low expression levels of FcεRI on the RBL cells may result in most of the IgE receptors being occupied with non-specific IgE antibodies. A number of serum factors, as well as the degree of severity of the allergic subjects clinical response have been considered important for inducing mediator release with these cell-lines (Ladics *et al.*, 2008). A recent study showed that IgE affinity purification of sera increased the reproducibility and sensitivity of RBL SX-38 cells, which suggested that some factors in the serum may hamper the binding of IgE to the FcεRI (Blanc *et al.*, 2009).

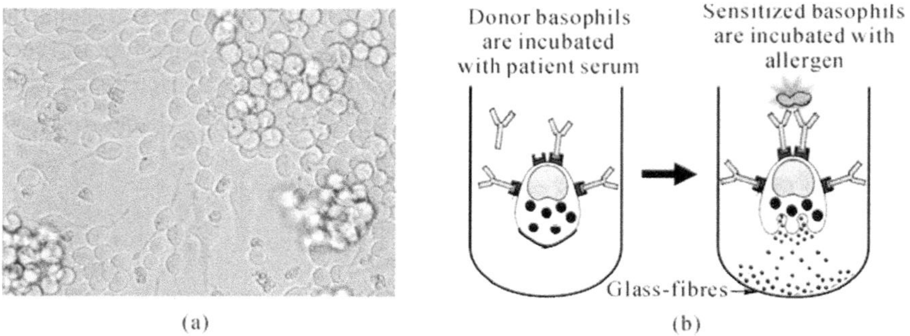

Fig. 12.1. (a) Microscopic picture of RBL-cells in a culture flask. Cells are not confluent yet, therefore some cells are still stretched. Furthermore it can be observed that the cells grow in layers; (b) Simplified graph of the histamine release test using glass-fibre coated microtiter plates. (adapted from Roitt *et al.*, 1998) (With Permission of Elsevier)

12.3.3 T-Cell Polarisation Assays

T-cell polarization assays are used to investigate T-cell mediated reactivity towards allergens and allergenic proteins. Allergic diseases result from an aberrant T-cell response to allergens dominated by long-lived Th2 cells (Akdis *et al.*, 2004). Allergen-specific CD4+ Th2 cells secrete high amounts of IL- 4 and IL-13, which induce the production of allergen-specific IgE antibodies that mediate immediate allergic symptoms (Christensen *et al.*, 2008; Akdis, 2009). In addition to this indirect involvement in immediate reactions, allergen-specific Th2 cells have been demonstrated to be directly involved in clinical late-phase reactions in target organs such as the lung and the skin (Haselden *et al.*, 1999; Bohle *et al.*, 2006).

12.3.3.1 Peripheral Blood Mononuclear Cells (PBMC)

The cells in human blood consist of erythrocytes (red blood cells), leukocytes (white blood cells) and thrombocytes (blood platelets). The leukocytes part is made up of granulocytes (neutrophils (54%–62%), eosinophils (1%–6%) and basophils (<1%)), monocytes (2%–10%) and lymphocytes (25%–33%) (T-cells, B-cells and NK cells). The monocytes and lymphocytes together form the peripheral blood mononuclear cell fraction (PBMC). PBMC can be isolated from the peripheral blood by density gradient separation and these cells can be used to assess cell-mediated immunity in general or, via antigen-specific stimulation, to detect previous exposure to a variety of antigens/allergens and to monitor, for example, the response to immunotherapy.

In the blood, 60%–70% of T-cells are CD4+ and 30%–40% represent CD8+. CD4+ T cells are generally designated helper cells and activate both humoral immune responses (B-cell help) and cellular responses (delayed-type hypersensitivity responses and others). CD8+ cells show a major cytotoxic activity against cells infected with intracellular microbes, and against tumour cells. A portion of the circulating CD4+ cells play an important regulatory role that acts to down modulate immune responses. These regulatory Treg cells consist of natural occurring Treg cells (CD4+ CD25+ Treg) and adaptive or induced Treg cells (Tr1). Treg cells are able to inhibit the development of allergen-specific Th2 and Th1 cell responses and therefore play an important role in a healthy immune response to allergens (Ozdemir *et al.*, 2009).

Both CD4+ and CD8+ T-cells differentiate into functionally distinct subsets after exposure to antigenic peptides processed and presented by antigen presenting cells (APCs), like dendritic cells, B-cells and monocytes/macrophages. This is best described for the transition of CD4+ T-cells from naïve to effector populations. Resting naïve CD4+ T-cells (designated Th cells) release very low levels of cytokines. Soon after stimulation by antigen and APC, the Th cells begin to produce IL-2 and are designated Th0. As the Th cells continue to respond to the activating signal, they progress toward polar extremes of differentiation designated Th1, Th2, and Th17 depending on the nature of the cytokines present at the site of activation (Sallusto and Lanzavecchia, 2009). IL-12 produced by macrophages or NK cells induces differentiation toward Th1; IL-4 produced by NK1.1$^+$ T-cells, basophils, or mast cells induces differentiation toward Th2; and TGF-β and IL-6 produced by yet to be defined cells induce differentiation toward Th17. Th1 cells are characterized by the production of IL-2, IFN-γ, and lymphotoxin, whereas Th2 cells produce IL-4, IL-5, IL-9, IL-13, and granulocyte-macrophage colony stimulating factor (GM-CSF) and Th17 cells produce the cytokines IL-6 and IL-17 (Bonilla and Oettgen, 2010; Chaplin, 2010). Recently, a new population of T helper cells (Th9 cells) have been distinguished. Th9 cells produce IL-9 and IL-10 and seem to lack suppressive function and constitute a distinct population of effector T-cells that promote tissue inflammation (Dardalhon *et al.*, 2008; Veldhoen *et al.*, 2008).

In T-cell polarisation assays based on hPBMC-cultures, blood is obtained from allergic persons and healthy controls and PBMC are isolated by means of Ficoll separation. After performing density gradient separation, the upper fraction consists of the plasma and the thrombocytes, the second fraction is the PBMC fraction, the third fraction consists of the Ficoll and the granulocytes and the fourth fraction, the pellet, consists of the erythrocytes (Fig. 12.2). Because antigen-specific cultures require the presence of APCs, PBMC providing autologous APCs, instead of purified T-cells are a better model to use. Antigen presentation in *in vitro* PBMC cultures is mainly accounted for by monocytes.

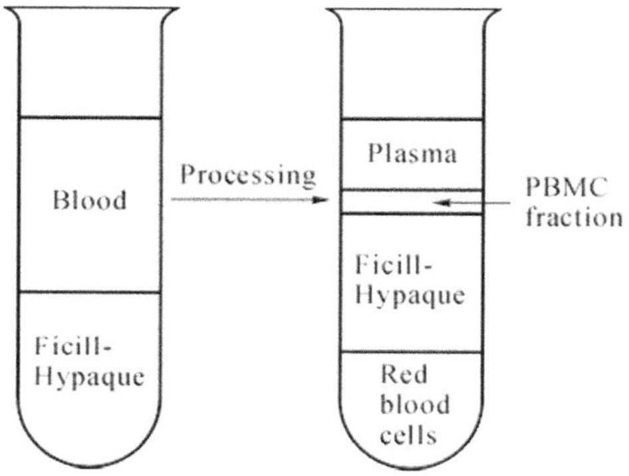

Fig.12.2. Overview of Ficoll separation of PBMC using density gradient separation

The PBMC consisting of T- and B-cells, natural killer cells and monocytes are cultured and stimulated with the allergen of interest. In the cultures, the allergens are presented by APCs to the adaptive immune cells via presentation of small peptide fragments of the allergens (T-cell epitopes) on the surface of the APCs. T-cell activation requires at least two signals. The major histocompatibility complex (MHC)-peptide complex associating with the T-cell receptor (TCR/CD3) provides signal 1, whereas signal 2 is delivered via the binding of CD28 to B7-1/2 (CD80/CD86) expressed on antigen presenting cells (also called co-stimulation). Subsequently, an immunological synapse is formed, resulting in the activation of several tyrosine kinases and recruitment of adapter proteins and specific downstream signalling leading to T-cell activation and specific cytokine production (Jeurink *et al.*, 2008).

By using combinations of surface marker sets and performing multicolour flow cytometric stainings, the phenotype of the cells in the culture can be analysed. A control which allows correction for possible differences between individual donors regarding proliferation or cytokine production is to perform a surface marker staining of the PBMC fraction immediately after isolation of the PBMC (Fig. 12.3). Furthermore, the cytokine values could even be corrected for the number of T-cells which are brought into the culture.

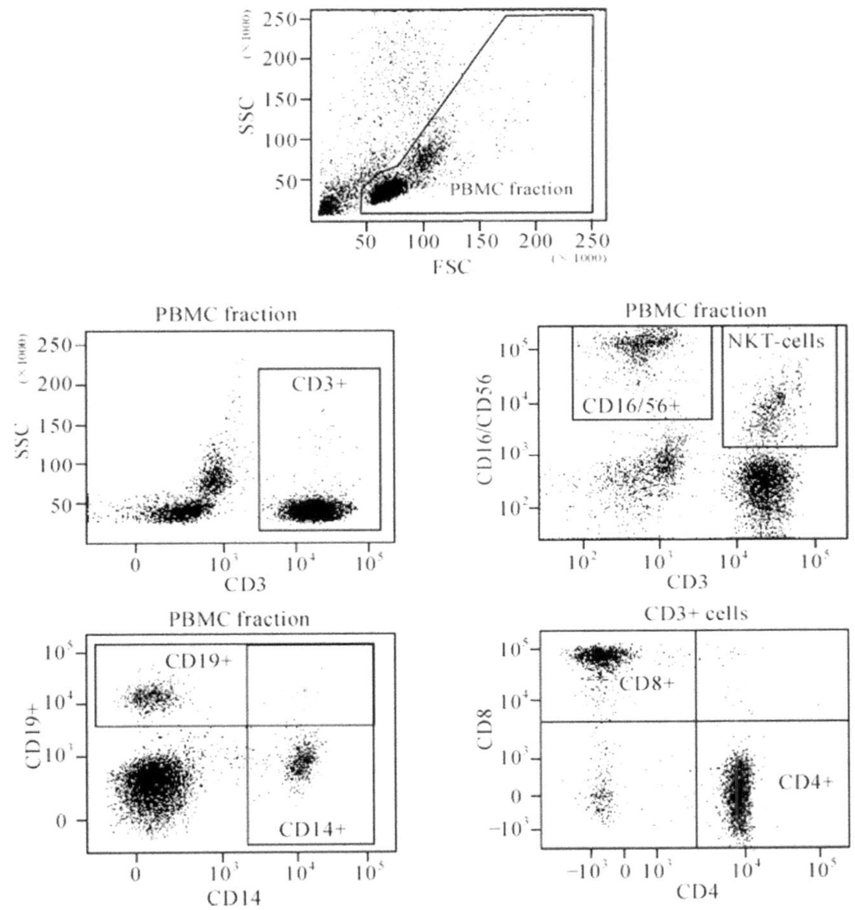

Fig. 12.3. Typical data of fresh human peripheral blood mononuclear cells (hPBMC) isolated from a donor and stained directly with a mixture of α-hCD3 (T-cells), α-hCD4 (T helper cells), α-hCD8 (cytotoxic T-cells), α-hCD25 (activated cells), α-hCD16/α-hCD56 (NK cells), α-hCD14 (monocytes) and α-hCD19 (B-cells). First the PBMC fraction was gated (debris was gated out) and within this PBMC fraction the percentages of T-cells (61±8%), NK cells (10%±3%), NKT cells (natural killer T-cells) (5%±3%), monocytes (16%±7%) and B-cells (9%±3%) were determined. Within the T-cell fraction the percentage T-helper cells (63%±9%) and cytotoxic T-cells (31%±10%) were assessed. As practically no activated cells (CD25+) were present, this results is not shown. Average ± SD values were obtained from the average values of PBMC stained directly after isolation from the blood of 21 patients. Seven-colour flowcytometric acquisition was performed on a FACSCanto II (BD Biosciences), using the BD FACSDiva software. The following monoclonal antibody mixture was used: α-hCD3 (V450), α-hCD4 (PerCP Cy 5.5), α-hCD8 (PE-Cy7), α-hCD25 (APC-H7) α-hCD14 (APC), α-hCD16 (PE), α-hCD19 (FITC) and α-hCD56 (PE)

12.3.3.2 Cytokine Production and Proliferation

Activated and differentiated cells will produce a characteristic cytokine production profile. T-cells of allergic individuals will typically produce IL-4, IL-5 and IL-13, which are signature Th2 type cytokines. Cytokine production can be measured in several ways. Most often the cytokine production is measured in the supernatant of the cultured cells. Cytokine enzyme-linked immunoabsorbent assay (ELISA) or flow cytometric techniques like the multiplexed bead assay (Cytometric Bead Array Flex sets, BD Biosciences) are commonly used to detect cytokines from cell supernatant or serum. Real-time quantitative polymerase chain reaction (RT-qPCR) technique can be implemented to assess mRNA expression of the cytokines in T-cell fractions. However, these methods do not allow for determining which proportion of the cells is responsible for the production of specific cytokines. Multiple staining using combinations of cell surface markers and staining for intracellular cytokines permit the assessment which type of cell produces what cytokines. This approach can give valuable extra information at the single-cell level. Optimizing experiments by assessing the optimal stimulation period for different cytokines is important as each stimulus results in different kinetics and these kinetics are crucially different for the individual cytokines. For example, in a study of Jeurink *et al.* (2008) monocyte derived cytokines such as TNF-α and IL-1β were present in cell cultures already after 1 day of culture and amounts decreased significantly over time, while T-cell derived cytokines such as IL-5, IL-13 and IFN-γ showed a slower kinetics and were highest on day 7 of the culture.

This optimization is also important when measuring proliferation of cells. Proliferation of PBMC can be measured using different methods, such as incorporation of ^{3}H-thymidine, staining with Ki-67 antibodies, loading with BrdU (5-bromo-2′-deoxyuridine), and CFSE (carboxyfluorescein diacetate succinimidyl ester) staining, each provide slightly different information. The ^{3}H-thymidine and BrdU methods determines the S phase of the cell cycle activity by measuring the thymidine incorporation into newly synthesized DNA of replicating cells. The Ki-67 antigen however, is expressed in all active stages of the cell cycle (all phases except the G_0 phase) and cells which are still in the G_1 phase of the cell cycle will be detected by a Ki-67 staining and not by a ^{3}H-thymidine incorporation assay or a BrdU assay. Upon cell division, the CFSE is covalently bound to free amine groups of intracellular macromolecules in the cell, and is distributed uniformly between daughter cells. CFSE staining can be the method of choice especially when the rate of the divisions of the cells is studied and when working with a uniformly sized cell population (such as resting T-or B-lymphocytes). However, hPBMC consists of a non-homogeneous population of cells (not uniformly sized and therefore not uniformly stained) which can make it difficult to differentiate between undivided and divided cells. As a result, this could decrease the sensitivity of the test. Furthermore, care should be taken when using CFSE because of its toxicity for dividing cells and effect on the expression of activation markers like CD69 and CD25. A disadvantage of ^{3}H-thymidine incorporation is that it does not allow identification of the phenotype of the proliferating cells, as it can only measure bulk cell divisions, a limitation overcome by flow cytometry (Ki-67 and CFSE staining), which enables analysis at a single-cell level. The use of selected CD markers in the same staining will make it possible to specify the phenotype of proliferating cells. It is difficult to directly compare the methods as they use different techniques (flowcytometry using fluorochromes versus scintillation using radio-activity) and furthermore, they also give different types of information. Therefore, it is advisable to use the same method in all experiments to be able to compare results.

12.3.3.3 Stimuli

The frequency of an antigen-specific lymphocyte in the blood is generally low. The frequency of Bet v 1-specific CD4+ T-cell was estimated to be in the range of 0.0001% to 0.001% of circulating CD4+ T-cells outside the birch pollen season, for both allergic and nonallergic individuals and up to 0.1% of allergen-specific T-cells for both groups during the pollen season. A significant expansion of specific T-cells occurred during exposure to seasonal allergens (van Overtvelt *et al.*, 2008; Wambre *et al.*, 2008). The frequency of antigen-specific T-cells is observed to differ per allergen. The frequency of peanut-specific circulating CD4+ precursors was found to be 0.6044%±0.0432% in peanut allergic patients and about 10 times lower in both non-allergic and peanut sensitized donors. Frequencies of house dust mite-specific T-cells in the blood of patients with atopic dermatitis varied between 0.0016%–0.026% (Neumann *et al.*, 1996). As a result of the low frequency of antigen-specific T-cells, antigenic stimulation typically activates only a very small fraction of T-cells. In mildly sensitized patients in particular, allergen-specific stimulation will reveal hardly any significant proliferation and cytokine synthesis. Therefore, amplification of the response by the use of polyclonal stimuli is often used. These polyclonal stimuli can also be used as a positive control.

Besides performing longer-term (several days) cultures that are stimulated with the allergen of interest, it is advisable to include a positive control. As a positive control, polyclonal stimuli can be used to assess the maximal stimulation capacity of the PBMC *in vitro* and thereby assess possible differences in the intrinsic immune responsiveness of antigen-specific cell cultures of different donors. Examples of mitogens are concanavalin A (ConA), phytohemaglutinin (PHA), phorbol myristate acetate (PMA) plus calcium ionophore (Ca-I) and lipopolysaccharide (LPS). Each stimulus acts at different sites and uses a different way to activate the cells. Anti-CD3 and αCD28 have been widely used to provide all T-cells with the required activation signals, and is therefore the most physiological way of stimulating all T-cells in a polyclonal fashion to assess the maximal stimulation capacities of the PBMC of the different donors. A microscopic picture of unstimulated and αCD3/αCD28 stimulated cells is depicted in Fig.12.4. Furthermore, αCD3/αCD28 can also be used as *in vitro* polyclonal activation stimulus for T-cells in long-period cultures to re-activate and increase the antigen-specific signal. A disadvantage is that non-specific T-cells that survived the long-term culture but are not allergen-specific will also react upon polyclonal activation which makes it difficult to see the specific effects of an allergen-specific stimulation.

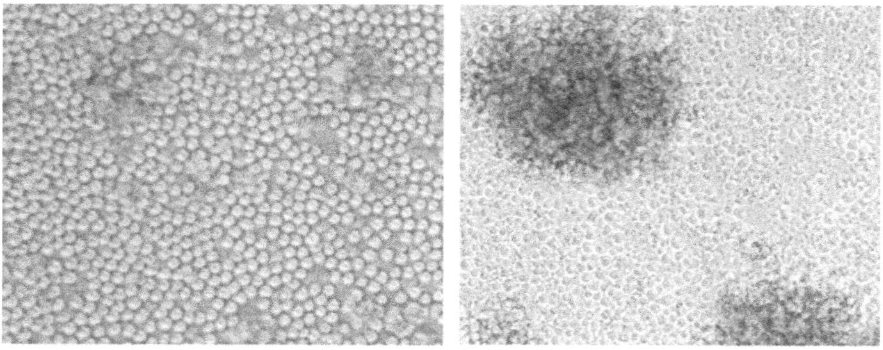

(a) (b)

Fig. 12.4. Microscopic picture of cell clones after the stimulation condition. (a) Unstimulated cells after 4 days of culture; (b) αCD3αCD28 stimulated cells after 4 days of culture

12.3.3.4 Culture Medium and the Use of Fresh Versus Cryopreserved PBMC

The culture medium seems to be important and could be responsible for differences in activation capacity. Laan *et al.* (1998) showed that proliferation was optimal when using RPMI as culture medium, whereas PBMC cultured in Yssel medium showed better cytokine responses. Yssel medium was first used as a serum-free medium based on IMDM that induced proliferative responses in mixed lymphocyte cultures that are comparable to those obtained in medium containing serum (Yssel *et al.*, 1984). IMDM contains many energy sources and in combination with the glucose in the human AB serum and the sodium pyruvate in IMDM, the cultured cells are provided with a much larger energy source than RPMI-1640 medium alone.

Another important aspect when dealing with *in vitro* stimulation of PBMC is the use of fresh cells compared to cryopreserved cells. Cryopreservation is in some cases unavoidable when working in the field or when handling large batches of patient samples. An advantage of cryopreservation is that thawing and culturing can be performed on a convenient moment and more samples can be thawed at once. This minimizes the operator-dependent interassay variability. However, cryopreservation could cause a small delay in the activation of the cryopreserved PBMC and could reduce the cytokine levels when stimulating with certain stimuli (Jeurink *et al.*, 2008). Monocytes can be successfully cryopreserved with different protocols, with recoveries ranging from 50% to 75% and afterwards stimulated to cytokine production (e.g., IL-10) which appeared not significantly altered compared to fresh PBMC (Best *et al.*, 2007). It is therefore advisable to choose the same methods for all analyses as this enables better comparison of the data obtained for individual PBMC donors, and allergens.

An advantage of the PBMC-model is that the results can be ascribed to innate and adaptive immune responses, based on cell specific responses. The response of the cells can be measured over time. After 1–2 days the response of the innate immune system (monocyte-specific) and after 5–7 days the response from the adaptive immune system (T-cell specific) can be measured. A disadvantage is that the low frequency of an antigen-specific lymphocyte in the blood which makes it difficult to see effects of an allergen-specific stimulation. In addition, the PBMC population can differ considerably among persons, but also within one person on a day-to-day basis, which should be controlled carefully. It is also advisable to use either only fresh or only cryopreserved cells, the same culture medium and the same read-out systems for assessing the cytokine production and proliferation in the different experiments and perform the cellular analysis on the same time points which have previously been optimized.

12.4 Conclusion

It can be concluded that gaining more knowledge on the effect of the food matrix and the role of food processing and digestion on individual allergen proteins is desirable to better understand their clinical impact (in terms of severity of reactions). It is also of importance in diagnostic procedures, in particular when food challenges are part of the diagnostic process.

In vitro digestion and *in vitro* processing of allergens can be performed effectively and these allergens can be analyzed in the presence or absence of different food matrices and processing procedures. The cellular measurements can be used to assess whether these different circumstances can have an influence on the potential allergenicity of the protein. T-cell polarisation assays based on hPBMC-cultures, histamine release tests and mediator release from RBL cells are examples of such cellular measurements that can be considered crucial in the provision of a comprehensive analysis of the effects of food matrix and processing methods on individual allergenic proteins.

References

Adel-Patient, K., Wal, J. M. (2008). Devenir des allergènes dans le tube digestif fate of food allergens in the digestive tract. Revue française d'allergologie et d'immunologie clinique, 48, 335-343.

Akdis, M. (2009). Immune tolerance in allergy. Curr Opin Immunol, 21(6), 700-707.

Akdis, M., Verhagen, J., *et al.* (2004). Immune responses in healthy and allergic individuals are characterized by a fine balance between allergen-specific T regulatory 1 and T helper 2 cells. J Exp Med, 199(11), 1567-1575.

Best, A., Hidalgo, G., *et al.* (2007). Issues concerning the large scale cryopreservation of peripheral blood mononuclear cells (PBMC) for immunotherapy trials. Cryobiology, 54(3), 294-297.

Blanc, F., Adel-Patient, K., *et al.* (2009). Capacity of purified peanut allergens to induce degranulation in a functional *in vitro* assay: Ara h 2 and Ara h 6 are the most efficient elicitors. Clin Exp Allergy, 39(8), 1277-1285.

Bohle, B., Zwolfer, B., *et al.* (2006). Cooking birch pollen-related food: Divergent consequences for IgE- and T-cell-mediated reactivity *in vitro* and *in vivo*. J Allergy Clin Immunol, 118(1), 242-249.

Bonilla, F.A., Oettgen H.C. (2010). Adaptive immunity. J Allergy Clin Immunol, 125, (Suppl 2), S33-40.

Burks, W., Ballmer-Weber, B.K. (2006). Food allergy. Mol Nutr Food Res, 50(7), 595-603.

Chaplin, D.D. (2010). Overview of the immune response. J Allergy Clin Immunol, 125, (Suppl 2), S3-23.

Chicon, R., Lopez-Fandino, R., *et al.* (2008). Proteolytic pattern, antigenicity, and serum immunoglobuline binding of beta-lactoglobulin hydrolysates obtained by pepsin and high-pressure treatments. Journal of Dairy Science, 91(3), 928-938.

Christensen, L.H., Holm, J., *et al.* (2008). Several distinct properties of the IgE repertoire determine effector cell degranulation in response to allergen challenge. J Allergy Clin Immunol, 122(2), 298-304.

Chung, S.Y., Champagne, E.T. (1999). Allergenicity of Maillard reaction products from peanut proteins. J Agric Food Chem, 47(12), 5227-5231.

Chung, S.Y., Maleki, S.J., *et al.* (2004). Allergenic properties of roasted peanut allergens may be reduced by peroxidase. J Agric Food Chem, 52(14), 4541-4545.

Dardalhon, V., Awasthi, A., *et al.* (2008). IL-4 inhibits TGF-beta-induced Foxp3+ T-cells and, together with TGF-beta, generates IL-9+ IL-10+ Foxp3(-) effector T cells. Nat Immunol, 9(12), 1347-1355.

Davis, P.J., Williams S.C. (1998). Protein modification by thermal processing. Allergy 53(Suppl 46), 102-105.

Dibbern, D.A., Palmer, G.W.Jr., *et al.* (2003). RBL cells expressing human Fc epsilon RI are a sensitive tool for exploring functional IgE-allergen interactions: studies with sera from peanut-sensitive patients. J Immunol Methods, 274(1-2), 37-45.

Ebo, D.G., Bridts, C.H., *et al.* (2008). Basophil activation test by flow cytometry: present and future applications in allergology. Cytometry B Clin Cytom, 74(4), 201-210.

Eiwegger, T., Rigby, N., *et al.* (2006). Gastro-duodenal digestion products of the major peanut allergen Ara h 1 retain an allergenic potential. Clin Exp Allergy, 36(10), 1281-1288.

Garcia, A., Wichers, J.H., *et al.* (2007). Decrease of the IgE-binding by Mal d 1, the major apple allergen, by means of polyphenol oxidase and peroxidase treatments. Food Chemistry, 103(1), 94-100.

Gardner, M.L.G. (1988). Gastrointestinal absorption of intact proteins. Annual Review of Nutrition, 8, 329-350.

Gibbs, B.F., Rathling, A., *et al.* (2007). Substantial differences in the kinetics of histamine release from human basophils caused by varying strengths of IgE-dependent activators. Inflamm Res, 56(Suppl 1), S5-6.

Gilfillan, A.M., Kadofong, H., *et al.* (1992). Conservation of signal transduction mechanisms via the human Fc-Epsilon-Ri-Alpha after transfection into a rat mast-cell line, Rbl 2h3. J Immunol, 149(7), 2445-2451.

Grimshaw, K.E., King, R.M., *et al.* (2003). Presentation of allergen in different food preparations affects the nature of the allergic reaction–a case series. Clin Exp Allergy, 33(11),1581-1585.

Groschwitz, K.R., Hogan S.P. (2009). Intestinal barrier function: molecular regulation and disease pathogenesis. J Allergy Clin Immunol, 124(1), 3-20.

Gruber, P., Becker, W.M., *et al.* (2005). Influence of the maillard reaction on the allergenicity of rAra h 2, a recombinant major allergen from peanut (*Arachis hypogaea*), its major epitopes, and peanut agglutinin. J Agric Food Chem, 53(6), 2289-2296.

Gruber, P., Vieths, S., *et al.* (2004). Maillard reaction and enzymatic browning affect the allergenicity of Pru av 1, the major allergen from cherry (*Prunus avium*). J Agricul Food Chem, 52(12), 4002-4007.

Harrer, A., Egger, M., Gadermaier, G., *et al.* (2010). Characterization of plant food allergens: An overview on physicochemical and immunological techniques. Mol Nutri Food Res, 54, 93-112.

Haselden, B.M., Kay, A.B., *et al.* (1999). Immunoglobulin E-independent major histocompatibility complex-restricted T-cell peptide epitope-induced late asthmatic reactions. J Exp Med, 189(12), 1885-1894.

Hoffmann-Sommergruber, K., O'Riordain, G., *et al.* (1999). Molecular characterization of Dau c 1, the Bet v 1 homologous protein from carrot and its cross-reactivity with Bet v 1 and Api g 1. Clin Exp Allergy, 29(6), 840-847.

Ilchmann, A., Burgdorf, S., *et al.* (2010). Glycation of a food allergen by the Maillard reaction enhances its T-cell immunogenicity: Role of macrophage scavenger receptor class A type Ⅰ and Ⅱ. J Allergy Clin Immunol, 125(1), 175-83 e1-11.

Jeurink, P.V., Vissers, Y.M., *et al.* (2008). T-cell responses in fresh and cryopreserved peripheral blood mononuclear cells: kinetics of cell viability, cellular subsets, proliferation, and cytokine production. Cryobiology, 57(2), 91-103.

Kaul, S., Luttkopf, D., *et al.* (2007). Mediator release assays based on human or murine immunoglobulin E in allergen standardization. Clin Exp Allergy,

37(1), 141-150.

Keil, T., McBride, D., *et al.* (2010). The multinational birth cohort of EuroPrevall: Background, aims and methods. Allergy, 65(4), 482-490.

Kepley, C.L., Youssef, L., *et al.* (1999). Syk deficiency in nonreleaser basophils. J Allergy Clin Immunol, 104(2 Pt 1), 279-284.

Kopper, R.A., Odum, N.J., *et al.* (2004). Peanut protein allergens: gastric digestion is carried out exclusively by pepsin. J Allergy Clin Immunol, 114(3), 614-618.

Kummeling, I., Mills, E.N., *et al.* (2009). The EuroPrevall surveys on the prevalence of food allergies in children and adults: background and study methodology. Allergy, 64(10), 1493-1497.

Laan, M.P., Baert, M.R., *et al.* (1998). Differential mRNA expression and production of interleukin-4 and interferon-gamma in stimulated peripheral blood mononuclear cells of house-dust mite-allergic patients. Eur Cytokine Netw, 9(1), 75-84.

Ladics, G.S. (2009). Food processing and allergenicity. Mol Nutr Food Res, 53(8), 945.

Ladics, G.S., van Bilsen, J.H., *et al.* (2008). Assessment of three human FcepsilonRI-transfected RBL cell-lines for identifying IgE induced degranulation utilizing peanut-allergic patient sera and peanut protein extract. Regul Toxicol Pharmacol, 51(3), 288-294.

MacGlashan, D.W. (2007). Endocytosis, recycling, and degradation of unoccupied Fc epsilon RI in human basophils. J Leukocyte Bio, 82(4), 1003-1010.

Mackie, A.R., Macierzanka, A. (2010). Colloidal aspects of protein digestion treatments. Current Opinion in Colloid & Interface Science, 15, 102-108.

Maleki, S.J. (2007). Allergenicity of processed food. INFORM–International News on Fats, Oils and Related Materials, 18(3), 195-197.

Maleki, S.J., Chung, S.Y., *et al.* (2000). The effects of roasting on the allergenic properties of peanut proteins. J Allergy Clin Immunol, 106(4), 763-768.

Maleki, S.J., Kopper, R.A., *et al.* (2000). Structure of the major peanut allergen Ara h 1 may protect IgE-binding epitopes from degradation. J Immunol, 164(11), 5844-5849.

Mills, E.N., Mackie A.R. (2008). The impact of processing on allergenicity of food. Curr Opin Allergy Clin Immunol, 8(3), 249-253.

Mills, E.N., Mackie, A.R., *et al.* (2007). The prevalence, cost and basis of food allergy across Europe. Allergy, 62(7), 717-722.

Mills, E.N., Marigheto, N.A., *et al.* (2003). Thermally induced structural changes in glycinin, the 11S globulin of soya bean (*Glycine max*)–an *in situ* spectroscopic study. Biochim Biophys Acta, 1648(1-2), 105-114.

Mills, E.N., Sancho, A.I., *et al.* (2009). Impact of food processing on the structural and allergenic properties of food allergens. Mol Nutr Food Res, 53(8), 963-969.

Mondoulet, L., Paty, E., *et al.* (2005). Influence of thermal processing on the allergenicity of peanut proteins. J Agric Food Chem, 53(11), 4547-4553.

Moreno, F.J., Rubio, L.A., *et al.* (2006). Uptake of 2S albumin allergens, Ber e 1 and Ses i 1, across human intestinal epithelial Caco-2 cell monolayers. J Agric Food Chem, 54(22), 8631-8639.

Nakamura, S., Suzuki, Y., *et al.* (2008). Reduction of in vitro allergenicity of buckwheat Fag e 1 through the Maillard-type glycosylation with polysaccharides. Food Chem, 103(3), 538-545.

Neumann, C., Gutgesell, C., *et al.* (1996). Comparative analysis of the frequency of house dust mite specific and nonspecific Th1 and Th2 cells in skin lesions and peripheral blood of patients with atopic dermatitis. J Mol Med, 74(7), 401-406.

Nowak-Wegrzyn, A., Fiocchi, A. (2009). Rare, medium, or well done? The effect of heating and food matrix on food protein allergenicity. Curr Opin Allergy Clin Immunol, 9(3), 234-237.

Ozdemir, C., Akdis, M., *et al.* (2009). T regulatory cells and their counterparts: masters of immune regulation. Clin Exp Allergy, 39(5), 626-639.

Palmer, G.W., Dibbern, D.A.Jr., *et al.* (2005). Comparative potency of Ara h 1 and Ara h 2 in immunochemical and functional assays of allergenicity. Clin Immunol, 115(3), 302-312.

Paschke, A. (2009). Aspects of food processing and its effect on allergen structure. Mol Nutr Food Res, 53(8), 959-962.

Poulsen, L.K. (2001). *In vivo* and *in vitro* techniques to determine the biological activity of food allergens. J Chromatography, B 756(1-2), 41-55.

Renn, P.T., Sathe, S.K. (1997). Effects of pH, temperature, and reactant molar ratio on L-leucine and D-glucose Maillard browning reaction in an aqueous system. J Agric Food Chem, 45(10), 3782-3787.

Sallusto, F., Lanzavecchia, A. (2009). Heterogeneity of CD4+ memory T-cells: functional modules for tailored immunity. Eur J Immunol, 39(8), 2076-2082.

Sancho, A.I., Rigby, N.M., *et al.* (2005). The effect of thermal processing on the IgE reactivity of the non-specific lipid transfer protein from apple, Mal d 3. Allergy, 60(10), 1262-1268.

Sathe, S.K., Sharma G.M. (2009). Effects of food processing on food allergens. Mol Nutr Food Res, 53(8), 970-978.

Schimek, E.M., Zwolfer, B., *et al.* (2005). Gastrointestinal digestion of Bet v 1-homologous food allergens destroys their mediator-releasing, but not T-cell-activating, capacity. J Allergy Clin Immunol, 116(6), 1327-1333.

Sen, M., Kopper, R., *et al.* (2002). Protein structure plays a critical role in peanut allergen stability and may determine immunodominant IgE-binding epitopes. J Immunol, 169(2), 882-887.

Sicherer, S.H., Sampson, H.A. (2010). Food allergy. J Allergy Clin Immunol, 125(2 Suppl 2), S116-125.

Skov, P.S., Pelck, I., *et al.* (1997). Hypersensitivity to the diphtheria component in the Di-Te-Pol vaccine. A type I allergic reaction demonstrated by basophil histamine release. Pediatr Allergy Immunol, 8(3), 156-158.

Stahl Skov, P., Norn, S., *et al.* (1984). A new method for detecting histamine release. Agents Actions, 14(3-4), 414-416.

Stanciuc, N., van der Plancken, I., *et al.* (2008). Denaturation impact in susceptibility of beta-lactoglobulin to enzymatic hydrolysis: A kinetic study. revue Roumaine De Chimie, 53(10), 921-929.

Taheri-Kafrani, A., Gaudin, J.C., *et al.* (2009). Effects of heating and glycation of beta-lactoglobulin on its recognition by IgE of sera from cow milk allergy patients. J Agric Food Chem, 57(11), 4974-4982.

Takagi, K., Nakamura, R., *et al.* (2003). Application of human Fc epsilon RI alpha-chain-transfected RBL-2H3 cells for estimation of active serum IgE. Biological & Pharmaceutical Bulletin, 26(2), 252-255.

Untersmayr, E., Jensen-Jarolim E., (2008). The role of protein digestibility and antacids on food allergy outcomes. J Allergy Clin Immunol, 121(6), 1301-1308.

Van Overtvelt, L., Wambre, E., *et al.* (2008). Assessment of Bet v 1-specific CD4+ T cell responses in allergic and nonallergic individuals using MHC class II peptide tetramers. J Immunol, 180(7), 4514-4522.

van Wijk, F., Nierkens, S., *et al.* (2005). The effect of the food matrix on *in vivo* immune responses to purified peanut allergens. Toxicol Sci, 86(2), 333-341.

Veldhoen, M., Uyttenhove, C., *et al.* (2008). Transforming growth factor-beta 'reprograms' the differentiation of T helper 2 cells and promotes an interleukin 9-producing subset. Nat Immunol, 9(12), 1341-1346.

Vlieg-Boerstra, B.J., Bijleveld, C.M., *et al.* (2004). Development and validation of challenge materials for double-blind, placebo-controlled food challenges in children. J Allergy Clin Immunol, 113(2), 341-346.

Vogel, L., Luttkopf, D., *et al.* (2005). Development of a functional in vitro assay as a novel tool for the standardization of allergen extracts in the human system. Allergy, 60(8), 1021-1028.

Wal, J. M. (2003). Thermal processing and allergenicity of foods. Allergy, 58(8), 727-729.

Wambre, E., van Overtvelt, L., *et al.* (2008). Single cell assessment of allergen-specific T-cell responses with MHC class II peptide tetramers: methodological aspects. Int Arch Allergy Immunol, 146(2), 99-112.

Warshaw, A.L., Bellini, C.A., *et al.* (1977). Intestinal mucosal barrier to intact antigenic protein-difference between colon and small-intestine. American Journal of Surgery, 133(1), 55-58.

Wichers, H.J., De Beijer, T., *et al.* (2004). The major peanut allergen Ara h 1 and its cleaved-off N-terminal peptide; possible implications for peanut allergen detection. J Agric Food Chem, 52(15), 4903-4907.

Wiegand, T.W., Williams, P.B., *et al.* (1996). High-affinity oligonucleotide ligands to human IgE inhibit binding to Fc epsilon receptor I. J Immunol, 157(1), 221-230.

Yssel, H., De Vries, J.E., *et al.* (1984). Serum-free medium for generation and propagation of functional human cytotoxic and helper T-cell clones. J Immunol Methodsk, 72(1), 219-227.

Yu, L.C. (2009). The epithelial gatekeeper against food allergy. Pediatr Neonatol,

50(6), 247-254.
Zeece, M., Huppertz, T., *et al.* (2008). Effect of high-pressure treatment on *in-vitro* digestibility of beta-lactoglobulin. Innovative Food Science & Emerging Technologies, 9(1), 62-69.

Part IV

Therapies and Pharmacy

13

Immunotherapy of Asthma: From Basic Research to Clinical Practices

Hui-Ying Wang[1*] and Hua-Hao Shen[2]

[1] Department of Allergy, the Second Affiliated Hospital, School of Medicine, Zhejiang University, Hangzhou, 310009, China
[2] Department of Respiralogy, the Second Affiliated Hospital, School of Medicine, Zhejiang University, Hangzhou, 310009, China
[*] E-mail: w_huiying@yahoo.com

Abstract: Allergen specific immunotherapy (SIT) is believed the only causative treatment that could change the progress of asthma. The mechanisms of SIT are complex, including the very early desensitization effects, modulation of T and B cell responses and related antibody isotypes, and migration of eosinophils, basophils, and mast cells to tissues, as well as release of their mediators. Basic studies have shown that immunotherapy works mainly through expanding allergen-specific Th1 immunity and suppressing the Th2 responses resulting from clinical observation. Treg cells are considered play key roles in modulating the switch from Th2 cells to Th0 or Th1 cells, through cytokine IL-10 and TGF-ß, as well as the switch from production of IgE to blocking antibody IgG4 isotype. The clinical practice over 100 years demonstrated its unique advantage over the first line medicine inhaled corticosteroid. It has dual functions including causative and preventative therapy. It can significantly reduce the symptom score and airway hyperresponsiveness in asthmatic patients. Furthermore, its salutary effects continue for years after the completion of treatment course. Based on the administration route, conventional subcutaneous specific immunotherapy (SCIT) and sublingual specific immunotherapy (SLIT) are the most popular methods used in clinical work. However, new administration methods such as intralymphatic immunotherapy and epicutaneous allergen- specific immunotherapy are under active clinical trials. The future studies of immunotherapy will focus on the alternative therapy with modified extract to provide the least side effects and the best clinical efficacies.

Allergen specific immunotherapy (SIT), also called desensitization or hyposensitization, is the repeated administration of allergen vaccines to allergic individuals in order to provide long term relief of symptoms and improvement in quality of life during subsequent natural allergen exposure (Bousquet *et al.*, 1998).

Bronchial asthma is considered a type I allergic reaction characterized by Th2 cells domination and allergen specific IgE mediation. The immunotherapy targeting at the allergic reaction is the optimal treatment. Clinical trials showed though inhaled gluccocorticoid could control the symptoms of asthma, it could not block the inflammation of airway (Guilbert *et al.*, 2006). However, SIT not only improves the clinical symptoms with reduced symptom score and medication, but also has long-term effect and prevents the progress of inflammation. Thus SIT is believed the only causative treatment that could change the progress of asthma. The current chapter focuses on the mechanism of SIT and its clinical practice.

13.1 The History of Allergen Immunotherapy of Asthma

SIT was introduced to treat asthma about 90 years ago. Since Noon and Freeman successfully treated hay fever sufferers by injecting them with pollen extracts in 1911, immunotherapy has been enthusiastically adopted as the treatment of choice for allergic rhinitis and asthma in North America and Europe. In the UK and South Africa, the procedure has never become popular but is used as a treatment option for grass pollen allergic rhinitis and for bee or wasp sting anaphylaxis. However, the process responsible for the reaction was unclear. The clinical efficacy and safety lead to controversial issues. In 1986, the practice of immunotherapy was practically halted in the UK when the *British Medical Journal* published its Committee on the Safety of Medicines (CSM) Report. This report cautioned against the use of immunotherapy in general practice and cited 26 anaphylactic deaths over a 30-year period. These deaths arose mainly as a result of inappropriate and injudicious use of the procedure in treating uncontrolled asthma.

However, with a deeper recognition of the processes responsible for the reactions of allergic diseases and immunotherapy, a large amount of basic and clinical studies proved the efficacy and safety of SIT. In addition, the improvement of treatment and application of standard allergen reagents also enhance the efficacy of immunotherapy. In 1998, the allergen immunotherapy congress of World Health Organization (WHO), based on the summary report of immunotherapy of the European Academy of Allergy and Clinical Immunology (EAACI) confirmed the dual significance of preventive and causative therapy, with long lasting efficacy and minor side effects. At the same time, the WHO position statement, *Allergen*

Immunotherapy: Therapeutic Vaccines for Allergic Diseases, was released, which was the first worldwide-accepted global guideline for immunotherapy (Bousquet *et al.*, 1998). It formally named specific allergic immunotherapy as specific allergy vaccination therapy (SAVT), and changed allergen extract to allergen vaccine. The Global Initiative for Asthma (GINA) also confirmed the important position of SIT in addition to the inhaled glucocorticoid treatment for asthma, and included SIT as part of the therapeutic regulation for asthma.

As the practice of allergen immunotherapy approaches its 100th anniversary, major changes appear likely in the way that it is practiced. The scientific rationale for immunotherapy has never been stronger.

13.2 Mechanism of Allergen Specific Immunotherapy

SIT is specific to the antigen administered. The mechanisms of SIT are complex and may differ depending on the allergen (venoms or inhalant allergens) and the route of immunization.

The mechanisms of action of allergen-specific immunotherapy include the very early desensitization effects, modulation of T- and B-cell responses and related antibody isotypes, and migration of eosinophils, basophils, and mast cells to tissues, as well as release of their mediators (Akdis CA *et al.*, 2011). The basic theory of mechanism of SIT is as follows.

13.2.1 Regulation of T-Cell TH2 and TH1 Cytokine Production

The main immunologic change of asthma is the imbalance of Th1/Th2 ratio and the disorder of function. Mechanistic studies continue to support the idea that immunotherapy expands allergen-specific Th1 immunity and suppresses the Th2 responses resulting from clinical observation. In 1998, there were reports that peripheral blood mononuclear cells (PBMCs) taken after immunotherapy showed reduced IL-4 production after allergen stimulation and enhanced IFN-γ expression (Durham *et al.*, 1998). The following reports demonstrated PBMCs of treated patients secreted less IL-2, IL-4, and IL-5 after immunotherapy (van Bever *et al.*, 1998). Some studies showed immunotherapy for dust mites had a marked effect in increasing the IFN-γ/IL-4 ratio in peripheral blood CD4+ T-cells after 3 months of the dose-increase phase (Klimek *et al.*, 1999). Those *in vitro* studies indicated that SIT could selectively reverse the Th2 response to Th1 response advantage, down-regulate Th2 cells and up-regulate Th1 cells. Oda *et al.* (1998) confirmed this hypothesis directly using *in vivo* studies. They developed both T-cell lines and T-cell clones from house dust mite-sensitive asthmatic patients before and at 18 months after rush immuno- therapy (RI). Eighteen months after RI, *D. farinae* specific T-cell clones had shifted their cytokine-secreting patterns to those of Th1- or Th0-like cells. In contrast, the T-cell clones of control subjects not given RI showed Th2-type responses throughout. The authors suggest that Th2 cells disappeared early after starting RI and that Th1 and Th0 populations emerged slowly and selectively later. The result also suggests that Th2 suppression occurs first and Th1 stimulation occurs later.

This was proved by the following studies. Guerra *et al.* (2001) found that immunotherapy makes allergen specific IL-4 producing T cells susceptible to apoptosis on exposure to allergens. Sade *et al.* (2003) also reported that repeated allergen administration during immunotherapy caused the proliferation of suppressor T-cells that have an oligoclonal TCR-V use. Laaksonen *et al.* (2003) found that signaling lymphocytic activation molecule, which usually increased in Th1 mediated autoimmune condition, had its mRNA expression increased in immunotherapy-treated patients and was associated with IFN-γ mRNA expression after 1 year of immunotherapy. This was interpreted as additional evidence that immunotherapy initiates an increased Th1 response.

With the new developments in the understanding of the T cell functions, Treg cells are considered play key role in this switch from Th2 cells to Th0 or Th1 cells. Two subgroups of Treg cells have received particular attention: the naturally occurring forkhead box P3 (FOXP3)+ CD4+ CD25+ regulatory T cells and the inducible Treg cells. Treg cells also increased in the tissues of allergic organs after allergen-SIT of allergic rhinitis (Radulovic *et al.*, 2008). Treg cells are the main resource of IL-10 and transforming growth factor - ß (TGF- ß). Neutralization of cytokine activity indicated that T-cell suppression was induced by IL-10 and TGF-ß. Immunotherapy also showed suppressed T-cell proliferation accompanied by diminished Th1 (IFN-γ) and Th2 (IL-5 and IL-13) cytokine responses and increased IL-10 and TGF-ß secretion by allergen-specific T cells (O'Hehir *et al.*, 2009). IL-10 could block the co-stimulatory CD28-B7.1 interaction and subsequent signaling pathways in T cells (Akdis *et al.*, 2001; Jutel *et al.*, 2003). Recent studies demonstrated that SIT can up-regulate the function of FOXP3+ CD4+ CD25+ Treg cells and significantly lead to more IL-10 production (Wei *et al.*, 2010). Moreover, a recent grass pollen subcutaneous immunotherapy (SCIT) study (Francis *et al.*, 2008) with an updosing period of 8–10 weeks showed that 2 weeks after first vaccine exposure T cells stimulated with antigen expressed an IL-10 signal, which was statistically significant at 4 weeks. This study revealed that allergen-induced Treg might play a more prominent role in the initial phases of SIT.

13.2.2 B-Cell Response and Antibody Synthesis

Compared to the plenty studies of T cell tolerance induced during SIT, B cell tolerance in the early course lacks enough investigations. However, B cells are the only cell type that is capable of producing antibodies and therefore are the central cellular component of the humoral immune responses. The production of IgE, IgG from these cells could determine the direction of allergic reaction. In addition, B cells can modulate CD4+ T cell responses by presenting antigens, expressing costimulatory molecules or producing cytokines. It was considered that under the modulation of IL-10 from Treg cells B cells could switch its production of IgE to IgG antibody isotype, in particular IgG4. While the latest findings showed that allergen-activated memory B cells can modulate IgE production by autocrine IL-10 (Milovanovicl *et al.*, 2009).

Since its first discover in 1967, allergen-specific IgE is considered as the marker of type I allergy. Early studies showed that SIT could change the production and role of IgE. SIT in patients with hay fever is associated with transient early increases in allergen-specific IgE, followed by blunting of seasonal increases in IgE (Bousquet *et al.*, 1998; Van Ree *et al.*, 1997; Lichtenstein *et al.*, 1973). While, in a proportion of individuals who received birch pollen SIT, low concentrations of IgE antibodies to previously unrecognized proteins were identified (Gleich *et al.*, 1982).

Block antibody theory is the generally accepted process for SIT in 1970s. The theory is from the evidence obtained that the serum total IgG from untreated allergic patients are at a low level, but increased after 3 months' SIT, and reached it's peak after 6 months' SIT. Then it was found that clinical symptoms were alleviated due to the increase of allergen specific IgG, so allergen specific IgG was considered as the important marker to evaluate the efficacy of SIT. This theory proposed that allergen-specific IgG antibodies induced by immunotherapy can disrupt the formation of allergen-IgE complexes that bind to antigen-presenting cells and facilitate allergen presentation. A further study by Gallego MT *et al.* (2010) confirmed this response after 1 year of immunotherapy. The latest study shows that injection of IgG attenuates asthmatic features and the function of lung CD11c(+) DCs via Fcγ receptor IIb in allergic airway inflammation on murine model, thus suggesting the possible process of IgG's roles in inhibition of allergic inflammation (Yamamoto *et al.*, 2010).

Quantitative measurements of allergen-specific IgG subclasses in SIT treated patients have revealed increases in allergen-specific IgG1 and IgG4 antibody concentrations in the serum and in the local target organs (Gehlar *et al.*, 1999; Jutel *et al.*, 2003; Moverare *et al.*, 2002). IgG4 is a non-inflammatory isotype protecting from allergic reaction. It is thought to capture the allergen before reaching the effector cell-bound IgE and thus to prevent the activation of mast cells and basophils (Akdis *et al.*, 2011). IgG4 is unable to bind complements efficiently and contains two different antigen-binding sites on one molecule. The bi-specificity turns the antibody functionally monovalent, thus preventing it from forming complexes (Schroeder *et al.*, 2010). Allergen-specific IgG4 might be directed against different epitopes of the allergen than IgE, yet an inhibition of the IgE-allergen binding by certain IgG is observed resulting in a blocking effect (Denepoux *et al.*, 2000).

Immunoglobulin isotype class switching, in particular for IgG4 to IgE, is dependent on Th2 cytokines (IL-4 and IL-13) and cognate interaction of CD40 on

Th cells and the CD40 ligand on B cells (James *et al.*, 2011; Jabara *et al.*, 1993; Punnonen *et al.*, 1993). Furthermore, the induction of IL-10 immunoregulatory cytokine following SIT has a profound effect on isotype class switching. In the presence of IL-4, additional IL-10 induces a preferential class switch in favour of IgG4 and has been shown to suppress both total and allergen-specific IgE responses (Agresti *et al.*, 1999; Meiler *et al.*, 2008).

Elevated levels of allergen specific IgA2 antibodies and polymeric IgA2 also have been reported following grass pollen-specific injection immunotherapy. Passive sensitization of monocytes in vitro using purified polymeric IgA2 from IgA-containing serum obtained following allergen SIT, followed by cross-linking in vitro of IgA on monocytes by antigen or anti-IgA resulted in IL-10 production (Pilette *et al.*, 2007). This indirect production of IL-10 from accessory cells may in turn favours isotype class switching in favour of IgG4 antibody production. These findings implicate a possible role for IgA antibodies in the induction of tolerance following SIT.

13.2.3 Decrease the Airway Reaction Specific to Allergen

Early studies showed that SIT could reduce the skin response to allergens. In recent years, more and more experiments have shown that SIT could decrease all of the sensitivity of skin response, nasal and bronchial provocation, which is allergen specific. Usually it could not block the response completely, but only increase the threshold of response. This occurred earlier than the specific IgG, which indicated the role of cytokines.

13.2.4 Influence on Mast Cells and Eosinophils

SIT can also inhibit the proliferation, activation and release of inflammatory material from mast cells. Children allergic to mites took SIT and were found to have reduced nasal mast cells, and after pollen SIT, adults' skin mast cells were also reduced (Durham *et al.*, 1999). The histamine and prostaglandin D2 level in nasal secretion also decreased. Meanwhile, studies also revealed the reduced eosinophils in nasal secretion of patients after SIT (Wilson *et al.*, 2001).

13.2.5 Inflammatory Cytokines

During chronic airway inflammation, a number of inflammatory cells such as eosinophils, mast cells, macrophage and inflammatory mediators released histamine, eosinophilic cationic protein. After SIT, asthmatic patients had reduced inflammatory cells and mediators in airways and circulation. Several studies cited above showed significant reductions in the eosinophil and basophil stimulator IL-5. Neutrophil degranulation *in vitro* after stimulation with formylmethionyl- leucyl-phenylalanine was found to be higher in the asthmatic patients than in the healthy control subjects and returned to normal levels after immunotherapy (Monteseirin *et al.*, 2001). IL-1b and TNF-a levels were also reduced. IL-2 and IL-6 levels, on the other hand, were initially low but increased after immunotherapy. This data indicates a reduction in inflammatory responses.

13.3 Clinical Efficacy of Immunotherapy

The clinical efficacy of SIT for asthma has been assessed in many trials. Compared to the first line medicine ICS, SIT has its unique advantages. SIT is not only beneficial to the control of symptoms, but also has its long-term effect and inhibition of the progress of polydensitization, which could not be achieved by ICS.

13.3.1 Symptom Control

Successful SIT reduces the symptoms of allergic asthma and the need for medication. Before 1990 there were numerous clinical trials of SIT in asthma. The available largest reviews of clinical trials included 32 (Bousquet *et al.*, 1998) and 33 (Bonifazi and Bilo, 1997) randomized placebo-controlled trials. Although most of them were not specifically designed for asthma and included small numbers of patients, but had asthma symptoms reported.

Meta-analyses of both subcutaneous immunotherapy and sublingual immunotherapy demonstrated significant efficacy on reduction of symptoms scores and asthma medication (Passalacqua G et al., 2011). These studies included SIT for mites, pollen, cat and so on. Asthma symptom scores improved significantly (Abramson et al., 2003). But there was only a borderline reduction of symptoms with mite allergens, whereas the effect was highly significant with pollens. No significant improvement with cat, dog or multiple allergen extracts was observed. Nevertheless a significant reduction in allergen-specific bronchial responses was consistently shown, lung function (either FEV1 or peak expiratory flow rate (PEFR)) was measured only in few trials and with controversial results (Hedlin G et al., 1999).

13.3.2 Prevention of Disease Progression

A multicenter study in Europe with children of birch or grass pollen allergies confirms that specific immunotherapy during childhood can reduce the incidence of later asthma (Moller *et al.*, 2002). Besides, SIT could also prevent new allergies. Although the clinical response to immunotherapy has been proven to be allergen-specific, there is now good evidence that administration of appropriate monotherapy to monosensitized patients can reduce the likelihood of the patients developing additional sensitivities. In two large-scale studies, the likelihood of developing additional positive skin tests was reduced from about 2/3 to 1/4. Furthermore, the protection was shown to persist 3 years after the 3-year to 4-year course of specific immunotherapy was completed (Pajno *et al.*, 2001; Purello-D'Ambrosio *et al.*, 2001).

13.3.3 Long-Term Effect

SIT offers a long-term effect to prevent the progress of inflammation, which lasts many years even after stopping therapy. In the past 5 years, the report by Durham *et al.* (1999) on long-term effects is the most important article for clinicians. Relief of grass pollen hay fever continued 3–4 years after stopping immunotherapy of 3 years' duration. In this study, sixteen patients received maintenance injections for an additional 3 years, 16 patients received matched placebos, and 15 new patients were followed without immunotherapy. Over the 3 years of the study, the symptoms reported by the maintenance and placebo-treated patients were similarly suppressed, whereas the new patients reported more severe symptoms. Inhibition of late-phase skin responses continued in both the maintenance and placebo-treated patients. In the placebo-treated patients, there was no evidence of the return of CD3+ or IL–4+ cells in allergen-challenged skin biopsy sites. This study confirmed that immunotherapy for respiratory allergies afforded long-term improvement just as insect venom immunotherapy protects against anaphylaxis long after it has been discontinued.

Another long-term follow-up of childhood immunotherapy in a retrospective study was conducted on asthmatic patients allergic to either house dust mite or to both house dust mite and grass pollen and treated with specific immunotherapy during childhood and were re-evaluated in early adulthood after mean cessation of immunotherapy for 9 years (Cools *et al.*, 2000). A control group of asthmatic patients with comparable asthma features had been treated with appropriate anti-asthmatic drugs during childhood but never received immunotherapy. The risk of frequent asthmatic symptoms was 3 times higher in the control group than in the systemic immunotherapy-treated group.

13.4 Glossary of SIT

SIT has different administration methods. The main popular treatments are subcutaneously immunotherapy (SCIT) and sublingual immunotherapy (SLIT). Both of these methods are divided into two steps: build-up step and maintenance step. During the build-up step, the dose of immunotherapy was increased gradually until it reaches the maximal dose. Then the treatment comes to maintenance step. The patients need to keep the maximal dose for about 3–5 years. But there is still obvious differences between these two methods.

13.4.1 Subcutaneous Immunotherapy (SCIT)

SCIT currently is established as an effective treatment for patients with IgE-mediated reactions to hymenoptera venom, allergic rhinitis, and allergic bronchial asthma. It is the classic route of SIT. The clinical efficacy of SIT is mostly built on the studies of SCIT and depends on the standardization of allergen and the effective dosage of allergen extract.

Allergen standardization. The quality of the allergen vaccine is critical for both diagnosis and treatment. When possible, standardized vaccines of known potency and shelf-life should be used. The most common vaccines used in clinical allergy practice are now available as standardized products or are pending standardization. However, there are many vaccines currently being marketed (many of which are only used occasionally), and it is neither feasible nor economical to standardize all of them. The measurement of major allergens for standardization is now a realistic and desirable goal that should be encouraged (Dreborg *et al.*, 1993; American Academy of Allergy, Asthma and Immunology, 1997). Allergen vaccines should be distributed, provided their potency, composition, and stability have been documented as either (1) vaccines from a single source material; (2) mixtures of related, cross-reacting allergen vaccines, such as grass pollen vaccines, deciduous tree-pollen vaccines, related ragweed-pollen vaccines, and related mite vaccines; or (3) mixtures of other allergen vaccines, provided that data for its stability and clinical efficacy are available. Where mixtures are marketed, the relative amounts of each component of the mixture should be indicated.

Effective doses of allergen extract. The use of major allergen content to express extract potency and dosing remains the only international language currently available. The effective dosage may be expressed as the quantity of the major allergen delivered as a maintenance dose. Expressed in this way, the range of dosages is seen to be relatively narrow (3–20 mg major allergen), with the exception of Alternaria, for which significant improvement was achieved with a dosage containing only 1.6 mg of the major allergen Alt a 1. Information on dosage responses is available for most of the extracts. In general, a dosage of 1/5 to 1/10 of the proven effective dosage has been less effective or ineffective in eliciting the responses being assessed. The information on the mean major allergen content of an extract is often available from the extract manufacturer.

Dosing schedules. For safety concerns, the rate of dosage increase with conventional immunotherapy is gradual. With weekly injections, 5 to 6 months may be required to reach the projected maintenance dosages. Alternatively, more rapid build-up to maintenance may be achieved by either rush or cluster schedules. With rush dosaging, multiple injections are given daily, and maintenance is reached after 1 or several days. The price of this rapid achieving of maintenance is an increased rate of systemic reactions, which is reduced somewhat by premedication (Sharkey *et al.*, 1996). Cluster dosaging consists of 2 or more injections per day, but visits are typically 1 or 2 days per week. Maintenance is not achieved as rapidly as with rush, but in comparison with conventional schedules, there does not appear to be an increased incidence of systemic reactions (Tabar *et al.*, 2005).

13.4.1.1 Assessment of Clinical Efficacy of Different Allergen

The allergens to induce asthma are a lot. The clinical efficacy of SIT varies for the differences of causative allergens and vaccines. Recent studies have shown that the different vaccines play crucial roles in the efficacy of SIT. Among these, standard pollen vaccine has the best clinical efficacy. Double blind clinical trials demonstrated that pollen vaccine could improve the symptoms of asthma, reduce the airway hyperreactivity and medication (Blaiss *et al.*, 2011). It has also been shown that mite vaccine has a better effect in children. It could reduce the exacerbation of symptoms, but it has less effect on airway hyperreactivity and irreversible impairment of ventilatory function. As for the dander of animals, the best way is to avoid the animal. But for cat allergies of difficult-controlled asthma, Fel d 1, a peptide vaccine comprising the immunodominant regions of the allergen was safe and well tolerated when given to subjects with cat allergy as a single dose (Worm *et al.*, 2011). Fungi are hard to be standardized since there are many kinds. Currently standardized A alternata extract has good clinical efficacy with reducing symptoms of asthma and rhinoconjunctivitis in children and adolescents without serious side effects (Kuna *et al.*, 2011). Combinations of multiple allergens, commonly used with SCIT in the US, are not recommended in Europe and China. The efficacy of reducing symptoms and medication doesn't show any significant benefit more than monotherapy.

13.4.1.2 Consideration of Safety

SCIT has potential side effects in forms of local, systemic, and even fatal reactions. Minor local pain, itching or swelling occuring at the injection site is a common occurrence. Sometimes a generalized allergic reaction with hives or flushing may occur. Occasionally, patients may experience an anaphylaxis of generalized reactions with throat swelling, wheezing or a drop in blood pressure. There have been cases of death from allergic reactions caused by allergen injections. Local reactions have not been found to be predictive of the subsequent occurrence of systemic reactions. The exact incidence of reactions to immunotherapy is a result of patient sensitivity, the dose administered, and whether the extract is modified to delay absorption. Therefore, the multiple studies reporting the incidence of reactions have limited general applicability.

Two surveys by the Immunotherapy Committee of the American Academy of Allergy, Asthma & Immunology in the years from 1985 to 2001 reported 34 fatalities, 33 from immunotherapy and 1 from skin testing to foods (Reid *et al.*, 1993; Bernstein *et al.*, 2004). Notably, 28 of 32 (88%) patients who experienced a fatal reaction had asthma that was labile, poorly controlled, or treated with oral corticosteroids in 21 of 28 cases (75%). Additional findings were that 18 of 32 (52%) occurred during the build-up phase, there were dosing errors in 5 (15%) occurred in the first injection from a new vial, and 3 (9%) occurred with unsupervised injections.

13.4.1.3 Practical Immunotherapy

Immunotherapy should only be performed by trained staff in an allergy clinic with facilities including an appropriate observation area, vaccine storage at 4℃, and access to resuscitation. An out of work telephone contact number should be provided. Injections should only be given in the immediate presence of a physician.

Immunotherapy protocols, in general, involve weekly injections for 8–16 weeks during an up-dosing phase, followed by monthly maintenance injections (empirically this has been extended in some centers to 6–8 weeks) for a period of 3–5 years. "Cluster" immunotherapy up-dosing schedules may involve repeated injections at each clinic visit. "Rush" protocols which may involve repeated up-dosing injections in order to achieve maintenance doses within several hours are applicable to venom sensitive patients, although they are unsuitable for patients with inhalant allergies in view of the marked increased occurrence of side effects. In general, manufacturers' recommendations should be carefully followed, although tailored to individual patients' circumstances.

Peak flow examination before and after injection, 30 min close monitoring after injection, and adrenaline at hand is important for safety of therapy.

13.4.1.4 Management of Adverse Events

In general, local swelling following injections is to be expected. No treatment is required other than reassurance, although occasionally use of an antihistamine may be indicated. Systemic reactions should be recognized and treated promptly, according to recommended guidelines. In general, mild rhinitis or wheezing may be treated by an antihistamine or bronchodilator with continued observation. More severe reactions, including moderate asthma, urticaria, or angio-edema require intravenous hydrocortisone and antihistamine. Adrenaline 0.5 mg by the intra-muscular route is indicated in rapidly evolving systemic reactions which do not respond to these measures and in all patients where there is associated moderate/severe respiratory impairment or hypotension. In general, if in doubt, give adrenaline which is more effective when administered early during a systemic reaction. Delayed systemic reactions are usually mild, involving mild urticaria or asthma and respond to antihistamines and/or inhaled bronchodilator therapy. Patients should be supplied with a contact telephone number of the clinic nurse/doctor and advised that if reactions do not respond to this treatment, they should attend their local accident and emergency department. All immunotherapy clinics should have detailed standard operating procedures and regular review of practice and staff training in immunotherapy procedures and the early recognition and treatment of adverse events (Wilson *et al.*, 2001). Clinics should supply patient information sheets which detail the benefits and risks of immunotherapy and detailed information on practical aspects of immunotherapy.

13.4.2 Sublingual Immunotherapy

Though SCIT is confirmed as the effective treatment for asthma, the shortcoming of SCIT extension of the use of allergen immunotherapy to the many appropriate patients is hampered by two drawbacks: concerns for safety and the inconvenience of the treatment schedules used. In response to these drawbacks, attention is being given to developing safer and more convenient ways to administer the currently available allergen extracts or to make modifications in the allergen extracts, to render them inherently safer and more convenient. It will be important, however, not to sacrifice the effectiveness of the current method of administering immunotherapy while seeking greater safety and convenience.

SLIT is also an effective therapy for allergic rhinitis, including pediatric patients, and for allergic asthma. A meta-analysis of 21 trials involving 959 subjects with allergic rhinitis found a significant reduction in symptoms and medication use (Wilson *et al.*, 2005). SLIT also has long-term preventative effects. Studies show a significantly reduced development of new skin test reactions in monosensitized patients treated with SLIT compared with untreated controls. The pre-seasonal treatment of SLIT could prevent the occurrence of asthma. There are a few long-term follow-up studies after discontinuation of SLIT. One open study in children showed a significant reduction in asthma after 5 years of therapy with house dust mite extract. This reduction in prevalence of asthma persisted without change on follow-up 5 years after SLIT was discontinued.

13.4.2.1 Safety/Convenience

The major advantage offered by SLIT over SCIT is safety; no fatal or near fatal reactions have been reported (Cox *et al.*, 2006). This reputation for safety has allowed home administration of SLIT, thus avoiding the shortcoming of SCIT, and the inconvenience of frequent visits to a physician's office to receive the injections. The absence of fatal or life-threatening reactions with SLIT, however, does not mean that this form of treatment is without adverse reactions. The predominant reaction is oral pruritus or swelling. The rate of systemic reactions including ocular, cutaneous, and respiratory was similar in the low-dose and high-dose studies, about 0.5 per 100 doses (Gidaro *et al.*, 2005). The majority of serious adverse events are single episodes of uvula edema, urticaria, abdominal pain, and vomiting that resulted in hospitalization.

13.4.2.2 Doses/Frequency/Dose Response

One of the unresolved questions with SLIT is that of dosaging. The most common way to express the dosages used in SLIT is to compare the cumulative amount administered sublingually over a period of 1 month to that administered by the same investigators for monthly maintenance SCIT. This has the limitation that maintenance dosages used for SCIT may vary widely among investigators. The dosage-response study revealed that large dosages may be more effective in short-term studies; however, the frequency and duration of treatment are important determinants of the response in SLIT.

13.4.2.3 Multiple Antigen

Despite the popular use of multiple antigens in SCIT, it has not been used in controlled studies of SLIT. If, indeed, the absorptive capacity is limited in the sublingual area, the use of multiple allergen mixes must be examined in controlled studies before widespread use of SLIT with multiple allergens can be endorsed.

Compared to SCIT, though both could reduce the symptoms of asthma obviously, the degree of reduction is different, and the medication for asthma is significantly less in the patients treated with SCIT. It was demonstrated that a single pre-seasonal course of SLIT with a dosage 45–225 times more than that given by SCIT will be about 1/2 less effective (Torres-Lima *et al.*, 2002). That is also explained by the compliance of the two methods. SLIT was considered safe and convenient, thus it had better compliance. However, the study showed that the 3-year's compliance of SLIT was worse than SCIT. One reason is the loss of confidence of this treatment.

13.4.3 Other Routes of Immunotherapy

Besides the foresaid two immunotherapy routes, people still put more efforts into the development of more attractive SIT or needle-free and potentially self-administrable treatment routes. Nasal, oral immunotherapy and bronchial approach are the early trials. But the potential risks of bronchial spasm or obvious gastrointestinal reaction limited the application of these methods.

Whereas, recent appearing intralymphatic immunotherapy (ILIT) has it unique advantage to enhance the efficacy of SIT. Compared to SCIT with only small fractions of subcutaneously injected antigen reaching the draining lymph nodes, direct intralymphatic injection of the antigen increasing antigen availability in the SLOs (see the geographic concept of immunogenicity) (Zinkernagel *et al.*, 1997 and 2000). It has high therapeutic efficacy with considerably reduced treatment dose and substantially shortened treatment duration. Combined with its good safety profile, ILIT is therefore likely to increase treatment compliance and reduce socioeconomic costs.

And epicutaneous allergen-specific immunotherapy (EPIT) also has the potential to emerge as a promising needle-free treatment route for IgE-mediated allergies. Though it was known as early as 1921 year, it is not performed in clinical trials until the beginning of 21 century. The clinical trials and animal experiments have proved its efficacy in allergic diseases including asthma (Mondoulet *et al.*, 2010; Dupont *et al.*, 2010). Yet, there is still potential to enhance its clinical efficacy and to reduce treatment duration.

13.4.4 Alternative Therapy with Modified Extracts

Since the beginning of SIT, efforts to improve the quality of extract were tried by many scientists. With the development of immunologic technique, a number of approaches to decrease the allergenicity (i.e., safety), while maintaining immunogenicity (i.e., efficacy) of allergen vaccines, have been explored as following.

Allergoids/Polymerized extracts. Creation of an allergoid by treatment with formaldehyde and polymerization with glutaraldehyde both reduce allergenicity of the extract while retaining immunogenicity (Corrigan *et al.*, 2005). Research continues with both approaches in Europe, but not in the US.

Recombinant allergens with site-directed mutagenesis and deletion. Successful immunotherapy has been reported using a mixture of recombinant major allergens of grass (Jutel *et al.*, 2005). Although this produces a consistent extract, there is no reduction in allergenicity and hence there is a limited advantage over natural allergen extracts. Site-directed mutagenesis or deletion, on the other hand, can reduce the IgE epitopes on the molecule while not affecting its ability to react with T lymphocytes. This approach has been applied to human studies with fragments or trimers of the major allergen of birch, Bet v 1.

Allergen-derived peptides. Because the IgE epitopes are expressed on the 3-dimensional structure of the allergen while the T lymphocyte epitopes are short segments of adjacent amino acids, it is possible to devise peptides too small to react with IgE but with retention of the T lymphocyte epitopes. Thus, reduced allergenicity with retained immunogenicity is achieved (Alexander *et al.*, 2005). This approach is being actively pursued with peptides of the major cat allergen, Fel d 1.

Immunostimulatory sequences. Unmethylated cytosine phosphorothionate guanosine segments that characterize bacterial and viral DNA react with the Toll-like receptor 9 of antigen presenting cells. Synthetic cytosine phosphorothionate guanosine DNA chains, when covalently linked to allergens, have the advantage not only of directing the response to the allergen toward a regulatory and Th1 bias but also of sterically interfering with the reaction of IgE with the allergen, thus reducing its allergenicity (Creticos *et al.*, 2006).

Fusion proteins. Mast cells and basophils express FcgRIIb, which contains an immunoreceptor tyrosine-based inhibition motif within its cytoplasmic tail. Aggregating FcgRIIb to the major IgE receptor, FcεRI, leads to inhibition of FceRI signaling. A fusion protein consisting of human Fcg plus the major cat allergen (Feld 1) has been developed and tested as a new form of immunotherapy (Zhu *et al.*, 2005). Pretreatment of basophils from subjects with cat allergies and cat allergen-sensitized cord blood-derived mast cells with the fusion protein reduced histamine release on Fel d 1 challenge. Active immunization of mice with the fusion protein produced inhibition of systemic, lung, and cutaneous reactivity to Fel d 1.

13.5 Summary

SIT developed from a desire to inoculate patients with hay fever against a supposed toxin in pollen. Although the underlying concept has proved false, one century's use proved its clinical efficacy in patients with rhinitis or asthma whose symptoms are clearly driven by allergic triggers. The exact mechanism of its work is still unclear, but its capability to switch Th2 reaction to Th1 reaction and switch

IgE to IgG4 antibody isotype under the regulation of Treg cells and IL-10 is believed widely. The dual function of causative and preventive therapy outstands its unique advantages over ICS. Concerns about safety and clinical efficacy stimulate the development of new modified vaccines and new administrated routes. The next century of SIT may see the development of vaccines that are better standardized or the use of recombinant allergens, both of which should improve the safety profile of SIT, but the basic concept of SIT remains valid as it enters its second century.

References

Abramson, M., Puy, R., Weiner, J. (2003). Allergen immunotherapy for asthma. Cochrane Database Systemic Review, 4, CD001186.

Akdis, C.A., Akdis, M. (2011). Mechanisms of allergen-specific immunotherapy. J Allergy Clin Immunol, 127(1), 18-27.

Akdis, C.A., Blaser, K. (2001). Role of IL-10 in allergen-specific immunotherapy and normal response to allergens. Microbes and Infection, 3, 891-898.

Agresti, A., Vercelli, D. (1999). Analysis of gamma4 germline transcription in human B-cells. Int Arch Allergy Immunol, 118, 279-281.

Alexander, C., Tarzi, M., Larche, M., *et al.* (2005). The effect of Fel d 1-derived T-cell peptides on upper and lower airway outcome measurements in cat-allergic subjects. Allergy, 60, 1269-1274.

American Academy of Allergy, Asthma and Immunology (1997). The use of standardized allergen extracts. Journal of Allergy and Clinical Immunology, 99, 583-586.

Bernstein, D.I., Wanner, M., Borish, L., *et al.* (2004). Twelve-year survey of fatal reactions to allergen injections and skin testing: 1990–2001. Journal of Allergy and Clinical Immunology, 113, 1129-1136.

Blaiss, M., Maloney, J., Nolte, H., *et al.* (2011). Efficacy and safety of timothy grass allergy immunotherapy tablets in North American children and adolescents. J Allergy Clin Immunol, 1, 64-71.

Bonifazi, F., Bilo`, M.B. (1997). Efficacy of specific immunotherapy in allergic asthma: Myth or reality? Allergy, 52, 698-710.

Bousquet, J., Lockey, R., Malling, H., the WHO panel members (1998). Allergen immunotherapy: Therapeutic vaccines for allergic diseases A WHO position paper. Journal of Allergy and Clinical Immunology, 102(4), 558-562.

Bousquet, J., Lockey, R.F., Malling, H.J. (1998). Allergen immunotherapy: Therapeutical vaccines for allergic diseases. Allergy, 54, 1-33.

Cools, M., van Bever, H.P., Weyler, J.J., *et al.* (2000). Long-term effects of specific immunotherapy, administered during childhood, in asthmatic patients allergic to either house-dust mite or to both house-dust mite and grass pollen. Allergy, 55, 69-73.

Cox, L.S., Linnemann, D.L., Nolte, H. *et al.* (2006). Sublingual immunotherapy: a comprehensive review. Journal of Allergy and Clinical Immunology, 117, 1021-1035.

Corrigan, C.J., Kettner, J., Doemer, C., *et al.* (2005). Efficacy and safety of preseasonally-specific immunotherapy with an aluminumadsorbed six-grass pollen allergoid. Allergy, 60, 801-807.

Creticos, P.S., Schroeder, J.T., Hamilton, R.G., *et al.* (2006). Immunotherapy with a ragweedtoll-like receptor 9 agonist vaccine for allergic rhinitis. New England Journal of Medicine, 355,1445-1455.

Denepoux, S., Eibensteiner, P.B., Steinberger, P., *et al.* (2000). Molecular characterization of human IgG monoclonal antibodies specific for the major birch pollen allergen Bet v 1. Antiallergen IgG can enhance the anaphylactic reaction. FEBS Lett, 465, 39-46.

Dupont, C., Kalach, N., Soulaines, P. *et al.* (2010). Cow's milk epicutaneous immunotherapy in children: A pilot trial of safety, acceptability, and impact on allergic reacitivity. J Allergy Clin Immunol, 125, 1165-1167.

Durham, S.R., Till, S.J. (1998). Immunologic changes associated with allergen immunotherapy. Journal of Allergy and Clinical Immunology, 102, 157-164.

Durham, S.R., Varney, V.A., Gaga, M., *et al.* (1999). Grass pollen immunotherapy decreases the number of mast cells in the skin. Clinical & Experimental Allergy, 29,1490-1496.

Durham, S.R., Walker, S.M., Varga, E.M., *et al.* (1999). Long-term clinical efficacy of grass-pollen immunotherapy.New England Journal of Medicine, 341, 468-475.

Francis, J.N., James, L.K., Paraskevopoulos, G., *et al.* (2008). Grass pollen immunotherapy: IL-10 induction and suppression of late responses precedes IgG4 inhibitory antibody activity. J Allergy Clin Immunol, 121, 1120-1155.

Gallego, M.T., Iraola, V., Himly, M., *et al.* (2010). Depigmented and polymerised house dust mite allergoid: allergen content, induction of IgG4 and clinical response. Int Arch Allergy Immunol, 1, 61-69.

Gehlar, K., Schlaak, M., Becker, W.M., *et al.* (1999). Monitoring allergen Immunotherapy of pollen-allergic patients: The ration of allergen- specific IgG4 to IgG1 correlates with clinical outcome. Clin Exp Allergy, 29, 497-506.

Gidaro, G.B., Marcucci, F., Sensi, L., *et al.* (2005). The safety of sublingual-swallow immune therapy: An analysis of published studies. Clinical & Experimental Allergy, 35, 565-571.

Gleich, G.J., Zimmermann, E.M., Henderson, L.L., *et al.* (1982). Effect of immunotherapy on immunoglobulin E and immunoglobulin G antibodies to ragweed antigens: A six-year prospective study. J Allergy Clin Immunol, 70, 261-271.

Guerra, F., Carracedo, J., Solana-Lara, R., *et al.* (2001). TH2 lymphocytes from atopic patients treated with immunotherapy undergo rapid apoptosis after culture with specific allergens. Journal of Allergy and Clinical Immunology, 107, 647-653.

Guilbert, T.W., Morgan, W.J., Zeiger, R.S., *et al.* (2006). Long-term inhaled corticosteroids in preschool children at high risk for asthma. N Engl J Med, 354, 1985-1997.

Hedlin, G., Wille, S., Browaldh, L., *et al.* (1999). Immunotherapy in children with allergic asthma: Effect on bronchial hyperreactivity and pharmacotherapy. J Allergy Clin Immunol, 103, 609-614.

Jabara, H.H., Loh, R., Ramesh, N., *et al.* (1993). Sequential switching from mu to epsilon via gamma 4 in human B-cells stimulated with IL-4 and hydrocortisone. J Immunol, 151, 4528-4533.

James, L.K., Shamji, M.H., Walker, S.M., *et al.* (2011). Long-term tolerance after allergen immunotherapy is accompanied by selective persistence of blocking antibodies. J Allergy Clin Immunol, 2, 509-516.

Jutel, M., Akdis, M., Budak, F., *et al.* (2003). IL-10 and TGF-beta cooperate in the regulatory T cell response to mucosal allergens in normal immunity and specific immunotherapy. European Journal of Immunol, 33,1205-1214.

Jutel, M., Jaeger, L., Suck, R., *et al.* (2005). Allergenspecific immunotherapy with recombinant grass pollen allergens. J Allergy Clin Immunol, 116, 608-613.

Klimek, L., Dormann, D., Jarman, E.R., *et al.* (1999). Short-term preseasonal birch pollen allergoid immunotherapy influences symptoms, specific nasal provocation and cytokine levels in nasal secretions, but not peripheral T-cell responses, in patients with allergic rhinitis. Clinical & Experimental Allergy, 29, 1326-1335.

Kuna, P., Kaczmarek, J., Kupczyk, M. (2011). Efficacy and safety of immunotherapy for allergies to Alternaria alternata in children. J Allergy Clin Immunol , 2, 502-508.

Laaksonen, K., Junikka, M., Lahesmaa, R., *et al.* (2003). *In vitro* allergen-induced mRNA expression of signaling lymphocytic activation molecule by PBMC of patients with allergic rhinitis is increased during specific pollen immuno-therapy. J Allergy Clin Immunol, 112, 1171-1177.

Lichtenstein, L.M., Ishizaka, K., Norman, P.S., *et al.* (1973). IgE antibody measurements in ragweed hay fever. Relationship to clinical severity and the results of immunotherapy. J Clin Invest, 52, 472-482.

Malling, H.J., Dreborg, S., Weeke, B. (1986). Diagnosis and immunotherapy of mould allergy, V: Clinical efficacy and side effects of immunotherapy with Cladosporium herbarum. Allergy, 41, 507-519.

Meiler, F., Klunker, S., Zimmermann, M., *et al.* (2008). Distinct regulation of IgE, IgG4 and IgA by T regulatory cells and Toll-like receptors. Allergy, 63, 1455-1463.

Milovanovicl, M., Heinel, G., Zuberbier, T., *et al.* (2009). Allergen extract-induced interleukin-10 in human memory B-cells inhibits immunoglobulin E production. Clin Exp Allergy, 39, 671-678.

Monteseirin, J., Bonilla, I., Camacho, J., *et al.* (2001). Elevated secretion of myeloperoxidase by neutrophils from asthmatic patients: The effect of immunotherapy. J Allergy Clin Immunol, 107, 623-626.

Moller, C., Dreborg, S., Ferdousi, H.A., *et al.* (2002). Pollen immunotherapy reduces the development of asthma in children with seasonal rhinoconjunctivitis (the PAT-study). Journal of Allergy and Clinical Immunology, 109, 251-256.

Mondoulet, L., Dioszeghy, V., Ligouis, M., *et al.* (2010). Epicutaneous immunotherapy on intact skin using a new delivery system in a murine model of allergy. Clin Exp Allergy, 40, 659-667.

Moverare, R., Elfman, L., Vesterinen, E., *et al.* (2002). Development of new IgE specificities to allergenic components in birch pollen extract during specific immunotherapy studied with immunoblotting and Pharmacia CAP System. Allergy, 57, 423-430.

Oda, N., Yamashita, N., Minoguchi, K., *et al.* (1998). Long-term analysis of allergen-specific T cell clones from patients with asthma treated with allergen rush immunotherapy. Cell Immunology, 190, 43-50.

O'Hehir, R.E., Gardner, L.M., de Leon, M.P., *et al.* (2009). House dust mite sublingual immunotherapy: the role for transforming growth factor-beta and functional regulatory T cells. Am J Respir Crit Care Med, 10, 936-947.

Pajno, G.B., Barberio, G., de Luca, F., *et al.* (2001). Prevention of new sensitization in asthmatic children monosensitized to house dust mite by specific immunotherapy: A six-year, follow-up study. Clinical & Experimental Allergy, 31, 1392-1397.

Passalacqua, G., Canonica, G.W. (2011). Specific immunotherapy in asthma: Efficacy and safety. Clin Exp Allergy, 41, 1247-1255.

Pilette, C., Nouri-Aria, K.T., Jacobson, M.R., et al. (2007). Grass pollen immunotherapy induces an allergen-specific IgA2 antibody response associated with mucosal TGF-beta expression. J Immunol, 178, 4658-4666.

Purello-D'Ambrosio, F., Gangemi, S., *et al.* (2001). Prevention of new sensitization in monosensitized subjects submitted to specific immunotherapy or not: Aretrospective study. Clinical & Experimental Allergy, 31,1295-1302.

Punnonen, J., Aversa, G., de Vries, J.E. (1993). Human pre-B cells differentiate into Ig-secreting plasma cells in the presence of interleukin-4 and activated CD41T cells or their membranes. Blood, 82, 2781-2789.

Radulovic, S., Jacobson, M.R., Durham, S.R., *et al.* (2008). Grass pollen immunotherapy induces Foxp3-expressing CD4+ CD25+ cells in the nasal mucosa. J Allergy Clin Immunol, 121, 1467-1472.

Reid, M.J., Lockey, R.F., Turkeltaub, P.C., *et al.* (1993). Survey of fatalities from skin testing and immunotherapy 1985-1989. Journal of Allergy and Clinical Immunology, 92, 6-15.

Sade, K., Kivity, S., Levy, A., Fireman, E. (2003). The effect of specific immunotherapy on T-cell receptor repertoire in patients with allergy to house-dust mite. Allergy, 58, 430-434.

Schroeder, H.W. Jr., Cavacini, L. (2010). Structure and function of immunoglobulins. J Allergy Clin Immunol, 125, S41-S52.

Sharkey, P., Portnoy, J. (1996). Rush immunotherapy: Experience with a one-day schedule. Annals of Allergy, Asthma & Immunology, 76, 175-180.

Tabar, A.I., Echechipia, S., Garcia, B.E., *et al.* (2005). Double-blind comparative study of cluster and conventional immunotherapy schedules with Dermatophagoides pteronyssinus. J Allergy Clin Immunol, 116,109-118.

Torres Lima, M., Wilson, D., Pitki, L., *et al.* (2002). Grass pollen sublingual immunotherapy for seasonal rhinoconjunctivitis: A randomized controlled trial. Clinical & Experimental Allergy, 32, 507-514.

Vailes, L., Sridhara, S., Cromwell, O. *et al.* (2001). Quantification of the major fungal allergens, Alt a 1 and Asp f 1, in commercial allergenic products. J Allergy Clin Immunol, 107, 641-646.

van Bever, H.P., Vereecke, I.F., Bridts, C.H., *et al.* (1998). Comparison between the in vitro cytokine production of mononuclear cells of young asthmatics with and without immunotherapy (IT). Clinical & Experimental Allergy, 28,

943-949.

van Ree, R., van Leeuwen, W.A., Dieges, P.H., *et al.* (1997). Measurement of IgE antibodies against purified grass pollen allergens (Lol p 1, 2, 3 and 5) during immunotherapy. Clin Exp Allergy, 27, 68-74.

Till, S.J., Francis, J.N., Nouri-Aria, K., *et al.* (2004). Mechanisms of immunotherapy. J Allergy Clin Immunol, 113, 1025-1034.

Wachholz, P.A., Durham, S.R. (2004). Mechanisms of immunotherapy: IgG revisited. Curr Opin Allergy Clin Immunol, 4, 313-318.

Wei, W., Liu, Y.G., Wang, Y.H., *et al.* (2010). Induction of CD4+ CD25+ Foxp3+ IL-10+ T-Cells in HDM-Allergic Asthmatic Children with or without SIT. Int Arch Allergy Immunol, 153, 19-26.

Wilson, D.R., Irani, A.M., Walker, S.M., *et al.* (2001). Grass pollen immunotherapy inhibits seasonal increases in basophils and eosinophils in the nasal epithelium. Clinical & Experimental Allergy, 31, 1705-1713.

Wilson, D.R., Torres-Lima, M., Durham, S.R. (2005). Sublingual immunotherapy for allergic rhinitis: Systematic review and meta-analysis. Allergy, 60, 4-12.

Worm, M., Lee, H.H., Kleine-Tebbe, J., *et al.* (2011). Development and preliminary clinical evaluation of a peptide immunotherapy vaccine for cat allergy. J Allergy Clin Immunol, 1, 89-97.

Yamamoto, M., Kobayashi, K., Ishikawa, Y., *et al.* (2010). The inhibitory effects of intravenous administration of rabbit immunoglobulin G on airway inflammation are dependent upon Fcγ receptor IIb on CD11c(+) dendritic cells in a murine model. Clin Exp Immunol, 2, 315-324.

Zhu, D., Kepley, C.I., Zhang, K., *et al.* (2005). A chimeric human-cat fusion protein blocks cat-induced allergy. Nature Medicine, 11, 446-449.

Zinkernagel, R.M., Ehl, S., Aichele, P., *et al.* (1997). Antigen localisation regulates immune responses in a dose- and time-dependent fashion: A geographical view of immune reactivity. Immunol Rev, 156, 199-209.

Zinkernagel, R.M. (2000). Localization dose and time of antigens determine immune reactivity. Semin Immunol, 12, 163-171.

14

Pharmacotherapy in Common Allergic Diseases

Erina M. Lin

University of California, 10833 Le Conte Ave, Los Angeles, CA 90095, US
E-mail: erinatu@gmail.com

Abstract: Allergic diseases are immunologically mediated hypersensitivity reactions that are increasing in prevalence throughout the world. Symptoms associated with allergic diseases can significantly affect one's sense of well-being. Management of allergic diseases involves diagnostic testing to identify suspected allergens accompanied by subsequent avoidance measures and pharmacotherapy. Pharmacotherapy plays an important role in the control and alleviation of allergic symptoms, but does not cure the underlying allergy. Nevertheless, the use of medications in allergy treatment is an essential component of effective management of allergic disease and can greatly enhance one's quality of life (WHO *et al.*, 2002). A wide array of medications is available for the treatment of common allergic diseases such as allergic rhinitis, allergic conjunctivitis, and asthma. Classes of medications recom- mended for use in allergic diseases are reviewed, and include antihistamines, leukotriene antagonists, mast cell stabilizers, corticosteroids, anticholinergics, and beta-agonists. Specific indications, mechanism of action, and potential side effects of these medications are discussed.

14.1 Introduction

It is estimated that over 20% of the world population has an IgE-mediated allergic disease including such conditions as allergic asthma, allergic rhinitis, and allergic conjunctivitis (WHO *et al.*, 2002). Investigation of allergic diseases begins with a detailed clinical history and physical examination with a focus on symptoms and exposures. Diagnostic testing to detect the presence of allergen-specific IgE antibodies can be performed to verify sensitization. Management of allergic diseases begins with avoidance measures and environmental control precautions (de Shazo and Kemp, 2010). Often times, pharmacotherapy must be initiated when symptoms persist despite these measures. In this chapter, we focus on standardized recommended pharmacotherapy regimens for the treatment of common allergic conditions including allergic rhinitis, allergic conjunctivitis and allergic asthma. These medications include controller agents as well as those that can be used as needed. In those with persistent symptoms despite allergen avoidance and an adequate trial of medication, initiation of immunotherapy may be warranted.

14.2 Allergic Rhinitis

Rhinitis refers to inflammation of the nasal mucosa characterized by symptoms of nasal congestion, rhinorrhea, pruritus, and sneezing (Bousquet and Khaltaev, 2001). Allergic rhinitis is the induction of symptoms of rhinitis after allergen exposure leading to an IgE-mediated immune reaction. A new classification of allergic rhinitis based on the allergic rhinitis and its impact on asthma (ARIA) initiative has recently been introduced categorizing allergic rhinitis according to frequency of symptoms and severity of disease. Frequency of symptoms is divided into intermittent and persistent diseases. Intermittent allergic rhinitis is defined as symptom duration of less than 4 days per week or less than 4 weeks' duration, whereas persistent allergic rhinitis is characterized by symptom duration of more than 4 days per week or greater than 4 weeks' duration. Furthermore, severity of symptoms is classified into mild or moderate-severe depending on the impact of symptoms on the quality of life. Mild disease causes no impairment of daily activities or troublesome symptoms with normal sleep, school and work. Moderate-severe disease has at least abnormal sleep, or abnormal school/work, or impairment of daily activities, or troublesome symptoms (Bachert, 2003; Dykewicz *et al.*, 1998).

The approach to treatment of allergic rhinitis depends upon the classification of disease from mild intermittent to severe persistent that involves allergen avoidance, pharmacotherapy, and allergen immunotherapy. The primary focus of this discussion is on pharmacotherapy, in which treatment options include oral and intranasal antihistamines, decongestants, mast cell stabilizers, leukotriene modifiers, and intranasal glucocorticosteroids.

Oral antihistamines work systemically and are comprised of first, second, and third generation agents. They reduce symptoms by blocking mast cell and basophil release of histamine, an important mediator in allergic inflammation. Antihistamines are particularly effective in treating symptoms of nasal pruritus, sneezing, and rhinorrhea. Ability to alleviate nasal congestion is less pronounced (Bousquet and Khaltaev, 2001; Verster and Volkerts, 2004). First generation antihistamines are lipophilic and incomplete H1 histamine receptor blockers which cross the blood-brain barrier thereby leading to their main disadvantage of causing sedation. Central nervous system symptoms are reported by 20% or more of patients. They may also have anticholinergic effects (Berkowitz *et al.*, 2006). First generation antihistamines include diphenhydramine, chlorpheniramine, hydroxyzine, brompheniramine, and others. They are generally recommended as second-line agents given their potential for side effects. Second generation antihistamines are lipophobic and selective H1 histamine receptor blockers. Thus, they are less likely to cause CNS side effects such as sedation (Bousquet and Khaltaev, 2001; Verster and Volkerts, 2004). The onset of action is within 1 h for most agents, and peak serum levels are reached in 2–3 h. They have a longer duration of action usually requiring only once or twice daily dosing (Verster and Volkerts, 2004). Second generation antihistamines include loratadine, cetirizine, and azelastine. Third generation antihistamines such as fexofenadine, desloratadine, and levocetirizine are metabolites of second generation antihistamines. Levocetirizine is a purified isomer of a second generation antihistamine. Third generation antihistamines are designed to have fewer CNS effects than its previous generation (Berger *et al.*, 2006).

Intranasal antihistamines including azelastine and olopatadine have the advantage of being delivered directly to the nasal mucosa, and can be used to treat primary symptoms of allergic rhinitis including nasal congestion. There is a rapid onset of action of less than 15 min. However, a potential side effect is sedation (Bronsky *et al.*, 1995).

The combination use of non-sedating oral antihistamines and oral decongestants provides better symptomatic relief than oral antihistamines do alone. Commonly used decongestants include pseudoephedrine and phenylephrine. However, decongestants should be used with caution given their potential adverse effects of hypertension, insomnia, irritability, and headache. They are relatively contraindicated in those with hypertension. Chronic use of decongestants is not recommended (Welsh *et al.*, 1987).

Intranasal cromolyn sodium and nedocromil sodium are mast cell stabilizers otherwise known as chromones that inhibit the release of histamine and other inflammatory mediators. While mast cell stabilizers help to reduce symptoms associated with allergic rhinitis, they appear to be less effective than intranasal corticosteroids (Ratner *et al.*, 2003). Required frequent dosing of 3–4 times per day makes it less convenient to use. However, given that they are poorly absorbed systemically, they have an excellent safety profile. They may be used in pregnancy and in pediatric patients (Bousquet and Khaltaev, 2001).

Leukotriene receptor antagonists such as montelukast work by blocking the effects of potent inflammatory mediators called cysteinyl leukotrienes. Montelukast helps to decrease symptoms of nasal congestion, sneezing, and rhinorrhea. Studies have shown that montelukast may be as effective as oral antihistamines, but less effective than intranasal corticosteroids in alleviating symptoms of allergic rhinitis

(Wilson *et al.*, 2004; Philip *et al.*, 2004; Nayak and Langdon, 2007). It is particularly useful in patients with allergic rhinitis and concomitant asthma given its potential to alleviate nasal and bronchial symptoms (Nayak and Langdon, 2007). Montelukast is well-tolerated and has a favorable safety profile (Borish, 2003).

Intranasal glucocorticoids (INGCs) have multiple anti-inflammatory pharmacological actions. They may decrease local inflammatory responses of the nasal mucosa by binding to intracellular glucocorticoid receptors of inflammatory cells (LaForce, 1999; deShazo and Kemp, 2009). There are first, second, and third generation INGC agents. The approximate bioavailability of the different agents has been studied. Although systemic absorption is low, there is still concern of potential systemic effects (deShazo and Kemp, 2010). First generation agents include beclomethasone (bioavailability unknown) and flunisolide (40%–50% bioavailability). Budesonide (10%–34%) is an example of a second generation agent. Third generation medications include fluticasone propionate (<2%), mometasone furoate (undetectable), and fluticasone furoate (<1%). There is less risk of systemic effects with second and third generation agents. Most studies have not shown a significant effect on growth of children with the use of INGCs, but potential effects of long-term use are not yet known. Once daily dosing vs. twice daily dosing appears to decrease the risk of affecting the hypothalamic-pituitary-axis in children (deShazo and Kemp, 2010; Juniper *et al.*, 2005). INGCs have been shown to be more efficacious than oral antihistamines, mast cell stabilizers, and leukotriene receptor antagonists in decreasing nasal symptoms. INGCs effectively reduce symptoms of allergic rhinitis, and are particularly effective in alleviating nasal congestion (Pullerits *et al.*, 2002; van Bavel *et al.*, 1994; Welsh *et al.*, 1987). Symptoms of local irritation caused by use of INGCs include nasal dryness, epistaxis, and burning (Blaiss, 2007). The correct technique for usage of nasal sprays includes keeping the head tilted down slightly during spraying and pointing the spray away from the septum to prevent septal irritation.

Recommendations for use of specific or combination therapies for allergic rhinitis depends upon the severity and frequency of symptoms, individual tolerability of various side effects associated with each agent, primary nasal symptoms, age of patient, and comorbidities. Usually, beginning with a trial of a single individual agent is reasonable. The addition of a second agent for a combination treatment regimen can be recommended if there is no satisfactory reduction of symptoms or significant improvement of quality of life while using a single agent.

14.3 Allergic Conjunctivitis

Allergic conjunctivitis is an IgE-mediated ocular disease that primarily affects the conjunctiva. In sensitized individuals, the conjunctiva is vulnerable to airborne pollens, animal dander, and other environmental allergens (Bielory and Friedlaender, 2008). Clinically, signs and symptoms of allergic conjunctivitis include pruritus, tearing, hyperemia, conjunctival edema, burning, and photophobia (Dana, 2009). Fifty percent of patients with allergic conjunctivitis have a personal or family history of other allergic diseases including allergic rhinitis, atopic dermatitis, and asthma (Doshnik and Ehlers, 1994). Allergic conjunctivitis can be subdivided into seasonal allergic conjunctivitis (SAC) and perennial allergic conjunctivitis (PAC). The onset of symptoms in SAC usually corresponds to one or more specific pollen seasons. PAC is due to a year-round exposure to airborne indoor environmental allergens such as dust mites, animal dander, and molds. Of note, a significant portion of patients with PAC may experience seasonal exacerbation of symptoms as well (Dana, 2009; Doshnik and Ehlers, 1994).

The management of allergic conjunctivitis involves the diagnosis of specific causative allergens, avoidance measures, and symptomatic relief. If pharmacologic treatment is needed, several classes of allergy medications are available. The use of oral and topical antihistamines, mast cell stabilizers, and nonsteroidal anti-inflammatory drugs can be effective. The use of corticosteroids is more potent, but given their increased risk for development of cataracts, glaucoma, and ocular infections are best prescribed and monitored by the ophthalmologist (Bielory and Friedlaender, 2008).

Oral and topical antihistamines can be used for symptomatic relief in allergic conjunctivitis. Nonsedating oral antihistamines including fexofenadine, loratadine, desloratadine, cetirizine, and levocetirizine have been shown to have efficacy in the treatment of allergic conjunctivitis. However, they may be associated with drying of the ocular mucosal membranes and decreased tear production (Ousler, 2007). Topical antihistamines include levocabastine, emedastine, and epinastine. They are especially effective in relieving symptoms of pruritus and erythema. Topical antihistamines are considered more effective than oral antihistamines given their faster onset of action with their immediate histamine receptor antagonism. Another advantage is local deliverance to the ocular tissues in higher concentration. Of note, topical vasoconstrictors such as naphazoline are widely available and sometimes combined with topical antihistamines, and may be used for short-term relief of symptoms. However, chronic use of these agents is generally not recommended due to their short duration of action and propensity for rebound hyperemia (Schultz, 2006).

Topical mast cell stabilizers have been a useful addition to the treatment regimens available for allergic conjunctivitis. Mast cell stabilizers include sodium cromoglycate, nedocromil, and lodoxamide. By inhibiting mast cell degranulation, they block the release of preformed allergic mediators. They do require a loading period, but are effective for long-term control. Their safety profile is excellent (Bielory and Friedlaender, 2008).

Dual-action agents combining topical antihistamines and mast cell stabilizers have become the new first class of choice in the treatment of allergic conjunctivitis. These combination agents include olopatadine, ketotifen, and azelastine. Due to their antihistamine and mast cell stabilizer components, they have both immediate and prophylactic activity rendering them more effective than other classes of agents. This symptomatic relief may also improve patient compliance (Schultz, 2006).

Topical nonsteroidal anti-inflammatory drugs (NSAIDs) such as ketorolac and diclofenac may be a useful adjunctive therapy for allergic conjunctivitis. NSAIDs inhibit the cyclooxygenase pathway thereby blocking the production of prostaglandins. By interfering with prostaglandin synthesis, they are effective in reducing conjunctival erythema and pruritus. Potential side effects include burning and stinging upon topical application (Manzouri *et al.*, 2006). Clinically, they are not considered to be as effective as other medications such as dual-action agents in controlling symptoms of allergic conjunctivitis (Deschenes *et al.*, 1999).

The primary focus of the medical management of allergic conjunctivitis is to alleviate symptoms. Total avoidance of the specific causative agents is often not possible. Therefore, the use of oral and topical ocular medications can be of great importance. Other ancillary control measures to consider include application of cool compresses, use of artificial tears, and refrigeration of topical ocular medications before use (Barney *et al.*, 2008).

14.4 Asthma

Asthma is a chronic inflammatory disease of the airways. It is characterized by recurrent episodes of wheezing, shortness of breath, chest tightness, and nighttime or early morning coughing. The pathophysiology of asthma involves components of airway obstruction, reversible hyperresponsiveness, and airway inflammation. These components are all potential targets for therapy (NHLBI and WHO, 1995). Airway remodeling may occur even before symptoms become evident, and the prevention of such remodeling would be ideal although how this can be effectively achieved remains to be determined (Bergeron and Boulet, 2006). The strongest identifiable predisposing factor for developing asthma is atopy, the genetic predisposition for the development of an IgE-mediated response to common aeroallergens (NHLBI, 2007). As there is no known cure for asthma, the goal of treatment is to control symptoms, prevent exacerbations, maintain good pulmonary function, and prevent mortality in order to live a normal, active life (Boulet, 2008).

Asthma management recommendations according to the National Heart, Lung, and Blood Institute (NHLBI) that partners with the Global Initiative on Asthma (GINA) were most recently updated in 2007. These guidelines provide a vehicle to help primary care physicians utilize a structured, stepwise approach in asthma care. Management. This involves determining asthma severity and degree of asthma control as measured by objective tests (spirometry and peak expiratory flow rate) and patient history. Asthma severity is assessed by quantifying impairment and risk of exacerbations as shown in Fig. 14.1. Asthma severity classifications consist of intermittent, mild persistent, moderate persistent, and severe persistent disease. Components of severity include reported symptoms over the past 4 weeks, lung function as measured by spirometry, and number of exacerbations requiring oral glucocorticoids per year. Severity is assigned according to the most severe category of impairment (Barnes, 1998). Management also entails control of environmental factors, treatment of comorbid conditions, and finally pharmacologic intervention.

Components of Severity		Classification of Asthma Severity ≥12 years of age			
			Persistent		
		Intermittent	Mild	Moderate	Severe
Impairment Normal FEV_1/FVC: 8–19 yr 85% 20–39 yr 80% 40–59 yr 75% 60–80 yr 70%	Symptoms	≤2 days/week	>2 days/week but not daily	Daily	Throughout the day
	Nighttime awakenings	≤2x/month	3–4x/month	>1x/week but not nightly	Often 7x/week
	Short-acting $beta_2$-agonist use for symptom control (not prevention of EIB)	≤2 days/week	>2 days/week but not daily, and not more than 1x on any day	Daily	Several times per day
	Interference with normal activity	None	Minor limitation	Some limitation	Extremely limited
	Lung function	• Normal FEV_1 between exacerbations • FEV_1 >80% predicted • FEV_1/FVC normal	• FEV_1 >80% predicted • FEV_1/FVC normal	• FEV_1 >60% but <80% predicted • FEV_1/FVC reduced 5%	• FEV_1 <60% predicted • FEV_1/FVC reduced >5%
Risk	Exacerbations requiring oral systemic corticosteroids	0–1/year (see note)	≥2/year (see note) →		
		Consider severity and interval since last exacerbation. Frequency and severity may fluctuate over time for patients in any severity category. Relative annual risk of exacerbations may be related to FEV_1.			
Recommended Step for Initiating Treatment		Step 1	Step 2	Step 3	Step 4 or 5
				and consider short course of oral systemic corticosteroids	
		In 2–6 weeks, evaluate level of asthma control that is achieved and adjust therapy accordingly.			

Fig. 14.1. Assessing asthma severity. EIB = exercise-induced bronchospasm; FEV1 = forced expiratory volume in 1 second; FVC = forced vital capacity; ICU = intensive care unit. (Adapted from NHLBI, 2007) (With permission of NHLBI)

Medications used in the treatment of asthma can be classified into either quick-relief or long-term control agents. Recommendations given by the United States National Asthma Education and Prevention Program (NAEPP) for use of specific quick-relief and long-term control medications are based on asthma severity. Specific medications are recommended using the stepwise approach for pharmacologic treatment of patients older than 12 years of age. These medication regimens are outlined in Fig. 14.2. There are also specific recommendations for the pediatric population as well. In each step of treatment, there is a recommended preferred treatment as well as an alternative treatment. The recommended step for initiating treatment of patients with mild intermittent asthma is Step 1. Those with mild intermittent asthma are best treated with a short-acting $beta_2$-agonist taken on an as needed basis for relief of symptoms. In those with mild persistent asthma (Step 2), the recommendation by the NAEPP is to begin a daily controller medication of a low-dose inhaled corticosteroid. In moderate persistent asthma (Step 3), the recommended medications include either a low-dose corticosteroid with the use of a long-acting beta agonist, or the use of a medium-dose corticosteroid. In severe persistent asthma (Step 4 or 5), a medium-dose or high-dose steroid in combination with a long-acting beta agonist is recommended. If necessary, Step 6 treatment can be initiated with the use of a high-dose corticosteroid in combination with a long-acting beta agonist with the addition of an oral corticosteroid. Assessment of how well asthma is controlled should occur every 3 months. Depending on the level of asthma control, treatment regimens can be either be stepped up if needed, or stepped down if possible (NHLBI, 2007).

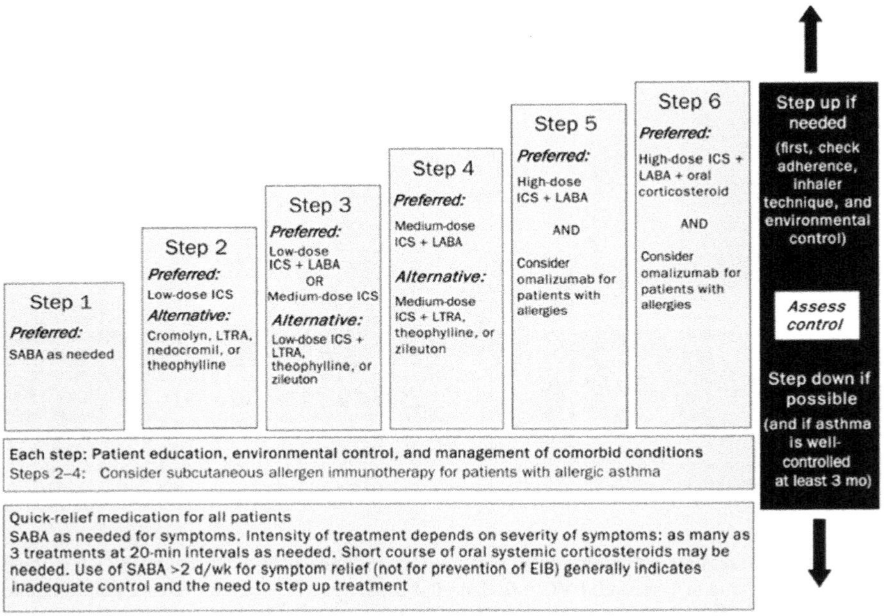

Fig. 14.2. Stepwise approach for managing asthma in patients aged 12 years or older (NHLBI, 2007). EIB = exercise-induced bronchospasm; ICS = inhaled corticosteroid; LABA = long-acting β-agonist; LTRA = leukotriene receptor antagonist; SABA = short-acting β-agonist (With permission of NHLBI)

Quick- relief medications include short-acting selective beta$_2$-agonists (SABAs), anticholinergics, and systemic corticosteroids. SABAs are bronchodilators that relax bronchial smooth muscle; they include albuterol, levalbuterol, and pirbuterol. SABAs are administered via inhalation through metered-dose inhalers (MDIs), hydrofluoralkanes (HFAs) or in solutions for nebulization. Albuterol can also be administered via an oral route, but this is not preferred. The inhaled route is recommended over oral administration given the faster onset of action, greater potency, and reduced side effects. Common side effects reported with SABAs include tachycardia and jitteriness. Levalbuterol is a purified preparation of the R isomer of albuterol which is believed to have fewer side effects. SABAs have a rapid onset of action (5 min) and an intermediate duration of effectiveness (4–6 h). They should be used on a needed basis. They can also be administered 10–15 min prior to exposure to known triggers such as exercise in order to prevent the development of symptoms. Anticholinergics inhibit muscarinic cholinergic receptors and reduce intrinsic vagal tone of the respiratory airway. Anticholinergics such as ipratropium bromide may provide an additive benefit to SABAs in moderate-to-severe asthma exacerbations. Ipratropium bromide's peak effect occurs at the point of 30 min, and lasts for 4–6 h. Ipratropium bromide may also be used as a single agent in cases in which an SABA is not tolerated. However, the bronchodilator effects of ipratropium bromide are not as pronounced as that with SABAs. Systemic corticosteroids can be used in combination with SABAs in moderate and severe exacerbations of asthma to improve symptoms and prevent further exacerbations (Boulet, 2008; NHLBI, 2007).

Long-term control medications commonly used in asthma consist of inhaled corticosteroids, long-acting $beta_2$ agonists (LABAs), leukotriene modifiers, mast cell stabilizers, and methylxanthines. Inhaled corticosteroids (ICSs) are the most effective long-term therapy available for the treatment of persistent asthma (Boulet, 2008; NHLBI, 2007). ICSs are potent broad anti-inflammatory agents. They suppress the release of inflammatory mediators, decrease the production of cytokines, and inhibit the recruitment of eosinophils (NHLBI, 2007). The use of ICSs has been shown to reduce airway hyperresponsiveness, prevent asthma exacerbations, reduce asthma-related mortality, and improve quality of life (Barnes, 1998). Dosages of ICSs are classified into low-dose, medium-dose, and high-dose groups. ICSs include beclomethasone dipropionate, budesonide, ciclesonide, flunisolide, fluticasone, mometasone furoate, and triamcinolone acetonide. At recommended dosages, there are minimal side effects associated with the use of ICSs. Most of the potential side effects are local including oropharyngeal candidiasis and hoarseness of voice. These local effects may be diminished with rinsing the mouth after administration, or using a spacer device with the MDI (Boulet, 2008). Studies on the potential of ICSs to interfere with linear growth in children suggest that there may be a small, possibly reversible effect on decreasing growth velocity. However, the small risk of adverse events appears to be outweighed by their positive benefits (CAMP Research Group 2000; Guilbert *et al.*, 2006; Leone *et al.*, 2003).

Long-acting beta agonists such as salmeterol and formoterol have a duration of action longer than 12 h. Formoterol has a faster onset of action (3–15 min) than salmeterol (30–48 min). LABAs are used in conjunction with inhaled corticosteroids for treatment of moderate or severe persistent asthma. LABAs are not recommended for use as a monotherapy (Boulet, 2008). The available LABA/ICS combination medications are salmeterol/fluticasone, formoterol/mometasone and formoterol/ budesonide. The addition of the LABA in these medication combinations likely improves asthma control by enhancing the anti-inflammatory effect of the corticosteroid (Eickelberg *et al.*, 1999). These combination medications are recommended for treatment of asthma cases in which a low dose or medium dose ICS is not sufficient to control symptoms (NHLBI, 2007). However, recent analyses have raised questions about their safety in long-term use.

Leukotriene modifiers used in the treatment of asthma include leukotriene receptor antagonists and a 5-lipoxygenase inhibitor. Montelukast and zafirlukast are leukotriene receptor antagonists (LTRAs) which block the action of cysteinyl leukotrienes (cysLTs) at the cysLT1 receptor. They work to inhibit bronchoconstriction and decrease airway inflammation. LTRAs are generally well-tolerated and side effects are uncommon. Of note, however, there have been a few reported cases of mood changes associated with the use of montelukast. LTRAs are considered as alternative therapy in mild or moderate persistent asthma. Zileuton is a 5-lipoxygenase pathway inhibitor which inhibits the formation of leukotrienes (LTs) including LTB4 and the cysLTs (LTC4, LTD4, LTE4) (NHLBI, 2007; Berger *et al.*, 2007). In a number of clinical trials, zileuton has been shown to improve airway function and inflammation, as well as quality of life in asthmatics. It is recommended as an alternative therapy in moderate or severe persistent asthma. Given the possibility of liver toxicity with the use of zileuton, liver enzymes must be monitored every 3 months. Although leukotriene modifiers have been shown to reduce the frequency of asthma exacerbations, they are not as effective as ICSs (Devillier *et al.*, 1999; Ducharme, 2003). The combination use of an LTRA and an ICS may help to improve asthma control (Vaquerizo, 2003).

Mast cell stabilizers include cromolyn sodium and nedocromil. They can be administered via MDI, HFA or a nebulizer. They are recommended as an alternative treatment for mild persistent asthma. As stabilizers of airway mast cells, they work to decrease inflammation and prevent early and late asthmatic responses to inhaled allergens and irritants. Prophylactic administration can inhibit acute bronchocon- striction if administered 10–15 min before known exposure to an asthma trigger including exercise (Kemp, 2009). Mast cell stabilizers can be used as a long-term controller therapy as well, but are not as effective as ICSs (Kemp, 2009; CAMP Research Group, 2000; Guevara, 2006). Their major advantage is that they have minimal side effects. Local irritation of the throat or cough may occur (Kemp, 2009).

Theophylline is a methylxanthine which was previously a mainstay long-term controller medication for asthma. It is no longer widely used because of its lower efficacy as compared to newer medications, and its risks for toxicity (Boushey *et al.*, 2005). However, it has still been shown to be potentially useful in patients whose asthma is not well-controlled on an ICS. The addition of theophylline to an ICS may improve lung function and symptom control (Peters *et al.*, 2009; Rivington *et al.*, 1995; Evans *et al.*, 1997). It works as a mild to moderate bronchodilator and has anti-inflammatory as well as immunomodulatory effects. It is recommended as an alternative treatment for mild, moderate, or severe persistent asthma (Hendeles *et al.*, 2009; Kidney *et al.*, 1995). It has an advantage of being inexpensive, and differs from many controller asthma medications in that it is an oral or intravenous formulation. The difficulty in the use of theophylline lies in its narrow therapeutic index and pharmacokinetic variability among individuals. Initiating and titrating oral therapy requires frequent measurements to obtain a satisfactory peak concentration. It is metabolized predominantly in the liver by the cytochrome P450 system. Potential side effects associated with theophylline toxicity include seizures and tachyarrythmias. Careful monitoring is required to achieve maximal efficacy and safety (Boushey *et al.*, 2005).

In patients with severe persistent asthma, the use of the anti-IgE treatment omalizumab may be considered. Omalizumab is a subcutaneously injected humanized monoclonal antibody that prevents allergen-induced activation of mast cells by blocking the binding of IgE to its receptors on mast cells as well as other inflammatory cells (Barnes, 2000). Omalizumab is administered every 2–4 weeks and dosed depending on body weight and the level of circulating IgE (Boushey *et al.*, 2005; Wenzel, 2009). Its use has been shown to decrease the need for oral and inhaled corticosteroids. Studies have also shown that the use of omalizumab reduces emergency room visits and hospitalizations in patients with asthma (Milgrom *et al.*, 1999; Soler *et al.*, 2001; Corren *et al.*, 2003). Its use may be limited by the high cost of treatment, variability in individual response, and the requirement for subcutaneous injections (Wenzel, 2009). Omalizumab is generally well-tolerated, but there have been reports of anaphylaxis following injection. There is also a risk of injection site reactions, serum sickness, and urticarial rash (Barnes, 2009).

The goal in the treatment of asthma is to control symptoms and prevent recurrent exacerbations of asthma in hopes of maintaining near normal pulmonary function and normal activity levels. Given the nature of asthma as a chronic disease, the risks versus benefits of treatment must be periodically assessed. Treatment decisions for adjusting pharmacotherapy are based on assessing the level of asthma control. Factors to consider before adjusting pharmacotherapy include unaddressed adverse environmental exposures, poor compliance and comorbidities. Overall, effective asthma management should encompass a preventative approach with regularly scheduled visits, pulmonary function monitoring, ongoing education, and frequent assessment of efficacy of pharmacotherapy (NHLBI, 2007).

14.5 Conclusion

Pharmacotherapy plays an essential role in the management of common allergic diseases including allergic rhinitis, allergic conjunctivitis and allergic asthma. The use of medications in conjunction with allergen avoidance measures can help to reduce symptoms and greatly enhance one's quality of life. Effective long-term management involves a delicate balance of controlling symptoms with the risks of side effects. Pharmacotherapy recommendations must be tailored to meet each specific individual's needs. Immunotherapy which modifies the immune response to allergies can be used when standard environmental control precautions and pharmacotherapy are inadequate. New directions in immunotherapy are being actively researched. Novel candidates include the use of T-cell reactive peptides, plasmid vaccines, and cytokine receptor modulation. These forms of immunotherapy show potential as valuable techniques on the horizon.

References

Adkinson, N.F., Busse, W., Bochner, B., *et al.* (2008). Middleton's Allergy: Principles and Practice (7th ed.). Mosby: Elsevier, pp. 1575-1583.

Bachert, C., van Cauwenberge, P. (2003). The WHO ARIA (allergic rhinitis and its impact on asthma) initiative. Chem Immunol Allergy, 82, 119-126.

Barnes, P.J. (1998). Efficacy of inhaled corticosteroids in asthma. J Allergy Clin Immunol, 102, 531-538.

Barnes, P.J. (2000). Anti-IgE therapy in asthma: Rationale and therapeutic potential. Int Arch Allergy Immunol, 123, 196-204.

Barnes, P.J. (2009). Anti-IgE therapy. Up to date. http://www.uptodate.com/patients/content/.

Barney, N.P., Graziano, F.M., Cook, E.B., *et al.* (2008). Treatment of allergic conjunctivitis. In: Adkinson: Middleton's Allergy: Principles and Practice (7th ed.). Mosby: Elsevier, p.64.

Berger, W., Hampel Jr., F., Bernstein, J., *et al.* (2006). Impact of azelastine nasal spray on symptoms and quality of life compared with cetirizine oral tablets in patients with seasonal allergic rhinitis. Ann Allergy Asthma Immunol, 97(3), 375-381.

Berger, W., De Chandt, M.T., Cairns, C.B. (2007). Zileuton: Clinical implications of 5-Lipoxygenase inhibition in severe airway disease. Int J Clin Pract, 61(4), 663-676.

Bergeron, C., Boulet, L.P. (2006). Structural changes in airway diseases: Characteristics, mechanisms, consequences, and pharmacologic modulation. Chest, 129, 1068-1087.

Berkowitz, R.B., McCafferty, F., Lutz, C., *et al.* (2006). Onset of action of fexofenadine hydrochloride 60 mg/pseudoephedrine hydrochloride 120 mg in subjects aged 12 years with moderate to severe seasonal allergic rhinitis: A pooled analysis of two single-dose, randomized, double-blind, placeboControlled allergen exposure unit studies. Clin Therapy, 28(10), 1658-1669.

Bielory, L., Friedlaender, M.H. (2008). Allergic conjunctivitis. Immunology and Allergy Clinics of North America, 28(1), 43-58.

Blaiss, M.S. (2007). Safety considerations of intranasal corticosteroids for the treatment of allergic rhinitis. Allergy Asthma Proc, 28(2), 145-152.

Borish, L.J. (2003). Allergic rhinitis: Systemic inflammation and implications for management. J Allergy Clin Immunol, 112(6), 1021-1031.

Boulet, L.P. (2008). Approach to adults with asthma. In: Adkinson: Middleton's Allergy: Principles and Practice (7th ed.). Mosby: Elsevier, p. 76.

Boushey Jr., H.A., Corry, D.B., Fahy, J.V., *et al.* (2005). In: Mason: Murray & Nadel's Textbook of Respiratory Medicine (4th ed.). Philadelphia: Elsevier, p. 37.

Bousquet, J., Khaltaev, N. (2001). ARIA Workshop Group, World Health Organization: Allergic rhinitis and its impact on asthma. J Allergy Clin Immunol, 108(Suppl 5), S147-S334.

Bousquet, J., Clark, T. J. H., Hurd, S. *et al.* (2007). GINA guidelines on asthma and beyond. Allergy, 62, 102-112.

Bronsky, E., Boggs, P., Findlay, S., *et al.* (1995). Comparative efficacy and safety of a once-daily loratadine-pseudoephedrine combination versus its components alone and placebo in the management of seasonal allergic rhinitis. J Allergy Clin Immunol, 96(2), 139-147.

Childhood Asthma Management Program (CAMP) Research Group (2000). Long-term effects of budesonide or nedocromil in children with asthma. N Engl J Med., 343(15), 1054-1063.

Corren, J. , Casale, T. , Deniz, Y. , *et al.* (2003). Omalizumab, a recombinant humanized anti-IgE antibody, reduces asthma-related emergency room visits and hospitalizations in patients with allergic asthma. J Allergy Clin Immunol, 111, 87-90.

Dana, M.R. (2009). Allergic conjunctivitis. www.uptodate.com.

Deschenes, J., Discepola, M., Abelson, M.B. (1999). Comparative evaluation of olopatadine ophthalmic solution (0.1%) using the provocative antigen challenge model. Acta Ophthalmol Scand, 77, 47-52.

deShazo, R.D., Kemp, S.F., Corren, J., Feldweg, A.M. (2010). Pharmacotherapy of allergic rhinitis. Up to Date. 2010 (18.1) February (cited 2010 February 10). http://www.uptodate.com/.

Devillier, P., Baccard, N., Advenier, C. (1999). Leukotrienes, leukotriene receptor antagonists and leukotriene synthesis inhibitors in asthma: an update. Part I. Synthesis, receptors and role of leukotrienes in asthma. Pharmacol Res, 40, 3-13.

Doshnik, P.C., Ehlers, W.H. (1994). Clinical immunologic diseases: Ocular allergy. In: The cornea: Scientific foundation and clinical practice (3rd ed.). Boston: Little Brown.

Ducharme, F.M. (2003). Inhaled glucocorticoids versus leukotriene receptor antagonists as single agent asthma treatment: Systematic review of current evidence. BMJ, 326, 621.

Dykewicz, M.S., Fineman, S., Skoner, D.P., *et al.* (1998). Diagnosis and management of rhinitis: Complete guidelines of the joint task force on practice parameters in allergy, asthma and immunology. American Academy of Allergy, Asthma, and Immunology. Ann Allergy Asthma Immunol, 81(5 Pt 2), 478-518.

Eickelberg, O., Pansky, A., Mussmann, R., *et al.* (1999). Transforming growth factor-beta1 induces interleukin-6 expression via activating protein-1 consisting of JunD homodimers in primary human lung fibroblasts. J Biol Chem, 274, 12933-12938.

Evans, D.J., Taylor, D.A., Zetterstrom, O., *et al.* (1997). A comparison of low-dose inhaled budesonide plus theophylline and high-dose inhaled budesonide for moderate asthma. N Engl J Med, 337(20), 1412-1418.

Guevara, J., Ducharme, F., Keren, R., *et al.* (2006). Inhaled corticosteroids versus sodium cromoglycate in children and adults with asthma. This is a reprint of a Cochrane review, prepared and maintained by The Cochrane Collaboration and published in The Cochrane Library 2009, Issue 3 http://www.thecochranelibrary.com.

Guilbert, T.W., Martinez, F.D., Zeiger R.S., *et al.* (2006). Long-term inhaled corticosteroids in preschool children at high risk for asthma. N Engl J Med, 354(19), 1985-1997.

Hendeles, L., Weinberger, M. (2009). Theophylline use in asthma. http://www.uptodate.com/ contents/theophylline-use-in-asthma.

Juniper, E.F., Stahl, E., Doty, R.L., *et al.* (2005). Clinical outcomes and adverse effect monitoring in allergic rhinitis. J Allergy Clin Immunol, 115(3 Pt 2), S390-413.

Kemp, J.P. (2009). The use of chromones (cromoglycates) in the treatment of asthma. http://www.uptodate.com/contents/the-use-of-chromones-cromoglycates-in-the-treatment-of-asthma.

Kidney, J., Dominguez, M., Taylor, P.M., *et al.* (1995). Immunomodulation by theophylline in asthma. Demonstration by withdrawal of therapy. Am J Respir Crit Care Med, 151(6), 1907-1914.

LaForce, C. (1999). Use of nasal steroids in managing allergic rhinitis. J Allergy Clin Immunol, 103(3 Pt 2), S388-S394.

Leone, F.T., Fish, J.E., Szefler, S.J., West, S.L. (2003). Systematic review of the evidence regarding potential complications of inhaled corticosteroid use in asthma: collaboration of American College of Chest Physicians, American Academy of Allergy, Asthma, and Immunology, and American College of Allergy, Asthma, and Immunology. Chest, 124(6), 2329-2340.

Manzouri, B., Flynn, T.H., Larkin, F., *et al.* (2006). Pharma- cotherapy allergic eye disease. Expert Opinion on Pharmacotherapy, 7, 1191-1200.

Milgrom, H., Fick Jr., R.B., Su, J.Q., *et al.* (1999). Treatment of allergic asthma with monoclonal anti-IgE antibody. N Engl J Med, 341, 1966-1973.

National Asthma Education and Prevention Program Science Base Committee and Expert Panel on the Management of Asthma (2002). Appendix A-2. Usual dosages for asthma medications. J Allergy Clin Immunol, 110(5, Suppl 1), S212-S219, doi: 10.1016/S0091-6749(02)80013-6.

National Heart, Lung, and Blood Institute (NHLBI)/World Health Organizatiom (WHO) (1995). Global Initiative for Asthma Management and Prevention. Workshop Report, US Department of Health and Human Services. National Institutes of Health, Bethesda, MD. Pub #95-3659.

National Heart, Lung, and Blood Institute (NHLBI) (2007). National Asthma Education and Prevention Program: Expert panel report III: Guidelines for the diagnosis and management of asthma. Bethesda, MD. , 74, 305.

Nayak, A., Langdon, R.B. (2007). Montelukast in the treatment of allergic rhinitis: An evidence-based review. Drugs, 67(6), 887-901.

Ousler, G.W., Workman, D.A., Torkildsen, G.L. (2007). An open-label, investigator-masked, crossover study of the ocular drying effects of two antihistamines, topical epinastine and systemic loratadine, in adult volunteers with seasonal allergic conjunctivitis. Clin Ther, 29(4), 611-616.

Peters, S., McCallister, J.W., Pascual, R. (2009). Treatment of moderate persistent asthma in adolescents and adults. http://www.uptodate.com/contents/treatment-of-moderate-persistent-asthma-in-adolescents-and-adults.

Philip, G., Nayak, A.S., Berger, W.E., *et al.* (2004). The effect of montelukast on rhinitis symptoms in patients with asthma and seasonal allergic rhinitis. Curr Med Res Opin, 20, 1549-1558.

Pullerits, T., Praks, L., Ristioja, V., *et al.* (2002). Comparison of a nasal glucocorticoid, antileukotriene, and a combination of antileukotriene and antihistamine in the treatment of seasonal allergic rhinitis. J Allergy Clin Immunol, 109(6), 949-955.

Ratner, P.H., Howland, W.C., Arastu, R., *et al.* (2003). Fluticasone propionate aqueous nasal spray provided significantly greater improvement in daytime and nighttime nasal symptoms of seasonal allergic rhinitis compared with montelukast. Ann Allergy Asthma Immunol, 90(5), 536-542.

Rivington, R.N., Boulet, L.P., Cote, J., *et al.* (1995). Efficacy of Uniphyl, salbutamol, and their combination in asthmatic patients on high-dose inhaled steroids. Am J Respir Crit Care Med, 151(2 Pt 1), 325-332.

Schultz, B.L. (2006). Pharmacology of ocular allergy. Cur Opinion Allergy Clin Immunol, 6(5), 383-389.

Soler, M., Matz, J., Townley, R., *et al.* (2001).The anti-IgE antibody omalizumab reduces exacerbations and steroid requirement in allergic asthmatics. Eur Respir J, 18, 254-261.

van Bavel, J., Findlay, S.R., Hampel F.C., Jr., *et al.* (1994). Intranasal fluticasone propionate is more effective than terfenadine tablets for seasonal allergic rhinitis. Arch Intern Med, 154(23), 2699-2704.

Vaquerizo, M.J., Casan, P., Castillo, J., *et al.* (2003). Effect of montelukast added to inhaled budesonide on control of mild to moderate asthma. Thorax, 58(3), 204-210.

Verster, J.C., Volkerts, E.R. (2004). Antihistamines and driving ability: Evidence from on-the-road driving studies during normal traffic. Ann Allergy Asthma Immunol, 92(3), 294-303; quiz 303-305, 355.

Welsh, P.W., Stricker, W.E., Chu, C.P., *et al.* (1987). Efficacy of beclomethasone nasal solution, flunisolide, and cromolyn in relieving symptoms of ragweed allergy. Mayo Clin Proc, 62(2), 125-134.

Wenzel, S. (2009). Treatment of severe asthma in adolescents and adults. http://www.uptodate.com/contents/treatment-of-severe-asthma-in-adolescents-and-adults.

Wilson, A.M., O'Byrne, P.M. (2004). Parameswaran KLeukotriene receptor antagonists for allergic rhinitis: A systematic review and meta-analysis. Am J Med, 116(5), 338-344.

15

Immunotherapy in Allergic Skin Disease

Hao Cheng

Department of Dermatology, Sir Run Run Shaw Hospital, School of Medicine, Zhejiang University, Hangzhou, 310016, China
E-mail: hz_chenghao@126.com

Abstract: Allergic skin diseases, such as atopic dermatitis (AD) and urticaria are common diseases with increasing incidence in recent decades that are responsible for a serious burden on patients and their families, highlighting the need for devising effective therapeutic and preventive strategies. During the past few years, there have been significant advances in our understanding of the cellular and immunologic mechanisms underlying allergic skin diseases. Allergen-specific immunotherapy (SIT) that involves the administration of increasing concentrations of crude allergen extracts over a period of time, in an attempt to switch the indi- vidual's allergic response to that of a non-allergic individual, is widely recognized as an effective treatment in asthma and allergic rhinitis, while its efficacy in most common allergic skin diseases, AD has been a matter of debate for a long time. Some novel immunotherapeutic approaches including monoclonal antibodies, CpG DNA, peptides, and recombinant allergen vaccines has significantly increased our treatment options. In this review, we provide some of the recent advances in our understanding of the immunotherapy in allergic skin disease AD.

15.1 Introduction

Allergic diseases range from potentially life-threatening allergic reactions (e.g., anaphylaxis or severe asthma) to chronic allergic diseases associated with significantly reduced quality of life (e.g., eczema, urticaria or allergic rhinitis). The prevalence of allergic skin diseases has been rising over the last few decades, resulting in morbidity and mortality in all ages of people. Atopic diseases include AD, allergic rhinoconjunctivitis, and allergic asthma. A subgroup of atopic patients usually goes through various atopic stages during their lifetime, starting with AD in early childhood, followed by the development of allergic bronchial asthma and allergic rhinoconjunctivitis later(Leung *et al.*, 2004). AD is a highly pruritic, chronic, and relapsing inflammatory skin disease, with a strong genetic predisposition, complex interactions of deficient innate and adaptive immune responses, and skin barrier dysfunction (Maintz, *et al.*, 2007; Boguniewicz, *et al.*, 2006) . Sensitizations to allergens, without any doubt, play a central role as one of the most common trigger factors leading of AD. Since we know that complete elimination of allergen exposure can be difficult or even impossible, the importance of allergen avoidance or elimination cannot be over-emphasized for treatment. There have been multiple modalities offering the opportunity for effective treatment in allergic skin disease, including pharmacotherapy (antihistamines, leukotriene inhibitors, topical and/or oral corticosteroids), and immunotherapy. Allergen-specific Immunotherapy (SIT), as a conventional attractive treatment option, represents a long-lasting efficacy and preventive effect in allergic rhinitis, conjunctivitis, stinging insect and asthma (Ozdemir, 2009). Currently, novel therapeutic immunomodulating approaches, such as selective targeting of allergen-specific T-cells (especially regulatory T-cells), recombinant allergens lacking IgE reactivity, small T-cell epitope-based peptides, monoclonal anti-IgE antibodies, anti-IL-5 antibodies, CpG DNA, and adjuvants (either co-administered or incorporated into a recombinant allergen vaccine to target tolerogenic dendritic cells) are being trialed clinically in allergic diseases with evidence of efficacy.

15.2 Definition of Allergen-Specific Immunotherapy

Allergen-specific immunotherapy, which is a medical procedure involving the administration of known allergen extracts, prevents symptoms in order to provide long term relief and improvement in quality of life during subsequent natural allergen exposure.

Immunotherapy, as a kind of desensitization or hyposensitization treatment, may induce an immune tolerance to allergens, followed by reducing the severity of allergic disorders or eliminating hypersensitivity. Conventional allergen-specific immunotherapy, i.e., subcutaneous allergen-specific immunotherapy (SCIT), involves injection of increasing doses of allergen extract followed by repeated injection of allergen at a maximum tolerated dose over 3–5 years in the maintenance phase. The diseases which can be treated by SCIT are mainly allergic rhinitis, allergic asthma, allergic conjunctivitis, and insect sting hypersensitivity (Abramson *et al.*, 1995; Calderon *et al.*, 2007; Álvarez-Cuesta *et al.*, 2006). The systematic studies on the effectiveness of SIT in patients with skin allergic disease, i.e., AD, are relatively few, with less data on the immunologic changes. The immunotherapy for food allergies is thought to be different from that of other allergies, though there is a fairly long history of oral immunotherapy for food allergies.

15.3 Historical Data on Allergen-Specific Immunotherapy

Immunotherapy started in the early 1900s. The concept of vaccinating patient to prevent allergies was initially described in 1911 by an English physician and immunologist Noon at St. Mary's Hospital in London (Noon, 1911). He injected subcutaneously aqueous extract of grass pollen into hay fever patients to prevent symptoms during the grass pollen season.

In 1903, Dunbar immunized patients' nasal mucosa for those who had hay fever with animal derived (horse and goose) grass pollen antisera (Dunbar, 1913), and Noon and Cantab introduced *in vivo* pollen doses and quantitation of individual sensitivity. In 1914, Freeman and Koessler reported that immunotherapy could produce long-lasting clinical effects (Gordon, 1995). In 1915, Robert A Cooke, a physician scientist in New York, developed the basic methods of allergen standardization and introduced immunotherapy into the USA with a treatment report of pollen immunization for 114 patients who had hay fever and asthma (Finegold, 2001). Since then, subcutaneous allergen-specific immunotherapy has developed as a routine therapy and the best practice protocols have been defined. Conventional SCIT is administered by subcutaneous injection of unfractionated allergen extract in either alum-precipitated or aqueous form. To minimize adverse reaction, allergen extracts are standardized as "Bioequivalent Allergy Units" or by quantitation of the dominant allergen.

15.4 The Route of Allergen Immunotherapy

Conventional allergen specific immunotherapy has been given by subcutaneous injection for nearly one century and has proven its effectiveness already. Subcutaneous immunotherapy (SCIT) would be uncomfortable for patients. So the inconvenience and adverse events of injection restrict the administration of SCIT. Therefore in an attempt to improve safety and enhance patients' acceptance, alternative methods for local administration of allergen immunotherapy have been developed since the mid-1980s, of which the most promising is the sublingual absorption (sublingual-swallow, or sublingual immunotherapy, SLIT) (Scadding *et al.*, 1986; Canonica *et al.*, 2003). Others are nasal insufflation (nasal immunotherapy, LNIT) (Dunbar, 1913) and bronchial inhalation. The intranasal route has also been shown to be effective, but less attractive because of inconvenience has not been widely accepted.

15.4.1 Subcutaneous Immunotherapy (SCIT)

In this seetion, mechanisms of SCIT, the clinical efficacy and side effects of SCIT will be introduced.

15.4.1.1 Mechanisms of SCIT

Allergen immunotherapy has been a valuable tool in treating allergic disorders for more than 90 years, whereas the exact mechanisms underlying SLIT are less well-understood. The discovery of IgE and the progress in T- and B-cell biology have directed the approaches to pathomechanisms of allergic reactions during the last 30 years. Successful SCIT reduces symptoms of allergic disease, characterized as decreasing responses to local allergen challenge in the skin, the nasal mucosa, and in the asthmatics airways. Local tissue changes demonstrated reduced recruitment and activation of effector cells including mast cells, eosinophils, and basophils in the allergic respiratory mucosa of the nose and the bronchi, and reduced release of inflammatory mediators such as eosinophil cationic protein and platelet activating factors. The switch to a Th1 response was first proposed by Cooke and his colleagues in 1935, which caused the production of IgG blocking antibodies (Cooke, 1935). To understand the underlying mechanisms of immunotherapy, multiple studies have been conducted to evaluate the effects on both T-cell and B-cell immune responses, cytokines production, and mucosal immunity as well. (Guerra *et al.*, 2001; Akdis *et al.*, 2007; Jutel *et al.*, 2008; James *et al.*, 2008).

SCIT effects on B-cells include changes in antibody production after immunotherapy: (1) Significant induction of allergen-specific IgG antibodies (IgG1, IgG4, and IgG2), especially protective IgG4 subclass which has been described as IgE blocking antibody by inhibiting the release of inflammatory mediators from mast cells and basophils; (2) Decreases of allergen-specific IgE antibodies; (3) Increases of IgA and IgM antigen-specific B lymphocytes (Gleich, *et al.*, 1982). Alteration of allergen specific T-cell reactivity correlates well with clinical improvement, which includes decreased antigen-specific T-cell proliferation (Akdis *et al.*, 1996; O'Brien, 1997), and changes of secreted cytokines profile such as decreased production of IL-4 and increased production of IL-10 and TGF-β. Increased levels of IL-10 will play a role in inducing T-cell tolerance, characterized as a shift in allergen-specific IgE to allergen-specific IgG4 and in dampening down both Th2 and Th1 cytokine-mediated inflammation at sites of allergen exposure. Both T-cells and DC are sources of IL-10, which can be induced by regulatory

T-cells with important cross-regulatory potential (Akdis *et al.*, 2004). In support of these mechanisms, increased IL-10 is detected earlier during immunotherapy for allergies. Using higher dose stimulation *in vitro*, decreased T-cell proliferative response to allergen with increased apoptosis of IL-4+ T-cells can be demonstrated. In this case, T-cells may undergo activation-induced cell death by allergen-specific effector T-cells apoptosis, which would be consistent with increased Treg activity in immunized individuals (Gardner *et al.*, 2004). Evidence is accumulating for a role of Treg cells in inducing tolerance in allergen-specific T-cells in healthy individuals and in allergic subjects following SIT. As we know that Th2 lymphocytes typically produce cytokines interleukin IL-4 and IL-5 which are needed for IgE production and eosinophil survival, while Th1 lymphocytes produce interferon gamma (IFN-) (Jutel *et al.*, 1995). Immunotherapy could induce a shifting or deviation from "allergic" lymphocyte (Th2) immune responses to a "non-allergic" lymphocyte (Th1) response by decreasing in the ratio of IL-4/IL-5 to IFN- levels. In agreement with this mechanism, decreased production of IL-4 and increased expression of mRNA for IFN- and IL-12 has been reported following immunotherapy (Durham *et al.*, 1996). Associated with these effects on T-cell and B cell immune responses, SCIT suppresses the number and activation of effector cells, including mast cells, basophils, and eosinophils in allergic tissues.

15.4.1.2 The Clinical Efficacy

Subcutaneous injection immunotherapy is usually recommended for treating certain severe or potentially life-threatening allergic conditions, and has been shown to be highly effective in seasonal allergic rhinitis, in particular, in patients with seasonal pollinosis due to grass, tree and weed pollens.

The long-term clinical efficacy of immunotherapy in significantly reducing symptoms of the above allergic disorders demonstrated the advantages over pharmaco-therapy. A Cochrane meta-analysis of SCIT in allergic rhinitis (51 double-blind placebo-controlled studies of 2871 subjects) showed a mean reduction in symptoms of 73% and a mean reduction in medication use of 57% (Abramson *et al.*, 2003). After 3–5 years' SCIT, patients have shown long-term remission of allergic rhinitis symptoms for at least 3–5 years following discontinuation of SCIT. A 3-year course of SIT with standardized allergen extracts has shown long-term clinical effects and the potential of preventing development of asthma in children with allergic rhinoconjunctivitis up to 7 years after treatment (Jacobsen *et al.*, 2007). Another Meta-analysis of SCIT in asthma (75 randomized controlled studies of 3,188 subjects) showed that SCIT also induces a significant reduction in asthma symptoms, medication and bronchial hyper-reactivity (Abramson, 2003). Allergen immunotherapy may also prevent progression of rhinitis to asthma in children, by decreasing the onset of new allergen sensitivities (Durham *et al.*, 1999; Des Roches A *et al.*, 1997). In a preventive allergy treatment study, 205 children aged 6–14 years with grass or birch pollen allergy from 6 pediatric allergy centers in Northern Europe were studied. The result showed that the patients who underwent specific immunotherapy had significantly less asthma than the untreated control group. Taken together, allergen immunotherapy indicated the important prophylactic value, in contrast to pharmacotherapy in which symptom relapse occurs immediately following discontinuation (Möller *et al.*, 2002).

Although the benefits of subcutaneous immunotherapy are apparent in both asthma and allergic rhinitis, there is only a few studies that have been conducted on the effectiveness of SCIT for treating allergic skin disease, especially atopic eczema. Double-blind placebo-controlled studies on the efficacy of subcutaneous specific immunotherapy in atopic dermatitis are relatively rare, as is the data on immunological changes induced by SCIT in AD patients. However, the most recent literature of controlled studies on specific immunotherapy in atopic dermatitis provides some promising results. Bussmann identified a total of 23 studies about subcutaneous or sublingual allergen SIT in AD with different allergens based on a systematic search in several databases such as MEDLINE, EMBASE, and CENTRAL (Bussmann *et al.*, 2006). Because of the heterogeneity of these studies, only the results of 5 placebo-controlled studies on subcutaneous allergen SIT were comparable, of which 4 studies reported proportions of patients with an improvement of symptoms. Czarnecka-Operacz *et al.* (2006) in their study found that SIT with house dust mite (HDM) was an efficacious and safe method of treatment for patients with AD and IgE-mediated airborne allergies, concluding that SIT may be a highly promising method of controlling skin inflammation in AD with the potential to prevent the development of AD into respiratory allergy. One multicenter study of 89 adult patients with severe atopic dermatitis administered HDM allergen extract subcutaneously demonstrated that patients received benefits from SIT (Werfel *et al.*, 2006). Bussmann *et al.* (2007) performed an open pilot study to assess clinical changes and objective laboratory parameters and evaluate the benefit of HDM SCIT in 25 AD patients. Subjective and objective SCORAD improved significantly within only 4 weeks of treatment, with an increased IgG4 and tolerogenic cytokine IL-10 level, decreased allergen specific IgE, CCL17 and IL-16 in the sera of the patients during SCIT. In this open-label study, SCIT with HDM extract led to a significant improvement of AD mirrored by a reduction of SCORAD as well as serological and immunological changes.

15.4.1.3 Side Effects of SCIT

Many patients develop a localized swelling at the injection site, which can be treated with oral antihistamines or ice packs. If the swelling is large, reducing the dosage is necessary. More serious reactions (such as wheezing, rash, dizziness or even anaphylaxis) are uncommon. Approximately 0.1% subjects receiving SCIT develop significant systemic reactions, which are associated with a risk of IgE-mediated adverse events, including systemic anaphylaxis (Mellerup *et al.*, 2000; Mauro *et al.*, 2007). Careful selection of patients, establishing a reporting system of dilutions and establishing appropriate dosing is essential for minimizing the risks. To reduce the allergenicity of allergen immunotherapy is very tough work, because allergoids as reduced allergenicity may reduce immunogenicity and clinical effectiveness.

15.4.2 Sublingual Immunotherapy

Sublingual immunotherapy (SLIT), termed as oral immunotherapy, oral desensitization or oral hyposensitization, involves the administration of allergen extracts containing drops or rapid-dissolving tablets under the tongue for 2–3 min, usually followed by swallowing. Because of the ease of administration, this therapy has a longer history of use and currently accounts for 40% of allergy treatment in Europe, where SLIT is applied more commonly than injected immunotherapy. Oral immunotherapy for food allergies started to get a wide range of attention only during the past few years. Unlike injection immunotherapy administrated with a gradual increase in dosage, sublingual immunotherapy is started usually at the full maintenance dosage. The major current drawback of sublingual immunotherapy is the high cost. As allergen doses required for effective treatment are at least 100-fold more than those needed for subcutaneous immunotherapy, the medication costs are at least three times higher than that for SCIT (Cox *et al.*, 2006; Frew, 2008).

15.4.2.1 The Clinical Efficacy

In order to prevent symptoms from perennial allergens such as dust mites, SLIT has been administered either prior to the spring or fall pollen allergy season, or continuously throughout the year to induce oral immune tolerance to inhaled allergens. A meta-analysis of 22 double blind placebo controlled trials of 979 subjects have confirmed the efficacy of this method of therapy for seasonal and perennial rhinitis, in terms of reduction in symptoms and rescue medication (Wilson *et al.*, 2005). Patriarca, as one of the pioneers in this field, reported clinical data repeatedly, and indicated that sublingual-oral-specific desensitization has been used for the treatment of food-allergic patients with a high percentage of success (Patriarca *et al.*, 2009). Comparison of results of meta-analysis studies of SLIT and SCIT in allergic rhinitis suggests that SCIT is clinically more effective but associated with fewer systemic adverse reactions. At present, however, no double-blind studies have demonstrated that SLIT, like SCIT, can prevent the development of sensitization to new allergens, nor long-lasting immunomodulating effects that result in sustained clinical remission after therapy is discontinued (Calamita *et al.*, 2006) . Although recommended as an alternative in international guidelines, further studies of potential long-term benefits, comparative studies and, in particular, more studies in children, in who sublingual-swallow treatment cannot currently be recommended, is a routine practice outside clinical trials.

So far, there is less data regarding efficacy and safety of specific sublingual immunotherapy (SLIT) in patients with AD. Mastrandrea's cohort study of 35 AD patients suggested that sublingual allergen-specific immunotherapy for Atopic Dermatitis treatment is safe and well-tolerated, and may favorably affect the natural course of the disease (Mastrandrea *et al.*, 2000). In an open non-controlled trial, SLIT with HDM extracts in patients with mild to moderate AD was effective in reducing the SCORAD after 1 year of SLIT treatment. Furthermore, SLIT could allow a gradual and relevant reduction of concomitant therapies with topical corticosteroids or immunosuppressants (Cadario *et al.*, 2007). To assess the effect of sublingual immunotherapy (SLIT) in children with atopic dermatitis, a randomized, double-blind, placebo controlled study of 56 children with atopic dermatitis was performed (Pajno *et al.*, 2007). The results suggested that sublingual immunotherapy to dust mite improves mild-moderate atopic dermatitis. Similarly, there was a significant reduction in the use of medications in the active group. This implies that sublingual immunotherapy may represent an additional therapeutic tool for the treatment of extrinsic atopic dermatitis in properly selected children.

15.4.2.2 Immune Response Induced by SLIT

The fact that SLIT induces local mechanisms in oral mucosa and/or regional lymph nodes are important, as we know that oral mucosa is a natural site of immune tolerance (Langerhans cells, Fc_R1, IL-10, IDO [indoleamine 2,3-dioxygenase]). During SLIT, the allergen is captured within the oral mucosa by Langerhans-like dendritic cells expressing high-affinity IgE receptors, producing IL-10 and TGF-b, and up-regulating indoleamine dioxygenase (IDO) (Moingeon *et al.*, 2006). Sublingual immunotherapy is considered to be associated with retention of allergen in sublingual mucosa for several hours, marked early increases in antigen-specific IgE, blunting of seasonal IgE, modest increases in antigen-specific IgG4 and IgE blocking activity, inhibition of eosinophils, reduction of adhesion molecules in target organ, and some evidence of increases in peripheral T-cell IL-10 (O'Hehir *et al.*, 2007).

The effect of SLIT on B-cell immune response in terms of IgG4 response is more limited than that induced by SCIT and is dependent on the dosage and duration of SLIT administration. Recent studies of SLIT in birch pollen allergic subjects, indicated that SLIT has effects on T-cells by inducing Tregs within a month. SLIT induces regulatory T-cell suppression through IL-10. After 1-year's therapy, SLIT induced immune deviation of allergen-specific T-cells characterizing as inhibition of Th2 and induction of Th1 immune response as well (Bohle *et al.*, 2007). Cosmi *et al.* (2006) in their study suggested that allergen-driven enhancement of IL-10- and IFN-γ-producing T-cells precedes and associates with SLIT-induced down-regulation of specific IgE, providing a rationale to explain the clinical benefit of SLIT in allergic patients.

15.4.2.3 Side Effects of Sublingual Immunotherapy

SLIT appears generally safe, with the main local side effects of local (oral) itching, salty or unpleasant taste, and swelling in the mouth, in general, trivial and requiring no treatment and only rarely results in discontinuation of therapy. In most studies, 5%–10% patients feel irritation or mild itching inside the mouth which is self-resolving, and does not frequently cause patients to discontinue SLIT. In rare cases (3%–5%) has the stomach become upset. The risk of potentially dangerous side effects, such as difficult breathing or rashes, is considered to be extremely uncommon. Based on a review of SLIT in 3,984 patients, 14 SLIT-related serious adverse events (mainly asthma exacerbations) were reported. There are three reported cases of anaphylaxis associated with SLIT administration. No fatal adverse events caused by SLIT have been reported in the past 20 years, and those were predominantly in Europe (Gidaro *et al.*, 2005; Kleine-Tebbe *et al.*, 2006).

15.5 New Approaches to Immunotherapy

In an effort to increase efficacy, as well as circumvent negative side effects of immunotherapy, new technology and compartments of the immune system which can be targeted in new SIT regimens are being tried to take the place of conventional immunotherapy. Application of recombinant DNA technology, almost complete repertories of the disease-eliciting allergens have been produced as recombinant molecules and are already in clinical use for the diagnosis and monitoring of allergic diseases (Fig. 15. 1)

15.5.1 Anti-IgE Therapy

Anti-IgE therapy is being explored as a strategy for minimizing IgE medicated side effects of allergen immunotherapy especially during induction. Intravenous injection of monoclonal anti-IgE antibodies is very effective in treating several types of atopy. The first agent is humanized monoclonal anti-IgE antibody Omalizumab, which binds to free and B-cell associated IgE, signaling IgE destruction, followed by free serum IgE decreasing. But the anti-IgE antibody does not seem to bind to IgE already bound to the Fc receptor on basophils and mast cells, as this would stimulate the allergic inflammatory response (Beck *et al.*, 2004).

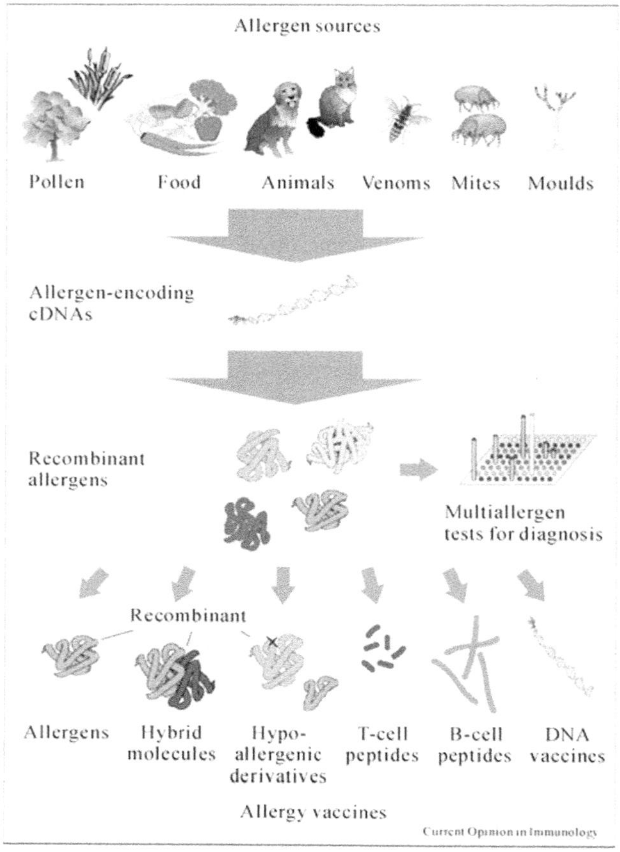

Fig. 15.1. From allergen genes to allergy vaccines (Linhart and Valenta, 2005) (With permission of Elsevier)

There are clinical trials indicating that omalizumab treatment can effectively control recurrent reactions during allergen immunotherapy for patients. Further clinical studies are ongoing to fully validate the role of anti-IgE antibody therapy in SIT. This agent should not be used in treating food allergies. Recently, new indications for omalizumab were suggested, which include AD characterized by elevated levels of IgE. Incorvaia, *et al.* (2008) reported a case with severe asthma and severe AD, both resistant to conventional drug treatment. The patient was treated with omalizumab every two weeks, which induced a rapid improvement of asthma along with a progressive decline of severity of AD. Another study was performed to assess the efficacy of omalizumab in 21 patients with allergic asthma and AD. The results indicated that omalizumab is effective in treating AD in patients with moderate to severe persistent allergic asthma (Sheinkopf *et al.*, 2008).

15.5.2 T-Cell Epitope-Based Peptides Immunotherapy

Peptide-based immunotherapy, i.e., short synthetic linear peptides corresponding to T-cell epitopes of allergens, is a novel attractive strategy for allergic disease (Larche, 2007). T-cell epitopes are the small peptides that result from digestion of the allergen by the antigen presenting cells. The administration of T-cell epitopes would alleviate the symptoms of allergy. However, it would not become sensitized to other components of intact allergen which conventional SIT has shown. The precise mechanism of immunologic tolerance induced by T-cell peptide epitopes remains incompletely defined. T-cell peptide epitopes for targeting allergen-specific T-cells, have reduced capability to cross-link mast cell- and basophil-bound IgE, and induced a population of antigen-specific regulatory T-cells, thereafter, can induce immunologic tolerance without adverse IgE mediated effects (Ali *et al.*, 2005). Patients treated with phospholipase A2 T-cell epitopes (bee venom allergen) have shown an increase in IgG4 antibodies, and an improvement of symptoms (Muller *et al.*, 1998). Eighteen cases treated with T-cell epitopes specific for Fel d 1 (the major cat allergen) showed a reduction in IL-4 release by cultured PBMCs from patients (Pene *et al.*, 1998). Other studies with Fel d 1 T-cell epitopes, however, show no effects on cytokine production (Simons *et al.*, 1996).

15.5.3 TLR9-Based Immunotherapy

The bacterial-derived products used for immunotherapy against allergies have a long history. The development of the "hygiene hypothesis", which proposes that microbial exposure in early childhood protects against the development of allergies made this therapy more reasonable. Recently, several molecularly defined bacterial products, especially bacterial immunostimulatory DNA sequences recognized by different toll-like receptors (TLRs) expressed by cells of the innate immune system, have been investigated as immunomodulators in the therapy of allergies. There is preliminary data from controlled trials which support the use of these bacterial adjuvants in allergen extracts for seasonal allergic rhinitis (Mothes *et al.*, 2003; Creticos *et al.*, 2004).

Cytosine phosphorothioate guanosine (CpG) DNA is a non-coding six-base-pair sequence of DNA which is highly enriched in bacteria and binds to TLR-9. CpG DNA has strong Th1 adjuvant properties, promoting IFN-γ production by natural killer cells and inducing type I interferons and IL-12 production by plasmacytoid dendritic cells (pDCs). Activation of TLR-9 CpG DNA in dendritic cells leads to activation of intracellular signaling pathways, which stimulates innate immunity and influences the adaptive immune response characterized by inhibition of APC-mediated Th2 cell activation and a long-lasting Th1 biasing effect, resulting in a more appropriate Th1/Th2 balance (Klinman *et al.*, 2004).

Conjugating the allergen to TLR- 9 ligand CpG DNA (as compared to administering the allergen and the CpG DNA separately) may enhance the immune response to allergen by approximately a hundred folds. The enhancement of immunogenicity is presumed to be due to the allergen and the CpG DNA localizing to the same antigen-presenting cell when administered as a conjugate, and localizing to different antigen-presenting cells when administered separately. On one hand, TLR9-based immunotherapy is an allergen-specific immunotherapy mode, which provides long-term inhibition of allergen-specific hypersensitivities. TLR9-based immunomodulation, on the other hand, is independent of allergens, hence, may have a potential therapeutic advantage in a broad spectrum of allergic diseases.

A large amount of data indicates that TLR 9-based immunotherapy is effective in the prevention and treatment of animal models of allergic disorders during the past few years. Synthetic CpG-rich oligodeoxynucleotide analogues (ODN) demonstrated impressive adjuvant effects in the treatment of anaphylaxis, asthma and allergic conjunctivitis in murine models (Kline *et al.*, 2007). Administration of CpG DNA in mouse models of allergy and asthma could inhibit Th2 cytokine responses, eosinophilic airway inflammation, mucus secretion, airway remodeling, and airway hyperreactivity. Current phase Ⅰ/Ⅱ clinical trials with TLR9-based immunotherapy demonstrate high immunogenic and therapeutic efficacy, and safety when compared with conventional allergen desensitization.

Inoue *et al.* (2005) have reported that administration of CpG-ODN through the barrier-disrupted skin may shift the immune response from type Th2 to Th1 and drastically attenuated the production of IgE in mice undergoing an IgE-type immune response. Later they examined the alterations in immune response in conventional NC/Nga mice, which spontaneously develop AD-like symptoms and high Th2-immune responses following an application of CpG-ODN to the skin. They found that CpG-ODN remarkably changed the immune response from type Th2 to Th1, which is characterized as decreased serum IgE level and up-regulated production of IgG2a. CpG-ODN also decreased the inflammatory infiltration of mast cells with the generation of Tregs in the skin and improved the lesions. These results suggest that CpG-ODN could be effective for immunotherapy in patients with AD (Inoue *et al.*, 2007). Additional studies are required to determine whether clinical trials that use CpG-ODN to reduce serum IgE levels and improve skin lesions with the generation of Tregs. Commercial preparations at least from 2 companies, either alone or with allergen immunotherapy (Aventis: AVE7279 and AVE0675 and Dynavax: 1018 ISS) are under clinical evaluation for treatment of allergic diseases.

15.5.4 Anti-IL-5 Antibody

IL-5 is a lineage specific growth factor for eosinophils (Lopez *et al.*, 1988). Increased levels of IL-5 and eosinophils are noted in the airways of asthmatics.

Administration of IL-5 by the inhalation route induces sputum eosinophilia in asthmatics (Leckie *et al.*, 2000; Cho *et al.*, 2004). Building on these findings, IL-5 provides a novel therapeutic target to specifically reduce eosinophilic inflammation in allergic diseases. Studies administering anti-IL-5 antibody to human subjects have demonstrated that anti-IL-5 antibody did reduce blood and sputum eosinophils in mild asymptomatic asthmatics. IL-5 is also thought to play. In a double-blind placebo-controlled study of human asthmatics, anti-IL-5 treatment significantly reduced airway biopsy eosinophils, and the levels of the extracellular matrix proteins tenascin and lumican, which are important in remodeled airways (Flood-Page *et al.*, 2003). The results suggest that anti-IL-5 reduces selected features of airway remodeling. In addition, the role of anti-IL-5 antibody in treating idiopathic hypereosinophilic syndrome (HES) has been examined, as we know that eosinophil proliferation is induced by IL-5. In a double-blind placebo-controlled study, HES patients receiving anti-IL-5 were able to taper off their dosage of prednisone easily, suggesting that anti-IL-5 therapy may be effective as a corticosteroid sparing agent in HES patients (Garrett *et al.*, 2004). However, further studies with anti-IL-5 are needed to determine its role in other eosinophil associated diseases such as eosinophilic esophagitis, airway remodeling in asthma, Churg Strauss syndrome, and eosinophilic pneumonia. One randomized, placebo controlled parallel study of 43 patients with AD indicated that mepolizumab (humanized monoclonal antibody to IL-5) did not result in clinical success in patients with AD, despite a significant decrease in peripheral blood eosinophils (Oldhoff *et al.*, 2005).

15.5.5 Allergen Modification and Anti-allergy DNA Vaccine

In this section, allergen structure, function and anti-allergy DNA vaccine will be introdued.

15.5.5.1 Allergen Structure and Function

The major allergens of most clinically important allergen sources have been cloned and sequenced, which provide an important foundation on considering structural modifications for optimal immunomodulation in clinical practice. The Allergen Nomenclature Subcommittee of the World Health Organization and International Union of Immunological Societies maintains a Protein Database providing a register of validated information (http://www.allergen. org/Allergen.aspx). More recently, the importance of structural biology and the function of allergen molecules is recognized in determining the response to allergen encounters. Crystal structures have been obtained for many well-characterized allergen molecules revealing sites of IgE epitopes on exposed outer surfaces of tertiary and quaternary structure. Higher order structures may also influence allergen processing by APC and subsequent presentation of peptides for T-cell recognition.

Gerstmayr *et al.* (2007) constructed rSbsC-Bet v 1, a recombinant fusion protein of a bacterial surface (S-layer) protein of Geobacillus stearothermophilus ATCC 12980 and a major birch pollen allergen Bet v 1, which exhibited reduced allergenicity and induced IFN-γ and IL-10 synthesis. They further found that mdDC responded to rSbsC-Bet v 1 with a significant up-regulation of costimulatory molecules, functional maturation, and the synthesis of IL-10 and IL-12, suggesting that the conjugating allergens to bacterial agents might be a promising approach to improve vaccines for specific immunotherapy of atopic allergies.

15.5.5.2 Anti-allergy DNA Vaccine

DNA vaccine is an attractive alternative for the prevention and treatment of allergic diseases. Recently, Dendritic cell-targeted allergen gene vaccination, using fascin gene promoter, was found to inhibit IgE production and allergic inflammation. Vaccination with DNA-encoded Ag85B (a microbial derivative from Mycobacterium kansasii) and AIMP1 (aminoacyl tRNA synthetase-interacting multifunctional protein 1) as adjuvants could down-regulate established Th2-mediated allergic responses and induce a strong Th1-type immune response in mice as well as in humans (Takatsu *et al.*, 2003; Kim *et al.*, 2006; Kim *et al.*, 2008). Mori *et al.* (2009) found that intraperitoneal injection of plasmid DNA encoding Ag85B on atopic dermatitis induced by oxazolone application in mice could suppress the dermal cell infiltration accompanied with the reduced IL-4 and augmented IFN-g, IL-10, and TGF-b mRNA expression in skin. Furthermore, Foxp3+ regulatory T-cells increased in the Ag85B treated mice. The results suggest that Ag85B DNA vaccine might be a potential therapy for Th2 type dermatitis.

15.5.6 Hypoallergenic Preparations for Immunotherapy

It is generally accepted that side effects of allergen specific immunotherapy are largely IgE mediated. Thus, to improve the safety and efficacy of allergen extract-based immunotherapy, development and application of hypoallergens have been proposed. The term "hypoallergenic allergen derivative" describes a recombinant allergen that has been modified to reduce the allergenic activity (i.e., with reduced IgE binding capacity) and to preserve T-cell epitopes reactivity and immunogenicity which can still induce IgG antibodies that recognize wild-type allergen and can block the recognition of the wild-type allergen by IgE antibodies of allergic patients. The approaches include chemical modification of allergen molecules, DNA shuffling on allergen genes, preparation of recombinant allergens with altered critical residues for IgE-binding and synthetic peptides based on dominant T-cell epitopes (Drew *et al.*, 2004; Gafvelin, 2007). Hypoallergenic allergoids were trialed with improved symptoms and safety. But till now, the preparation of hypoallergenic extracts is poorly standardized and the characterization of allergoids is difficult. The generation of hypoallergenic recombinant molecules depends on the particular features of the allergen molecule. The recombinant hypoallergen will retain multiple T-cell epitopes and thus target the polyclonal allergen-specific T-cell population in patients. Potential hypoallergenic latex preparations identified include modified non-IgE-reactive allergen molecules and short T-cell epitope peptides (Rolland *et al.*, 2008).

Several clinical trials of recombinant birch and grass pollen allergen-based vaccines have been undertaken with promising results. The first injection immunotherapy trials conducted with recombinant vaccines for birch pollen and grass pollen allergies show that recombinant allergen-based immunotherapy has vaccination characteristics and is clinically effective (Valenta *et al.*, 2007). Another approach to reduce the risk of Th2 sensitization after allergen gene vaccination is to link the allergen genes to ubiquitin, and to skew the allergen-specific immune response toward a Th1 type with minimal antibody production. Bauer *et al.* (2006) showed that one hypoallergenic DNA vaccine encoding major birch pollen allergen Bet v 1 covalently linked to ubiquitin could effectively induce antiallergic T-cell responses without detectable antibody responses in a mouse model for type Ⅰ allergy.

Recently targeted allergen presentation pathway in APC by recombinant technology has been explored for improving allergen presentation to T-cells. This strategy could potentially make T-cell responses more efficient without administering increased doses of allergen to patients. Fusion proteins, consisting of a [HIS]6-tag for protein purification, a transactivator of transcription (TAT) peptide for converting extracellular to cytoplasmic protein, a truncated invariant chain (Ii) peptide for protein targeting to endosomal/lysosomal compartments and an allergen were created. With these fusion proteins, T-cell proliferation was induced at 10–100 folds lower dosages than that for the control proteins, and a shift in cytokine profile from Th2- to Th1-type with concurrent IL-10 production was also observed. Thus these recombinant fusion proteins might be promising, safe and effective vaccines against allergies, though further research to validate the effectiveness of these preparations is needed in clinical trials (Crameri *et al.*, 2007).

15.6 Perspective

Allergic diseases are a major health concern worldwide. To date, allergen specific immunotherapy is a clinically effective treatment for allergy disorders. The successful development of safe and easily administered immunomodulator agents holds significant potential for early intervention. Research activity focuses on

T-cell epitope mapping, Treg induction, allergen modifications, adjuvants and adjunct therapies and hypoallergenic approaches.

Conventional subcutaneous immunotherapy is the most established form but increasingly sublingual immunotherapy (SLIT) is becoming popular due to increased safety and easy administration and proven efficacy. However, the optimal dosage and duration of SLIT therapy, as well as the use of multiple allergens in SLIT still needs to be investigated. New knowledge on mechanisms has prompted the development of modified vaccines, such as recombinant allergens, which can undoubtedly better standardize the allergen extracts and affords the opportunity for individualized treatment (Valenta *et al.*, 2007).

Newer immunotherapeutic approaches, such as anti-IgE (omalizumab) therapy, may help improve the efficacy and safety profile of immunotherapy. In addition, the short synthetic allergen-derived peptides, corresponding to T-cell peptides, have been recently evaluated in clinical trials for the safety and efficacy. The main advantage of this approach is the reduction in systemic IgE-mediated adverse events compared with whole allergen immunotherapy. CpG DNA combined with allergens are under investigation with early promising trials. These approaches offer exciting opportunities to further refine allergen related approaches to immunotherapy with the goal of preventing allergic skin diseases.

References

Abramson, M.J., Puy, R.M., Weiner, J.M. (2003). Allergen immunotherapy for asthma. Cochrane Database Syst Rev, (4), CD001186.

Abramson, M.J., Puy, R.M., Weiner, J.M. (1995). Is allergen immunotherapy effective in asthma? A meta-analysis of randomized controlled trials. Am J Respir Crit Care Med, 151, 969-974.

Akdis, M., Akdis, C.A. (2007). Mechanisms of allergen-specific immunotherapy. J Allergy Clin Immunol, 119(4), 780-791.

Akdis, C. Akdis, A., Blesken, M. *et al.*, (1996). Epitope-specific T-cell tolerance to phospholipase A2 in bee venom immunotherapy and recovery by IL-2 and IL-15 *in vitro*. J Clin Invest, 98(7), 1676-1683.

Akdis, M., Verhagen, J., Taylor, A. *et al.* (2004). Immune responses in healthy and allergic individuals are characterized by a fine balance between allergen-specific T regulatory 1 and T helper 2 cells. J Exp Med, 199 (11), 1567-1575.

Ali, F.R., Larché, M. (2005). Peptide-based immunotherapy: A novel strategy for allergic disease. Expert Rev Vaccines, 4(6), 881-889.

Álvarez-Cuesta, E., Bousquet, J., Canonica, G.W., *et al.* (2006). Standards for practical allergen-specific immunotherapy. Allergy, 61 (Suppl 82), 1-20.

Bauer, R., Scheiblhofer, S., Kern, K., *et al.* (2006). Generation of hypoallergenic DNA vaccines by forced ubiquitination: preventive and therapeutic effects in a mouse model of allergy. J Allergy Clin Immunol, 118, 269-276.

Bousquet, J., Demoly, P., Michel, F. (2001). Specific immunotherapy in rhinitis and asthma. J Allergy Clin Immunol, 87(1), 38-42.
Beck, L. A., Marcotte, G. V., MacGlashan, D., *et al.* (2004). Omalizumab-induced reductions in mast cell Fce psilon RI expression and function. J Allergy Clin Immunol, 114(3), 527-530.
Boguniewicz, M., Leung, D.Y. (2006). Atopic dermatitis. J Allergy Clin Immunol, 117, S475-S480.
Bohle, B., Kinaciyan, T., Gerstmayr, M., *et al.* (2007). Sublingual immunotherapy induces IL-10-producing T regulatory cells, allergen specific T-cell tolerance, and immune deviation. J Allergy Clin Immunol, 120(3), 707-713.
Bussmann, C., Bockenhoff, A., Henke, H., *et al.* (2006). Does allergen-specific immunotherapy represent a therapeutic option for patients with atopic dermatitis? J Allergy Clin Immunol, 118(6), 1292-1298.
Bussmann, C., Maintz, L., Hart, J. (2007). Clinical improvement and immunological changes in atopic dermatitis patients undergoing subcutaneous immuno-therapy with a house dust mite allergoid: A pilot study. Clin Exp Allergy, 37(9), 1277-1285.
Cadario, G., Galluccio, A.G., Pezza, M. (2007). Sublingual immunotherapy efficacy in patients with atopic dermatitis and house dust mites sensitivity: A prospective pilot study. Curr Med Res Opin, 23(10), 2503-2506.
Calamita, Z., Saconato, H., Pela, A.B. (2006). Efficacy of sublingual immunetherapy in asthma: Systematic review of randomized-clinical trials using the Cochrane collaboration method. Allergy, 61(10), 1162-1172.
Calderon, M.A., Alves, B., Jacobson, M. (2007). Allergen injection immunotherapy for seasonal allergic rhinitis. Cochrane Database Syst Rev, 24 (1), CD001936.
Canonica, G.W., Passalacqua, G. (2003). Noninjection routes for immunetherapy. Review J Allergy Clin Immunol, 111 (3), 437-448.
Cho, J.Y., Miller, M., Baek, K.J., *et al.* (2004). Inhibition of airway remodeling in IL-5-deficient mice. J Clin Invest, 113, 551-560.
Cooke, R.A., Barnard, J.H., Hebald, S., Stull, A. (1935). Serological evidence of immunity with coexisting sensitization in a type of human allergy (hay fever). J Exp Med, 62, 733-751.
Cosmi, L., Santarlasci, V., Angeli, R., *et al.* (2006). Sublingual immunotherapy with Dermatophagoides monomeric allergoid down-regulates allergen-specific immunoglobulin E and increases both interferon-gamma- and interleukin-10-production. Clin Exp Allergy, 36(3), 261-272.
Cox, L.S., Linnemann, DL., Nolte, H., *et al.* (2006). Sublingual immunotherapy: a comprehensive review. J Allergy Clin Immunol, 117(5), 1021-1035.
Crameri, R., Flückiger, S., Daigle, I, *et al.* (2007) Design, engineering and *in vitro* evaluation of MHC class- II targeting allergy vaccines. Allergy, 62(2), 197-206.
Creticos, P.S., Schroeder, J.T., Hamilton, R.G., *et al.* (2006). Immunotherapy with a ragweed-toll-like receptor 9 agonist vaccine for allergic rhinitis. N Engl J

Med, 355(14), 1445-1455.
Czarnecka-Operacz, M., Silny, W. (2006). Specific immunotherapy in atopic dermatitis-four-year treatment in different age and airborne allergy type subgroups. Acta Dermatovenerol Croat, 14(4), 230-240.
Des Roches, A., Paradis, L., Menardo, J.L., *et al.* (1997). Immunotherapy with a standardized dermatophagoides pteronyssinus extract. VI. Specific immunotherapy prevents the onset of new sensitizations in children. J Allergy Clin Immunol, 99 (4), 450-453.
Drew, A.C., Eusebius, N.P., Kenins, L., *et al.* (2004). Hypoallergenic variants of the major latex allergen Hev b 6.01 retaining human T lymphocyte reactivity. J Immunol, 173(9), 5872-5879.
Dunbar, W.P. (1913). The present state of knowledge of hayfever. J Hyg, 13, 105.
Durham, S.R., Walker, S.M., Varga, E.M., *et al.* (1999). Long-term clinical efficacy of grass-pollen immunotherapy. N Engl J Med, 341(7), 468-475.
Durham, S.R., Ying, S., Varney, V.A., *et al.* (1996). Grass pollen immunotherapy inhibits allergen-induced infiltration of CD4+ T lymphocytes and eosinophils in the nasal mucosa and increases the number of cells expressing messenger RNA for interferon-gamma. J Allergy Clin Immunol, 97(6), 1356-1365.
Finegold, I. (2001). Immunotherapy historical perspective. Annals of Allergy, Asthma & Immunology, 87(1), 3-4.
Flood-Page, P.T., Menzies-Gow, A.N., Kay, A.B., *et al.* (2003). Eosinophil's role remains uncertain as anti-interleukin-5 only partially depletes numbers in asthmatic airway. Am J Respir Crit Care Med, 167, 199-204.
Frew, A.J. (2008). Sublingual immunotherapy. N Engl J Med, 358(21), 2259-2264.
Gafvelin, G., Parmley, S., Neimert-Andersson, T., Blank, U., Eriksson, T.L.J., van Hage, M., Punnonen, J. (2007). Hypoallergens for Allergen-specific Immunotherapy by Directed Molecular Evolution of Mite Group 2 Allergens. J Bio Chem, 282, 3778-3787.
Gardner, L.M., O'Hehir, R.E., Rolland, J.M. (2004a). High dose allergen stimulation of T cells from house dust mite-allergic subjects induces expansion of IFN gamma+T cells, apoptosis of CD4+IL-4+ T-cells and T-cell anergy. Int Arch Allergy Immunol, 133(1), 1-13.
Gardner, L.M., Thien, F.C., Douglass, J.A., *et al.* (2004c). Induction of T "regulatory" cells by standardized house dust mite immunotherapy: An increase in CD4+ CD25+ interleukin-10+ T-cells expressing peripheral tissue trafficking markers. Clin Exp Allergy, 34(8), 1209-1219.
Garrett, J.K., Jameson, S.C., Thomson, B., *et al.* (2004). Anti-Interleukin-5 (mepolizumab) therapy for hypereosinophilic syndromes. J Allergy Clin Immunol, 113, 115-119.
Gerstmayr, M., Ilk, N., Schabussova, I., *et al.* (2007). A novel approach to specific allergy treatment: The recombinant allergen-Slayer fusion protein rSbsC-Bet v 1 matures dendritic cells that prime Th0/Th1 and IL-10-producing regulatory T-cells. J Immunol, 179(11), 7270-7275.
Gidaro, G.B., Marcucci, F., Sensi, L., *et al.* (2005). The safety of sublingualswal-

low immunotherapy: an analysis of published studies. Clin Exp Allergy, 35(5), 565-571.

Gleich, G.J., Zimmermann, E.M., Henderson, L.L., *et al.* (1982). Effect of immunotherapy on immunoglobulin E and immunoglobulin G antibodies to ragweed antigens: A six-year prospective study. J Allergy Clin Immunol, 70(4), 261-271.

Gordon, B.R. (1995). Immunotherapy basics. Otolaryngology–Head and Neck Surgery, 113,597-602.

Guerra, F., Carracedo, J., Solana-Lara, R., *et al.* (2001). TH2 lymphocytes from atopic patients treated with immunotherapy undergo rapid apoptosis after culture with specific allergens. J Allergy Clin Immunol, 107(4), 647-653.

Incorvaia, C., Pravettoni, C., Mauro, M., *et al.* (2008). Yacoub MR, Tarantini F, Riario-Sforza GG. Effectiveness of omalizumab in a patient with severe asthma and atopic dermatitis. Monaldi Arch Chest Dis, 69(2), 78-80.

Inoue, J., Yotsumoto, S., Sakamoto, T., *et al.* (2005). Changes in immune responses to antigen applied to tape-stripped skin with CpG oligodeoxynucleotide in NC/Nga mice. Pharm Res, 22, 1627-1633.

Inoue, J., Aramaki, Y. (2007). Suppression of skin lesions by transdermal application of CpG-oligodeoxynucleotides in NC/Nga mice, a model of human atopic dermatitis. J Immunol, 178, 584-591.

Jacobsen, L., Niggemann, B., Dreborg, S., *et al.* (2007). Specific immunotherapy has long-term preventive effect of seasonal and perennial asthma: 10-year follow-up on the PAT study. Allergy, 62(8), 943-948.

James, L.K., Durham, S.R. (2008). Update on mechanisms of allergen injection immunotherapy. Clin Exp Allergy, 38(7), 1074-1088.

Jutel, M., Akdis, M., Blaser, K., *et al.* (2006). Mechanisms of allergen specific immunotherapy–T-cell tolerance and more. Allergy, 61(7), 796-807.

Jutel, M., Pichler, W.J., Skrbic, D., *et al.* (1995). Bee venom immunotherapy results in decrease of IL-4 and IL-5 and increase of IFN-gamma secretion in specific allergen-stimulated T cell cultures. J Immunol, 154(8), 4187-4194.

Kim, E., Kim, S.H., Kim, S., *et al.* (2006). The novel cytokine p43 induces IL-12 production in macrophages via NF-kB activation, leading to enhance IFN-g production in CD4t T cells. J Immunol, 176, 256-264.

Kim, E., Kim, S.H., Kim, S., *et al.* (2008). AIMP1/p43 protein induces the maturation of bone marrow-derived dendritic cells with T helper type 1-polarizing ability. J Immunol, 180, 2894-2902.

Kleine-Tebbe, J., Ribel, M., Herold, D.A. (2006). Safety of a SQ-standardized grass allergen tablet for sublingual immunotherapy: A randomized, placebo-controlled trial. Allergy, 61(2), 181-184.

Kline, J.N. (2007). Eat dirt: CpG DNA and immunomodulation of asthma. Proc Am Thorac, 4(3), 283-288.

Klinman, D.M. (2004). Immunotherapeutic uses of CpG oligodeoxynucleotides. Nat Rev Immunol, 4, 249-258.

Larche, M. (2007). Peptide immunotherapy for allergic diseases. Allergy, 62(3),

325-331.
Leckie, M., Brinke, A., Khan, J., *et al.* (2000). Effects of an interleukin-5 blocking monoclonal antibody on eosinophils, airway hyper-responsiveness, and the late asthmatic response. Lancet, 356, 2144-2148.
Leung, D.Y., Boguniewicz, M., Howell, M.D., *et al.* (2004). New insights into atopic dermatitis. J Clin Invest, 113, 651-657.
Linhart, B., Valenta, R. (2005). Molecular design of allergy vaccines. Current Opinion in Immunology, 17, 646-655.
Lopez, A.F., Sanderson, C.J., Gamble, J.R., *et al.* (1998). Recombinant human interleukin 5 is a selective activator of human eosinophil function. J Exp Med, 167, 219-224.
Maintz, L., Novak, N. (2007). Getting more and more complex: the pathophysiology of atopic eczema. Eur J Dermatol, 17, 267-283.
Mauro, M., Russello, M., Incorvaia, C., *et al.* (2007). Comparison of efficacy, safety and immunologic effects of subcutaneous and sublingual immunotherapy in birch pollinosis: A randomized study. Eur Ann Allergy Clin Immunol, 39(4), 119-122.
Mastrandrea, F., Serio, G., Minelli, M., *et al.* (2000). Specific sublingual immunotherapy in atopic dermatitis. Results of a 6-year follow-up of 35 consecutive patients. Allergol Immunopathol (Madr), 28(2), 54-62.
Mellerup, M.T., Hahn, G.W., Poulsen, L.K. *et al.* (2000). Safety of allergen-specific immunotherapy. Relation between dosage regimen, allergen extract, disease and systemic side-effects during induction treatment. Clin Exp Allergy, 30(10), 1423-1429.
Moingeon, P., Batard, T., Fadel, R., *et al.* (2006). Immune mechanisms of allergen-specific sublingual immunotherapy. Allergy, 61(2), 151-165.
Möller C., Dreborg, S., Ferdousi, H.A., *et al.* (2002). Pollen immunotherapy reduces the development of asthma in children with seasonal rhinoconjunctivitis (the PAT-study). J Allergy Clin Immunol, 109(2), 251-256.
Mori, H., Yamanaka, K., Matsuo, K., *et al.* (2009). Administration of Ag85B showed therapeutic effects to Th2-type cytokine-mediated acute phase atopic dermatitis by inducing regulatory T-cells. Arch Dermatol Res, 301,151-157.
Mothes, N., Heinzkill, M., Drachenberg, K.J., *et al.* (2003). Allergen-specific immunotherapy with a monophosphoryl lipid A-adjuvanted vaccine: Reduced seasonally boosted immunoglobulin E production and inhibition of basophil histamine release by therapy-induced blocking antibodies. Clin Exp Allergy, 33 (9), 1198-1208.
Muller, U., Akdis, C.A., Fricker, M., *et al.* (1998). Successful immunotherapy with T-cell epitope peptides of bee venom phospholipase A2 induces specific T-cell anergy in patients allergic to bee venom. J Allergy Clin Immunol, 101, 747-754.
Noon, L. (1911). Prophylactic inoculation against hayfever. Lancet, 1, 1572-1573.
O'Brien, R.M., Byron, K.A., Varigos, G.A., *et al.* (1997). House dust mite immunotherapy results in a decrease in Der p 2-specific IFN-gamma and IL-4 expression by circulating T lymphocytes. Clin Exp Allergy, 27, 46-51.

O'Hehir, R.E., Sandrini, A., Anderson, G.P., *et al.* (2007). Sublingual allergen immunotherapy: Immunological mechanisms and prospects for refined vaccine preparation. Curr Med Chem, 14(21), 2235-2244.

Oldhoff, J.M., Darsow, U., Werfel, T., *et al.* (2005). Anti-IL-5 recombinant humanized monoclonal antibody (Mepolizumab) for the treatment of atopic dermatitis. Allergy, 60, 693-696.

Ozdemir, C. (2009). An immunological overview of allergen specific immunotherapy – subcutaneous and sublingual routes. Ther Adv Respir Dis, 3, 253-262.

Pajno, G.B., Caminiti, L., Vita, B., *et al.* (2007). Sublingual immunotherapy in mite-sensitized children with atopic dermatitis: a randomized, double-blind, placebo-controlled study. J Allergy Clin Immunol, 120, 164-170.

Patriarca, G., Schiavino, D., Pecora, V., *et al.* (2009). Food allergy and food intolerance: Diagnosis and treatment. Intern Emerg Med, 4, 11-24.

Pene, J., Desroches, A., Paradis, L. (1998). Immunotherapy with Fel d 1 peptides decreases IL-4 release by peripheral blood T-cells of patients allergic to cats. J Allergy Clin Immunol, 102, 571-578.

Rhyner, C., Kundig, T., Akdis, C.A., *et al.* (2007). Targeting the MHC II presentation pathway in allergy vaccine development. Biochemical Society Transactions, 35(4), 833-834.

Rolland, J.M., O'Hehir, R.E. (2008). Latex allergy: A model for therapy. Clin Exp Allergy, 38(6), 898-912.

Scadding, K., Brostoff, J. (1986). Low dose sublingual therapy in patients with allergic rhinitis due to dust mite. Clin Exp Allergy, 16, 483-491.

Sheinkopf, L.E., Rafi, A.W., Do, L.T. (2008). Efficacy of omalizumab in the treatment of atopic dermatitis: A pilot study. Allergy Asthma Proc, 29(5), 530-537.

Simons, F.E., Imada, M., Li, Y., *et al.* (1996). Fel d 1 peptides: effect on skin tests and cytokine synthesis in cat-allergic human subjects. Int Immunol, 8, 1937-1945.

Takatsu, K., Kariyone, A. (2003). The immunogenic peptide for Th1 development. Int Immunopharmacol, 3, 783-800.

Valenta, R., Niederberger, V. (2007). Recombinant allergens for immunothe- rapy. J Allergy Clin Immunol, 119(4), 826-830.

Werfel, T., Breuer, K., Rueff, F., *et al.* (2006). Usefulness of specific immunotherapy in patients with atopic dermatitis and allergic sensitization to house dust mites: A multicentre, randomized, dose-response study. Allergy, 61, 202-205.

16

Traditional Chinese Medicine for Treating Food Allergy and Associated Eczema: From Research to Practice Perspective

Xiu-Min Li

Division of Allergy and Immunology, Center for Chinese Herbal Therapy for Asthma and Allergy, the Mount Sinai School of Medicine, New York, USA
E-mail:xiu-min.li@mssm.edu

Abstract: Food allergy is a growing problem in modern society that deserves an ongoing effort to develop safe and effective treatment. Traditional Chinese medicine (TCM) has a long human use history and is beginning to play a role in health care in the US, mainly via licensed practitioners. There is also increasing scientific evidence demonstrating the safety and efficacy of TCM for allergic diseases. We developed a herbal formula named food allergy herbal formula-2 (FAHF-2) derived from a classical herbal formula Wu-Mei-Wan that has been used in TCM to treat intestinal parasite infections and food allergy-like symptoms. Over the past years, we have generated a number of publications showing that FAHF-2 can prevent and reverse established peanut allergies in an animal model of peanut anaphylaxis, and that the effect is long lasting. These studies suggest that FAHF-2, and perhaps other Chinese herbal medicines, may have a potential for treating food allergies. FAHF-2 is the first botanical drug that has entered clinical trials as a United States Food and Drug Administration Investigational New Drug. Our phase Ⅰ studies showed that FAHF-2 is safe and well-tolerated. Food allergies are often associated with other allergic conditions such as eczema (atopic dermatitis), particularly recalcitrant eczema. Given the growing interest in alternative and complementary medicine (CAM) therapies from both families and physicians, we have established a TCM/integrative medicine clinical program to help children and adults with recalcitrant eczema associated with food allergies. The clinical outcomes have been well received. TCM treatment improved quality of life, reduced food (including peanut) specific IgE levels, can be used long term, and no side effects were observed. TCM as monotherapy or integrative medicine may be an important approach for treating food allergies and associated eczema. It is possible that some of these herbal remedies will progress from dietary supplements to prescription drugs via clinical studies.

16.1 Background

Allergic diseases are significant health problems in westernized countries, and represent a tremendous burden with respect to both quality of life and health care expenditures (O'Connell, 2004; Isolauri *et al.*, 2004; Bousquet *et al.*, 2005). This chapter will focus on food allergies and new trend of increased use of complementary and alternative medicine, especially traditional Chinese medicine.

16.1.1 Food Allergy and Related Conditions

The prevalence of childhood peanut allergies has been significantly increasing. The prevalence of peanut and tree nut allergies in children younger than 18 years is 2.1%, compared with 1.2% in 2002 and 0.6% in 1997 (Sicherer *et al.*, 2010). Concurrent with food allergies over the past several decades, the number of individuals with asthma, atopic dermatitis and allergic rhinitis has increased dramatically in industrialized countries (Eichenfield *et al.*, 2003). The reasons for the food allergy epidemic are unknown, but there are several hypothesis, including the hygiene hypothesis (Pali-Scholl *et al.*, 2009), the overall allergy march (Worth and Sheikh, 2010) and epigenetics (Bjorksten, 2005; van Panhuys *et al.*, 2008; Pali-Scholl *et al.*, 2009; Prescott and Clifton, 2009). Clinically, food allergic children often suffer from other allergic conditions such as eczema, and environmental allergies. It has been shown that 40% of recalcitrant eczema is associated with food allergies (Worth *et al.*, 2010). A recent international cohort study of 2,222 infants with atopic eczema aged 11.5–22.5 months found that 64% of those who developed the condition before 3 months of age also had a high-risk of IgE-mediated food sensitivity to egg and/or cow's milk and/or peanut. In some clinical settings, this number is even higher. In our recent clinical study, of persons 12–45 years of age–with peanut or tree nut allergies, 94% also suffer from other allergies including eczema, asthma and allergic rhinitis (Wang *et al.*, 2010). This study also showed that among peanut and tree nut allergic subjects, 79% also suffer from other food allergies. Planning diets to accommodate avoidance of multiple foods is challenging, and negatively impacts nutritional status (miz-Echevarria *et al.*, 2008; Noimark and Cox, 2008). In addition, the potential for accidental ingestion is greater in multiple food allergies, thereby creating higher anxiety levels for patients and their families (Ostblom *et al.*, 2008; Sicherer *et al.*, 2001). Conventional therapies are not satisfactory for multiple allergic conditions and many families are also concerned about side effects of chronic steroids use for eczema and asthma. This information highlights the need to develop more effective health care for patients with food allergies as well as other allergies.

16.1.2 Complementary and Alternative Medicine (CAM) in the US

A National Institutes of Health (NIH)/National Center for Complementary and Alternative Medicine (NCCAM) survey reported that more than 50% of Americans have used some forms of CAM treatment. The chronic nature of allergic diseases and the lack of satisfactory treatment lead many families of children with allergies and asthma to seek CAM treatments (Slader *et al.*, 2006). Although the results of surveys on CAM use by children vary between the studies, the reported rates by children with asthma ranges from 33% to 89% (Slader *et al.*, 2006). A recent survey reported that a majority of pediatricians believe that their patients are using CAM and have a positive attitude towards CAM. Because CAM therapies are generally considered to be safe and effective by patients, pediatricians would also consider referring to a CAM practitioner and would like more education on CAM (Sawni and Thomas, 2007), so that they might better discuss the implications of using these therapies and potentially improve adherence to the prescribed medication regimen and improve allergy and asthma management (Mark, 2007).

Increased use of CAM by patients is also a driving force for allergists to gain more insight into CAM. The American Academy of Allergy, Asthma and Immunology (AAAAI), Complementary & Alternative Practice in Allergy Committee conducted a survey of allergic patients in 2007. Of 450 responders, 80% were interested in learning more about complementary or alternative medicine. The participants clearly reflected their interest in learning more about various CAM modalities with 88% extremely interested in Herbal and Botanical Medicine. This highlights the need to develop safe and effective CAM therapy. In 2008, Dr. Hugh Sampson gave a presidential address at the AAAAI meeting, which outlined a plan to promote research on food allergies, and CAM. Since then, there have been a number of milestones. In 2009, the Journal of Allergy and Clinical Immunology (JACI) published a special issue on CAM. The same year, the AAAAI held a seminar on CAM research with outstanding attendance, including the director of NCAAM. The AAAAI Natural Products website provides assistance for allergists in their practice. We recently organized the first East-West Scientific Conference on Allergy and Traditional Medicine in China. This conference further contributed to efforts to accelerate global collaboration to develop safe and effective CAM therapy for food allergies and related conditions.

16.1.3 Traditional Chinese Medicine Research and Food Allergy

Traditional Chinese medicine (TCM) has a long human use history in China and other Asian countries, such as Korea and Japan, for treating and preventing disease, and is part of main stream medicine in these countries. Main components of TCM include herbal therapy, acupuncture, acupressure/massage, mind-body therapy and dietary therapy. TCM is part of the mainstream of medicine and the cost is covered by insurance in these countries. TCM began to play a role in the health care system in the US. Acupuncture has been approved by the FDA as a medical device; Herbal medicine is still viewed as a dietary supplement and the cost is not covered by insurance. However, this situation may change in favor of herbal medicines. In recent years the US FDA has provided guidance for investigating botanical drug products, including complex formulas containing several herbs, focusing on efficacy, safety and consistency (the US Food and Drug Administration (FDA) & Center for Drug Evaluation and Research, 2004). The FDA recently issued guidelines for investigating herbal medicine as a botanical drug-prescription drug. The National Institutes of Health (NIH)/ NCCAM defines TCM as Whole Medical Systems(NCCAM, 2008a). The NCCAM/NIH provides grants to support clinical and basic research on CAM. Several publications including ours indicate that TCM has potential for treating asthma (Li *et al.*, 2002), managing allergic rhinitis (Jindal *et al.*, 2008) and improving quality of life of atopic dermatitis patients(Hon *et al.*, 2007). Therefore some of the TCM remedies may become botanical drugs, i.e., prescription drugs via clinical investigation. This chapter will focus on the research into the investigation of TCM for treating food allergy and food allergy associated with eczema.

16.2 Development of TCM Formula for Treating Food Allergy

The NIH/NCCAM defines TCM as whole medical systems. The NIH/ NCCAM provides grants to support clinical and basic research on CAM. TCM practice does not only focus on the disease or a single organ. Rather, it focuses on establishing and maintaining the balance of yin-yang (two opposite, but complementary forces), the homeostasis of organ systems in the body, and interactions with the environment. TCM used to be associated with the lack of knowledge of allergic diseases, it was recently postulated as resulting from the loss of homeostasis in interactions between human organs, such as the lungs, skin, and gut, with the environment and foods. Formulas, a mixture of many herbs, are commonly used in TCM practice. It is believed that formulas maximize (synergize) benefits and minimize side effects. Treatment is customized for each patient. Some traditional formulations are fairly standard. Wu Mei Wan is a classical Chinese formula for treating intestinal parasite infection, first documented by Dr. Zhang Zhongjing in a classical Chinese medical text compiled ~2,000 years ago. This formula, along with other classical TCM formulas, has been translated into English as *Chinese Herbal Medicine: Formulas and Strategies* by Bensky *et al.*(1993). The prescriptions of Wu Mei Wan include intermittent attacks of abdominal pain, a stifling sensation, irritability, and warmth in the chest (anxiety), accompanied by vomiting, diarrhea after eating, and cold hands and feet. Based on our understandings from different angles of food allergy clinical expression, immunopathological mechanisms, TCM formula principle and updated research in TCM and allergy, we developed food allergy herbal formula-1 (FAHF-1), containing 11 herbs and then refined that formula to make FAHF-2 which contains 9 herbs. The following are the major findings in preclinical and clinical studies related to FAHF-2.

16.2.1 FAHF-2 Prevents and Reverses Peanut Allergy in an Animal Model of Peanut Anaphylaxes

Peanut allergies are potentially life threatening. There is no curative therapy for this disorder. We previously found that a herbal formula, food allergy herbal formula (FAHF)-1, blocked peanut-induced anaphylaxis in a murine model when challenged immediately after post therapy. In this study we tested whether FAHF-2, an improved herbal formula, from which 2 herbs, Zhi Fu Zi (Radix Lateralis Aconiti Carmichaeli Praeparata) and Xi Xin (Herba Asari), were eliminated, is equally effective to FAHF-1, and if so, whether protection persists after therapy is discontinued. The raw herb quality was ascertained according to the standards required by the *Pharmacopoeia of the People's Republic of China* (The State Pharmacopoeia Commission of The People's Republic of China, 2005). Based on organoleptic and microscopic examination, the raw herbs used in FAHF-2 were identified as *Prunus mume*, *Zanthoxylum schinifolium*, *Angelica sinensis*, *Zingiber officinalis*, *Cinnamomum cassiae*, *Phellodendron chinense*, *Coptis chinensis*, *Panax ginseng*, and *Ganoderma lucidum*. The quality and ratio of herbs used has been described previously (Kattan *et al.*, 2008). Product quality was monitored by HPLC fingerprinting according to the FDA's Guidance for Industry Botanical Drug Products (the US Food and Drug Administration (FDA) & Center for Drug Evaluation and Research, 2004). In this study, mice allergic to peanut were treated with FAHF-2 for 7 weeks and were challenged 1, 3, or 5 weeks post therapy. The results showed that after challenges, all sham-treated mice developed severe anaphylactic signs, significant decrease in rectal temperatures, significantly increased plasma histamine levels, and had marked vascular leakage. In contrast, no sign of anaphylactic reaction, decrease in rectal temperatures, or elevation of plasma histamine levels was observed in FAHF-2-treated mice in 5 separate experiments (Fig.16.1). IgE levels were significantly reduced by FAHF-2 treatment and remained significantly lower as long as 5 weeks post therapy. Splenocytes from FAHF-2-treated mice showed significantly reduced IL-4, IL-5, and IL-13, and enhanced IFN-g production to recall peanut stimulation *in vitro*.

FAHF-2 also showed high safety margin, mice fed 24 times of the dosage were safe.

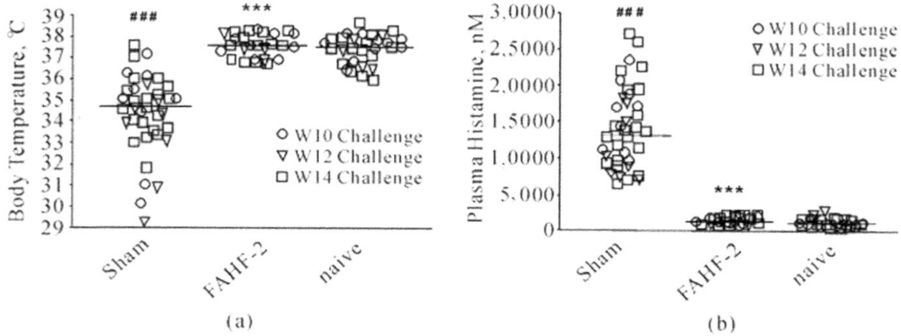

Fig. 16.1. FAHF-2 prevent peanut allergy in murine model (Srivastava *et al.*, 2005). (a) FAHF-2 prevented a drop in core body temperatures. Rectal temperatures were measured 25 min following i.g. peanut challenge in mice challenged at wks 10 (○), 12( ), and 14 (□). Bars indicate the medians of temperature from 5 sets of experiments (sham, *n*=38; FAHF-2, *n*=26; naive, *n*=29). ###, *P*<0.001 vs naive; and ***, *P*< 0.001 vs sham. (b) *FAHF-2 blocked histamine release.* Data are plasma histamine levels 30 min post challenge the same mice-as that in (a) ###, *P*<0.001 vs naïve; and ***, *P*< 0.001 vs sham (With permission of Elsevier)

16.2.2 Pharmacologic and Immunologic Effects of Individual Herbs in the Food Allergy Herbal Formula-2 (FAHF-2) on Peanut Allergy

In this study, we investigated the pharmacological actions of individual herbs comprising FAHF-2 on peanut-induced anaphylactic reactions in a murine model of peanut allergies, and determined if all nine herbs are necessary to prevent anaphylactic reactions, or if a simplified formula containing fewer herbs would be equally effective. Our results showed that some individual herbs reduced peanut-induced anaphylactic symptoms but no single herb offered full protection from anaphylactic symptoms equivalent to FAHF-2. The herbs had highly variable effects on histamine release, as well as peanut-specific serum IgE and IgG2a levels. The herbs also had variable effects on IL-4, IL-5 and IFN- levels. A simplified formula comprised of only the most efficacious individual herbs showed only partial efficacy and was not able to reproduce the effects of FAHF-2 (Table 16.1). This finding suggested that component herbs of FAHF-2 may be working synergistically to produce the curative therapeutic effects produced by the whole formula, which appears to be the best option for future clinical trials (Kattan *et al.*, 2008).

Table 16.1 Anaphylactic reaction rates and median scores obtained 30 min after oral PN challenge (Kattan *et al.*, 2008) (With permission of Wiley-Blackwell)

Herbal tratment	Anaphylactic reactions *n*/total	Percent of mice with a reaction (%)	Anaphylactic score Median (range), pvalues vs Sham
Sham	9/11	81	2(0–4)
FAHF-2	0/8	0	0(0), P<0.001
HB	1/4	25	0(0–2), P=0.008
GJ	3/5	60	2(0–2), P=0.058
HL	3/5	60	2(0–4), P=0.058
CJ	5/8	63	2(0–4), P=0.140
LZ	3/4	75	2.5(0–3), P=0.590
WM	6/7	86	2(0–4), P=0.387
GZ	6/7	86	2(0–3), P=0.330
DG	5/5	100	3(0–3), P=0.640
RS	4/4	100	3(2–3), P=0.386
Naïve	0/15	0	0(0–0), P<001

Sham, peanut allergic mice fed water; FAHF-2, food allergy herbal formula-2; HB, Huang Bai; GJ, Gan Jang; HL, Huang Liang; CJ, Chuan Jiao; LZ, Ling Zhi; WM, Wu Mei; GZ, Gui Zhi; DG, Dang Gui; RS, Ren Shen

16.2.3 FAHF-2 Silences Peanut-Induced Anaphylaxis for a Prolonged Post-Treatment Period via IFN- Producing CD8+ T-Cells

While tremendous strides have been made in food allergy awareness over the past 10 years, there is no satisfactory therapy to prevent or reverse this disease. Other than immediate access to post-anaphylactic rescue medications, strict avoidance is the only way to manage this condition. Unfortunately, accidental ingestion is common. Conventional allergen-specific immunotherapy is not satisfactory because of the unacceptable number of adverse reactions and poor maintenance of tolerance afterwards (Oppenheimer *et al.*, 1992; Nelson *et al.*, 1997). Anti-IgE therapy for food allergies has shown only modest and transient benefits(Leung *et al.*, 2003). Several new therapies for food allergies, such as oral immunotherapy (OIT) and sublingual immunotherapy (SLIT) are under investigation. OIT and SLIT for egg, milk, hazelnut and PN allergies appear to desensitize the majority of patients while on therapy and allow them to ingest some amount of the allergen without allergic reactions (Sicherer and Sampson, 2007; Jones *et al.*, 2009). However, there is, as yet, no evidence that these therapies induce long-term tolerance. One study found that significant allergic reactions occurred when therapy was discontinued for 1–3 weeks and followed by additional therapy using the same dosage, or by accidental ingestion of the food allergen (Rolinck-Werninghaus *et al.*, 2005). Effective, safe, convenient and long-lasting therapies for food allergies are urgently needed.

Since FAHF-2 treatment re-established tolerance to peanut after peanut hypersensitivity was fully established in mice and this effect was shown to persist for 4 weeks post therapy (Qu *et al.*, 2007), we sought to determine if FAHF-2 mediated protection could be extended long term and explored the mechanisms underlying its persistent immunomodulatory effects. In this study, peanut-allergic mice received FAHF-2 daily oral by gavage for 7 weeks, and then received 7 oral peanut challenges at 4–10 week intervals over 36 weeks. For mechanistic studies, some mice received CD4+ or CD8+ T-cell-depleting antibodies or interferon-neutralizing antibodies. Anaphylactic symptoms, body temperatures and plasma histamine levels were recorded following each challenge, and PN-specific immunoglobulin levels and cytokine profiles of splenocytes, mesenteric lymph node cells and purified CD4+ and CD8+ T-cells were determined.

The results showed that FAHF-2 treatment protected mice from anaphylaxis for over 36 weeks after discontinuing treatment. Peanut-specific IgE levels were reduced up to 50%, whereas IgG2a levels were increased up to 60%, and these effects persisted overtime. At week 50, the experiment was terminated and mice were sacrificed. Splenocytes and mesenteric lymph node cells showed significant suppression of Th2 cytokines (IL-4, IL-5 and IL-13) and enhancement of IFN-ã (Fig. 16.2). Furthermore, Th2 cytokine production by CD4+ T-cells from FAHF-2-

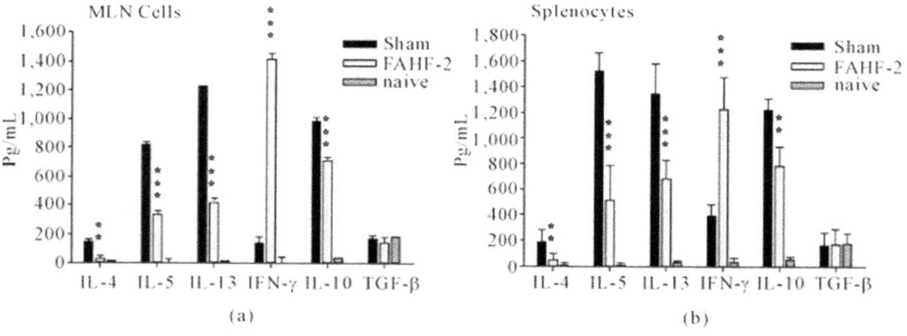

Fig. 16.2. FAHF-2 treatment reduced Th2 cytokine levels and increased IFN- secretion by specific modulation of CD4+ and CD8+ T-cells: MLN cells and SPC were collected from each group of mice immediately following evaluation of clinical effect and blood drawing following the 7th challenge (week 50). Single MLN cells and SPCs were prepared and stimulated with crude CPE for 72 h. Cytokines in MLN culture supernatants (a) and SPC culture supernatants (b) were measured by ELISA. Data are shown as Means ± SEM of pooled cultures from representative of one of two experiments measured in triplicate (n=5 mice per group). **, P<0.01; ***, P<0.001 vs sham (Srivastava *et al.*, 2009) (With permission of Elsevier)

treated mice was reduced up to 75%, whereas CD8+ T-cell IFN- production was increased by up to 85% at the final challenge. Neutralization of INF- and depletion of CD8+ T-cells markedly attenuated FAHF-2 efficacy(Srivastava et al., 2009). This study showed that FAHF-2 provides long-term protection from anaphylaxis by inducing a beneficial shift in allergen-specific immune responses mediated largely by elevated CD8+ T-cell interferon- production.

16.2.4 Food Allergy Herbal Formula-2 Protection against Peanut Anaphylactic Reaction Is via Inhibition of Mast Cells and Basophils

Mast cells and basophils are key effector cells of IgE-mediated anaphylactic reactions. Using a well-established murine model of peanut (PN) anaphylaxis, we found that the food allergy herbal formula-2 (FAHF-2) abrogated anaphylactic reactions in mice and this protection persisted for at least 36 weeks after therapy was discontinued (Srivastava *et al.*, 2005; Qu *et al.*, 2007; Srivastava *et al.*, 2009). The increased IFN- production by CD8+ T-cells, as shown previously, may be an important mechanism underlying FAHF-2 modulated potent and long-term protection (Srivastava *et al.*, 2005; Qu *et al.*, 2007). However, neutralization of IFN- or depletion of CD8+ T-cells blocked the suppression of IgE and Th2 cytokine production induced by FAHF-2, while protection from anaphylaxis still existed up to 4 weeks post therapy (Srivastava *et al.*, 2009). We therefore hypothesized that FAHF-2 may also directly inhibit mast cells/basophils. To confirm this possibility, we investigated whether FAHF-2 inhibits mast cell/basophil numbers and IgE mediated activation. In this study, peanut allergic mice (PNA mice) were treated with FAHF-2 intragastrically (i.g.) for 7 weeks and challenged (i.g.) with peanut 1 day and 4 weeks post treatment. Peripheral blood basophil numbers and peritoneal mast cell numbers and Fc RI expression were determined. Direct effects of FAHF-2 on murine mast cell line MC/9, and effects of 4 fractions and 3 compounds isolated from FAHF-2 on rat basophilic leukemia cells (RBL-2H3) and human skin mast cells degranulation, and on the IgE mediated Syk signaling pathway were determined. The results showed that while all sham-treated PNA mice developed anaphylaxis, FAHF-2-treated PNA mice were protected against anaphylaxis following peanut challenge at 1 day and 4 weeks post therapy. Reduction of peripheral blood basophil numbers began after 1 week of treatment and continued for at least 4 weeks post therapy. The number of and Fc RI expression by peritoneal mast cells were also significantly decreased 4 weeks post therapy. FAHF-2 treated MC/9 cells showed significantly reduced IgE-induced Fc RI expression, Fc RI γ mRNA subunit expression, proliferation, and histamine release upon challenge. Fraction 2 (F2) from FAHF-2 inhibited RBL-2H3 cell and human mast cell degranulation. Three compounds from F2-berberine, palmatine and jatrorrhizine were inhibited of RBL-2H3 cell degranulation via suppressing Syk phosphorylation (Song *et al.*, 2010). This study demonstrates that FAHF-2 reduction of basophils and mast cell numbers as well as suppression of IgE-mediated mast cell activation may contribute to FAHF-2 persistent protection against peanut anaphylaxis.

16.2.5 Clinical Safety and Immunological Effects Clinical Study

Because FAHF-2 has been shown to have potent therapeutic effect in animal model of peanut anaphylaxis and high safety profile in preclinical study, and all individual herbs in FAHF-2 have long human use history and have been lawfully used in the US, FAHF-2 received approval as an investigational new drug (IND) by the US FDA for the study of its safety and efficacy in patients ages 12–45 years old with peanut and/or tree nut, fish and shellfish food allergies. This clinical study included both phase Ⅰ (safety study) and phase Ⅱ (efficacy study). The phase Ⅰ study included acute phase Ⅰ study and extended phase Ⅰ study. We have completed both phase Ⅰ studies. The acute phase Ⅰ is a randomized, double-blinded, placebo-controlled, dosage escalation trial to evaluate the safety and tolerability of FAHF-2 in subjects with food allergies. In this study, subjects received one of three doses of FAHF-2 or placebo: 2.2 grams (4 tablets), 3.3 grams (6 tablets) or 6.6 grams (12 tablets) 3 times/d for 7 d. Four active and two placebo subjects were treated at each dosage level. Vital signs, physical examination, laboratory data, pulmonary function tests and electrocardiogram data were monitored. Immunomo- dulatory studies were also performed. Nineteen food allergic participants were included in the study. Two subjects (one FAHF-2 and one placebo) reported mild gastrointestinal symptoms. One patient withdrew from the study due to an allergic reaction that was unlikely related to the study medication. No significant differences were found in vital signs, physical examination, laboratory data, pulmonary function tests and electrocardiogram data obtained at pre- and post-treatment visits. Significantly decreased IL-5 levels were found in the active treatment group after 7 days *in vitro* studies of peripheral blood mononuclear cells (PBMCs) cultured with FAHF-2 also demonstrated a significant decrease in IL-5 and increase in IFN-γ and IL-10 levels. This data showed that FAHF-2 is safe and well tolerated by subjects with food allergies and also showed beneficial immnunoregulatory effects, which is favorable for induction of food allergy tolerance. These results were published in *Annals Allergy Asthma Immunolology* (Wang *et al.*, 2010). The extended phase Ⅰ study was a 6-month open label study. The results also showed that FAHF-2 is safe and well-tolerated and showed inhibitory effect on basophil activation after 6 months of treatment The phase Ⅱ study is underway.

16.3 TCM for Food Allergy Associated Eczema

Atopic dermatitis (AD) is a chronic, inflammatory and pruritic skin disease that affects up to 20% of children. Regardless of dietary modifications, daily hydration therapy, and topical corticosteroids, some of these children continue to suffer from severe skin lesions-recalcitrant eczema. Clinically, 40% of refractory, moderate-to-severe atopic dermatitis is associated with food hypersensitivity. TCM has been used for centuries to treat eczema. However, an effect on recalcitrant eczema has not been previously reported. We issued a report on a case series of 14 patients [median age 5.4 years (IQR 0.5–52)] with persistent recalcitrant AD who were treated with TCM at Ming Qi Natural Health Center in Manhattan between August 2006 and May 2008, titled a retrospective analysis of patients with AD who received TCM. The TCM consisted of Shizheng Herbal Tea, herbal bath additive, herbal creams, and acupuncture. The SCORAD index for AD severity was calculated, with scores >40–103 considered severe, 15–40 moderate, and <15 mild. The Dermatology Life Quality Index (DLQI) was calculated on a scale of 0–30, with a score of 30 representing the highest impairment to life quality. Baseline median (IQR) SCORAD and DLQI scores were 89 (42–103) and 17 (10–30), compared to 11 (0–62) and 1 (0–14) after a median of 8 months (3–24) treatment (*t*-test, $P<0.0001$ and $P<0.0001$). Good improvement (60%–90% reduction in SCORAD) was reached in 13/14 patients after 3.3 mo (1.6–8.6). Greater than 50% improvement in DLQI scores was achieved in 13/14 patients in 2.4 mo (0.7–5.9). Peripheral eosinophilia decreased from 1000±700 mc/L to 500±200 ($P=0.03$) with no change in total blood counts. No abnormality of liver and kidney function was observed. Patients reported a reduction in use of steroids, antibiotics and antihistamines within 3 months of TCM treatment. These results showed that TCM is both a safe and effective treatment for patients with persistent AD, especially those with severe disease and significant life quality impairment (Wisniewski J *et al.*, 2010).

16.4 Conclusion

TCM is beginning to play a role in the US health care system. There is increasing scientific evidence demonstrating that TCM has potential for treating food allergies and associated eczema. Although the clinical effects are promising, continued research is required to ensure that some of these herbal medicines progress from dietary supplements to prescription drugs in the US. Our goals are to accelerate clinic trials and to identify the active compounds that target on specific food allergy aspects. We are also building up collaborations with experts in the field of allergy and traditional Chinese medicine in Western countries and Eastern countries to accelerate the efforts to develop safe and effective natural products for treating food allergies and other allergic conditions.

Acknowledgements

These studies are supported by grants 1 P01 AT002625-01 and R011R01AT 001495-01A1 and 2 R01 AT001495-05A1 from the National Institutes of Health/the National Center for Complementary and Alternative Medicine/a grant from the Food Allergy Initiative and Winston Wolkoff Children's Holistic Medicine for Allergy Foundation to Dr. Xiumin Li, and part by a grant from the National Institutes of Health/National Institute of Allergy and Infectious Diseases, AI083883 to Julie Wang.

I would like to give a great thanks to my colleagues and collaborators including Hugh Sampson, MD, Kamal Srivastava Mphil, Jullie Wang MD, Sangita Patil Ph.D, Nan Yang Ph.D, Wei Zhao MD, Ph.D, Mia Oh MD, Julia Wisniewski MD, Anna Nowak-Wegrzyn MD Joseph Goldfarb Ph.D, Jacob Kattan MD,Ying Song MD, Sylvan Wallenstein Ph, Sally Noone RN, Jingjing Mei RN, Brian Schofield JD, Scott Sicherer MD and all FAHF-2 study team.

Author's Disclosure: Xiu-Min Li, as a consultant for the Food Allergy Initiative, receives grant support from the Food Allergy Initiative and the NIH Xiu-Min Li and Hugh Sampson have shares of US Patent PCT/US 05/08600 on FAHF-2, PCT/US05/08600 on ASHMI, and Herbal Springs LLC.

References

Bensky, D., Barolet, R., Bensky, D., *et al.* (1993). Chinese Herbal Medicine: Formulas & Strategies. Seattle, WA: Eastland Press, pp. 466-467.

Bjorksten, B. (2005). Genetic and environmental risk factors for the development of food allergy. Curr Opin Allergy Clin Immunol, 5, 249-253.

Bousquet, J., Bousquet, P. J., Godard, P., *et al.* (2005). The public health implications of asthma. Bull. World Health Organ, 83, 548-554.

Eichenfield, L.F., Hanifin, J.M., Beck, L.A., *et al.* (2003). Atopic dermatitis and asthma: parallels in the evolution of treatment. Pediatrics, 111, 608-616.

Hon, K.L., Leung, T.F., Ng, P.C., *et al.* (2007). Efficacy and tolerability of a Chinese herbal medicine concoction for treatment of atopic dermatitis: A randomized, double-blind, placebo-controlled study. Br J Dermatol 157(2), 357-363.

Isolauri, E., Huurre, A., Salminen, S., *et al.* (2004). The allergy epidemic extends beyond the past few decades. Clin Exp Allergy, 34, 1007-1010.

Jindal, V., Ge, A., Mansky, P.J. (2008). Safety and efficacy of acupuncture in children: a review of the evidence. J. Pediatr. Hematol. Oncol., 30, 431-442.

Jones, S. M., Pons, L., Roberts, J.L., *et al.* (2009). Clinical efficacy and immune regulation with peanut oral immunotherapy. J Allergy Clin Immunol, 124, 292-300.

Kattan, J.D., Srivastava, K.D., Zou, Z.M., *et al.* (2008). Pharmacological and immunological effects of individual herbs in the Food Allergy Herbal Formula-2 (FAHF-2) on peanut allergy. Phytother. Res., 22, 651-659.

Leung, D.Y., Sampson, H.A., Yunginger, J., *et al.* (2003). Effect of anti-IgE therapy in patients with peanut allergy. N Engl J Med, 348, 986-993.

Li, X.M., Teper, A., Zou Z.M., *et al.* (2002). The Chinese herbal formula, MSSM-002, can reverse the established Th2 phenotype, which is accompanied with down regulation of the Th2 transcription factor GATA-3. J Allergy Clin Immunol (Abstract), 109, S155.

Mark, J.D. (2007). Integrative medicine and asthma. Pediatr Clin North Am, 54, 1007-1023.

miz-Echevarria, L., Bilbao, A., Andrade, F., *et al.* (2008). Fatty acid deficiency profile in children with food allergy managed with elimination diets. Acta Paediatr, 97, 1572-1576.

NCCAM. http://nccam.nih.gov/health/backgrounds/wholemed.htm.Publication No. D347, updated November 2010.

Nelson, H.S., Lahr, J., Rule, R., *et al.* (1997). Treatment of anaphylactic sensitivity to peanuts by immunotherapy with injections of aqueous peanut extract. J Allergy Clin Immunol, 99, 744-751.

Noimark, L., Cox, H. E. (2008). Nutritional problems related to food allergy in childhood. Pediatr. Allergy Immunol, 19, 188-195.

O'Connell, E. J. (2004). The burden of atopy and asthma in children. Allergy, 59(Suppl 78), 7-11.

Oppenheimer, J.J., Nelson, H.S., Bock, S.A., *et al.* (1992). Treatment of peanut allergy with rush immunotherapy. J Allergy Clin Immunol, 90, 256-262.

Ostblom, E., Egmar, A.C., Gardulf, A., *et al.* (2008). The impact of food hypersensitivity reported in 9-year-old children by their parents on health-related quality of life. Allergy, 63, 211-218.

Pali-Scholl, I., Renz, H., Jensen-Jarolim, E. (2009). Update on allergies in pregnancy, lactation, and early childhood. J Allergy Clin Immunol, 123, 1012-1021.

Prescott, S.L., Clifton, V. (2009). Asthma and pregnancy: emerging evidence of epigenetic interactions in utero. Curr Opin Allergy Clin Immunol, 9, 417-426.

Qu, C., Srivastava, K., Ko, J., Zhang, T.F., *et al.* (2007). Induction of tolerance after establishment of peanut allergy by the food allergy herbal formula-2 is associated with up-regulation of interferon-gamma. Clin Exp Allergy, 37, 846-855.

Rolinck-Werninghaus, C., Staden, U., Mehl, A., *et al.* (2005). Specific oral tolerance induction with food in children: transient or persistent effect on food allergy? Allergy, 60, 1320-1322.

Sawni, A.,Thomas, R. (2007). Pediatricians' attitudes, experience and referral patterns regarding Complementary/Alternative Medicine: A national survey. BMC. Complement Altern. Med., 7, 18.

Sicherer, S.H., Munoz-Furlong, A., Godbold, J.H., *et al.* (2010). US prevalence of self-reported peanut, tree nut, and sesame allergy: 11-year follow-up. J Allergy Clin Immunol, 125(6), 1322-1326.

Sicherer, S.H., Noone, S.A., Munoz-Furlong, A. (2001). The impact of childhood food allergy on quality of life. Ann. Allergy Asthma Immunol, 87, 461-464.

Sicherer, S.H., Sampson, H.A. (2007). Peanut allergy: Emerging concepts and approaches for an apparent epidemic. J Allergy Clin Immunol, 120, 491-503.

Slader, C.A., Reddel, H.K., Jenkins, C.R., *et al.* (2006). Complementary and alternative medicine use in asthma: Who is using what? Respirology, 11, 373-387.

Song, Y., Qu, C.F., Srivastava, K.D., *et al.* (2010). Food allergy herbal formula 2 protection against peanut anaphylactic reaction is via inhibition of mast cells and basophils. J Allergy Clin Immunol, 126, 1208-1217.

Srivastava, K.D., Kattan, J.D., Zou, Z.M., *et al.* (2005). The Chinese herbal medicine formula FAHF-2 completely blocks anaphylactic reactions in a murine model of peanut allergy. J Allergy Clin Immunol, 115, 171-178.

Srivastava, K.D., Qu, C., Zhang, T., *et al.* (2009). Food Allergy Herbal Formula-2 silences peanut-induced anaphylaxis for a prolonged posttreatment period via IFN-gamma-producing CD8+ T-cells. J Allergy Clin Immunol, 123, 443-451.

The State Pharmacopoeia Commission of the People's Republic of China (2005). Pharmacopoeia of the People's Republic of China (6 ed.) (Vol. 1). People's Medical Publishing House.

The US Food and Drug Administration (FDA) & Center for Drug Evaluation and Research (2004). Guidance for Industry Botanical Drug Products (Revised ed.). Volume 69, Number 111.

van Panhuys, N., Le Gros, G., McConnell, M.J. (2008). Epigenetic regulation of Th2 cytokine expression in atopic diseases. Tissue Antigens, 72, 91-97.

Wang, J., Patil, S., Yang, N., *et al.* (2010). Safety, tolerability, and immunologic effects of a food allergy herbal formula (FAHF-2) in food allergic individuals: A randomized, double-blinded, placebo-controlled, dose escalation phase Ⅰ study. Ann Allergy Asthma Immunol, 105, 75-84.

Wisniewski, J., Oh, M., Nowak-Wegrzyn, A., *et al.* (2010). Efficacy and Safety of Traditional Chinese Medicine for Treatment of Atopic Dermatitis (AD). J Allergy Clin Immunol (Abstract), 123, S37.

Worth, A., Sheikh, A. (2010). Food allergy and atopic eczema. Curr Opin Allergy Clin Immunol, 10(3), 226-230.

17

Immunomodulation by Food

Ingrid van der Meer[1], Jurriaan Mes[2], Amrah Weijn[2], Harry Wichers[3*]

[1] Plant Research International, Wageningen University and Research Center, P.O. Box 16, 6700 AA Wageningen, the Netherlands
[2] Agrotechnology and Food Sciences Group, Wageningen University and Research Center, P.O. Box 17, 6700 AA Wageningen, the Netherlands
[3] Cell Biology and Immunology group, Wageningen University and Research Center, P.O. Box 338, 6700 AH Wageningen, the Netherlands
[*] E-mail: harry.wichers@wur.nl

Abstract: An optimally active and balanced immune system is a pre-requisite for maintaining health. Immunity strongly varies in different life stages. This chapter describes the putative role that foods can play in achieving proper immune functioning. Different food components can exert effects on specific immune compartments. This offers potential for use in a variety of applications, depending on the characteristics of deviating metabolic processes and consumers or patients. It becomes more and more clear that the etiology of a given disease varies from person to person. Thus, also treatments, or supporting dietary measures, should be tuned more towards individual needs. This means that in the functional foods arena more knowledge is required of individual consumers' responses, and the deviations in metabolism that underlie the development of chronic anomalies. Therefore, functional foods should be regarded as food products with specific health benefits for specific groups of consumers. An overview is given of the current research published on immunomodulation by food type, with a focus on allergies, in relation to probiotics, beta-glucans, and fungal immunomodulatory proteins (FIPs). We conclude that immunomodulation via the diet has its merits in, at the least, specific immune-related anomalies such as allergies. It may well be that there is also potential in other application areas such as maintaining immune homeostasis.

17.1 Background

Proper immune functioning is pivotal to sustain health homeostasis. Since World War II, a decrease in the prevalence of "classical" infectious diseases, such as mumps, measles or tuberculosis, was observed. Concurrently, the prevalence of immune-related disorders, such as multiple sclerosis, irritable bowel disease, type 1 diabetes and various allergic diseases, strongly increased. Also, disease incidence decreased in the North-South direction in the Northern hemisphere, and vice versa in the Southern hemisphere (Bach *et al.*, 2002). Changes in health care and medical practice, living environment and life style, welfare levels, dietary habits, and the like are thought to be causally connected to this. A relationship with increased hygienic practice and subsequent reduced exposure to microbes was observed for the strongly increased occurrence of asthma and allergic disease (Strachan *et al.*, 1989). Chronic inflammatory processes are involved in many chronic diseases, for instance cardiovascular diseases, tumor formation, and metabolic syndrome (Wisse, 2004; Després and Lemieux, 2006; Kabingu *et al.*, 2007; Valdes *et al.*, 2007; Oda *et al.*, 2008).

Immunity, described as activity of its various compartments, is strongly varying in distinguishable life stages. There is, for instance, a decline in the functional capacity to elicit generalized and specific immune responses, and regulatory cells show a decrease in production and response to regulatory signals. Overall, these developments result in impairment, with age, of innate and adaptive immune responses, increased self-antigen reactivity, increased incidence of infection and increased vulnerability for the development of neoplastic malformations. Overall implications of such an aging phenomena of immunity are that susceptibility for specific health conditions may vary with age, immune impairment eventually leading to increased risk of mortality (Gardner and Murasko, 2002; Burns and Goodwin, 2004; Derhovanessian *et al.*, 2008; Gorczynski and Terzioglu, 2008; Srivastava *et al.*, 2008).

Dietary components and immunity. Many food components have an impact on immune functioning, although some perhaps have stronger effects than others. For instance, protein energy malnutrition (PEM) affects many aspects of immune functioning. The functioning of T-cell and B-cell subsets and their functions, and of innate immunity, is related to protein nutritional status. Relieving this protein shortage can restore diminished immune responses. In patients that suffer from inflammatory processes this happens more slowly. The imbalance between normal macrophage functions and decreased T-cell functions is, in part, responsible for chronic inflammatory processes in vulnerable patients. This explains why acute phase responses are more detrimental for nutritional status and nutrient reserves in elderly patients than in adults (Keusch, 2003; Lesourd, 2004). Not only absolute levels, but also dietary protein-carbohydrate ratio appears important in maintaining immune responsiveness. A low protein-high carbohydrate for longer intervals seems beneficial, as was shown in a rat model (Pal and Poddar, 2008).

Dietary components that influence immune functioning are being applied in so-called Immunonutrition. These are specific nutritional preparations for clinical use in critically ill patients with a jeopardized immune functioning. For instance, amino acids such as arginine and glutamine, nucleotides, and specific poly-unsaturated fatty acids, notably ω-3 PUFAs are incorporated into these preparations (Calder, 2007; Fernandes, 2008).

Immune functioning is a target for the development of functional foods. In particular, vitamins and minerals as Zn and Se, PUFAs such as docosahexaenoic acid (22:6n-3; DHA) and eicosapentaenoic acid (20:5n-3; EPA) gain attention in this respect. The reader is referred to publications that deal with these concepts (Harbige, 1996; Calder and Kew, 2002; Lopez-Varela *et al.*, 2002; Calder, 2007; Fernandes, 2008; Hoyles and Vulevic, 2008). For some other nutrients and products, research is still in a less advanced stage, e.g., at best in the stage of *in vitro* or animal models that will be dealt with below.

The conclusion that modulation of immune response may well serve as a basis for the development of functional foods appears obvious. To realize such paradigms, it is important to realize that functional foods should be regarded as food products with specific health benefits for targeted groups of consumers. Needs of individuals or groups of individuals, must be matched with immunomodulatory properties of specific products. Products that may be of benefit to allergic consumers do not necessarily have a beneficial effect on other consumers with different requirements for immune mitigation. The objective of these researches is to develop immunity-related markers for health effects of food, and to develop dietary or food products to support vital and balanced immune functioning throughout life.

17.2 Research into Immunomodulation by Food, with a Focus on Allergy

17.2.1 Probiotics

Probiotic bacteria are non-pathogenic micro-organisms that are thought to exert positive health effects on the host. In the majority of cases, *Lactobacillus* or *Bifidobacterium* spp. are studied for such beneficial effects. Relationship between the composition of intestinal microbiota and the occurrence and incidence of allergic diseases was noted approximately a decade ago (Bjorksten *et al.*, 2000; Bottcher *et al.*, 2000; Kalliomaki *et al.*, 2001). A comparison between Estonian and Swedish school children made clear that allergic children were less often colonized with *Lactobacilli* and more often with aerobic coliforms and *Staphylococcus aureus* (Bjorksten *et al.*, 2000; Bottcher *et al.*, 2000). Also, lower *Bifidobacterium* and higher *Clostridium* colonization numbers were observed (Kalliomaki *et al.*,

2001). These phenomena were thought to be relevant for proper maturation of immunity in the intestine, and to have effects on the balance between Th1- and Th2-cells (Bottcher *et al.*, 2000; Kalliomaki *et al.*, 2001). This suggests a perspective for nutritional and dietary intervention, via probiotics.

Lactobacilli (LGG), were given to 159 pregnant women, who were at risk for allergies, for 2 weeks and to their babies for 6 months. This resulted in 50% less infants in the LGG group that demonstrated atopic eczema at year 2 (Kalliomaki *et al.*, 2003). LGG and *Bifidobacterium lactis* Bb12 were shown to moderate severity of atopic eczema (Isolauri *et al.*, 2000). In a similar study serum IL-10 was elevated significantly in the LGG group (Pessi *et al.*, 2000). However, sometimes such results were reproducible by others (Rosenfeldt *et al.*, 2003), but sometimes this was not the case, e.g., in the development of atopic dermatitis (Brouwer *et al.*, 2006). In a group of teenagers and young adults, application of *Lactobacillus rhamnosis* failed to diminish symptoms of birch-pollen allergies (Helin *et al.*, 2002). Also, LGG-supplementation via diet a few weeks before delivery was found not to have an effect on proliferation of PBMCs from atopic or control neonates (Kopp *et al.*, 2008).

In conclusion, a strict experimental design is required to better understand and compare different studies. Results in allergy prevention are still variable. *In vitro* observations suggest that a possible mechanism for positive effects on allergy-related immunity is via restoring imbalances between T-cell subsets. This may proceed indirectly via TLR-signaling in innate immune cells. *In vivo* results may be strongly strain-dependent, as in *in vitro* screening of probiotics strain collections a variety of immune-effects, e.g., IL-10 inducing or Th1-skewing, may be observed (Vissers *et al.*, 2010; 2011). At this stage, however, effects of probiotics on gut permeability or intestinal allergen processing cannot be excluded. Responses likely also depend on patient characteristics; recently, interactions between *Lactobacillus*-S-layer protein and gut-DCs were described, which may contribute to the mechanism of action (Konstantinov *et al.*, 2008), in particular primarily on innate immunity. In addition to allergic disease, probiotics are being investigated in celiac disease, gastroenteritis, IBD, colon and colorectal cancer, antibiotic associated diarrhea etc. (de Vrese and Schrezenmeir, 2008).

17.2.2 β-Glucans

β-(1→3)-(1→6) glucans are widely occurring polymers of glucose with a β-(1→3) backbone and β-(1→6) cross-links. Typical Mr's are in the range of several hundreds of kDa's. This indicates Dp's of several thousands (Kataoka *et al.*, 2002; Volman *et al.*, 2008). β-Glucans can be found in bacteria, yeasts, fungi (Igor *et al.*, 2006), seaweed (Vetvicka *et al.*, 2007) and cereals (the latter mainly containing (1→4)-cross-links (Muralikrishna *et al.*, 2007).

Orally administered β-glucans (0.5%–1.0% from *Aureobasidium pullulans*) moderated food allergic reactions in ovalbumin-allergic Balb/C mice (Kimura *et al.*, 2007). Finely dispersed shiitake β-glucan (Lentinan), applied orally, significantly alleviated symptoms of Japanese cedar pollen-induced rhinitis in human patients. Sneezing, nasal congestion, and conjunctivitis were reduced, and oral uptake of the β-glucan before symptom onset suggested preventive effects. Allergic symptoms were not only relieved for seasonal allergies to cedar pollen but also for perennial allergies. Oral ingestion of β-1,3-glucan in allergic individuals reduced the spontaneous increase in allergen-specific and in total IgE titers. This clinical response to treatment correlated with the capacity of monocytes to bind to β-1,3-glucan (Yamada *et al.*, 2007). The route of application, and the degree of dispersion appeared important: non-dispersed lentinan did not give the desired effects (Kimura *et al.*, 2007; Yamada *et al.*, 2007) .

The immunomodulatory effect of β-glucans seems to result from activation of innate pathways, e.g., in macrophages (Kataoka *et al.*, 2002; Volman *et al.*, 2008). β-Glucans stimulated the production of TNF-α, IFN-γ and IL-12 in ICR-mice (Tada *et al.*, 2008). Similar observations were made for human peripheral blood mononuclear cells in which various fungal extracts were screened (Lull-Noguera *et al.*, 2005). For *Aureobasidium pullulans*, in particular, IL-8 production in PBMCs and a monocyte cell line was detected (Ikewaki *et al.*, 2007). Activation of T-cells in PBMC-cultures is probably an indirect effect of prior stimulation of innate cells. Reactions appear to depend on molecular characteristics of β-glucans such as Mr, branching and secondary structure.

A number of receptors for β-glucans (dectin-1, complement receptor 3, TLR2 and TLR6, scavenger receptors and lactosylceramide) have been identified. Dectin-1 is expressed on dendritic cells, macrophages, monocytes, neutrophils, eosinophils, some T-cells and in humans also on B-cells. Possible expression of dectin-1 on intestinal cells is under debate (Volman *et al.*, 2008). Interaction of dectin-1 and TLR2 upon β-glucan-binding causes synergistic effects on TNF-α and IL12 production, In this reaction cascade, NF-kB (Meyer-Wentrup, 2005) and tyrosine kinase are thought to be involved (Brown, 2006).

Peanut allergies relieving effects were observed in C3H/HeJ-mice, for an orally applied complex herbal preparation, in which β-glucans (from *Ganoderma lucidum*) were present (Li *et al.*, 2001). Orally applied *G. lucidum* preparations were effective in a mouse model for house dust mite allergies (Liu *et al.*, 2003).

Taken altogether, orally applied β-glucans or glucan-containing preparations can cause mitigation of various allergy symptoms. Details on mode of action, structure-function studies and research on application methods, for instance incorporation into food products or supplements, are required.

17.2.3 Fungal Immunomodulatory Proteins (FIPs)

FIPs are 15 kDa proteins of fungal origin. The FIPs form *Flammulina velutipes* (golden needle mushroom; FIP-fve), *Volvariella volvacea* ((paddy) straw mushroom; FIP-vvo), *Ganoderma lucidum* and *G. tsugae* (Japanese lacquer mushroom; LZ-8 and FIP-gts, resp.) have been described to share high sequence homology (Hsu *et al.*, 1997).

The X-ray structure of FIP-fve is known (Seow *et al.*, 2003; Paaventhan *et al.*, 2003). The protein was a homodimer. Each subunit contains an N-terminal secondary structural element, an α-helix followed by a β-strand, linked to a domain consisting almost exclusively of β-sheets adopting an Ig-like fold (Seow *et al.*, 2003). The α-helices H_A and H_B are amphipathic and the side chains of the amino acids on the hydrophobic face of one helix pack against those of the other α-helix. This results in binding via hydrophobic interactions. Lin *et al.* (1997) predicted that the N-terminal residues 1-13 of FIP-Gts formed a α-helix. Recombinant mutants of FIP-Gts lacking residues 1-13 (the N-terminal α-helix) were not able to dimerise. Triple mutants in which Leu5, Phe7 and Leu9 were deleted lost the amphipathic character, the ability to form dimers with itself and with the wild type FIP-Gts protein. This indicates that this is important for immunomodulatory activity (Lin *et al.*, 1997; Paaventhan *et al.*, 2003). Fve has lectin-like properties, being able to agglutinate erythrocytes, and were proposed to be classified as FIPs (Paaventhan *et al.*, 2003).

The biological relevance of FIPs for allergy management became apparent when they were administered, either orally or nasally, in food allergic, as well as respiratory allergic, mouse models. Feeding of FIP-fve every other day to ovalbumin-allergic Balb/c-mice (at 200 mcg/mouse (ca. 4 mg/kg), every other day), reduced the allergy symptom score, histamine release, and intestinal damage. The Th2-dominant phenotype was shifted towards a strongly elevated Th1-response, showing potential prophylactic properties (Hsieh *et al.*, 2003). FIP-fve could be applied in local nasal immune therapy to suppress allergic responses to house dust mite allergies in Balb/c mice (Liu *et al.*, 2003). Also here, a stronger Th1-response was observed.

FIP-fve appeared to stimulate IFN-γ production in human PBMCs via p38-MAPK-activation (Wang *et al.*, 2004). Our own observations (Jeurink *et al.*, 2008) and unpublished results indicated that natural FIPs or cloned FIPs from *Flammulina* or *Ganoderma* have a limited effect on NO-production in RAW 264.7 cells (a murine monocyte cell line), and a Th1-skewing effect in human PBMC-cultures from healthy volunteers. This was partly confirmed by observations for an immunomodulatory protein from *Auricularia* (Jew's ear) mushroom (Sheu *et al.*, 2004).

17.3 Directions for Future Research

Based on these observations, the option to modulate immune functioning via diet appears to be a realistic option. This approach appears to have merits in, at the least, specific immune-related aberrancies such as allergies. Likely, as there is no reason to assume beforehand that there are "immune compartments" whose activity can not be modified via diet, there is also potential in other application areas such as maintaining immune homeostasis. It appears as if different food components can exert effects on specific immune compartments, thus offering potential for use in a variety of applications, depending on the characteristics of deviating chronic processes and consumers, or patients characteristics. Foods can thus play a significant role in fortifying and balancing immune responses.

Considering the pivotal role of immune homeostasis in general health, and the very relevant and significant societal and economic perspectives, immunomodulation holds a significant potential for application. As it becomes much clearer that the etiology of a given "disease" varying from person to person, the conclusion appears at hand that also treatments, or supporting dietary measures, should be tuned towards individual needs. This emphasizes that, in the "functional foods" domain, more knowledge of individual consumers' responses, and their deviations in metabolism that underlie the development of chronic anomalies, is required. More insight in underlying mechanisms is a prerequisite to realize product development.

With respect to the impact of immunomodulatory foods and food components on allergic reactions, as summarized in this manuscript, in all described cases a skewing of cytokine profiles from Th2-dominance towards a more re-balanced state, with an enhanced Th1-response, could be observed. In all cases the active products were micro-organisms or products derived thereof. This leads to the hypothesis that, if indeed an overly hygienic living environment is at the basis of the increase in allergic reactions, this situation can perhaps be (partly) restored by exposure to specific microbial constituents, likely activating innate immunity, and subsequently skewing adaptive immunity towards a strengthened cell-mediated response.

References

Bach, J.F. (2002). The effect of infections on susceptibility to autoimmune and allergic diseases. New Engl J Med, 347, 911-920

Bjorksten, B., Naaber, P., Sepp, E., Mikelsaar, M. (2000). The intestinal microflora in allergic Estonian and Swedish 2-year-old children. Clin Exp Allergy, 29(3), 342-346.

Bottcher, M.F., Nordin, E.F., Sandin, A., *et al.* (2000). Microflora-associated characteristics in faeces from allergic and nonallergic infants. Clin Exp

Allergy, 30(11), 1590-1596.
Brown, G.D. (2006). Dectin-1: A signaling non-TLR pattern-recognition receptor. Nature Rev Immunol, 6, 33-43.
Brouwer, M.L., Wolt-Plompen, S.A., Dubois, A.E., *et al.* (2006). No effects of probiotics on atopic dermatitis in infancy: A randomized placebo-controlled trial. Clin Exp Allergy, 36(7), 899-906.
Burns, E.A., Goodwin, J.S. (2004). Effects of aging on immune function. J Nutr Health Ageing, 8(1), 9-18.
Calder, P.C. (2007). Immunonutrition in surgical and critically ill patients. Br J Nutr, 98 (S1), S133-9.
Calder, P.C., Kew, S. (2002). The immune system: A target for functional Foods? Br J Nutr, 88, S165-S176.
Derhovanessian, E., Solana, R., Larbi, A. *et al.* (2008). Immunity, ageing and cancer. Immunity & Ageing, 5, 11.
Després, J.P., Lemieux, I. (2006). Abdominal obesity and metabolic syndrome. Nature, 444(7121), 881-887.
de Vrese, M., Schrezenmeir, J. (2008). Probiotics, prebiotics, and synbiotics. Adv Biochem Engin/Biotechnol, 111, 1-66.
Fernandes, G. (2008). Progress in nutritional immunology. Immunol Res, 40, 244-261.
Gardner, E.M., Murasko, D.M. (2002). Age-related changes in Type 1 and Type 2 cytokine production in humans. Biogerontology, 3, 271-289.
Gorczynski, R.M., Terzioglu, E. (2008). Aging and the immune system. Int Urol Nephrol, 40(4), 1117-1125.
Harbige, L.S. (1996). Nutrition and immunity with emphasis on infection and autoimmune disease. Nutr Health, 10, 285-312.
Helin, T., Haahtela, S., Haahtela, T. (2002). No effect of oral treatment with an intestinal bacterial strain, Lactobacillus rhamnosus (ATCC 53103), on birch-pollen allergy: A placebo-controlled double-blind study. Allergy, 57(3), 243-246.
Hoyles, L., Vulevic, J. (2008). Diet, immunity and functional foods. Adv Exp Med Biol, 635, 79-92.
Hsieh, K.Y., Hsu, C.I., Lin, J.Y., *et al.* (2003). Oral administration of an edible-mushroom-derived protein inhibits the development of food-allergic reactions in mice. Clin Exp Allergy, 33, 1595-1602.
Hsu, H.C., Hsu, C.I., Lin, R.H. *et al.* (1997). Fip-vvo, a new fungal immunomodulatory protein isolated from Volvariella volvacea. Biochem J, 323, 557-565
Ikewaki, N., Fujii, N., Onaka, T., *et al.* (2007). Immunological actions of Sophy β-glucan (β-1,3-1,6 glucan), currently available commercially as a health food supplement. Microbiol Immunol, 51(9), 861-873.
Isolauri, E., Arvola, T., Sütas, Y., *et al.* (2000). Probiotics in the management of atopic eczema. Clin Exp Allergy, 30(11), 1604-1610.
Jeurink, P.V., Lull-Noguera, C., Savelkoul, H.F.J., Wichers, H.J. (2008). Immunomodulatory capacity of fungal proteins on the cytokine production of

human peripheral blood mononuclear cells. Int Immunopharmacol, 8(8), 1124-1133.

Kabingu, E., Vaughan, L., Owczarczak, B., *et al.* (2007). CD8+ T-cell-mediated control of distant tumours following local photodynamic therapy is independent of CD4+ T-cells and dependent on natural killer cells. Br J Cancer, 96, 1839-1848.

Kalliomäki, M., Kirjavainen, P., Eerola, E., *et al.* (2001). Distinct patterns of neonatal gut microflora in infants in whom atopy was and was not developing. J Allergy Clin Immunol, 107(1),129-134.

Kalliomäki, M., Salminen, S., Poussa, T., *et al.* (2003). Probiotics and prevention of atopic disease: 4-year follow-up of a randomised placebo-controlled trial. Lancet, 361(9372), 1869-1871.

Kataoka, K., Muta, T., Yamazaki, S., Takeshige, K. (2002). Activation of macrophages by linear (1→3)-β-D-glucans. J Biol Chem, 277(39), 36825-36831.

Keusch, G.T. (2003). The history of nutrition: malnutrition, infection and immunity. J Nutr, 133(1), 336S-340S. Kimura, Y., Sumiyoshi, M., Suzuki, T., *et al.* (2007). Inhibitory effects of water-soluble low-molecular-weight β-(1,3–1,6) D-glucan purified from Aureobasidium pullulans GM-NH-1A1 strain on food allergic reactions in mice. Int Immunopharmacol, 7, 963-972.

Konstantinov, S.R., Smidt, H., de Vos, W.M., *et al.* (2008). S layer protein A of Lactobacillus acidophilus NCFM regulates immature dendritic cell and T-cell functions. Proc Natl Acad Sci U S A,105(49), 19474-19479.

Kopp, M.V., Goldstein, M., Dietschek, A., *et al.* (2008). Lactobacillus GG has in vitro effects on enhanced interleukin-10 and interferon-gamma release of mononuclear cells but no *in-vivo* effects in supplemented mothers and their neonates. Clin Exp Allergy, 38(4), 602-610.

Lesourd, B. (2004). Nutrition: A major factor influencing immunity in the elderly. J Nutr Health Aging, 8(1), 28-37.

Li, X.M., Zhang, T.F., Huang, C.K., *et al.* (2001). Food Allergy Herbal Formula-1 (FAHF-1) blocks peanut-induced anaphylaxis in a murine model. J Allergy Clin Immunol, 108, 639-646.

Lin, W.H., Hung, C.H., Hsu, C.I., *et al.* (1997). Dimerization of the N-terminal amphiphatic α-helix domain of the fungal immunomodulatory protein from Ganoderma tsugae (Fip-gts) defined by a yeast two-hybrid system and site-directed mutagenesis. J Biol Chem, 272, 20044-20048.

Liu, Y.H., Tsai, C.F., Kao, M.C., *et al.* (2003). Effectiveness of Dp2 nasal therapy for Dp-2 induced airway inflammation in mice: Using oral Ganoderma lucidum as an immunomodulator. J Microbiol Immunol Infect, 36, 236-242.

Liu, Y.H., Kao, M.C., Lai, Y.L., Tsai, J.J. (2003). Efficacy of local nasal immunotherapy for Dp2-induced airway inflammation in mice: Using Dp2-peptide and fungal immunomodulatory peptide. J Allergy Clin Immunol, 112(2), 301-310.

Lull-Noguera, C., Wichers, H.J., Savelkoul, H.F.J. (2005). Anti-inflammatory and immuno-modulating properties of fungal metabolites. Mediat Inflamm,

2005(2), 63-80.

López-Varela, S., González-Gross, M., Marcos, A. (2002). Functional foods and the immune system: A review. Eur J Clin Nutr, 56(S3), S29-S33.

Meyer-Wentrup, F., Cambi, A., Adema, G.J., Figdor, C.G. (2005). "Sweet Talk": Closing in on C Type Lectin Signaling. Immunity, 22, 399-400.

Muralikrishna, G. and Subba Rao, M.V.S.S.T. (2007) Cereal non-cellulosic polysaccharides: Structure and function relationship–an overview. Crit Rev Food Sci Nutr, 47(6), 599-610.

Oda, E. (2008). The metabolic syndrome as a concept of adipose tissue disease. Hypertens Res, 31(7), 1283-1291.

Paaventhan, P., Jospeh, J.S., Seow, S.V., *et al.* (2003). A 1.7Å structure of fve, a member of the new fungal immunomodulatory protein family. J Mol Biol, 322, 461-470.

Pessi, T., Sütas, Y., Hurme, M., *et al.* (2000). Interleukin-10 generation in atopic children following oral Lactobacillus rhamnosus GG. Clin Exp Allergy, 30(12), 1804-1808.

Pal, S., Poddar, K. (2008). Dietary protein-carbohydrate ratio: exogenous modulator of immune response with age. Immunobiol, 213, 557-566.

Rosenfeldt, V., Benfeldt, E., Nielsen, S.D., *et al.* (2003). Effect of probiotic Lactobacillus strains in children with atopic dermatitis. J Allergy Clin Immunol, 111(2), 389-395.

Schepetkin, I.A., Quinn, M.T. (2006). Botanical polysaccharides: Macrophage immunomodulation and therapeutic potential. Int Immunopharmacol, 6, 317-333.

Seow, S.V., Kuo, I.C., Paaventham, P., *et al.* (2003). Crystallisation and preliminary X-ray crystallographic studies on the fungal immunomodulatory protein Fve from the golden needle mushroom (*Flammulina velutipes*). Acta Crystallog, 59, 1487-1489.

Sheu, F., Chien, P.J., Chien, A.L., *et al.* (2004). Isolation and characterization of an immunomodulatory protein (APP) from the Jew's Ear mushroom Auricularia polytricha. Food Chem, 87(4), 593-600.

Strachan, D.P. (1989). Hay fever, hygiene, and household size. Br Med J, 299(6710), 1259-1260.

Srivastava, S., Lundqvist, A., Childs, R. (2008). Natural killer cell immunotherapy for cancer: a new hope. Cytotherapy, 10(8), 775-783.

Tada, R., Tanioka, A., Iwasawa, H., *et al.* (2008). Structural characterisation and biological activities of a unique type β-D-glucan obtained from Aureobasidium pullulans. Glycoconj J, 25(9), 851-861.

Valdés-Ramos, R., Benítez-Arciniega, A.D. (2007). Nutrition and immunity in cancer. Br J Nutr, 98(Suppl 1), S127-132.

Vetvicka, V., Dvorak, B., Vetvickova, J., *et al.* (2007). Orally administered marine (1→3)-β-D-glucan Phycarine stimulates both humoral and cellular immunity. Int J Biol Macromol, 40, 291-298.

Vissers, Y.M., Snel, J., Zuurendonk, P.F., *et al.* (2010). Differential effects of

Lactobacillus acidophilus and Lactobacillus plantarum strains on cytokine induction in human peripheral blood mononuclear cells. FEMS Immunol Med Microbiol, 59(1), 60-70.

Vissers, Y.M., Snel, J., Zuurendonk, P.F., *et al.* (2011). Lactobacillus strains differentially modulate cytokine production by hPBMC from pollen-allergic patients. FEMS Immunol Med Microbiol, 61, 28-40.

Volman, J.J., Ramakers, J.D., Plat, J. (2008). Dietary modulation of immune function by beta-glucans. Physiol Behav, 94(2), 276-284.

Wang, P.H., Hsu, C.I., Tang, S.C., *et al.* (2004). Fungal immunomodulatory protein from Flammulina velutipes induces interferon-γ production through p38 mitogen-activated protein kinase signaling pathway. J Agric Food Chem, 52, 2721-2725.

Wisse, B.E. (2004). The inflammatory syndrome: The role of adipose tissue cytokines in metabolic disorders linked to obesity. J Am Soc Nephrol, 15(11), 2792-800.

Yamada, J., Hamuro, J., Hatanaka, H. *et al.* (2007). Alleviation of seasonal allergic symptoms with superfine β-1, 3-glucan: A randomized study. J Allergy Clin Immunol, 119, 1119-1126.

18

Prospective Studies on Mite (*Dermatophagoides farinae*) Allergen Immunotherapy in China

Ting-Huan Wen

Medical Acarology Laboratory, Shanghai Medical College, Fudan University, Shanghai, 200032, China
E-mail: thwen26@gmail.com

Abstract: Mite allergy study was founded in China beginning in the 1970s and conducted at Shanghai First Medical College. The highest allergenic contents had been detected in comparing the results with those of foreign literature . Around 80% of allergic cases were sensitive to mite allergen by the skin prick tests in concordance with the nasal provocation test and sIgE assay results. Seasonal classic immunotherapy for allergic patients by product *Injectio Dermatophagoidei Farinae*, the only commercial allergen licensed by the Chinese government, achieved very good results in improving symptoms of allergic disorder in a majority of the cases, in addition, long-lasting effects of mite immunotherapy was also certified with minimal adverse reactions. Studies on sublingual mite vaccine for mite allergic disorders have been synchronized with identified foreign trends since 1992. Sublingual drops are further acceptable for children's cases almost without age restriction with a high efficacy rate. It has also been shown that rush mite immunotherapy can achieve quick relief of allergic symptoms and long-lasting curative effects.

18.1 Introduction

A lot of local native resources are involved in Chinese daily life as remedies for allergic disorders. House dust and soil have been used as a popular secret recipe for a visitor suffering from allergic attack during his visits in foreign territories.

The contemporary allergen immunotherapy (IT) in mainland China was first identified in 1956, and at the same time, two allergy schools were also founded. One was the Beijing Union Medical College Hospital, in the ENT Department led by late Prof. Qing-Song Zhang processing subcutaneous injection for allergic rhinitis with polyvalent allergen desensitization assigned specifically for the individual patient twice a week and subsequently maintained for many years. The other one was the Shanghai First Medical College Huashan Hospital, in the Asthma Section of the Department of Internal Medicine led by late Prof. Min-Gang Wang processing subcutaneous injections for asthmatics with univalent *Platanus hispanica* pollen extract in spring and *Ricinus communis* pollen extract in autumn once a week maintaining for a 4-month period alternatively every year, and several seasonal courses of desensitization were needed for long-lasting treatment of the symptomatic attacks.

18.2 Mite Allergen

Voorhorst and his Department of Allergology of Leiden University Hospital revived the mite allergy during the 1950s, and proved that the mite allergies are a worldwide problem (Voorhorst *et al.*, 1969). Since 1970's, studies on mite allergies have been led by the Medical Acarology Laboratory, Shanghai First Medical College (Wen, 2005), In which high concentrations of the major allergens Der f 1 and Der f 2 had been detected by Heymann *et al.* (1989) and Wen *et al.* (1990). The highest ratio contained in the Chinese pure culture of the dust mite *Dermatophagoides farinae* (*Df*) had been achieved in comparing with that of the US FDA reference and foreign commercial extracts, reaching 6.9 mg/mL total protein, and 621 μg/mL of Der f, 87 μg/mL of Der f 2.

18.3 Skin Prick Test with *Df* Extract

Skin prick tests (SPT)with *Df* extract for diagnosis of patients sensitive to mite and for epidemiological studies were conducted by this laboratory for the first time and foremost in China in early 1973. More than 80% of allergic asthmatic patients were sensitive to the mite allergen by SPT which was in concordance with the nasal provocation test and sIgE assay results(Peng *et al.*, 1985; Wen *et al.*, 1992) (Table 18.1). The result indicated that mite is a potent major indoor allergen in this country.

Table 18.1 Comparison between skin prick test, nasal provocation test and sIgE assay with *Dermatophagoides farinae* (*Df*) extract

Group	Age (year)	SPT		NPT		*Df* sIgE	
		Case	Positive	Case	Positive	Case	Positive
Asthma	16–39	82	67 (81.7%)	65	54 (83.1%)	76	60 (78.9%)
Control	17–38	163	20 (12.8%)	151	16 (10.6%)	47	1 (2.1%)

Asthma: Controlling $P<0.001$

SPT (skin prick test): NPT (nasal provocation): *Df* sIg E $P>0.05$

Concordance rate SPT: NPT 81.5%; SPT: sIgE 89.5%; NPT: sIgE 78.5%

18.4 Classic Immunotherapy with Mite Extract

The classic seasonal immunotherapy had been conducted in Shanghai First Medical College Huashan Hospital in 1973, promising curative efficacy, up to 76.5% after two or more seasonal desensitizations (Chu *et al.*, 1981) (Tables 18.2 & 18.3), and even more efficacy were shown in the patients of atopic dermatitis (88.7%), meanwhile minimal adverse reactions were achieved by *Df* allergen IT (Wen *et al.*, 1998). Then *Injectio Dermatophagoidei Farinae* became the first government licensed commercial extract for allergy IT in China in 1978 and approved as the state pharmaceuticals by SFDA in 2003.

Table 18.2 Efficacy of first course of seasonal *Df* SIT

Efficacy \ Year	1973	1974	1975	1976	Total (%)
Excellent	5	7	5	2	19 (4.0)
Good	8	29	13	11	61 (12.9)
Fair	44	133	36	22	235 (49.6)
Nay	50	64	29	16	159 (33.5)
Grand Total	107	233	83	51	474 (100)

Excellent+Good+Fair 315/474 (66.5%) effective cases

Table 18.3 Efficacy of *Df* SIT after second seasonal courses

Efficacy \ Year	1974	1975	1976	Total (%)
Excellent	6	6	38	50 (9.8)
Good	53	24	69	146 (28.5)
Fair	60	66	69	195 (38.1)
Nay	18	58	45	121 (23.6)
Grand Total	139	154	221	512 (100)

Excellent+Good+Fair 393/514 (76.5%) effective cases

Extended studies on various modes of *Df* extract IT had been conducted in Shanghai, such as aluminum hydroxide modified *Df* extract injections with slow releasing and long effect for two-week intervals (Wang *et al.*, 1994; Jin *et al.*, 1995) (Table 18.4) and *Df* percutaneous smear drops for daily use (Zhu *et al.*, 1992) (Table 18.5)

Table 18.4 Efficacy of $Al(OH)_3$ absorbed *Df* extract in comparing with crude mite (*Df*) IT with asthmatics

Allergen IT	Case No.	Dosage (BU)	Effect. (%)	SPT 20 wk	SPT 10 mon	sIgG	sIgE	Adv. Reaction (swelling)
$Al(OH)_3$ -*Df*	18	16220*(1000/2 wk)	94.3	↓	↓	↑	↑	(5×1.5)d(Ø3.2 m)
Df	18	6820 (200/wk)	88.4	↓	↓	↑	↑	(14×4)d(Ø6.1 cm) Wang *et al.*, 1994
$Al(OH)_3$ -*Df*	22	6,220 (1,000/2wk)	91.0	↓		↑	↑	-
Df	19	2,220 (200/wk)	84.2	↓		↑	↑	Jin *et al.*, 1995

* $P<0.05$

Table 18.5 Efficacy of percutaneous IT with mite extract for allergic rhinitis compared with classic injections

IT	Case No. F	Case No. M	Age Aver. (year)	Rhinitis Anamn. (year)	Effective Cases No. (%)	Dosage (BU)
Df Pc Rubbing	23	27	30.5	4.5	38 (76)	350×4 wk
Df Sc Injection	23	27	34.4	3.5	37 (74)*	200×10 wk

Pc: percutaneous; Sc: subcutaneous; * $P>0.05$

Retrospective study on the long-lasting clinical efficacy of mite specific immunotherapy with asthmatics had been carried out for a 11-year period, the achievement of the classic IT was acceptable, and 97.1% cases were effective (Su *et al.*, 2001) (Table 18.6).

Table 18.6 Long-term efficacy of *Df* SIT with asthmatics sensitive to mite

Grade	7-11 years SIT (*n*=30) Case	%	2-5 years SIT (*n*=151) Case	%	4-10 years Long-term efficacy(*n*=140) 4	5	6	7	8	9	10	Total	%
Excel.	18	60.0	96	64.8	8	10	17	17	17	3	2	74	52.9
Good	10	33.3	34	22.5	0	4	14	10	7	1	0	36	25.7
Fair	2	0.7	10	6.6	0	5	7	12	2	0	0	26	18.6
Nay	0	0	11	7.3	1	0	2	1	0	0	0	4	2.9
Effect.	30	100	140	92.7	8	19	38	39	26	4	2	136	97.1

18.5 Mite Sublingual Vaccination

Since 1992, sublingual drops of *Df* extract have been studied once again in the same laboratory in Shanghai. Further excellent (>90%) clinical efficacy was obtained with *Df* sublingual immunotherapy (SLIT) for allergic patients and it is comparable with that of *Df* injections in this country and as well as that SLIT reported in foreign literature. Moreover it has been shown that there is almost no age restriction in using *Df* SLIT for asthma, rhinitis, atopic dermatitis, chronic urticaria and infantile eczema in this country (Wen *et al.*, 1997; Zhao *et al.*, 1999; Zheng *et al.*, 2000; Pan *et al.*, 2001; Zhang *et al.*, 2003; Huang *et al.*, 1997; 2003; Wen, 2006) (Table 18.7). The achievement of *Df* allergen SLIT now is being adopted freely by WOLWO Biotech and licensed by SFDA.

Df extract tablet SLIT for allergic asthma and/or rhinitis and all had achieved the same clinical efficacy as that of commercial *Df* injection. Specific and non-specific clinical efficacies have been shown in *Df* extract IT for the patients sensitive to different allergenic categories.

Table 18.7 Curative efficacies of SLIT with SMU-*Df* extract

Author	Case No.	Symptom	Course dur.	Dosage (BU)	Efficacy
Wen *et al.*, 1997	33	asthma	23 wk	92,210	93.9%
Huang *et al.*, 1997	50	bronchiolitis	12–20 wk	50,000–70,000	94.4%
Zhao *et al.*, 1999	13	asthma	12 mon	50,000–70,000	100%
Zheng *et al.*, 2000	60	rhinitis/asthma	12 mon	500,000	88.6%
Pan *et al.*, 2001	32	asthma	13 wk	100,000	93.0%
Zhang *et al.*, 2003	35	asthma	12 mon	500,000	100%
Huang *et al.*, 2003	28	asthma	28 wk	80,000	92.9%
Yu *et al.*, 2004	24	atopic dermatitis	8-15 mon	70,000–130,000	87.5%

18.6 Rush Schedule *Df* Vaccine IT

Evidence has been shown that a rush schedule with high dosage (5–20 μg of major allergen) IT can help to alleviate the allergic symptoms as quickly as one week after starting IT (Bousquet *et al.*, 1998; Akdis and Akdis, 2007), and 1–2 months intermittent maintenance IT could be an improved approach for IT (Peng *et al.*, 1992). Improvement of the patient's quality of life and lowering the ratio of cost-effectiveness of IT could then be achieved. The risk factor of adverse reactions during a rush schedule IT could be prevented as in concordance with antihistamine or topical steroid at initial injections. A trial of rush schedule IT for the urticaria patients has been shown to have better efficacy than that of the classic injections (Table 18.8).

Table 18.8 Efficacious comparison between two *Df* IT schedules and antihistamine treatment for urticaria

Treatment	Excel. (%)	Good (%)	Fair (%)	Nay (%)	Total
Classic schedule IT	34 (28.3)	32 (26.7)	32 (26.7)	22 (18.3)	120
Rush schedule IT	50 (39.1)	50 (39.1)	30 (23.4)	0 (0)	130
Subtotal of IT	84 (33.8)	82 (32.3)	62 (25.0)	22 (8.9)	250
Antihistamine	0 (0)	0 (0)	20 (12.7)	138 (87.3)	158

State standards of the mite allergens including *D. farinae* and *D. pteronyssinus* (*Dp*), *Blomia tropicalis* (*Bt*), etc. now are in processing. Recombinants of the major allergens of Der f and Der p and other mites have also been producing in China

References

Akdis, M., Akdis, C.A. (2007). Mechanisms of allergen-specific immunotherapy. J Allergy Clin Imunol, 119(4), 780-789.

Bousquet, J., Lockey, R.F., Malling, H.J. (1998). WHO position paper–Allergen immunotherapy, therapeutic vaccines for allergic diseases. Allergy, 53(Suppl 44), 1-42.

Chu, J.C.L., Wen, T.H., Chen, X.J. (1981). Treatment of asthmatic patients sensitive to mites (*Dermatophagoides farinae*). A four-year study of immunotherapy with *Dermatophagoides farinae*. Ann Allergy, 47, 107-109.

Heymann, P.W., Chapman, M.D., Aalberse, R.C., *et al.* (1989). Antigenic and structural analysis of group II allergens (Der f II and Der p II) from house dust mites (*Dermatophagoides* spp.). J Allergy Clin Immunol, 83960, 1055-1067.

Huang, Y., Chen, K.H., Dong, F., *et al.* (1997). Studies on mite oral extract depressing the incidence of bronchial asthma after bronchiolitis. Pract Paed Clin J, 12(6), 367-369 (in Chinese).

Huang, Y., Li, J.B., Li, F.R., *et al.* (2003). Clinical observations on child asthma

control with mite oral extract. Chongqing Med, 32, 427-429 (in Chinese).
Jin, M.L., Cai, Y.Y., Wang, B.L., *et al.* (1995). Efficacy of slow releasing mite extract in treatment of allergic asthma. Shanghai Med, 18, 151-152 (in Chinese).
Pan, X.L., Geng, Y.C., Wang, S.A., *et al.* (2001). Comparison of the curative efficacies of child asthma between oral and injection dust mite vaccines. Chin J Appl Immun Allergy Asthma, 5(3), 225-227 (in Chinese with English abstract).
Peng, Z., Wen, T., Hong, S.S., *et al.* (1985). Skin prick test, nasal provocation test and measurement of specific IgE with dust mite extract in allergic asthmatic patients. Shanghai J Immunol, 5, 29-33 (in Chinese with English abstract).
Peng, Z., Nacierio, R.M., Norman, P.S., *et al.* (1992). Quantitative IgE and IgG-subclass responses during and after long-term ragweed immunotherapy. J Allergy Clin Immunol, 89(2), 519-529.
Su, S.L., Zhang, Y.P., Wen, T.H. (2001). Long and short immunotherapy effect upon bronchial asthma using the fluid extract of dust mite. Chin J Appl Immun Allergy Asthma, 5(2), 4-50 (in Chinese with English abstract).
Voorhorst, R., Spieksma, F.T.M., Varekamp, N. (1969). House Dust Atopy and the House Dust Mite *Dermatophagoides pteronyssinus* (Trouessart, 1877). Leiden: Stafleu's Scientific Publishing Co., pp. 159.
Wang, B.L., Cai, Y.Y., Cheng, X.L., *et al.* (1994). Introduction of a new slow releasing *Dermatophagoides farinae* extract in treatment of allergic asthma. Acta Acad Med Shanghai, 21, 480 (in Chinese).
Wen, T., Wang, E., Shen, S., *et al.* (1990). Allergenic potency of SMU *Df* extract in comparison with VUS *Df* extract, and diagnosis and immunotherapy for atopic dermatitis and rhinitis with SMU-*Df* extract in China. Arb Paul-Ehrlich-Institut, 85, 217-227.
Wen, Y.Y., Wen, T.H., Hua, Z.Y., *et al.* (1992). Evaluation of skin prick test with dust mite sensitive asthma. Chin J Pharmaceut, 23(7), 315-317 (in Chinese with English abstract).
Wen, T., Cai, Y.Y., Chen, X.J., *et al.* (1998). Safety analysis of diagnosis and immunotherapy with *Dermatophagoides farinae* extract in allergic patients in China. Syst Appl Acarol, 3, 105-112.
Wen, T., Wang, B.L., Jin, M.L., *et al.* (1997). Oral versus subcutaneous immunotherapy with dust mite allergen in asthmatics in China. The 5th WPAS and the 7th Korea-Japan Joint Allergy symposium (Seoul), Abstract, p.115.
Wen, T.H. (2005). Dust mite sensitivity and asthma. In: Li, M.H., *et al.* (ed). Asthmology (2nd ed). Beijing: People's Medical Publishing House, pp. 936-960.
Wen, T.H. (2006). Vaccines for child allergic asthma and immunotherapy per os. Chinese Medical Association–Subassociation of Allergology (ed). All China Symposiums on Diagnosis and Management of Child Allergic Diseases in Advance. Chengdu, pp. 38-40 (in Chinese).
Zhang, Z.X., Zhao, Y., Li, H.W., *et al.* (2003). Clinical observation on the curative

efficacy of child asthma controlled by Manyi or, sublingual mite vaccine. Chin J Appl Immun Allergy Asthma, 7(1), 74-75 (in Chinese with English abstract).

Zhao, J., Chen, Y.Z., Ding, Y., *et al.* (1999). Oral mite desensitization solution as supplementary therapy in treating children asthma. Chin Pract Paed J, 14(11), 661-663 (in Chinese with English abstract).

Zheng, Y., Yin, R., Liang, C.Y., *et al.* (2000). Oral mite vaccine in immunotherapy of allergic rhinitis and asthma. West China Med J, 15(3), 294.

Zhu, J.R., Wen, T.H., Chen, N.Z. (1992). Curative efficacy of allergic rhinitis by percutaneous application of glycerinated *Dermatophagoides farinae* extract. Clin ENT J, 6(1), 50 (in Chinese).

19

The Influence of Pollen Concentration on the Dispension of Antihistaminics and Corticosteroids to Hay Fever Patients

Arnold J.H. van Vliet[1], **Hilde Tobi**[2]

[1] Environmental Systems Analysis Group, Wageningen University, P.O. Box 47, 6700 AA Wageningen, the Netherlands & Foundation for Sustainable Development, P.O. Box 570, 6700 AN Wageningen, the Netherlands
[2] Research Methodology Group –ECS, Wageningen University, P.O. Box 8130, 6700 EW Wageningen, the Netherlands
E-mail: [1]Arnold.vanvliet@wur.nl; [2] hilde.tobi@wur.nl

Abstract: Climate change may induce alterations in the start and duration of the pollen season. Future changes in climate are likely to significantly advance the start of the pollen season and change the pollen concentration in the atmosphere. These changes may have consequences for the use and costs of medication for allergic rhinitis (AR). The aim of the study presented here was to investigate the effects of changes in the pollen concentration on medication dispensing and medication costs. Weekly pharmacy dispensing data on medication for hay fever and the associated costs were modeled using weekly pollen counts for different species during the period 2001–2005. Both the pollen counts and the pharmacy data have been collected from the Netherlands. The majority of the annually dispensed AR-medication appeared not to be directly related to pollen counts and was dispensed outside the pollen season. However, the large weekly variation in AR medication dispersion and costs could be attributed to a substantial degree to grass, Birch and Alder pollen concentration in the atmosphere. This significant impact of weekly pollen counts on the variation in medication dispensing suggests the importance of changes in pollen concentration in determining the incidence of symptoms. Climate change induced changing pollen seasons will make it more difficult for patients and health professionals to anticipate the need for AR medication.

19.1 Introduction

Allergic rhinitis (AR) is a significant societal problem (Beasley, 1998; Skoner, 2001; Asher *et al.*, 2006). The prevalence of AR in 6 European countries was shown to range from 17% in Italy to 29% in Belgium with an average of 23% (Bauchau and Durham, 2004). Several studies show that the health costs of hay fever are substantial (Gupta *et al.*, 2004; Schramm *et al.*, 2003; Weiss and Sullivan, 2001). Although AR is a complex disease with many factors which influence its course, it will only come to the forefront after exposure of patients to allergens via, for example, pollen.

The start and duration of the pollen season varies from year to year due to the relationship between plant phenology and climate characteristics such as temperature (Burr *et al.*, 2003; Rasmussen, 2002; Spieksma and Nikkels, 1998; van Vliet *et al.*, 2002). Increases in temperature and CO_2 concentrations can also increase the total amount of pollen produced and the pollen concentration in the atmosphere (Damialis *et al.*, 2007; Rasmussen, 2002; Teranishi *et al.*, 2000; Ziska and Caulfield, 2000; Ziska *et al.*, 2003). Hence the start, duration and intensity of the pollen seasons are likely to change significantly with the projected likely increase of temperature of 1.8–4.0℃ in the 21st century (Solomon *et al.*, 2007). Furthermore, air pollutants such as ozone and particulate matter affect both airborne allergenic pollen and the airways of exposed subjects (D' Amato *et al.*, 2007).

Given the high prevalence of hay fever, the expected changes in pollen concentration due to climate changes are likely to have a large socio-economic impact. The main objective of the study presented here was to investigate the effects of changes in the pollen concentration on medication dispensed, and medication costs, for the period 2001–2005 in the Netherlands.

19.2 Research Design

Prescription drugs for the treatment of allergic rhinitis as mentioned in the formulary Farmacotherapeutisch Kompas (published by the Health Care Insurance Board (Health Care Insurance Board)) were identified using the Anatomical Therapeutic Chemical (ATC) classification system of the World Health Organization (the WHO Collaborating Centre for Drug Statistics Methodology and the WHO Collaborating Centre for Drug Statistics Methodology. For this study, all prescriptions with the following ATC codes were included: R01AC and R01AD (nasal preparations for topical use with antiallergic agents), S01GX (ophthalmological decongestants and antiallergics other than sympathomimetics), R06AE and R06AX (antihistamines for systematic use: piperazine derivatives and other histamines) and V01AA (allergen abstracts).

In this study, the focus was on prescriptions from general practitioners and medical specialists that can only be provided in pharmacies. Stichting Farmaceutische Kengetallen (SFK) provided pharmacy dispensing data from about 40% of all community pharmacies throughout the Netherlands. Previous studies have shown that pharmacy dispensing data gave accurate information on the use of prescription drugs (Lau *et al.*, 1997; Monster *et al.*, 2002). The SFK provided data on the daily total number of hay fever prescriptions dispensed, the total costs of the dispensed AR medication (excluding pharmacists' fees) and the number of pharmacy visits made for hay fever medication. In addition, for each half year during the period 2001–2005, the total number of patients that got AR medication from the pharmacies was also available.

For pollen counts, the daily counts were used of grass (*Poaceae*), Birch (*Betula*) and Alder (*Alnus*) as measured with a Burkard trap (Driessen *et al.*, 1988) by both national pollen monitoring stations (Leiden University Medical Center, LUMC, and Elkerliek Hospital in Helmond). The pollen counts of these stations were averaged to give a better representation of pollen concentration at any given site.

Because of pharmacies having limited opening hours in the weekends, the relationship between pollen counts and dispensing of medicines were analyzed per weekday only. Both the medication data and the pollen counts were aggregated for each week by summing the daily numbers. When needed, weekly numbers were standardized by subtracting the weekly year average and dividing by the standard deviation of that year.

A time series auto-regression analysis was performed to determine to what extent the weekly variation in the dependent variables (the number of prescriptions and the total costs of medicines) could be explained by the pollen concentration of the different species and by the variable "year" to account for non-seasonal changes over time.

All calculations and graphical representation were performed in Microsoft Excel (version 2003) and SPSS (version 12).

19.3 The Influence of Pollen Concentration on the Dispensing of Antihistaminics and Corticosteroids

Table 19.1 shows the numbers of prescriptions, unique patients, costs involved and the pollen counts per 6 months. Compared to 2001, in 2005, the number of prescriptions increased 14% in the first half of the year and only 10% in the second half of the year. The number of unique patients to whom AR medication was dispensed also increased 9% for the first half of the year and only 6% for the second half of the year. The number of prescriptions per patient increased slightly from 1.9 for the first half of 2001 to 2.0 for the first half of 2005, where the number of prescriptions per patient for the second half of the year remained about 1.8. Medication costs showed an increase of 30% for the first half of the year and 33% for the second half of the year between 2001 and 2005. Medication costs per patient increased 19% and 26% for the first and the second half of the year, whereas costs per prescription increased 14% and 21% respectively.

For pollen-induced allergic rhinitis, the results for the first half of the year were most important (Table 19.1). The 2-year cycle of Birch pollen is well-known and described (Ranta *et al.*, 2008), but otherwise, the year-to-year counts showed no significant trend over the 5-year period. The weekly amount of prescriptions and costs involved in AR-medication can be found in Fig. 19.1. Fig. 19.1 clearly shows a minimum number of weekly prescriptions and costs throughout the whole year on top of which a seasonal pattern became visible. Fig. 19.2 depicts the standardized weekly values for prescriptions and grass, Birch and Alder pollen counts. Visual inspections suggest that the peaks in medication dispensing occurred especially after days with high pollen loads of Birch and grass.

Table 19.1 Numbers of prescriptions, patients and costs in the study period 2001–2005.

Year	Months	Prescriptions	Unique Patients	Visits	Costs (Euro)	Pollen counts*		
						Grass	Birch	Alder
2001	Jan–June	1,037,758	542,403	904,696	17,125,072	2715	1382	1317
	July–Dec	756,988	427,859	696,692	14,518,758	994	5	3
2002	Jan–June	1,189,371	591,228	1,022,644	21,119,847	2185	4479	1785
	July–Dec	801,688	436,236	738,183	16,408,922	1069	7	94
2003	Jan–June	1,161,096	594,014	1,000,687	21,370,154	2839	1098	3530
	July–Dec	841,015	445,845	771,021	17,356,923	896	7	175
2004	Jan–June	1,209,627	582,314	1,046,831	21,444,407	1400	5229	3751
	July–Dec	816,558	451,097	750,595	17,898,843	623	5	44
2005	Jan–June	1,181,784	589,514	1,019,141	22,206,361	2253	1935	1736
	July–Dec	832,499	453,044	760,360	19,354,806	634	3	21

* Averaged over the two pollen collection stations.

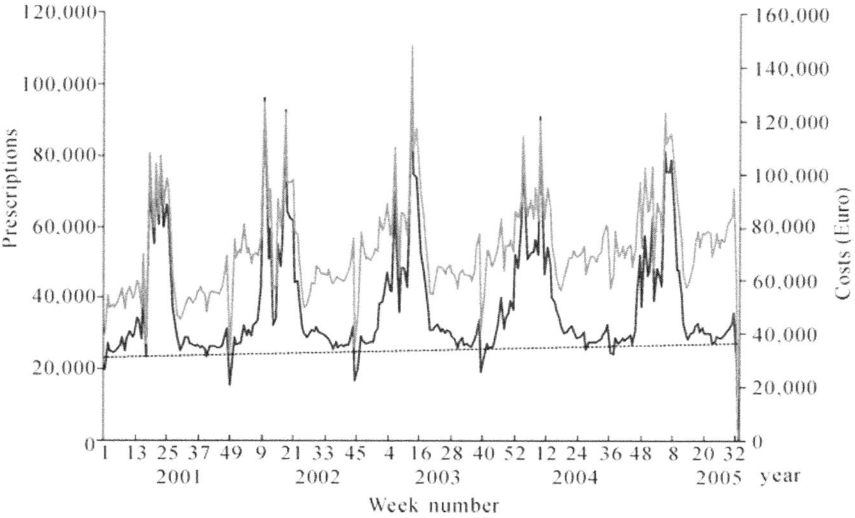

Fig. 19.1. Overview of the weekly number of prescriptions and the costs (Euro) associated with those prescriptions. Black line: Number of prescriptions; Gray line: Costs; Dashed line: the minimum number of prescriptions per week

This visible association between pollen counts and AR-medication dispensing was supported by the linear regression with autocorrelation analysis (Table 19.2). Based on these models, it was estimated that weeks with a maximum Birch pollen count as high as 1,770 would yield about 46,020 additional prescriptions in the pharmacies studied and 490×10^3 Euro additional costs for AR-medication per week. A maximum weekly count of 1,049 grass pollen, would be followed by 81,822 additional prescriptions and 880×10^3 Euro additional costs per week. Extrapolating to the whole of the Netherlands, the weekly additional prescriptions and costs in the species' pollen season, is estimated at 106,000 prescriptions and $1{,}128 \times 10^3$ Euro for Birch and 188,000 prescriptions and $2{,}024 \times 10^3$ Euro for grass. The variable "Year" also contributed significantly to the models. For the dependent variable costs, this year effect incorporates inflation, but it is unclear what it would represent for the dependent variable prescriptions.

Based on the regression equations in Table 19.2, it was estimated that about 16% of the prescriptions and 10% of the costs can be directly explained to pollen. This relative limited contribution of pollen to the total amount of prescriptions and the total amount of costs was due to the high minimum number of prescriptions over the year (the dashed line in Fig. 19.1).

Table 19.2 Parameter estimates and variation explained in linear regression models with autocorrelation

Dep. variable	Independent variables										
	intercept	Grass		Birch		Alder		Year		Var explained	
		B	p	B	p	B	p	B	p	R2	R2adj
Prescriptions	27,507	78	0.000	26	0.000	–	–	1,337	0.023	0.61	0.61
Visits	25,510	58	0.000	19	0.000	–	–	1,094	0.016	0.59	0.58
Costs	513,658	839	0.000	277	0.000	106	0.011	46,365	0.000	0.56	0.56

19.4 Conclusion and Discussion

The effect of pollen counts on the number of prescriptions dispensed and associated costs can be clearly shown, as pollen counts can explain 16% of the annual prescriptions and 19% of the annual costs. Between 2001 and 2005, the cost per prescription increased 14% and 21% (the first and the second half of the year respectively), such values below the 22% threshold one would have expected based on an economic discount rate of 4%.

Our results also suggest that a large proportion of the patients use AR medication year round. This may be explained by AR medication being used for other diseases like asthma. In addition, the medication may have been used for treating reactions to year-round available allergens such as house dust mite, cats, dogs or foods. House dust mite allergy is a common allergy in the Netherlands with a prevalence of 22% among people younger than 44 years which is about twice the prevalence of grass pollen allergy and Birch pollen allergy (Smithuis *et al.*, 2000). Another explanation could be that patients use medication year round as a precautionary approach or because of ignorance of the actual cause of the allergy as allergic rhinitis remains frequently undiagnosed (Bauchau and Durham, 2004).

High pollen counts of Birch and grass were followed by an increased number of prescriptions meaning that many patients started medication days after exposure and experiencing allergenic effects. Although Alder pollen is potentially allergenic, the amount of Alder pollen only contributed to the explanation of the weekly variation in costs and not the amount of prescriptions. Why Alder pollen counts only contributed to costs requires further investigation, as well as identifying detailed information on the diagnosis of AR and allergies. A possible source of the absence of Alder pollen in the model, and the observed time lag between exposure to Birch and grass pollen and starting to take medication is the lack of knowledge about when the pollen season starts. It is likely that patients do not know that AR problems can occur as early as in February when Alder trees release their pollen. People might also be misguided about the duration of the pollen season. For example, the Dutch national hay fever forecasts are only distributed from 15th,

May until 15th, July, while the duration of the pollen season is much longer. Several studies have shown that health complaints are lower when AR-patients started using their medication several weeks before the start of the pollen seasons (Bousquet *et al.*, 2006; Kurowski *et al.*, 2004) and that complaints decreased with an increase in duration of using medication (Bachert *et al.*, 2004; Bousquet *et al.*, 2006; Harvey *et al.*, 1996; Kurowski *et al.*, 2004; Potter, 2005). Thus, there is a major health gain possible with the timely start of medication.

This study investigated medication dispensed by community pharmacies. Consequently, the results are likely to underestimate the total costs involved in medication for allergic rhinitis. However, there is no reason to suspect that this invalidates in any way the trends and associations described in this paper. People who experience complaints severe enough to warrant a visit to a medical doctor will receive prescription medication, and all these people were incorporated into the analysis. These are also the patients who are likely to benefit most from knowing which pollen or other allergens provoke allergic responses, and from having more information and knowledge on the start, duration and intensity of the pollen season. To our knowledge, there is no study published that has investigated whether patients and medical doctors know when these pollens can be expected.

The anticipated increase in temperature, more extreme climate events (e.g., warm periods in February), and the increase in CO_2 concentration will advance the start of pollen release, make this start more variable, and will increase the pollen concentration (Beggs, 2004; D'Amato *et al.*, 2007; Emberlin *et al.*, 2002; Wayne *et al.*, 2002; Ziska and Caulfield, 2000). The changing pollen dynamics and concentrations are likely to trigger allergic reactions in more people especially in combination with air pollution. The projected changing pollen seasons will make it more difficult for patients and health professionals to anticipate the rises and falls of the need for SAR medication concluding, the pattern of medication dispensing after high pollen counts showed an opportunity for improvement of health care by means of a timely treatment by health professionals. The changing pollen seasons emphasize the importance of health promotion and educating people on pollen and pollen season.

Acknowledgements

We thank the organizations that funded the research: the Ministry of Agriculture, Nature, and Food Quality; Siemens Healthcare Diagnostics; The Climate Changes Spatial Planning Programme; and the Space for Geo-Information Research Programme. Furthermore, we thank Leiden University Medical Center and Elkerliek Hospital for providing the pollen counts.

References

Asher, M.I., Montefort, S., Bjorksten, B., Lai, C.K.W., Strachan, D.P., Weiland, S.K. *et al.* (2006). Worldwide time trends in the prevalence of symptoms of asthma, allergic rhinoconjunctivitis, and eczema in childhood: ISAAC Phases One and Three repeat multicountry cross-sectional surveys. The Lancet, 368(9537), 733-743.

Bachert, C., Bousquet, J., Canonica, G.W., Durham, S.R., Klimek, L., Mullol, J. *et al.* (2004). Levocetirizine improves quality of life and reduces costs in long-term management of persistent allergic rhinitis. J Allergy Clin Immunol, 114(4), 838-844.

Bauchau, V., Durham, S.R. (2004). Prevalence and rate of diagnosis of allergic rhinitis in Europe. Eur Respir J, 24(5), 758-764.

Beasley, R. (1998). Worldwide variation in prevalence of symptoms of asthma, allergic rhinoconjunctivitis, and atopic eczema: ISAAC. The Lancet, 351(9111), 1225-1232.

Beggs, P.J. (2004). Impacts of climate change on aeroallergens: Past and future. Clin Exp Allergy, 34, 1507-1513.

Bousquet, J., van Cauwenberge, P., Ait Khaled, N., Bachert, C., Baena-Cagnani, C.E., Bouchard, J., *et al.* (2006). Pharmacologic and anti-IgE treatment of allergic rhinitis ARIA update (in collaboration with GA2LEN). Allergy, 61(9), 1086-1096.

Burr, M.L., Emberlin, J.C., Treu, R., Cheng, S., Pearce, N.E. (2003). Pollen counts in relation to the prevalence of allergic rhinoconjunctivitis, asthma and atopic eczema in the International Study of Asthma and Allergies in Childhood (ISAAC). Clin Exp Allergy, 33(12), 1675-1680.

D'Amato, G., Cecchi, L., Bonini, S., Nunes, C., Annesi-Maesano, I., Behrendt, H. *et al.* (2007). Allergenic pollen and pollen allergy in Europe. Allergy, 62(9), 976-990.

Damialis, A., Halley, J.M., Gioulekas, D., Vokou, D. (2007). Long-term trends in atmospheric pollen levels in the city of Thessaloniki, Greece. Atmos Environ, 41(33), 7011-7021.

Driessen, M.N.B.M., Derksen, J.W.M., Spieksma, F.T.M., Roetman, E. (1988). Pollenatlas van de Nederlandse Atmosfeer. Hilversum: Drukkerij Onkenhout B.V.

Emberlin, J., Detandt, M., Gehrig, R., Jaeger, S., Nolard, N., Rantio-Lehtimäki, A. (2002). Responses in the start of Betula (birch) pollen seasons to recent changes in spring temperatures across Europe. Int J Biometeorol, 46(4), 159-170.

Gupta, R., Sheikh, A., Strachan, D.P., Anderson, H.R. (2004). Burden of allergic disease in the UK: Secondary analyses of national databases. Clin Exp Allergy, 34(4), 520-526.

Harvey, R.P., Comer, C., Sanders, B., Westley, R., Marsh, W., Shapiro, H., *et al.* (1996). Model for outcomes assessment of antihistamine use for seasonal

allergic rhinitis. J Allergy Clin Immunol, 97(6), 1233-1241.
Health Care Insurance Board. Health Care Insurance Board. Available: www.fk.cvz.nl.
Kurowski, M., Kuna, P., Gorski, P. (2004). Montelukast plus cetirizine in the prophylactic treatment of seasonal allergic rhinitis: Influence on clinical symptoms and nasal allergic inflammation. Allergy, 59(3), 280-288.
Lau, H.S., de Boer, A., Beuning, K.S., Porsius, A. (1997). Validation of pharmacy records in drug exposure assessment. J Clin Epidemiol, 50(5), 619-625.
Monster, T.B.M., Janssen, W.M.T., de Jong, P.E., de Jong-van den Berg, L.T.W. (2002). Pharmacy data in epidemiological studies: An easy to obtain and reliable tool. Pharmacoepidemiol Drug Safety, 11, 279-284.
Potter, P.C. (2005). Efficacy and safety of levocetirizine on symptoms and health-related quality of life of children with perennial allergic rhinitis: A double-blind, placebo-controlled randomized clinical trial. Annals Allergy, Asthma Immunol, 95(2), 175-180.
Ranta, H., Hokkanen, T., Linkosalo, T., Laukkanen, L., Bondestam, K., Oksanen, A. (2008). Male flowering of birch: Spatial synchronization, year-to-year variation and relation of catkin numbers and airborne pollen counts. Forest Ecology and Management, 255(3-4), 643-650.
Rasmussen, A. (2002). The effects of climate change on the birch pollen season in Denmark. Aerobiologia, 18(3), 253-265.
Schramm, B., Ehlken, B., Smala, A., Quednau, K., Berger, K., Nowak, D. (2003). Cost of illness of atopic asthma and seasonal allergic rhinitis in Germany: 1-yr retrospective study. Eur Respir J, 21, 116-122.
S.F.K. (2008). Stichting Farmaceutische Kengetallen. Available: www.sfk.nl.
Skoner, D.P. (2001). Allergic rhinitis: Definition, epidemiology, pathophysiology, detection, and diagnosis. J Allergy Clin Immunology, 108(1, Part 2), S2-S8.
Smithuis, L.O.M.J., Haan, G.J.H., van der Laan, L.R., Pekelharing, J.M., Rikken, S.A.J.J., Rutten, W.P.F. (2000). Wetenschappelijke verantwoording van het landelijk model van een probleemgeoriënteerd aanvraagformulier voor laboratoriumonderzoek door huisartsen. Nederlands Tijdschrift voor Klinische Chemie December, (Special), 14-19.
Solomon, S., Qin, D., Manning, M., Chen, Z., Marquis, M., Averyt, K.B. *et al.* (2007). Climate Change (2007): The Physical Science Basis. Contribution of Working Group I to the Fourth Assessment Report of the Intergovernmental Panel on Climate Change. Cambridge: Cambridge University Press.
Spieksma, F.T.M., Nikkels, A.H. (1998). Airborne grass pollen in Leiden, the Netherlands: Annual variations and trends in quantities and season starts over 26 years. Aerobiology, International Journal of Aerobiology, 14, 347-358.
Teranishi, H., Kenda, Y., Katoh, T., Kasuya, M., Oura, E., Taira, H. (2000). Possible role of climate change in the pollen scatter of Japanese cedar Cryptomeria japonica in Japan. Climate Research, 14(1), 65-70.
van Vliet, A.J.H., de Groot, R.S., Overeem, A., Jacobs, A.F.G., Spieksma, F.T.M. (2002). The influence of temperature and climate change on the timing of pollen release in the Netherlands. International J Climatol, 22, 1757-1767.

Wayne, P., Foster, S., Connolly, J., Bazzaz, F., Epstein, P. (2002). Production of allergenic pollen by ragweed (*Ambrosia artemisiifolia* L.) is increased in CO_2-enriched atmospheres. Annals of Allergy, Asthma and Immunology, 8, 279-282.

Weiss, K.B., Sullivan, S.D. (2001). The health economics of asthma and rhinitis. I. Assessing the economic impact. J Allergy Clin Immunol, 107(1), 3-8.

WHO Collaborating Centre for Drug Statistics Methodology. WHO Collaborating Centre for Drug Statistics Methodology. Available:www.whocc.no/atcddd/.

Ziska, L.H., Caulfield, F.A. (2000). Rising CO_2 and pollen production of common ragweed (*Ambrosia artemisiifolia*), a known allergy-inducing species: Implications for public health. Australian J Plant Physiol, 27(10), 893-898.

Ziska, L.H., Gebhard, D.E., Frenz, D.A., Faulkner, S., Singer, B.D., Straka, J.G. (2003). Cities as harbingers of climate change: Common ragweed, urbanization, and public health. J Allergy Clin Immunol, 111(2), 290-295.

Part V

Hypoallergenic Products

20

In Search of Hypoallergenic Birch Trees: Characterization of PR-10 Genes from Eight Betula Species and Detection of Bet v 1 Isoforms in Birch Pollen Using a Combined Genomics-Proteomics Approach

M.J.M. Smulders*, Martijn F. Schenk, Luud J.W.J. Gilissen

Allergy Consortium Wageningen and Plant Research International, Wageningen UR, P.O. Box 16, NL-6700 AA Wageningen, the Netherlands
* E-mail: rene.smulders@ wur.nl

Abstract: Bet v 1 is an important cause of hay fever in the northern hemisphere. Bet v 1 isoforms from the European white birch (*Betula pendula*) have been investigated extensively, but the allergenic potency of other birch species is unknown. PR-10 genes were cloned and sequenced from 8 different birch species, representing all major groups within the genus, to establish the presence of these genes. In total, 134 unique sequences were discovered. Sequences were attributed to different PR-10 genes or pseudogenes that were subdivided into 7 subfamilies. Five subfamilies were common to all birch species. Q-TOF LC-MSE was applied to identify which PR-10/Bet v 1 genes are expressed in pollen and to determine relative abundances of individual isoforms in the pollen proteome. This showed that genes of two subfamilies were expressed in pollen, while each species expressed a mixture of isoforms with at least 4–5 different isoforms. Q-TOF LC-MSE allowed fast screening of Bet v 1 isoforms in pollen by determining the presence and relative abundance of isoforms similar to those with a known high IgE-reactivity (Bet v 1a =PR-10.01A01) and to the hypoallergenic isoform Bet v 1d (=PR-10.01B01). *Betula pendula* contains a Bet v 1 mixture in which both isoforms with a high and low IgE-reactivity are abundant. The presence of isoforms with high IgE-reactivity is a condition for determining the influence for the allergenicity of this species. Birch species that express isoforms similar to Bet v 1a are predicted to be allergenic as well.

20.1 Introduction

Birch pollen is a major cause of Type I allergies. The major birch allergen is a pollen protein from the European white birch *(Betula pendula* syn. *B. verrucosa)* termed Bet v 1 (Breiteneder *et al.*, 1989; Jarolim *et al.*, 1989). Bet v 1 and other pathogenesis-related class 10 (PR-10) proteins constitute the largest group of aeroallergens (Breiteneder and Ebner, 2000). Individuals who have been sensitized to birch pollen are particularly prone to develop Oral Allergy Syndrome (OAS) due to an IgE-mediated cross-reaction between Bet v 1 and PR-10 food proteins (Wensing *et al.*, 2002; Ferreira *et al.*, 2004). As a result, PR-10 proteins are also among the four most common groups of food allergens (Breiteneder and Ebner, 2000). Allergic diseases have a negative impact on the patient's quality of life and are associated with high economic costs (Meltzer, 2001). Consequently, allergies to birch pollen and OAS represent relevant targets for prevention.

B. pendula is the most common birch species in Europe and it has been widely investigated in relation to hay fever, but it is not the only *Betula* species present. *B. nana* has a circumpolar distribution and is primarily found at northern latitudes. *B. pubescens* and *B. pendula* trees cover large parts of the land and form stable climax forests in Scandinavia (Atkinson, 1992). As a result, the birch pollen load is very high in Scandinavia (Rasmussen, 2002). *B. pubescens* and *B. pendula* are also present in north-east Europe as is *B. humilis*. Pollen counts in this area are similar to those in Scandinavia (Latalowa *et al.*, 2002). Towards the center of their distribution range, *B. pendula* and *B. pubescens* are colonists in primary or secondary succession (Atkinson, 1992). Pollen counts for several West European cities show that birch pollen is abundant and occurs in the same order of magnitude as grass pollen (Spieksma *et al.*, 2003). Continuing southwards, birch becomes limited in cooler mountainous areas. Birch is present in the alpine foothills and around the northern Italian lakes. Here, *B. pubescens* reaches the southernmost limit of its range, while *B. pendula* is found in scattered populations throughout the rest of Italy (Atkinson, 1992). Birch allergy is also reported in Eastern Asia, Japan, Canada and the US, where *B. pendula* does not occur (Abe *et al.*, 1997; Eriksson *et al.*, 1998). Other birch species do occur in these areas and are likely candidates for causing allergic complaints.

The prevalence of sensitization to birch shows a distinct geographical pattern in Europe. There is a good correlation between the degree of exposure throughout the main flowering season of birch and the occurrence of sensitization among atopic patients. Relatively high numbers of birch-sensitized patients are found in areas with many birch trees and high pollen counts. The "hot-spot" of sensitization to birch lies in Scandinavia (Burney *et al.*, 1997; Eriksson *et al.*, 1998; Gislason *et al.*, 1999), while the occurrence of sensitization to birch is also high among atopic patients in Estonia, Lithuania and the Russian Federation (Eriksson *et al.*, 1998; Jõgi *et al.*, 1998; Raukas-Kivioja *et al.*, 2003; Vartiainen *et al.*, 2002). In these areas, sensitization to grass occurs at a similar level to birch. The prevalence of sensitization to birch is also high throughout north-west Europe and the Alps, although sensitization to grass has a higher prevalence (Burney *et al.*, 1997; Horak *et al.*, 2002; Nowak *et al.*, 1996; Riedler *et al.*, 2000; von Mutius *et al.*, 1998). Where exposure levels are intermediate, such as in France or the UK, this leads to intermediate levels of sensitization to birch (Burney *et al.*, 1997). Continuing southwards, birch is a relevant allergenic tree in the alpine and Po region in Italy, where over 30% of the atopic patients are sensitized to birch, as opposed to 2% in the most southern regions (Corsico *et al.*, 2000). An 18%–20% share of sensitization to birch/hazel was recorded around Genoa (north Italy), while sensitization to olive pollen, grass and *Parietaria* was higher in this area (Crimi *et al.*, 1999; Silvestri *et al.*, 1996). Birch represents a minor source of allergens in Spain and Greece, compared to grass or olive pollen (Gioulekas *et al.*, 2004; Sunyer *et al.*, 2000).

20.2 The Major Birch Pollen Allergen Bet v 1 Is a PR-10 Protein

PR-10 proteins are reported as a multigene family present in Gymnosperms, Monocots and Dicots (Ekramoddoullah *et al.*, 2000; Gao *et al.*, 2005; Huang *et al.*, 1997). The classification as PR-proteins is based on the induced expression in response to pathogen infections by viruses, bacteria or fungi (McFadden *et al.*, 2001; Poupard *et al.*, 2003; Pühringer *et al.*, 2000; Robert *et al.*, 2001; Swoboda *et al.*, 1995c), to wounding (Liu *et al.*, 2003; Poupard *et al.*, 1998) or to abiotic stress (Moons *et al.*, 1997; Srivastava *et al.*, 2004; Utriainen *et al.*, 1998). Other PR-10 family members are constitutively expressed during plant development (Pinto *et al.*, 2005; Walter *et al.*, 1996) or in specific tissues (Huang *et al.*, 1997; Liu and Ekramoddoullah, 2003). No general function has been described and PR-10 proteins are thought to act as ribonuclease (Bantignies *et al.*, 2000; Moiseyev *et al.*, 1997; Park *et al.*, 2004), as cytokine-binding proteins (Fujimoto *et al.*, 1998), as storage proteins (Flores *et al.*, 2002; Richard-Molard *et al.*, 2004), as plant steroid carriers (Markovic-Housley *et al.*, 2003) or as cryoprotective proteins (Ukaji *et al.*, 2004). PR-10 proteins are homogeneous within species (Wen *et al.*, 1997), which may be maintained by concerted evolution. Arrangements of duplicated *PR-10* genes into clusters as found for Mal d 1 genes in apples (Gao *et al.*, 2005a), may facilitate this evolutionary process.

The *B. pendula* genome contains multiple *PR-10* genes with varying expression patterns. mRNAs of PR-10 genes are detected in several tissues, including birch pollen (Breiteneder *et al.*, 1989; Friedl-Hajek *et al.*, 1999; Swoboda *et al.*, 1995b), roots (Feugey *et al.*, 1999; Koistinen *et al.*, 2002; Poupard *et al.*, 2001; Poupard *et al.*, 1998; Utriainen *et al.*, 1998), and leaves (Utriainen *et al.*, 1998; Valjakka *et al.*, 1999). The true Bet v 1 allergens are the PR-10 proteins that are expressed in pollen (Schenk *et al.*, 2009). The first Bet v 1 isoform was identified by immunoscreening of a cDNA expression library from pollen with serum of birch allergic patients (Breiteneder *et al.*, 1989). Other Bet v 1 alleles have subsequently been sequenced (Friedl-Hajek *et al.*, 1999; Hoffmann-Sommergruber *et al.*, 1997; Son *et al.*, 1999; Swoboda *et al.*, 1995b). Bet v 1 is estimated to encompass 10% of the total protein content of birch pollen (Larsen, 1995). A mixture of pollen from multiple birch trees was found to contain multiple Bet v 1 isoforms (Swoboda *et al.*, 1995b).

20.3 Occurrence and Diversity of Bet v 1 Isoforms

Measures to prevent hay fever may focus on the development of hypoallergenic food or hypoallergenic varieties of pollen-producing plants or trees. Variations in the concentration of allergens has been found for foods such as apple, peach, nectarine and peanut, and for pollen from olive trees (Ahrazem *et al.*, 2007; Castro *et al.*, 2003; Koppelman *et al.*, 2001; Marzban *et al.*, 2005). Variations in the allergenicity among different protein variants involved in the allergic reaction (Ferreira *et al.*, 1996) may also prove to be a source of hypoallergenic products. Plant varieties with a reduced allergenicity can be selected within the natural range of variations. For example, high and low allergenic apple cultivars were identified by using SPTs (Bolhaar *et al.*, 2005; Carnés *et al.*, 2006). The apple cultivar Santana has been identified as hypoallergenic in these tests, and a clinical trial on this cultivar confirmed the results (Bolhaar *et al.*, 2005; Kootstra *et al.*, 2007). Breeding for hypoallergenicity may further reduce the allergenic properties of apples.

To establish the number of PR-10/Bet v 1 genes and the isoform diversity within a single tree, PCR amplification, cloning and sequencing of PR-10 genes was performed on several diploid *B. pendula* cultivars and on trees from 7 other diploid and polyploid *Betula* species, representing the 4 subgenera recognized within the genus Betula by Schenk *et al.* (2008). Sequences were attributed to putative genes based on sequence identity and intron length. Thirty two different *PR-10* sequences were recovered from *B. pendula* and assigned to thirteen putative genes. Alleles from ten identified genes had previously been described (Breiteneder *et al.*, 1989; Friedl-Hajek *et al.*, 1999; Hoffmann-Sommergruber *et al.*, 1997; Son *et al.*, 1999; Swoboda *et al.*, 1995b), while three were new genes

(Schenk *et al.*, 2006). The genes could be classified into 5 subfamilies (I-V) based on identities of the coding regions and on distinct intron lengths. The average similarity between alleles within each subfamily was 95% or more (Schenk *et al.*, 2006). An organization of PR-10 genes into subfamilies was also reported for *Malus domestica* (Gao *et al.*, 2005) and for *Pinus monticola* (Liu and Ekramoddoullah, 2004).

PR-10 genes were found in all birch species we subsequently examined. In total, 134 unique PR-10 sequences were recovered from the 8 Betula species. When sequences were attributed to different genes or pseudogenes and clustered into subfamilies, two additional subfamilies were recognized, which were only present in a limited number of species. Genes from the 5 subfamilies identified in *B. pendula* are present in all birch species examined.

Sequence homology to database sequences of cDNAs obtained from specific tissues suggests that 7 genes are transcribed in pollen and 3 genes in somatic tissue, while the transcription of 3 other genes remains unknown (Schenk *et al.*, 2006). Differences in transcription coincide with the division between subfamilies. Of the 14 different Bet v 1 isoforms encoded by these 7 genes in the *B. pendula* cultivars, 9 were entirely new. The allergenicity of the new isoforms can be examined in the future by expressing the isoforms as recombinant proteins and use these in a SPT or T-cell activation tests. Ferreira *et al.* (1996) divided the Bet v 1 isoforms into 3 groups according to their IgE-reactivity and confirmed the division between high, moderate, or low IgE-reactivity in a Skin Prick Test (SPT). One high and one low IgE-reactive isoform from their analysis were 100% identical to isoforms that we have obtained from a single tree, while 2 intermediate IgE-reactive isoforms differed only by one amino acid from the alleles of 2 other identified genes in *B. pendula.* Investigations towards sensitization and immunotherapy should anticipate that patients are exposed to a mixture of Bet v 1 isoforms of different IgE-reactivity, even if pollen originates from a single birch tree.

20.4 Detecting Hypoallergenic and Allergenic Bet v 1 Proteins in Pollen

The recently developed Q-TOF LC-MS^E method enables peptide identification, but has the additional advantage of determining relative abundances of peptides (Silva *et al.*, 2005). The quantification of isoforms with a known IgE-reactivity (Ferreira *et al.*, 1996) can be used to predict the allergenicity of particular birch species and potentially to identify species with a reduced allergenicity (Schenk *et al.*, 2010). Q-TOF LC-MS^E was applied to identify which PR-10/Bet v 1 genes are expressed in pollen, and to determine the relative abundance of individual isoforms in the pollen proteome (Schenk *et al.*, 2009; 2011a). Protein analysis of pollen from *B. nigra*, *B. chichibuensis*, *B. lenta*, *B. costata* and *B. pendula* (representing all 4 subgenera) revealed that the genes of 2 subfamilies were expressed in pollen, while each species expressed a mixture of isoforms with at least 4–5 different isoforms, which were similar to isoforms with a high IgE-reactivity and were abundant in all species except *B. lenta*. Isoforms similar to those with a low IgE-reactivity were detected in *B. pendula* and were close relatives of this species only. Q-TOF LC-MS^E therefore allowed fast screening of birch pollen by determining the presence and relative abundance of Bet v 1 isoforms. *B. pendula* pollen contains a Bet v 1 mixture in which both isoforms with a high and low IgE-reactivity are present. Unfortunately, Bet v 1a, which is known to have a high IgE-reactivity, was estimated to encompass ~40% of the total amount of Bet v 1 and thereby was the most abundant isoform in the mixture (Schenk *et al.*, 2009; 2011a). Other birch species that express variants similar to Bet v 1a are predicted to be allergenic as well.

The involvement of multiple Bet v 1 isoforms has to be considered with regard to applications in allergy diagnostics. Natural mixtures of allergens are difficult to standardize, in contrast to recombinant allergens. This is especially so when allergens are unstable, which is the case for several food allergens (Bohle and Vieths, 2004). The use of recombinant allergens also allows for determining the sensitization profile of individual patients on the level of individual allergens rather than on food products as a whole. For diagnostic purposes, recombinant isoforms should be recognized by as many patients as possible. rBet v 1a seems an excellent candidate for diagnosis of birch pollen allergies (Pauli *et al.*, 1996; Valenta *et al.*, 1998; van Ree *et al.*, 1999). This isoform has a high IgE-reactivity. Bet v 1a was found in high quantities in all examined *B. pendula* cultivars. As a consequence, every individual sensitized against *B. pendula* is likely to react to Bet v 1a. *B. populifolia* (a North-American species) also contains the Bet v 1a allele, while *B. plathyphylla* and *B. costata* (Asian species) express isoforms that are highly similar and will contain similar epitopes, suggesting that when sensitization takes place against the latter species, patients will react to Bet v 1a as well. Other abundant species on which no sequencing data were acquired, such as *B. pubescens* (Europe), *B. papyrifera* (North-America) and *B. ermanii* (Asia) take an intermediate phylogenetic position between *B. pendula* and *B. costata* (Schenk *et al.*, 2008) and can be expected to contain similar isoforms as well.

Birch trees that have a reduced allergenicity may be more suitable for planting within the urban environment. A reduced allergenicity of birch trees can be based on a reduced quantity of allergens or on the absence of allergen isoforms with a high IgE-reactivity. Our results (Schenk *et al.*, 2009; 2011a) indicate that birch trees that produce pollen with substantial reduction levels of Bet v 1 are unlikely to be identified, given the high amount of Bet v 1-type proteins in pollen of each of the examined birch species. Only *B. lenta* (subgenus *Betulenta*) lacked isoforms that were similar to known isoforms with a high IgE-reactivity. This species and related species represent the most promising candidates for further screening of allergenicity by, for example, skin prick tests or a nasal challenge. Alternatively, genetic modification (GM) can be applied to reduce the allergenicity of the plants involved (Bhalla and Singh, 2004; Le *et al.*, 2006). However, developments which apply this technique should take societal concerns about GM into account (Frewer *et al.*, 2004; Schenk *et al.*, 2008b; 2011b; Zechendorf, 1994). Another approach may be the application of male sterility, which occurs in some interspecific hybrids.

References

Abe, Y., Kimura, S., Kokubo, T., *et al.* (1997). Epitope analysis of birch pollen allergen in Japanese subjects. J Allergy Clin Immunol, 17, 485-493.

Ahrazem, O., Jimeno, L., López-Torrejón, G., *et al.* (2007). Assessing allergen levels in peach and nectarine cultivars. Ann Allergy Asthma Immunol, 99, 42-47.

Atkinson, M.D. (1992). Biological flora of the British Isles: *Betula penduala* Roth (*B. verrucosa* Ehrh.) and *B. pubescens* Ehrh. J Ecol, 80, 837-870.

Bantignies, B., Séguin, J., Muzac, I., *et al.* (2000). Direct evidence for ribonucleolytic activity of a PR-10-like protein from white lupin roots. Plant Mol Biol, 42, 871-881.

Bhalla, P.L., Singh, M.B. (2004). Knocking out expression of plant allergen genes. Methods, 32, 340-345.

Bohle, B., Vieths, S. (2004). Improving diagnostic tests for food allergy with recombinant allergens. Methods, 32, 292-299.

Bolhaar, S.T.H.P., van de Weg, W.E., van Ree, R., *et al.* (2005). *In vivo* assessment with prick-to-prick testing and double-blind, placebo-controlled food challenge of allergenicity of apple cultivars. J Allergy Clin Immunol, 116, 1080-1086.

Breiteneder, H., Ebner, C. (2000). Molecular and biochemical classification of plant-derived food allergens. J Allergy Clin Immunol, 106, 27-36.

Breiteneder, H., Pettenburger, K., Bito, A., *et al.* (1989). The gene coding for the major birch pollen allergen Bet v I, is highly homologous to a pea disease resistance response gene. EMBO Journal, 8, 1935-1938.

Burney, P., Malmberg, E., Chinn, S., *et al.* (1997). The distribution of total and specific serum IgE in the European Community Respiratory Health Survey. J Allergy Clin Immunol, 99, 314-322.

Carnés, J., Ferrer, A., Fernández-Caldas, E. (2006). Allergenicity of 10 different apple varieties. Ann Allergy Asthma Immunol, 96, 564-570.

Castro, A.J., Alche, J.D., Cuevas, J., *et al.* (2003). Pollen from different olive tree cultivars contains varying amounts of the major allergen Ole e 1. Int Arch Allergy Immunol, 131, 164-173.

Corsico, R., Falagiani, P., Ariano, R., *et al.* (2000). An epidemiological survey on the allergological importance of some emerging pollens in Italy. J Invest Allerg Clin, 10, 155-161.

Crimi, E., Voltolini, S., Gianiorio, P., *et al.* (1990). Effect of seasonal exposure to pollen on specific bronchial sensitivity in allergic patients. J Allergy Clin Immunol, 85, 1014-1019.

Ekramoddoullah, A.K.M., Yu, X.S., Sturrock, R., *et al.* (2000). Detection and seasonal expression pattern of a pathogenesis-related protein (PR-10) in Douglas-fir (*Pseudotsuga menziesii*) tissues. Physiol Plant, 110, 240-247.

Eriksson, N.E., Holmén, A., Möller, C., *et al.* (1998). Sensitization according to skin prick testings in atopic patients with asthma or rhinitis at 24 allergy clinica in Northern Europe and Asia. Allergol Int, 47, 187-196.

Ferreira, F., Hawranek, T., Gruber, P., *et al.* (2004). Allergic cross-reactivity: From gene to the clinic. Allergy, 59, 243-267.

Ferreira, F., Hirtenlehner, K., Jilek, A., *et al.* (1996). Dissection of immunoglobulin E and T lymphocyte reactivity of isoforms of the major birch pollen allergen Bet v 1: Potential use of hypoallergenic isoforms for immunotherapy. J Exp Med, 183, 599-609.

Feugey, L., Strullu, D.G., Poupard, P., *et al.* (1999). Induced defense responses limit Hartig net formation in ectomycorrhizal birch roots. New Phytol, 144, 541-547.

Flores, T., Alape-Girón, A., Flores-Díaz, M., *et al.* (2002). A novel tuber storage protein from the Andean tuber crop oca with antibacterial and antifungal activities. Plant Physiol, 128, 1291-1302.

Frewer, L., Lassen, J., Kettlitz, B., *et al.* (2004). Societal aspects of genetically modified foods. Food Chem Toxicol, 42, 1181-1193.

Friedl-Hajek, R., Radauer, C., O'Riordain, G., *et al.* (1999). New Bet v 1 isoforms including a naturally occurring truncated form of the protein derived from Austrian birch pollen. Mol Immunol, 36, 639-645.

Fujimoto, Y., Nagata, R., Fukasawa, H., *et al.* (1998). Purification and cDNA cloning of cytokinin-specific binding protein from mung bean (*Vigna radiata*). Eur J Biochem, 258, 794-802.

Gao, Z.S., van de Weg, W.E., Schaart, J.G., *et al.* (2005). Genomic cloning and linkage mapping of the Mal d 1 (PR-10) gene family in apple (*Malus domestica*). Theor Appl Genet, 111, 171-183.

Gioulekas, D., Papakosta, D., Damialis, A., *et al.* (2004). Allergenic pollen records (15 years) and sensitization in patients with respiratory allergy in Thessaloniki, Greece. Allergy, 59, 174-184.

Gislason, D., Björnsson, E., Gislason, T., *et al.* (1999). Sensitization to airborne and food allergens in Reykjavík (Iceland) and Uppsala (Sweden) – a comparative study. Allergy, 54, 1160-1167.

Hoffmann-Sommergruber, K., Vanek-Krebitz, M., Radauer, C., *et al.* (1997). Genomic characterization of members of the Bet v 1 family: Genes coding for allergens and pathogenesis-related proteins share intron positions. Gene, 197, 91-100.

Horak, F., Studnicka, M., Gartner, C., *et al.* (2002). Parental farming protects childen aganist atopy: Longituinal evidence involving skin prick test. Clin Exp Allergy, 32, 1155-1159.

Huang, J.C., Chang,, F.C., Wang, C.S. (1997). Characterization of a lily tapetal transcript that shares sequence similarity with a class of intracellular pathogenesis-related (IPR) proteins. Plant Mol Biol, 34, 681-686.

Jarolim, E., Rumpold, H., Endler, A.T., *et al.* (1989). IgE and IgG antibodies of patients with allergy to birch pollen as tools to define the allergen profile of Betula verrucosa. Allergy, 44, 385-395.

Jõgi, R., Janson, C., Björnsson, E., *et al.* (1998). Atopy and allergic disorders among adults in Tartu, Estonia compared with Uppsala, Sweden. Clin Exp Allergy, 28, 1072-1080.

Koistinen, K.M., Hassinen, V.H., Gynther, P.A.M., *et al.* (2002). Birch PR-10c is induced by factors causing oxidative stress but appears not to confer tolerance to these agents. New Phytol, 155, 381-391.

Kootstra, H.S., Vlieg-Boerstra, B.J., Dubois, A.E.J. (2007). Assessment of the reduced allergenic properties of the Santana apple. Ann Allergy Asthma, Immunol, 99, 522-525.

Koppelman, S.J., Vlooswijk, R.A.A., Knippels, L.M.J., *et al.* (2001). Quantification of major peanut allergens Ara h 1 and Ara h 2 in the peanut varieties Runner, Spanish, Virginia, and Valencia, bred in different parts of the world. Allergy, 56, 132-137.

Larsen, J.N. (1995). Isoallergens – significance in allergen exposure and response. ACI News, 7, 141-160.

Latalowa, M., Mietus, M., Uruska, A. (2002). Seasonal variations in the atmospheric *Betula* pollen count in Gdańsk (southern Baltic coast) in relation to meteorological parameters. Aerobiologia, 18, 33-43.

Le, L.Q., Mahler, V., Lorenz, Y., *et al.* (2006). Reduced allergenicity of tomato fruits harvested from Lyc e 1–silenced transgenic tomato plants. J Allergy Clin Immunol, 118, 1176-1183.

Liu, J.J., Ekramoddoullah, A.K.M. (2004). Characterization, expression and evolution of two novel subfamilies of Pinus monticola cDNAs encoding pathogenesis-related (PR)-10 proteins. Tree Physiol, 24, 1377-1385.

Liu, J.J., Ekramoddoullah, A.K.M., Yu, X.S. (2003). Differential expression of multiple PR10 proteins in western white pine following wounding, fungal infection and cold-hardening. Physiol Plant, 119, 544-553.

Markovic-Housley, Z., Degano, M., Lamba, D., *et al.* (2003). Crystal structure of a hypoallergenic isoform of the major birch pollen allergen Bet v 1 and its likely biological function as a plant steroid carrier. J Mol Biol, 325, 123-133.

Marzban, G., Puehringer, H., Dey, R., *et al.* (2005). Localisation and distribution of the major allergens in apple fruits. Plant Sci, 169, 387-394.

McFadden, H.G., Chapple, R., de Feyter, R., *et al.* (2001). Expression of pathogenesis-related genes in cotton stems in response to infection by *Verticillium dahliae*. Physiol Mol Plant Pathol, 58, 119-131.

Meltzer, E.O. (2001). Quality of life in adults and children with allergic rhinitis. J Allergy Clin Immunol, 108, S45-S53.

Moiseyev, G.P., Fedoreyeva, L.I., Zhuravlev, Y.N., *et al.* (1997). FEBS Letters, 407, 207-210.

Moons, A., Prinsen, E., Bauw, G., *et al.* (1997). Antagonistic effects of abscisic acid and jasmonates on salt stress-inducible transcripts in rice roots. Plant Cell, 9, 2243-2259.

Nowak, D., Heinrich, J., Jörres, R., *et al.* (1996). Prevalence of respiratory symptoms, bronchial hyperresponsiveness and atopy among adults: West and East Germany. Eur Respir J, 9, 2541-2552.

Park, C.J., Kim, K.J., Shin, R., *et al.* (2004). Pathogenesis-related protein 10 isolated from hot pepper functions as a ribonuclease in an antiviral pathway. Plant J, 37, 186-198.

Pauli, G., Oster, J. P., Deviller, P., *et al.* (1996). Skin testing with recombinant allergens rBet v 1 and birch profilin, rBet v 2: Diagnostic value for birch pollen and associated allergies. J Allergy Clin I mmunol, 97, 1100-1109.

Pinto, M.P., Ribeiro, A., Regalado, A.P., *et al.* (2005). Expression of *Lupinus albus* PR-10 proteins during root and leaf development. Biol Plant, 49, 187-193.

Poupard, P., Brunel, N., Leduc, N., *et al.* (2001). Expression of a Bet v 1 homologue gene encoding a PR 10 protein in birch roots: induction by auxin and localization of the transcripts by in situ hybridization. Aust J Plant Physiol, 28, 57-63.

Poupard, P., Parisi, L., Campion, C., *et al.* (2003). A wound- and ethephon-inducible PR-10 gene subclass from apple is differentially expressed during infection with a compatible and an incompatible race of *Venturia inaequalis*. Physiol Mol Plant Pathol, 62, 3-12.

Poupard, P., Strullu, D.G., Simoneau, P. (1998). Two members of the Bet v 1 gene family encoding birch pathogenesis-related proteins display different patterns of root expression and wound-inducibility. Aust J Plant Physiol, 25, 459-464.

Pühringer, H., Moll, D., Hoffmann-Sommergruber, K., *et al.* (2000). The promoter of an apple Ypr10 gene, encoding the major allergen Mal d 1, is stress- and pathogen-inducible. Plant Sci, 152, 35-50.

Rasmussen, A. (2002). The effects of climate change on the birch pollen season in Denmark. Aerobiologia, 18, 253-265.

Raukas-Kivioja, A., Raukas, E., Loit, H.M., *et al.* (2003). Allergic sensitization among adults in Tallinn, Estonia. Clin Exp Allergy, 33, 1342-1348.

Richard-Molard, C., Brugière, N., Moille, M., *et al.* (2004). Molecular characterization of a gene encoding a vegetative storage protein (CiVSP) from *Cichorium intybus* and its expression in the root and shoot in relation to nitrogen status and pathogen resistance. Physiol Plant, 121, 568-577.

Riedler, J., Eder, W., Oberfeld, G., *et al.* (2000). Austrian children living on a farm have less hay fever, asthma and allergic sensitization. Clin Exp Allergy, 3, 194-200.

Robert, N., Ferran, J., Breda, C., *et al.* (2001). Molecular characterization of the incompatible interaction of Vitis vinifera leaves with *Pseudomonas syringae* pv. pisi: Expression of genes coding for stilbene synthase and class 10 PR protein. Eur J Plant Pathol, 107, 249-261.

Schenk, M..F, Gilissen, L.J.W.J., Esselink, G.D., *et al.* (2006). Seven different genes encode a diverse mixture of isoforms of Bet v I, the major birch pollen allergen. BMC Genomics, 7, 168.

Schenk, M.F., Thienpont, C.N., Koopman, W.J.M., *et al.* (2008a). Phylogenetic relationships in *Betula* (Betulaceae) based on AFLP markers. Tree Genet Genom, 4, 911-924.

Schenk, M.F., Fischer, A.R.H., Frewer, L.J., *et al.* (2008b). The influence of perceived benefits on acceptance of GM applications for allergy prevention. Health, Risk Soc, 10, 263-282.

Schenk, M.F., Cordewener, J.H.G., America, A.H.P., *et al.* (2009). Characterization of PR-10 genes from eight Betula species and detection of Bet v 1 isoforms in birch pollen. BMC Plant Biol, 9, 24.

Schenk, M.F., Gilissen, L.J.W.J., *et al.* (2010). Editorial. Mass spectrometry and pollen allergies. Exp Rev Proteom, 7, 627-630.

Schenk, M.F., Cordewener, J.H.G., America, A.H.P., *et al.* (2011a). Proteomic analysis of the major birch allergen Bet v 1 predicts allergenicity for 15 birch species. J Proteom, 74, 1290-1300.

Schenk, M.F., van der Maas, M.P., Smulders, *et al.* (2011b). Consumer attitudes towards hypoallergenic apples that alleviate mild apple allergy. Food Qual Pref, 22, 83-91.

Silva, J.C., Denny, R., Dorschel, C.A., *et al.* (2005). Quantitative proteomic analysis by accurate mass retention time pairs. Anal Chem, 77, 2187-2200.

Silvestri, M., Oddera, S., Rossi, G.A., *et al.* (1996). Sensitization to airborne allergens in children with respiratory symptoms. Ann Allergy Asthma Immunol, 76, 239-244.

Son, D.Y., Scheurer, S., Hoffmann, A., *et al.* (1999). Pollen-related food allergy: cloning and immunological analysis of isoforms and mutants of Mal d 1, the major apple allergen, and Bet v 1, the major birch pollen allergen. Eur J Nutrition, 38, 201-215.

Spieksma, F.T.M., Corden, J.M., Detandt, M., *et al.* (2003). Quantitative trends in annual totals of five common airborne pollen types (*Betula*, *Quercus*, *Poaceae*, *Urtica*, and *Artemisia*), at five pollen-monitoring stations in western Europe. Aerobiologia, 19, 171-184.

Srivastava, S., Fristensky, B., Kav, N.N.V. (2004). Constitutive expression of a PR10 protein enhances the germination of *Brassica napus* under saline conditions. Plant Cell Physiol, 45, 1320-1324.

Sunyer, J., Soriano, J., Antó, J.M., *et al.* (2000). Sensitization to individual allergens as risk factors for lower FEV1 in young adults. Int J Epidemiol, 29, 125-130.

Swoboda, I., Jilek, A., Ferreira, F., *et al.* (1995b). Isoforms of Bet v 1, the major birch pollen allergen, analyzed by liquid-chromatography, mass-spectrometry, and cDNA cloning. J Biol Chem, 27, 2607-2613.

Swoboda, I., Scheiner, O., Heberle-Bors, E., *et al.* (1995c). cDNA cloning and characterization of three genes in the Bet v 1 gene family that encode pathogenesis-related proteins. Plant Cell Environ, 18, 865-874.

Ukaji, N., Kuwabara, C., Takezawa, D., *et al.* (2004). Accumulation of pathogenesis-related (PR) 10/Bet v 1 protein homologues in mulberry (*Morus bombycis* Koidz.) tree during winter. Plant Cell Environ, 27, 1112-1121.

Utriainen, M., Kokko, H., Auriola, S., *et al.* (1998). PR-10 protein is induced by copper stress in roots and leaves of a Cu/Zn tolerant clone of birch, *Betula pendula.* Plant Cell Environ, 21, 821-828.

Valenta, R., Vrtala, S., Laffer, S., *et al.* (1998). Allergy reviews series II : Recombinant allergens. J Allergy Clin Immunol, 53, 552-561.

Valjakka, M., Luomala, E.M., Kangasjärvi, J., Vapaavuori, E. (1999). Expression of photosynthesis- and senescence-related genes during leaf development and senescence in silver birch (*Betula pendula*) seedlings. Physiol Plant, 106, 302-310.

van Ree, R., van Leeuwen, W.A., Akkerdaas, J.H., *et al.* (1999). How far can we simplify in vitro diagnostics for Fagales tree pollen allergy? A study with three whole pollen extracts and purified natural and recombinant allergens. Clin Exp Allergy, 29, 848-855.

Vartiainen, E., Petäys, T., Haahtela, T., *et al.* (2002). Allergic diseases, skin prick test responses, and IgE levels in North Karelia, Finland, and the Republic of Karelia, Russia. J Allergy Clin Immunol, 109, 643-648.

von Mutius, E., Weiland, S.K., Fritzsch, C., *et al.* (1998). Increasing prevalence of hay fever and atopy among children in Leipzig, East Germany. Lancet, 351, 862-866.

Walter, M.H., Liu, J.W., Wünn, J., *et al.* (1996). Bean ribonuclease-like pathogenesis-related protein genes (Ypr10) display complex patterns of developmental, dark-induced and exogenous-stimulus-dependent expression. Eur J Biochem, 239, 281-293.

Wen, J., Vanek-Krebitz, M., Hoffmann-Sommergruber, K., *et al.* (1997). The potential of Betv1 homologues, a nuclear multigene family, as phylogenetic markers in flowering plants. Mol Phylogenet Evol, 8, 317-333.

Wensing, M., Akkerdaas, J.H., van Leeuwen, A., *et al.* (2002). IgE to Bet v 1 and profilin: Cross-reactivity patterns and clinical relevance. J Allergy Clin Immunol, 110, 435-442.

Zechendorf, B. (1994). What the public thinks about biotechnology–Better than synthetic food but worse than organ-transplantation–a survey of opinion polls. Biotechnology, 12, 870-875.

21

Hypoallergenic Soybean, from Genes to Cultivar

Rong-Xia Guan, Xu-Qian Fang, Ru-Zhen Chang, Li-Juan Qiu*

National Core Facility for Crop Gene Resources and Genetic Improvement / Key Laboratory of Soybean Biology (Beijing), Institute of Crop Sciences, Chinese Academy of Agricultural Sciences, Beijing, 100081, China
* E-mail: lijuan_qiu@263.net

Abstract: Soybean is one of the most important sources of edible vegetable oil and protein. However, soybean is considered as an allergenic food, particularly in western countries. At least 16 IgE-reactive proteins have been identified in soybean, among which Gly m Bd 30K, Gly m Bd 28K and Gly m Bd 60K appeared to be major allergens. These allergens account for less than 1% of total soybean protein. Soybean germplasm selection, breeding and genetic modification have been applied to develop soybean with low- or non-allergenicity. This chapter addresses the characterization of soybean allergens and possible approaches towards developing hypoallergenic soybean cultivars.

21.1 Soybean Protein and Allergy

Soybean [*Glycine max* (L.) Merr.] is one of the major sources of vegetable protein and lipid, and is utilized in the food industry as an ingredient for many food products. Soy protein is particularly important because of its valuable nutritional, functional and processing properties. Soybean originated from China with more than 4,000 years of cultivation history (Hymowitz and Newell, 1981). However, in western countries the history of soybeans as a human food source dates back only several decades. Yet, today, the United States is the leading soybean producer in the world, reaching 92.5 million tons in 2009 and US soybean meal represented 67% of the world total protein meal consumption (http//www.soystats.com). Although there is evidence that soybean originated in China where genetic diversity is the highest, soybean germplasms from America and Japan also showed great genetic differences at the molecular level (Qiu *et al.*, 1997, Guan *et al.*, 2010). As a relatively cheap source of amino acids and protein, soybean is also widely used in the animal feed industry, accounting for 77% of protein content (Kerley *et al.*, 2003). Soybean protein consists of 136 phytochemicals, and there is evidence that individuals who consume soybean-rich diets exhibit a lower prevalence of high cholesterol, cancer and obesity (Friedman and Brandon, 2001; Fang *et al.*, 2004; Duranti *et al.*, 2004). Soybean seeds contain about 40% protein, which is composed of two major storage protein components, β-conglycinin (7S globulin) and glycinin (11S globulin) (Derbyshire *et al.*, 1976).

However, soybean is not a safe food for everyone: It ranks among the "top eight" most allergenic food sources, next to cow's milk, hen's eggs, fish, crustaceans, peanuts, tree nuts and wheat (Taylor and Helfe, 2000; Kabourek and Taylor, 2003), which account for over 90% of the documented food allergies world-wide (Metcalfe *et al.*, 1996). The first report of soy allergenicity in humans was described in 1934 (Duke *et al.*, 1934). The exact prevalence of soy allergy in the general population is not fully clear because this is dependent on local feeding habits and exposure; however, a prevalence rate of 0.3%–0.4% in the general population has been quoted (Becker *et al.*, 2004). Soy allergy appears to affect 6% of atopic children and 14% of children with cow's milk allergy (Errahali *et al.*, 2002). The symptoms of soybean allergy range from skin, gastrointestinal and respiratory reactions to severe systemic reactions including anaphylaxis (Herian *et al.*, 1990).

Regarding the clinical manifestations, three main types of soy allergenic reactions have been reported. The first type is an IgE-mediated reaction that involves abnormal responses of the immune system, such as respiratory, cutaneous, and gastrointestinal symptoms, which are usually observed minutes to hours after ingestion (Taylor *et al.*, 1999). The most severe reaction is anaphylaxis which may result in death due to cardiac or respiratory collapse. The second type represents non-IgE-mediated reactions, including fever, vomiting, failure to thrive, and body stools and can often be outgrown over time (Burks *et al.*, 1994). Soybean allergic reactions are mostly transient and non-life-threatening, and usually outgrown by the age of 3 years old (Kabourek and Taylor, 2003). Although this allergy may be transient and usually outgrown, its severity and frequency have increased, particularly in adults. Therefore, it is necessary to make an effort to remove allergenic proteins from soy products whenever possible (Van, 2004).

21.2 Characterization and Genomic Analysis of Soybean Allergens

Four protein families and super-families account for the majority of legume allergens: Cupins (including the 7S and 11S globulins); prolamin superfamily (composed of nonspecific lipid transfer proteins, and the 2S storage albumins); profilins and the larger group of pathogenesis-related proteins (mostly composed of homologues of the major birch pollen allergen, Bet v 1). The clinical features of patients with food-dependent exercise-induced anaphylaxis (FDEIA) due to tofu consumption were caused by allergen β-conglycinin. But the β-conglycinin in soy milk did not cause allergic symptoms in the patient. The main reason was that the β-conglycinin in soy milk disappeared within 20 min of pepsin digestion, whereas β-conglycinin digestion in tofu needs more than 120 min (Adachi *et al.*, 2009).

Several research groups have reported immune responses to multiple fractions of soy protein, with no fraction being the most consistent antigen. To date, at least 16 IgE-binding soybean proteins have been described, of which several have been characterized in more detail, including Kunitz trypsin inhibitor, thiol-protease Gly m Bd 30k, the α subunit of the major storage protein β-conglycinin (BC), the acidic chain of the major storage protein glycinin (G) G1 subunit, and the basic chain of the G2 subunit. In addition, two 2S albumins have been newly identified as having potential allergenicity using a protein microarray approach (Lin *et al.*, 2006). Although numerous IgE reactive soy proteins have been identified, only a few allergens have been thoroughly characterized. Immuno-blot analysis using sera from soybean-allergic patients and matrix-assisted laser desorption ionization time-of-flight mass spectrometry (MALDI-TOF) analysis revealed that the α′-, α-, and β-subunits of β-conglycinin were all major allergenic proteins (Krishnan *et al.*, 2009).

To date, only a few soy proteins including the hydrophobic soybean hull protein Gly m 1, the soybean hull protein Gly m 2, soybean profilin Gly m 3, the Bet v 1-homologous pathogenesis-related PR-10 protein Gly m 4, Beta-conglycinin Gly m 5 and Glycinin Gly m 6 are officially recognized as soybean allergens based on the criteria of the International Union of Immunological Societies Allergen Nomenclature Subcommittee (IUIS) (Ballmer-Weber and Vieths, 2008). Among them, 3 soybean proteins, Gly m Bd 28K, Gly m Bd 30K (previously referred to as 34kDa oil-body-associated protein) (Kalinski *et al.*, 1990) and Gly m Bd 60K (the same as α-subunit of β-conglycinin) (Ogawa, 1995) have been identified as major allergenic soybean proteins. These will be described in more detail below.

21.2.1 Gly m Bd 28K

Gly m Bd 28K is a major soybean glycoprotein allergen, which was isolated and purified from the 7S-globulin fraction. The encoding gene for Gly m Bd 28K has an open reading frame (ORF) of 473 amino acids, and exhibits high homology with MP27/MP32 proteins in pumpkin seeds. Gly m Bd 28K was suggested to be biosynthesized as a pro-protein form just as MP27/MP32, which would be composed of a signal peptide, Gly m Bd 28K and the C-terminal peptide (the 23kDa peptide). The 23kDa peptide was further demonstrated as a new glycoprotein allergen (Tsuji *et al.*, 1997; 2001; Hiemori *et al.*, 2004).

Epitope mapping of Gly m Bd 28K showed a dominant linear C-terminal IgE binding epitope residing between residues S256 and A270 (Xiang *et al.*, 2004). Alanine scanning of this dominant epitope indicated that five amino acids, Y260, D261, D262, K264 and D266, contribute the most in IgE-binding. Examination of the IgE-binding ability of the glycan moiety of Gly m Bd 28K in the binding reaction with patients' sera showed that the N-linked glycan moieties may be one of the common IgE-reactive determinants (Hiemori *et al.*, 2000). Gly m Bd 28K is a relatively minor seed constituent comprising less than 1% of total soy protein.

21.2.2 Gly mBd 30K (P34)

A soybean band designated Gly m Bd 30K as a 7S soybean protein allergen was initially assigned as the major allergenic protein in soybean extracts that bound specific IgE in the serum of 65% of Japanese patients with atopic dermatitis (Ogawa *et al.*, 1993). The first 15 amino acids of Gly m Bd 30K showed considerable sequence similarity (15 N-terminal amino-acid residues) to the thiol proteinase of the papain family, including the allergenic thiol proteinase Der p 1 from house dust mites (Kalinski *et al.*, 1992). It is an outlying member of the papain superfamily that appears to be a pseudomonas defense protein.

The isolated cDNA clone encoding P34 contains 1,350 base pairs terminating in a poly(A)+ tail and an open reading frame of 1,137 base pairs in length. The open reading frame includes a deduced amino acid sequence which matches 23 of 25 amino-terminal amino acids as determined by automated Edman degradation of P34 and P32. The cDNA predicts a 26,641 Da mature protein of 257 amino acids (Kalinski *et al.*, 1990).

Epitope mapping of P34 protein using an overlapping peptide strategy revealed 5 distinct immunodominant epitopes of P34. Single-site amino acid substitution of the 5 immunodominant epitopes of Gly m Bd 30K with alanine revealed that IgE binding in sera from 6 soybean-sensitive patients could be reduced or eliminated in epitopes 6 and 16 (Helm *et al.*, 2000). This finding provided information necessary for the modification of the genes responsible for encoding an antigen recognized by IgE.

21.2.3 Gly m Bd 60K (β-Conglycinin)

Gly m Bd 60K was identified as α-subunit of β-conglycinin. β-conglycinin (7S globulin), which makes up about 85% of the total 7S fraction, is a glycoprotein with a molecular weight ranging from 140 to 170 kDa. It is a trimer composed of 3 prevalent types of subunits, α, α' and β, which have molecular masses of 76, 72 and 53 kDa, respectively (Nielsen, 1985).

Interestingly, although α-subunit of β-conglycinin has allergencity, known as Gly m Bd 60K (Ogawa *et al.*, 1995), the α'-subunit, which is highly homologous to the α-subunit, had no allergencity. Ogawa *et al.* (1995) showed that the epitope(s) of the α-subunit was located on the peptide 232–383, which had no sugar moiety, suggesting that the sugar moiety of α-subunit was not essential for the binding of IgE antibodies. Beardslee *et al.* (2000) have suggested that the IgE epitopes are confined to regions of the 170 residue N-terminal insert found in the α-subunit.

21.2.4 Other Soybean Allergen (Gly m 1, Gly m 2, Gly m 3, Gly m 4)

The hull from soybean seeds has been noted as a source of aeroallergens which causes soybean-induced asthma. Three new allergenic proteins, Gly m 1A, 1B and 2, were identified from the soybean seed coat (Gonzalez *et al.*, 1992). The hull components responsible for respiratory allergies were identified and officially denominated by the International Union of Immunological Society (IUIS) as Gly m 1 and Gly m 2 (Codina *et al.*, 1997).

Rihs *et al.* (1999) found a soybean profilin, named rGly m 3, by a PCR-based cDNA cloning technique, and demonstrated that the profilin was reactive with IgE antibodies in the sera of soybean-sensitive patients. Epitope mapping with recombinant profiling indicated that IgE-binding to rGly m 3 was dependent on the integrity of a conformational structure. In view of the high allergic incidence, plant profilin is likely to be a panallergen in pollens and vegetable foods. Furthermore, soybean Gly m 3 was found to be homologous with a sequence identity of 73% to Bet v 2, a birch pollen allergen, and to 11 other plant profilins with a 69%–88% identity.

Gly m 4 was the major soy allergen for patients allergic to birch pollen and thus soybean is another birch pollen-related allergenic food (Mittag *et al.*, 2004). This type of allergy is quite severe in Central Europe. Gly m 4 has been found to be homologous to Bet v 1, a birch pollen allergen. Accounting for the high allergic incidence, Gly m 4 from soy is commercially available (Ballmer-Weber and Vieths, 2008) and is recommended for diagnostic use for birch pollen allergic patients with suspicion of soy allergy.

21.3 Selection and Genetic Modification to Create Hypoallergenic Germplasms

There are a few options for symptom treatment when an allergic reaction has occurred, but it is better to avoid the products that contain a certain allergenic ingredient because the threshold for allergens is very low. Key goals of research aimed at controlling soy allergy are to identify soy protein allergens and to decrease soy allergenic reactivity. The most efficient way to prevent soybean allergy is strict avoidance of food products containing soybean proteins. But this is very challenging as soy protein has been widely used in the food industry. So far, several approaches have been proposed and investigated to reduce the allergenicity of proteins, including enzymatic methods, physical treatment or development of soybean seeds that lack the allergenic protein (Samoto *et al.*, 1996).

Ogawa's group was able to eliminate beta-conglycinin and Gly m Bd 28K allergens by developing a mutant line induced by chemical mutation breeding techniques. Gly m Bd 30K, the main allergen, was almost completely eliminated using a salting out and centrifugation technique. In preliminary trials, about 80% of soybean-sensitive patients could ingest without adverse reaction the soybean products that were produced from the combined hypoallergenic soybean source. However, this preparative procedure has not been adopted because of limitations on further processing by the food industry.

Breeding hypoallergenic lines is an effective way to enhance food safety and to enable the consumption of soybean products by sensitive individuals. Two primary strategies have been used to obtain hypoallergenic soybean lines. First, germplasm is screened for the absence or reduced content of specific allergenic protein(s). Second, genetic transformation is used to silence native genes encoding allergenic proteins.

Germplasm screenings have been conducted at the protein and DNA levels. Protein screens have been performed using specific (monoclonal or polyclonal) antibodies that recognize a specific allergen in stained gels that evaluate the overall protein profile of the legume varieties of interest. For example, using a monoclonal antibody against Gly m Bd 28K and 30K, selection for hypoallergenic lines in soybean germplasm collections has been carried out both in Japan and the US (Yaklich, 1999). A series of Gly m Bd 28K null lines were identified; another testing showed that 80% of Japanese soybean varieties did not contain Gly m Bd 28K (Bando *et al.*, 1996). China is the center of origins of soybean with approximately 23,000 accessions currently preserved in the national gene bank. Zhang *et al.*, (2006) evaluated 60 wild soybean accessions and 421 cultivars for Gly m Bd 28K. The absence ratio of Gly m Bd 28K in cultivars (37.8%) was higher than that in wild soybean (13.3%). The Gly m Bd 28K protein null accessions accounted for 11.7%, 60.7% and 60.9% in north spring soybean, Huanghuai summer soybean, and south China soybean, respectively. Cultivars with protein content greater than 43% in these three ecotypes were 10.0%, 57.1% and 88.5%, respectively. The results also showed that cultivars lacking Gly mBd 28K allergen tended to be high in protein content, and perhaps associated with the habit of eating vegetable soybean in south China. But P34 null germplasms were not found among the Chinese soybean germplasm accessions (Guan *et al.*, 2004). Tsuji *et al.* (2001) isolated a 1,319 bp open reading frame of Gly m Bd 28K, from which genome we amplified the full sequence with a length of 3,183 bp including 4 introns and 5 exons. Eleven SNPs, of which 2 were located in exon1 and the others in introns, and a 2 bp Indel were found through alignment of the sequences from 16 cultivated and 9 wild soybean accessions. Only one SNP in exon1 generated an amino acid non-sysnonymous substitution (H>R). More research is needed to analyze the relationship of the sequence diversity with the protein function of Gly m Bd 28K (unpublished data). Recently, Wu *et al.* (2012) evaluated the temporal-spatial expression of the major allergens in soybean seed developing and germinating stage, they found that allergens accumulated sooner in cotyledons but degradated faster in embryonic axes, and Gly mBd 28K accumulated 7 to 14 d earlier than Gly m Bd 30K and 60K. These findings will be helpful for developing safer hypoallergenic soybean food.

Another effective way to remove allergens from soybean is genetic modification. By using transgene-induced gene silencing to prevent the accumulation of Gly m Bd 30K protein in soybean seeds, transgenic plants with suppression of Gly m Bd 30K-related peptides were developed. The results of IgE immunoblot analysis with a serum pool of soy-allergic individuals indicated that no additional allergenic proteins were produced in the modification of P34. The transgenic lines exhibited no phenotypes: Growth and development were identical to the wild type (Herman *et al.*, 2003).

Although genetic modification for removing P34 in soybean was successful, scientists continue to search for naturally reduced or modified germplasm with reduced or no P34 because of the public reluctance to consume genetically modified products. A total of 16,266 accessions from the USDA *Glycine* collection (Urbana, IL) consisting of soybean (*G. max*), its wild annual (*G. soja*), and wild *Glycine* perennial relatives were screened to identify accessions with low or no P34 expression. Seven *G. soja* accessions appeared to have low levels of the P34 protein, while 2 *G. max* and 3 wild perennial accessions seem to lack completely the allergenic protein. The 2D SDS/PAGE and immuno-blot analysis were performed on the two *G. max* null accessions PI603570 and PI567476, in which P34 content is less than 1% of the standard level. The results also showed that the mutation of P34 did not affect the expression of other seed proteins, indicating that the two lines have normal accumulation of seed proteins other than P34. Six single nucleotide alterations were identified in the 1.14 kb open reading frame in both P34-null accessions. These nucleotide mutations were likely the result of 4 amino acid substitutions (Joseph *et al.*, 2006). However, by sequencing the encoding genes of wild-type and low-P34 soybean lines, a four-base-pair insertion at the P34 start codon was found in low-P34 germplasm accessions. No other sequence differences were observed in the P34 genomic DNA region between the two low-P34 germplasm accessions and the genomic sequence of the standard cultivar Williams 82. This four-base-pair insertion has been proposed as being responsible for the reduced level of P34. The low-P34 phenotype in segregation populations was completely associated with the PI 567476 and PI 603570A P34 alleles. The molecular marker assays were capable of being used to incorporate the mutant P34 gene into elite soybean cultivars. But it still needs to be tested as to whether this minimal protein expression can be tolerated by soy sensitive patients.

β-Conglycinin (Gly m 5) contains important soybean allergens. The Japanese cultivars "Mo-shi-dou" (Gong 503) and "Keburi" were identified as having reduced levels of β-conglycinin (Kitamura *et al.*, 1981). "Mo-shi-dou" contains low levels of both the α- and β-subunits of β-conglycinin, whereas "Keburi" is null for the α' subunit. A cultivar, named Tohoku 124, which was selected by γ-ray irradiation of a breeding line derived from "Mo-shi-dou" and "Keburi" , has low levels of the β-subunit of β-conglycinin and is also lacking the α-subunit, the α' subunit and Gly m Bd 28K (Takahashi *et al.*, 1994). Guan *et al.*, (2004) selected a soybean accession lacking the β-subunit of β-conglycinin from the collection of Chinese soybean landraces. A β-conglycinin deficient (*cgdef*), lacking all 3 subunits, was developed by γ-ray irradiation of the soybean cultivar "Yumeyutaka". The *cgdef* mutant displays some unfavorable effects on plant growth, and the deficiency is controlled by a single recessive gene (Kitagawa *et al.*, 1991; Hayashi *et al.*, 2001). Recently, the *Cgdef* gene was mapped to soybean chromosome 19 between SSR marker Satt523 and Sat_388 (Hayashi *et al.*, 2009). Another 7S globulin deficient mutant QT2, developed from a cross of a natural mutant lacking β-conglycinin (QY2) and EnB1 lacking glycinin, was controlled by a single dominant gene *Scg-1*. This gene was mapped to soybean chromosome 20 in the same region as the α- and β-subunit genes tightly linked to each other (Hajika *et al.*, 1996; 1998; Tsubokura *et al.*, 2006). Although these two mutants had different regulation mechanisms, the molecular markers linked with the deficient genes could lead to the selection of elite soybean lines lacking β-conglycinin via marker assisted selection.

21.4 Conclusion

Significant progress has been made in the area of soybean allergenic proteins. The research advances are well reflected not only in the search for germplasm and development of elite lines with reduced allergenic proteins or their absence, but also in molecular technologies involved in genetic mechanisms aimed at reduction of allergenicity and at genetic modification of underling genes in breeding programs. Soybean germplasm accessions with decreased or null allergens of Gly m Bd 28K, Gly m Bd 30K and Gly m Bd 60K have been identified/developed and molecular markers tightly linked with the related genes are also mapped on the soybean chromosomes. These research findings have provided insights into factors affecting soybean allergenic protein composition and suggest ways to produce low allerginicity in soybean cultivars by marker assisted selection or by pyramiding all the imported genes into one elite line.

Acknowledgements

This work was supported by the State Key Basic Research and Development Plan of China (973) [Project No. 2009CB118400], the Ministry of Agriculture of China for Transgenic Research [2008ZX08009-003] and National Key Technologies R&D Program in the 10th 5-Year Plan [Project Nos. 20064AA100104, 2006AA10Z1B3 and 2008AA10Z153].

References

Adachi, A., Horikawa, T., Shimizu, H., *et al.* (2009). Soybean β-conglycinin as the main allergen in a patient with food-dependent exercise-induced anaphylaxis by tofu: Food processing alters pepsin resistance. Clin Exp Allergy, 39, 167-173.

Bando, N., Tsuji, H., Yamanashi, R., *et al.* (1996). Identification of the glycosolation site of a major soybean allergen, Gly m Bd 30K. Biosci Biotechnol Biochem, 60, 347-348.

Beardslee, T.A., Zeece, M.G., Sarath, G., *et al.* (2000). Soybean glycinin G1 acidic chain shares IgE epitopes with peanut allergen Ara h 3. Int Arch Allergy Immunol, 123(4)., 299-307.

Ballmer-Weber, B.K., Vieths, S. (2008). Soy allergy in perspective. Curr Opin Allergy Clin Immunol, 8, 270-275.

Becker, W., Brasseur, D., Bresson, J.L., *et al.* (2004). Opinion of the sScientific pPanel on dDietetic pProducts, nNutrition and aAllergies on a request from the Commission relating to the evaluation of allergenic foods for labeling purposes (Request No. EFSA-Q-2003-016). EFSA, 32, 1-197.

Burks, A.W., Casteel, H.B., Fiedorek, S.C., *et al.* (1994). Prospective oral food challenge study of two soybean protein isolates inpatients with possible milk or soy protein enterocolitis. Ped Allergy Immunol, 5, 40-45.

Codina, R., Lockey, R.F., Fernandez-Caldas, E., *et al.* (1997). Purification and characterization of a soybean hull allergen responsible for the Barcelona asthma outbreaks. II. Purification and sequencing of the Gly m 2 allergen. Clin Exp Allergy, 27(4), 424-430.

Derbyshire, E., Wright, D.B., Boulter, D. (1976). Legumin and vicilin, storage protein of legume seeds. Phytochemistry, 15, 3-24.

Duke, W. W. (1934). Soybean as a possible important source of allergy. J Allergy, 5, 300-302.

Duranti, M., Lovati, M.R., Dani, V., *et al.* (2004). The alpha-subunit from soybean 7S globulin lowers plasmalipids and upregulates liver beta-VLDL receptors in rats fed a hypercholesterolemic diet. J Nutr, 134, 1334-1339.

Errahali, Y., Morisset, M., Moneret-Vautrin, D.A., *et al.* (2002). Allergen in soy oils. Allergy, 57(7), 648-649.

Fang, N., Yu, S., Badger, T.M. (2004). Comprehensive phytochemicalprofile of soy protein isolate. J Agric Food Chem, 52, 4012-4020.

Friedman, M., Brandon, D.L. (2001). Nutritional and health benefits of soy proteins. J Agric Food Chem, 49, 1069-1086.

Gonzalez, R., Polo, F., Zapatero, L., *et al.* (1992). Purification and characterization of major in halant allergens from soybean hulls. Clin Exp Allergy, 22, 748-755.

Guan, R.X., Chang R.Z., Qiu L.J., *et al.* (2004). Analysis of protein subunit 7s/11s constitution and allergen lacking of soybean [*Glycine max* (L.). Merrill Cultivars. Acta Agron Sin, 30 (11), 1076-1079.

Guan, R.X., Chang R.Z., Li, Y.H. (2010). Genetic diversity comparison between Chinese and Japanese soybeans (*Glycine max* (L.). Merr.). revealed by nuclear SSRs.Genet Resour Crop Evol, 57, 229-242.

Hajika, M., Takahashi, M., Sakai, S., *et al.* (1996). A new genotype of 7S globulin (β-conglycinin). detected in wild soybean (*Glycine soja* Sieb. et Zucc.). Breed Sci, 46, 385-386.

Hajika, M., Takahashi, M., Sakai, S., *et al.* (1998). Dominant inheritance of a trait lacking β-conglycinin detected in a wild soybean line. Breed Sci, 48, 383-386.

Hayashi, M., Kitamura, K., Harada, K. (2009). Genetic mapping of *Cgdef* Gene controlling accumulation of 7S globulin (β-Conglycinin). subunits in soybean seeds. J Heredity, 100(6), 802-806.

Hayashi, M., Miyahara, A., Sato, S., *et al.* (2001). Construction of a genetic linkage map of the model legume *Lotus japonicus* using an intraspecific F_2 population. DNA Res, 8, 301-310.

Helm, R.M., Cockrell, G., Connaughton, C., *et al.* (2000). Mutational analysis of the IgE-binding epitopes of P34/Gly m Bd 30K. J Allergy Clin Immunol, 105(2 part 1), 378-384.

Herian, A.M., Taylor, S.L., Bush, R.K. (1990). Identification of soybean allergens by immunoblotting with sera from soy-allergic adults, Int Arch Allergy Appl Immunol, 92, 193-198

Herman, E.M., Helm, R.J., Kinney, A. (2003). Genetic modification removes an immunodominant soybean allergen. Plant Physiol, 132, 36-43.

Hiemori, M., Bando, N., Ogawa, T. (2000). Occurrence of IgE antibody-recognizing N-linked glycan moiety of a soybean allergen, Gly m Bd 28K. Allergy and Immunol, 122, 238-245.

Hiemori, M., Hitomi, I., Kimoto, M., *et al.* (2004). Identification of the 23-kDa peptide from the precursor of the Gly m Bd 28K, a major soybean allergen, as a new allergen. Biochim Biophys Acta, 1675, 174-183.

Hymowitz, T., Newell, C.A. (1981). Taxonomy of the genus Glycine, domestication and uses of soybeans. Econ Bot, 35, 272-288.

Joseph, L.M., Hymowitz, T., Schmidt, M.A., *et al.* (2006). Evaluation of Glycine germplasm for nulls of the immunodominant allergen P34/Gly m Bd 30K.

Crop Sci, 46, 1755-1763.
Kabourek, J.L. and Taylor, S.L. (2003). Soyfoods and allergies: Separating fact from fiction. The Soy Connection, 11, 1-6.
Kalinski, A., Melroy, D.L., Dwivedi, R.S., *et al.* (1992). A soybean vacuolar protein (P34). related to thiol proteases is synthesized as a glycoprotein precursor during seed maturation. J Biol Chem, 267, 12068-12076.
Kalinski, A., Weisemann, J.M., Matthews, B.F., *et al.* (1990). Molecular cloning of a protein associated with soybean seed oil bodies that is similar to thiol proteases of the papain family. J Biol Chem, 265(23), 13843-13848.
Kerley, M.S. and Allee, G.L. (2003). Modifications in soybean seed composition to enhance animal feed use and value: moving from a dietary ingredient to a functional dietary component. Ag Bio Forum, 6, 14-17.
Kitagawa, S., Ishimoto, M., Kikuchi, F., *et al.* (1991). A characteristic lacking or decreasing remarkably 7S globulin subunits induced with c–ray irradiation in soybean seeds. Jpn J Breed, 41(Suppl 2), 460-461.
Kitamura, K. and Kaizuma, N. (1981). Mutant strains with low levels of subunits of 7S globulin in soybean (*Glycine max Merr.*). seed. Japan. J. Breed., 31, 353-359.
Krishnan, H.B., Kim, W.S., Jang, S., *et al.* (2009). All three subunits of soybean β-conglycinin are potential food allergens. J Agric Food Chem, 57, 938-943.
Lin, J., Shewry, P.R., Archer, D.B., *et al.* (2006). The potential allergenicity of two 2S albumins from soybean (*Glycine max*): A protein microarray approach. Int Arch Allergy Immunol, 141, 91-102.
Metcalfe, D.D., Astwood, J.D., Townsend, R., *et al.* (1996). Assessment of allergenic potential of foods derived from genetically engineered crop plants. Crit Rev Food Sci Nutr, 36, S165-S186.
Mittag, D., Vieths, S., Vogel, L., *et al.* (2004). Soybean allergy in patients allergic to birch pollen: clinical investigation and molecular characterization of allergens. J Allergy Clin Immunol, 113(1), 148-154.
Nielsen, N.C. (1985). Structure of soy proteins. In: Altschul, A.M., and Wilcke, H.L. (eds.). New Protein Food, Vol. 5. Seed Storage Proteins. Orlando: Academic Press, pp. 27-64.
Ogawa, T., Tsuji, H., Bando, N., *et al.* (1993). Identification of the soybean allergenic protein, Gly m Bd 30K, with the soybean 34-kDa oil-body-associated protein. Biosci Biotechnol Biochem, 57, 1030-1033.
Ogawa, T., Bando, N., Tsuji, H., *et al.* (1995). α-Subunit of β-conglycinin, an allergenic protein recognized by IgE antibodies of soybean-sensitive patients with atopic dermatitis. Biosci Biotechnol Biochem, 59(5), 831-833.
Qiu, L.J., Nelson, R.L., Vodkin, L.O. (1997). Evaluation of soybean germplasm with random amplification polymorphic DNA (RAPD) markers. Scientia Agric Sin, 23, 408-417.
Rihs, H.P., Chen, Z., Rueff, F., *et al.* (1999). Binding of the recombinant allergen soybean profilin (rGly m 3). is mediated by conformational epitopes. J Allergy Clin Immunol, 104, 1293-1301.
Samoto, M., Takahaski, K., Fukuda, Y., *et al.* (1996). Substantially complete

removal of the 34 kDa allergenic soybean protein, Gly m Bd 30K from soy milk of a mutant lacking the α and α'-subunits of conglycinin. Biosci Biotech Biochem, 60, 1006-1010.

Takahashi, K., Banba, H., Kikuchi, A., *et al.* (1994). An induced mutant line lacking the R-subunit of β-conglycinin in soybean (*Glycine max* (L.). Merrill). Breed Sci, 46, 65-66.

Taylor, S.L., Hefle, S.L., Munoz-Furlong, A. (1999). Nutrition and the life cycle: food allergies and avoidance diets. Nutr Today, 34, 15-22.

Taylor, S.L., Helfe, S.L. (2000). Hidden triggers of adverse reactions to foods. Can J Allergy Clin Immunol, 5, 106-110.

Tsubokura, Y., Hajika, M., Harada, K. (2006). Molecular markers associated with β-conglycinin deficiency in soybean. Breed Sci, 56, 113-117.

Tsuji, H., Hiemori, M., Kimoto, M., *et al.* (2001). Cloning of cDNA encoding a soybean allergen, Gly m Bd 28 K. Biochim Biophys Acta, 1518(1-2), 178-182.

Tsuji, H., Bando, N., Heimori, M., *et al.* (1997). Purification of characterization of soybean allergen Gly m Bd 28K. Biosci Biotech Biochem, 61, 942-947.

van, R.R. (2004). Clinical importance of cross-reactivity in food allergy. Curr Opin Allergy Clin Immunol, 4, 235-240.

Wu, Y.M., Guan, R.X, Liu, Z.X., *et al.* (2012). Synthesis and degradation of the major allergens in developing and germinating soybean seed. J Integr Plant Biol, 54(1), 4-14.

Xiang, P., Haas, E.J., Zeece, M.G., *et al.* (2004). C-Terminal 23 kDa polypeptide of soybean Gly m Bd 28K is a potential allergen. Planta, 220, 56-63.

Yaklich, R.W., Helm, R.M., Cockrell, G., *et al.* (1999). Analysis of the distribution of the major soybean seed allergens in a core collection of *Glycine max* accessions. Crop Sci, 39(5), 1444-1447.

Zhang, Y.Q., Guan, R.X., Liu, Z.X., *et al.* (2006). Identification of Gly m Bd 28K and Gly m Bd 30K lacking soybean by using random sampling of core collection in soybean. Acta Agron Sin, 32(3), 324-329.

22

Fruit Allergy and Genetic and Genomic Tools to Select Hypoallergenic Fruit Cultivars

Zhong-Shan Gao[1*], Zhao-Wei Yang[1], Luud Gilissen[2]

[1] Department of Horticulture/Allergy Research Center, Zhejiang University, Hangzhou, 310058, China
[2] Plant Research International, Wageningen University and Research Centre, P.O. Box 619, 6700 AP, Wageningen, the Netherlands
[*] E-mail: gaozhongshan@zju.edu.cn

Abstract: Fruit allergy has become a health problem with regard to food safety and quality of life. Breeding and selection of hypoallergenic fruit cultivars is an efficient way to counteract allergy. This chapter reviews the prevalence of fruit allergy, common symptoms, and fruit species involved and their corresponding allergen proteins. Advancements in their linkage map positions and gene expression in fruit tissue are summarized, and genomic sequencing of apple and peach allergen genes. The genetic basis of cultivar differences in apple allergenicity has been illustrated in particular. Strategies to produce hypoallergenic fruits through conventional breeding and marker-assisted selection are discussed.

22.1 Introduction to Fruit Allergy

Fruit is probably one of the oldest edible raw foods for humans, and raw fruit is still highly recommended for improving and maintaining health. However, almost every kind of fruit has the ability to provoke an allergic reaction in some individuals (Gao and Gilissen, 2011), and the increasing prevalence of fruit allergy has drawn social attention and scientific interest (Marzban *et al.*, 2005). To our knowledge, the first record of the clinical connection of fruit allergy to hay fever was in 1942 (Tuft and Blumenstein, 1942). In 1982, apple allergy and its association with birch pollen allergy in Central and Northern Europe was reported (Eriksson *et al.*, 1982). Since then, the number of fruit allergy reports has increased explosively, and now allergenic fruit species are not limited to a few but involve almost all fruits in the world, although with very different allergenic profiles and relevance. The current literature includes more than 20 fruit genera (Table 22.1). Many major fruit crops in temperate zones belong to the Rosaceae family, so there is frequently more in-depth allergy research on fruits such as the apple, peach, sweet cherry, plum, apricot and pear. In addition, rapid globalization of the world market has led to an increased supply of fruits, especially in westernized societies. In parallel with the substantial increase in the consumption of tropical fruits over the last 20 years, allergies to these fruits have also increased, such as allergies to kiwi fruit (Fine, 1981; Palacin *et al.*, 2008), mango (Paschke *et al.*, 2001b) and lychee (Song *et al.*, 2007).

Fruit allergy symptoms are usually mild and the local reactions (itching and swelling in the oral cavity, throat and inner ear, running nose and eyes, coughing), collectively called oral allergy syndrome (OAS). Other fruit allergy-related symptoms have been reported in the skin, the gastrointestinal tract, nose and lungs and the cardiovascular system. Anaphylactic reactions to fruit do occur but only in rare cases. Fresh fruit has already become the most common cause of food allergy in patients over 5 years of age in Europe (Fernandez-Rivas *et al.*, 2008). Fruit allergy appears to be a multi-factorial immunological disease with highly individual features and related to a variety of environmental factors (Andersen *et al.*, 2010).

Most fruit allergy/allergen research has been done in Western Europe, especially in Austria, Spain, the Netherlands and Italy (Gao and Gilissen, 2011). The prevalence of apple allergy alone has been estimated as 2% in European countries (Hoffmann-Sommergruber, 2005). In China, fruit is one of the ten major causes of food allergy (Bai, 2005), with the most frequently reported allergenic fruit being peach and mango. The increasing prevalence of peach and mango allergy may be in line with the dramatic increase in peach production and consumption (Yang *et al.*, 2010a). Mild OAS does not completely deprive patients of fruit consumption or require medical treatment, but it does make fruit less attractive. Severe OAS has a direct impact on the patient through fruit avoidance, also resulting in reduction of fruit consumption by their family members. This will further decrease fruit intake that is already below the recommended level in Europe (Andersen, 2010).

As yet, there is no cure for fruit allergy other than avoidance of the specific fruits, so the development of low or non-allergenic fruits would be good news for all fruit allergy sufferers (Gilissen *et al.*, 2006). In Europe, fruit allergy has become a research theme in the last 10 years. In the EU-SAFE project on fruit allergy(funded by the European Commission and run from 2001 to 2003), the apple was chosen as the model (Hoffmann-Sommergruber, 2005). Follow-up

Table 22.1 Allergenic fruit crops in the literature (Gao and Gilissen, 2011) (With permission of John wiley & Sons)

Genus	Common names	References
Actinidia	Kiwi	Garcia *et al.*, 1989, Lucas *et al.*, 2003
Artocarpus	Jackfruit	Wuthrich *et al.*, 1997; Bolhaar *et al.*, 2004
Carica	Papaya	Tamburrini *et al.*, 2005
Castanea	Chestnut	Sanchez-Monge *et al.*, 2006
Citrullus	Watermelon	Jordanwagner *et al.*, 1993; Hoffmann-Sommergruber and Bruckmuller, 2009
Citrus	Mandarin, orange	Ebo *et al.*, 2007; Ahrazem *et al.*, 2005;
Cucumis	Melon, cucumber	Rodriguez *et al.*, 2000; Caballero and Martin- Esteban, 1998
Diospyros	Persimmon	Bolhaar *et al.*, 2005
Eriobotrya	Loquat	Hajime *et al.*, 2002
Ficus	Fig	Gandolfo *et al.*, 2001; Focke *et al.*, 2003
Fragaria	Strawberry	Karlsson *et al.*, 2004
Hylocereus	Dragon Fruit	Kleinheinz *et al.*, 2009
Litchi	Lychee	Song *et al.*, 2007
Lycopersicon	Tomato	Caballero and Martin-Esteban, 1998; Reche *et al.*, 2001; Pravettoni *et al.*, 2009
Malus	Apple	Ebner *et al.*, 1991
Mangifera	Mango	Duque *et al.*, 1999; Hegde and Venkatesh, 2007
Morus	Mulberry	Caiaffa *et al.*, 2003
Musa	Banana	Sanchez-Monge *et al.*, 1999; Saraswat and Kumar, 2005
Persea	Avocado	Blanco *et al.*, 1994; Diaz-Perales *et al.*, 2003
Phoenix	Date	Kwaasi *et al.*, 1999
Prunus	Peach, plum Apricot, cherry	Pastorello *et al.*, 1994; 1999; 2000; Scheurer *et al.*, 1997; 2004; Reuter *et al.*, 2005
Punicia	Pomegranate	Gaig *et al.*, 1999; Zoccatelli *et al.*, 2007
Pyrus	Pear	Karamloo *et al.*, 2001
Rubus	Raspberry	Sherson *et al.*, 2003
Vitis	Grape	Pastorello *et al.*, 2003

research on apple and peach allergy was carried out within the EU-ISAFRUIT project (Chen *et al.*, 2008; Botton *et al.*, 2008; 2009a; 2009b; Krath *et al.*, 2009), aiming at the increase in fruit consumption in Europe by reduction of allergenicity through selection, breeding and gene technology.

22.2 Identified Fruit Allergens and Genomic Research

With advanced technology and new analysis methods, most allergens have been characterized for their molecules and encoding genes. Here we summarize the main fruit allergens and the updated apple and peach allergen genes in linkage maps and the genomes.

22.2.1 Fruit Allergen Identification and Classification

Fruit allergens are generally identified by western blotting, using the extracted proteins from fruit and clinically confirmed sera, to the specific fruit, from allergic patients. With the development of molecular biology, more information has become available on the allergen amino acids sequences and encoding DNAs, and on the characterization of allergen protein structures and motifs (Gao and Gilissen, 2011). Currently, a very large number of ESTs and complete DNA data of fruit allergen genes are stored in the NCBI database for apple, peach, tomato, kiwi, grape and citrus fruits, and many of the proteins have been further identified by 2D and mass spectrography (Helsper *et al.*, 2002, Reuter *et al.*, 2005; Herndl *et al.*, 2007). By analyzing large data sets of the reported allergen protein sequences, researchers recently found common structures and properties of protein families over a wide range of plant species, genera and even families (Radauer *et al.*, 2008). Below, we give more details of the major allergen protein families.

PR10 Pathogeneses related family 10 proteins are small, acidic, intracellular proteins with molecular masses ranging from 15 to 18 kDa (van Loon and van Strien, 1999). PR10 proteins are generally unstable to heat and sensitive to proteolysis. Apple Mal d 1 and peach Pru p 1 belong to this PR10 family. In apple and peach, PR10 proteins are encoded by many homologous (multiplied and evolved) genes. The complicated expression pattern of these PR-10 proteins suggests that they have acquired diverse biological functions (Atkinson *et al.*, 1996; Liu and Ekramoddoullah, 2006). It was recently found that the peach Pru p 1 isoallergens Pru p 1.01 and Pru p 1.06D have RNA hydrolysis and cytokine binding activities (Zubini *et al.*, 2009).

nsLTP Non-specific lipid transfer protein belongs to pathogenesis-related protein family 14. nsLTP was discovered about 35 years ago as a multigene family encoding 9 kDa proteins (90–95 amino acids) that are distributed throughout the entire plant kingdom (Kader, 1996). The protein is very stable because of 4 internal disulphide bonds. nsLTP is functionally involved in lipid metabolism but lacks any specificity for fatty acids, phospholipids or cutin monomers (Douliez *et al.*, 2001). Peach Pru p 3 and apple Mal d 3 are representative allergens of this protein family in fruit, and Pru p 3 is thought of as the primary sensitizing allergen (Aseo *et al.*, 2006).

Profilins Profilins are small (12–15 kDa) cytosolic proteins that are found in all eukaryotic cells. In 1991, birch pollen profilin was identified as a relevant allergen (Valenta *et al.*, 1991). Plant profilin amino acids are typically 70% to 80% similar, and display striking features of conserved length (most consist of 131 to 134 amino acids), protein domains and structure (Radauer and Hoffmann-Sommergruber, 2004). Multiple profilin isoforms can be expressed in individual tissues. The potent allergenicity of Mal d 4 and Pru p 4, for example, has been frequently reported as a result of cross-reactivity with birch profilin Bet v 2.

Thaumatin-like protein (TLP) TLP is one of the major protein constituents of many mature edible fruits (Menu-Bouaouiche *et al.*, 2003). It is homologous to an intensely sweet tasting protein isolated from the fruit of *Thaumatococcus daniellii*. Most TLP amino acid sequences have 16 conserved cysteine residues that form eight disulfide bonds contributing to the protein's resistance to proteolysis and heat (Breiteneder, 2004). TLPs belong to the PR-5 family of pathogenesis-related proteins (van Loon and van Strien, 1999), and it has been suggested that they may be a new class of panallergens in food as well as in pollen (Breiteneder, 2004). The apple Mal d 2 allergen (Krebitz *et al.*, 2003), the kiwi fruit TLP (Gavrovic-Jankulovic *et al.*, 2008) and peach Pru p 2 (Palacin *et al.*, 2010) are three examples.

Chitinases Chitinases are digestive enzymes that break down glycosidic bonds in chitin. Some plant chitinases are classified as pathogenesis related proteins (PR-3) that are synthesized after systemic acquired resistance induction (Salzer *et al.*, 2000). Chestnut, avocado, banana and Indian date (Diaz-Perales *et al.*, 1998; Sanchez-Monge *et al.*, 1999; Lee *et al.*, 2006) have this type of allergenic protein.

Cysteine proteases Cysteine proteases have a common catalytic mechanism that involves a nucleophilic cysteine thiol in a catalytic triad. Cysteine proteases are commonly encountered in fruits including papaya, pineapple and kiwifruit. The amount of protease tends to be higher when the fruit is unripe (Tamburrini *et al.*, 2005).

Other allergenic plant fruit allergens include germin-like protein in citrus fruits (Cit s 1, Cit r 1) (Poltl *et al.*, 2007), cucumisin, a subtilisin-like endopeptidase (Cuesta-Herranz *et al.*, 2003), endo-β-1,3-glucanase of banana (Receveur-Brechot *et al.*, 2006), actinidin in kiwifruits (Palacin *et al.*, 2008), and a PR-1 protein in melon (Asensio *et al.*, 2004).

22.2.2 Genomics of Putative Allergen Genes

Once the allergen-encoding DNA sequences were available, localization of these allergen genes in established linkage maps became feasible. The four apple allergen gene families were mapped on the reference map developed from the Prima × Fiesta cross (Maliepaard *et al.*, 1998) and two alternative Fiesta x Discovery and Jonathan × Prima maps, all constructed at Plant Research International, Wageningen, the Netherlands. The basic strategies employed included four consecutive processes: (1) Search for the apple allergen sequences in the DNA databases (GenBank/EMBL) or other published databases and analyze these sequences to design PCR primers; (2) Clone and sequence the target genes on gDNA templates of Prima and Fiesta by PCR; (3) Analyze the polymorphisms of allelic sequences from the two parents and design primers to create markers; (4) Test these markers on the mapping population to obtain segregating data for mapping these new markers to known linkage maps.

The PCR-based agarose gel method to detect SNPs was the major technique used to develop allele specific makers for mapping the allergen gene members. The basis of this technique is that primers with a specific mismatch at the 3' end (SNPs) with an extra mismatch added within the last three bases of the primer will significantly reduce the PCR product of the mismatch allele, but have relatively little effect on the amplification of the correct allele (Kwok *et al.*, 1990). Simple primers were designed, empirically replacing one of the second or third base at the 3' end, and then gradient PCR was used to find the best annealing temperature to distinguish a specific allele (Fig. 22.1).

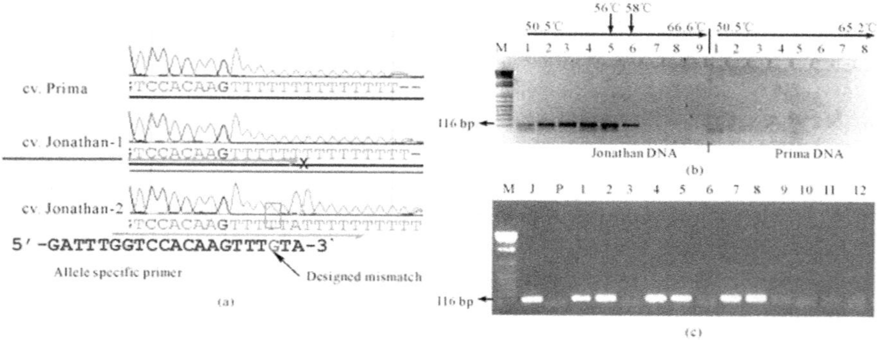

Fig. 22.1. Development of SNAP markers for gene mapping and allele testing (Gao *et al.*, 2005). (a) Sequencing reveals an SNP in two parents of an apple mapping population, Prima and Fiesta, used to design an allele specific forward primer; (b) Test of specificity using gradient PCR. Only Jonathan gives the expected 116-nt band in a wide range of Tm (lanes 2–6; 51.4–60.6℃); (c) Test for segregation in 12 descendants of JO × PM. (With permission of Springer)

22.2.3 Linkage Mapping of Putative Allergen Genes in Apple and Peach

Intensive gene cloning and mapping has given a comprehensive overview of the apple allergen genes on eight linkage groups (LGs) (Gao *et al.*, 2005b,c,d; Chen *et al.*, 2008). The Mal d 1 family consists of 18 gene members, of which 16 have been mapped as multiple gene clusters on the two homoeologous linkage groups (chromosomes) 13 and 16. A single Mal d 1.05 locus has been identified on LG 6, and 2 gene copies of Mal d 2 identified at the same position on LG 9. Other Mal d 2 genes are likely on the homoeologous LG 17, which was also indicated by similar EST sequences in GenBank. Mal d 3 genes have been located on LG 4 and 12. Genomic characterization of Mal d 4 revealed the existence of 4 genes of which 2 gene copies were found on LG 9 and 2 other single genes on LG 2 and LG 8.

Using similar genomic cloning and gene mapping in peach, allele specific markers have been developed for the 2 parents T (Texas almond) and E (Earlygold peach) of an F1 hybrid that gave rise to the well-known TxE reference map for the *Prunus* species (Howad *et al.*, 2005). In addition, direct sequencing of the 8 bin-mapping individuals revealed the SNP genotypes (Howad *et al.*, 2005). Eighteen putative peach allergen genes of 4 families similar to apple were later mapped on five linkage groups of the TxE reference map (Chen *et al.*, 2008). Apple and peach putative allergen genes have the expected syntenic and colinear regions: The Pru p 1 gene cluster on linkage group 1 (named G1 in peach) corresponds to regions with 2 clusters of homologous Mal d 1 genes on apple LG13 and LG16; Pru p/du 4.01 on the lower part of G1 and Mal d 4.03 on LG8, and Pru p 2.01A/B on G3 and Mal d 2.01A/B on LG9. Three more syntenic regions were detected in this study: Pru p 4.02 on G7 and the region of LG2 where Mal d 4.02 maps, Pru p 3.01–3.03 on G6 corresponding to Mal d 3.01 on LG12 and Mal d 3.02 on LG4, and Pru p 4.01 in the region of LG8, is syntenic with the upper part of LG15 where Mal d 4.02B and Mal d 4.03B maps.

Genomic sequences and linkage mapping of apple and peach allergen genes can be used as references for closely related fruits in the Maloideae and Prudeae subfamilies. Apple and peach putative allergen proteins are compared in Fig. 22.2. The peach genome sequence was made available in April, 2010 (http://www.rosaceae.org/peach/genome). For example, the exact sequence position of peach Pru p 1 isoallergen genes was illustrated in Fig. 22.3. BLASTing previously filed genomic sequences for physical position and neighbor regions is informative for gene expression and gene diversity surveys among diverse germplasms.

Other available fruit linkage maps which have been updated in recent years include kiwi (Fraser *et al.*, 2009), pear (Yamamoto *et al.*, 2007), raspberry (Stafne *et al.*, 2005), strawberry (Sargent *et al.*, 2009) and loquat (Gisbert *et al.*, 2009). The interesting allergen genes can be located in the same way and compared among species. Recent advances in whole genome sequencing of fruits will certainly provide much more information on sequence and physical map position.

For detailed apple and putative apple and peach allergen genes in the genomes, please go to the electronic supplementary materials in Yang *et al.* (2011).

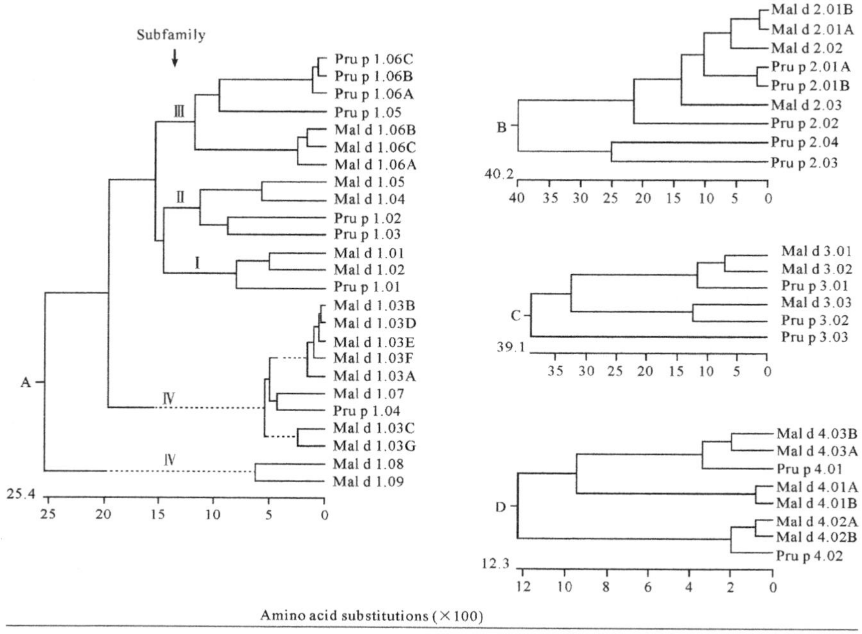

Fig. 22.2. Phylogenetic trees derived from the deduced PR-10, TLP, LTP and profilin (PRF) families in peach (Pru p 1–4) and apple (Mal d 1–4) based on our genomic sequences and ESTs retrieved from the NCBI (Chen *et al.*, 2008) (With permission of BioMed Central)

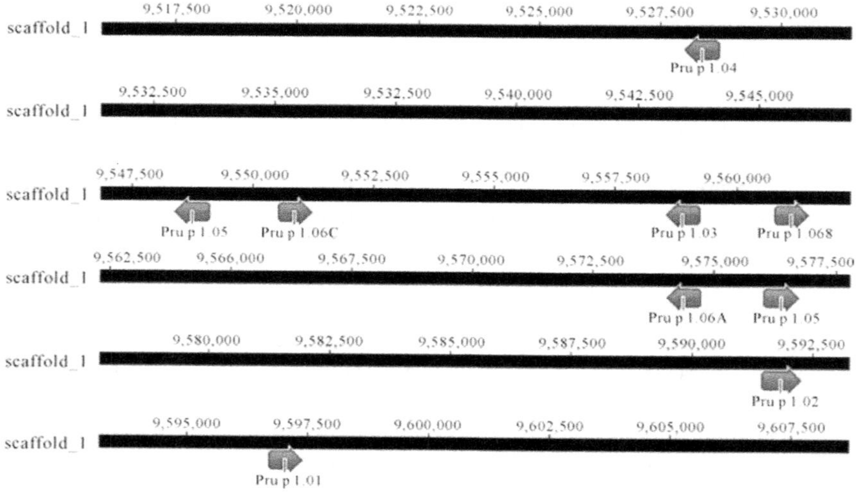

Fig. 22.3. Genome locations of previously mapped Pru p 1 isoallergen gene members

22.3 Expression Profiles of Putative Allergen Genes

The expression profiles of many homologous putative isoallergen genes are diverse. In a specific tissue, at a certain time or developmental stage, several proteins may be simultaneously present but their individual quantity and allergenicity may differ (Gao and Gilissen, 2011). Gene expression using RT-PCR in apple and peach (Botton *et al.*, 2008; 2009a; 2009b; Yang *et al.*, 2011), comprehensive immunology tests for apple (Herndl *et al.*, 2007), and synergistic combination of molecular biology and proteomics in cherry (Reuter *et al.*, 2005) have identified novel allergenic proteins. In apple fruit, a recent comprehensive gene expression study on 4 classes of putative allergens has given insight into the genetic and environmental factors affecting their allergenic potential (Botton *et al.*, 2008). Transcripts of some allergen gene members are accumulated to a high degree in ripe fruit, while others are undetectable. RT-PCR and EST sequence analysis have demonstrated the mRNA expression for 5 Mal d 1 allergen genes (at the genomic level, 18 genes can be identified) in mature fruit. These sequences represent 2 genes (Mal d 1.01 and -03E) on LG 13 and 3 genes (Mal d 1.02, -06A and -06B) on LG 16. Mal d 1.04 has only been found in mature leaves (Botton *et al.*, 2008; Beuning *et al.*, 2004; Gao *et al.*, 2008). At the protein level, only 3 Mal d 1 isoallergen proteins have been found in apple fruit, mainly Mal d 1.02 (Mal d 1b) (Helsper *et al.*, 2002; Herndl *et al.*, 2007) and, to a minor extent, the isoallergens Mal d 1.06 (Helsper *et al.*, 2002; Beuning *et al.*, 2004; Gao *et al.*, 2005a) and Mal d 1.03 (Zheng *et al.*, 2007). The majority of the Mal d 1 isoallergens in apple fruit are encoded by genes located on LG 16.

In peach, real-time PCR has shown that the most abundantly expressed member is Pru p 1.01, followed by Pru p 1.06 (Chen *et al.*, 2008; Zubini *et al.*, 2009; Yang *et al.*, 2011), with their expression levels increasing during fruit maturation and ripening. Differential expression of LTP1 (Pru p 3.01) and LTP2 (Pru p 3.02) has been observed (Botton *et al.*, 2002; 2006; 2009a). Pru p 3.01 and Mal d 3.01 have been found mainly in the peel of peach (Zuidmeer *et al.*, 2005) and apple (Sancho *et al.*, 2005), respectively. Tomato LTPs are present as different isoforms in pulp, peel and seeds (Pravettoni *et al.*, 2009). As for TLP genes in apple, the highest expression has been found for Mal d 2.01, followed by Mal d 2.02 and Mal d 2.03 (Botton *et al.*, 2008). A similar expression pattern was found in peach, with Pru p 2.01 and Pru p 2.04 expressed at a higher level than Pru p 2.02 and Pru p 2.03 (Yang *et al.*, 2011). Gene expression studies have helped to narrow down key candidate allergen-gene members, and a follow-up survey on genetic diversity and expression can be used in a wide array of cultivars or pre-breeding materials.

22.4 Selection of Hypoallergenic Cultivar

In nature, variation in allergenicity for some fruit species does occur. It will be worthwhile to identify and select those hypoallergenic cultivars for specific fruit allergy sufferers.

22.4.1 Different Allergenicity among Various Cultivars

The severity of allergic reactions to apple is not only related to the specific sensitivity of the individual, but also largely depends on the apple cultivar. For instance, Mal d 1 from the cultivar Golden Delicious has been found to be highly allergenic, whereas Mal d 1 from the cultivar Gloster is generally much less reactive (Vieths *et al.*, 1994). *In vivo* skin prick testing (SPT) with 21 different apple cultivars revealed a range of allergenic reactivity from very high to very low (Bolhaar *et al.*, 2005). On further investigation of the allergenicity of Santana in double-blind placebo-controlled food challenges (DBPCFC) and oral challenges of whole apple fruits, this new apple variety was found to be hypoallergenic for 75% of patients with a mild apple Mal d 1 allergy. On this basis, Santana has recently been marketed as "suitable for individuals with mild apple allergy" (Kootstra *et al.*, 2007) (Fig. 22.4). In Fig. 22.4, This study was reviewed and approved by the Ethics Committee of the University Medical Center, Utrecht under document number 01–050. All patients provided written, informed consent before enrolling in the study.

In vitro immunoassays to test the allergenicity of fruits of apple, peach, nectarine and kiwi have also been conducted. Allergenicity may depend on the total amount of Mal d 1 proteins, as suggested by Son *et al.* (1999). The Mal d 1 content measured in apple cultivars in Germany varied greatly from 1.3 to 20.1 μg/gFW at the time of harvest, and increased during storage (Matthes and Schmitz-Eiberger, 2009; Bolhaar *et al.*, 2005). Five cultivars from the Netherlands and Italy showed quite different contents (about 100 folds) for 3 cultivars, ranging from 6 to 455 μg/g (Zuidmeer *et al.*, 2006). These differences in Mal d 1 content could not readily be associated with the differences in SPT responses. For example, Mal d 1 content in Golden Delicious (grown in the Netherlands) was 135 μg/g determined by Bet v 1-ELISA, which is much lower than the 455μg/g of Jonathan, while the former had a much higher score in SPT (Bolhaar *et al.*, 2005). In 2010, the allergenicity of different apple cultivars in Italian apple-allergic patients was also assessed by SPT and sIgE level with rMal d 1 and rMal d 3 using UnmmuoCAP. Again, the Golden Delicious cultivar was in the most allergenic group (Ricci *et al.*, 2010).

A high level of variation in the content of Pru p 1 and Pru p 3 in peach and nectarine cultivars has been found by sandwich ELISA (Ahrazem *et al.*, 2007).

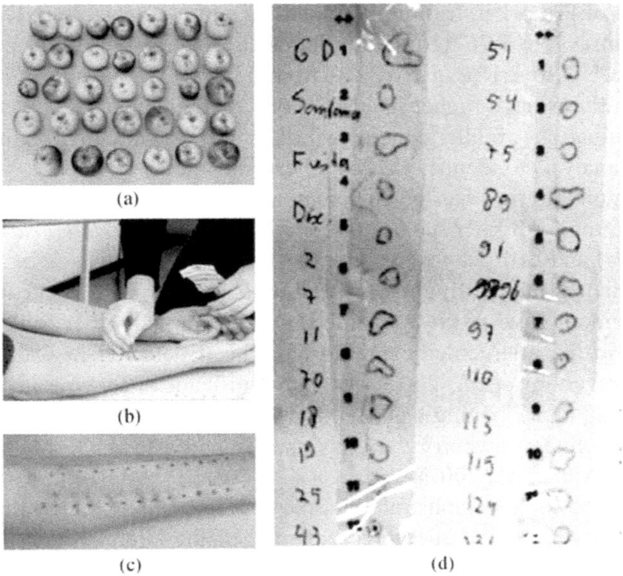

Fig. 22.4. Skin prick test of apple cultivars and accessions. (Gao and Gilissen, 2011) (a) Tested sample; (b) A doctor performs the skin prick test; (c) A positive reaction producing a wheal 20 min later; (d) Marked area for different cultivars and accessions. (With permission of John Wiley & Sons)

Fig. 22.5. Santana was bred as a low Mal d 1 allergenic apple cultivar after two generations of crossing. (a) Golden Delicious, parent of Elstar; (b) Elstar, parent of Santana; (c) Santana; (d) low (mild) allergenicity label on Santana in supermarket, the Netherlands. (Photos (b) and (c) were kindly provided by Dr. Henk Scholten and Bert Meulenbroek, Plant Research International, WUR)

The amount of Pru p 1 in fruit pulp ranged from non-detectable (ND) to 0.68 μg/g FW, and from ND to 1.76 μg/g FW in the peel. The amount of Pru p 3 in the pulp ranged from ND to 2.1 μg/g FW, and from 53.6 to 338.45 μg/g FW in the peel. Most US cultivars have higher levels of both allergens than the Spanish cultivars. Clinical evaluation of allergenicity of different peach cultivars is needed for association analysis to identify the low allergenic alleles.

IgE immunoblotting has given remarkably different protein profiles and IgE binding patterns for 3 kiwi species: green kiwi (*Actinindia deliciosa*), golden kiwi (*A. chinensis*) and hardy (*A. arguta*) (Chen *et al.*, 2006; Lucas *et al.*, 2007). For a specific patient group, it might be possible to identify and select low allergenic accessions among diverse kiwi germplasm collections. In other fruit, no particular differences in the allergenicity pattern have been detected in 6 different cherry cultivars (Pravettoni *et al.*, 2007) and no significant difference in the allergenic potency during fruit ripening has been found in mango (Paschke *et al.*, 2001b). In contrast, the white strawberry varieties usually had a lower Fra a 1 content than the red ones (Alm *et al.*, 2007).

The available data demonstrate that the degree of allergenicity may depend on the amount as well as the allergenic potential of the allergen, and that in several fruits the ripening stage and the storage conditions are determining factors.

22.4.2 Allelic Diversity Survey and Association Analysis on Apple Allergenicity

Following the mapping of the putative apple allergen genes, preliminary genetic analysis on specific alleles controlling the low Mal d 1 allergenicity was carried out (Gao *et al.*, 2008). The allelic diversity of the seven intron-containing Mal d 1 genes was assessed in a set of 14 apple cultivars by sequencing or, indirectly, by molecular marker tests. In contrast to what was assumed, sequencing of 10 cultivars gave only 2 variants of the main isoallergen Mal d 1.02 (with slight amino acid changes), making the validity of the 8 variants, which can be deduced from previous sequences in the database, doubtful (Table 22.3). Comparison of the Mal d 1 allelic composition between the high-allergenic cultivar Golden Delicious and the low-allergenic cultivar Santana, which are linked in pedigree, showed an association between the presence of specific Mal d 1.06A alleles (located on LG 16) and low allergenicity. This association was confirmed in 10 other cultivars. The deduced Mal d 1.06A allele dosage effects are also shown in Table 22.4. An SSR marker of 154 bp developed from Mal 1.06A can be used to screen germplasms and select for potential hypo-allergenic sources.

These findings indicate the need to reconsider the relevance of merely assessing total amounts of Mal d 1 protein in allergenicity research and diagnostic tests, and to focus future research on the association of specific Mal d 1 isoforms and allergenicity among a larger group of cultivars and allergy patients.

Table 22.3 SNPs in the Mal d 1.02 sequences found in the database (Gao, 2005a) (With Courtesy of Zhong-Shan Gao)

Sequen-ce allele [a]	Cultivars [b]	Nucleotide position in coding sequences [c]																									
		9	18	19	23	65	161	165	168	180	186	222	231	273	306	313	316	369	386	387	401	414	420	448	453	458	459
Mal d 1.02 consensus		C	T	T	C	C	T	G	G	T	C	G	G	T	G	T	G	G	A	G	A	G	T	T	T	G	C
Mal d 1.0201.01 (L42952, AF074721,AY428578)	GD, GA, IR, PM, FS, GD, JO, CO							A			T																
Mal d 1.0201.02	PM											A	A												C		
Mal d 1.0201.03	PS, IM, JO CO, RD, FJ												A												C		
Mal d 1.0201.04 (AF124826)	IR, GD, FJ																								C		
Mal d 1.0201.05	IM															C											
Mal d 1.0201.06	RD															C		A									
Mal d 1.0201.07	DS																					A	C				
AF124827	GA				*T*			A			T																
AF124828	GS													C		C											
AF124833	GD																			*A*					C		
AF124834	GL												A														
AF124836	GA							A					A														*G*
Mal d 1.0202 (AF124822)	GS							A		G	T																

(To be continued)

(Table 22.3)

Sequen-ce allele[a]	Cultivars[b]	Nucleotide position in coding sequences[c]																									
		9	18	19	23	65	161	165	168	180	186	222	231	273	306	313	316	369	386	387	401	414	420	448	453	458	459
Mal d 1.0203 (AF124824)	GD							A			T													*C*			
Mal d 1.0204 (AF124825)	GL												A				A								C		
Mal d 1.0205 (AF124835)	JB												A								T				C		
Mal d 1.0206 (AF020542)	MT					G							A		A										C		
Mal d 1.0207 (AF026911)	GD-seedling						C																		C		
Mal d 1.0208 (AF488060)	CO	*T*	*A*	*C*				A			T								*T*							*A*	
Mal d 1.0209	DS								T																C		

[a] Variant-alleles in bold type are from our sequences (accession numbers listed in Table 2 inGao et al., BMC Plant Biology 2008, 8:116); The same reference sequences are given in brackets;

[b] Golden Delicious (GD), Priscilla (PS), Ingrid Marie (IM), Cox (CO), Jonathan (JO), Red Delicious (RD), Fuji (FJ), Discovery (DS), Prima (PM), Fiesta (FS). Gala (GA), Jamba (JB), Gloster (GL), McIntosh (MT), IR-Idared, Granny Smith (GS): seedling offspring of GD apple cultivar (seedling-GD);Cultivars in bold type deduced from genomic sequences;

[c] Position refers to the coding sequence. SNP nucleotides in rectangular boundary were true, in italic and underlined were errors after cross-checking. Other SNPs are not certain.

Further analysis of the Mal d 1 gene cluster haplotypes of the allergenic trait demonstrated an interesting phenomenon. The Mal d 1.06A-SSR-154 marker and the Mal d 1.02-3PS/IM SNAP marker were associated with low allergenicity in one haplotype. In the case of Santana and Priscilla, their homozygous Mal d 1.0201.03 allele has a specific 231A residue (Table 22.3). The same 231A residue was found in Gloster (AF124825, AF124834), Jamba (AF124835), McIntosh (AF020542), and Gala (AF124836) (Table 22.3). Gloster and Jamba have been reported to be less allergenic than Golden Delicious (Son *et al.*, 1999). Another cultivar, Wijcik McIntosh, is also homozygous for Mal d 1.06A-ssr-154, so we therefore expect its Mal d 1 allergenicity to be low. If this holds true in other cultivars and selections, future breeding programs for low allergenic apples will be facilitated, enabling deliberated choices from parental cultivars and genetic markers for selection in young seedling populations.

Santana is a new, low-allergenic cultivar with high levels of resistance to scab, mildew and *Nectria* canker. Large-scale production of this cultivar has begun in the Netherlands. In addition, this cultivar may also have lower Mal d 3 content since its parents, Elstar and Priscilla, are both low in this allergen (Sancho *et al.*, 2008). Santana will, therefore, be a very good parent for further breeding of low-allergenic cultivars. Molecular markers from this research could be used to select favorable alleles in progeny.

Information on allelic diversity of the key allergen genes in other fruit is still very limited. In tomato, the ripening inhibitor (rin) mutant showed reduced allergenicity related to the proteins β-fructofuranosidase and polygalacturonase 2A which cross-react with Japanese cedar pollen allergens. In Japan, patients allergic to tomato showed lower levels of reactivity to the extract from this hybrid mutant (Kitagawa *et al.*, 2006).

Table 22.4 Mal d 1.06A gene SSR marker associated with allergenicity

Cultivar[a]	PS	ST	JO	EC	PM	ES	FJ	GA	EL	BE	FS	DE	PI	GD
Allergenicity score[b]	30	35	48	51	61	61	61	64	67	72	83	87	89	100
Mal d 1.06A-SSR −154[c]	++	++	+	+		+	+	+	+					
−156							+					+	+	+
−162			+			+		+	+	++	++	+	+	+
−164				+	+									
−174					+									

[a] Cultivars: Priscilla (PS), Santana (ST), Jonathan (JO), Ecolette (EC), Prima (PM), Elstar (ES), Fuji (FJ), Gala (GA), Elise (EL), Bellida (BE), Fiesta (FS), Delblush (DE), Pinova (PI), Golden Delicious (GD); GD is the parent of ES, the latter is the parent of ST;
[b] Responses are expressed in percentage relative to Golden Delicious (GD);
[c] homozygous ++.

Pru p 1 and Pru p 3 are the main allergens in peach. The encoding genes are located in clusters on 2 linkage groups, G1 and G6 (Chen *et al.*, 2008). Based on their bin map positions, SSR markers tightly linked or flanking these 2 major allergen genes were to be used for a diversity survey of a large number of genotypes (Chen *et al.*, 2008). However, preliminary cloning and sequencing of Pru p 1.01 and Pru p 3.01 in the encoding region for a number of peach cultivars from different types did not show any polymorphism, which is probably due to the low genetic diversity between the chosen peach varieties. It may be that there is very little variation in some fruits, as has been demonstrated in strawberry (Musidlowska-Persson *et al.*, 2007).

Recombinant food allergens have been frequently produced for research and diagnosis. Most are comparable to their natural counterparts. The generation of hypoallergenic isoforms and mutants is also useful for therapeutic purposes and for the determination of epitopes and cross-reactive structures (Lorenz *et al.*, 2001). Identification and quantification of the Mal d 1 isoallergen and variants in fruits of Santana and Golden Delicious will provide further solid evidence with regard to which Mal d 1 genes and alleles are expressed and related to allergenicity. A modified mutation of a Mal d 1 isoform (AJ41771) has already shown reduced allergenicity (Hoffmann-Sommergruber *et al.*, 2005).

MALDI-TOF or Q-TOF and peptide sequencing, together with available genomic information, can identify different proteins in the natural allergen mixtures, but accurate isoallergen and variant quantification will be difficult. One approach to solve this problem may be the development of monoclonal recombinant variants of all known Mal d 1 allergens by genome cloning, and their application for Mal d 1 isoallergen separation and allergenicity testing. Molecular modeling of deduced variants and 3D structural determination will also help to elucidate the molecular mechanism of allergenicity.

22.5 Conclusion

Natural variation among cultivars and pre-selected breeding materials can be tested by clinical skin prick tests and verified by double-blind placebo controlled food (fruit) challenges to find suitable low allergenic cultivars for fruit allergy patients. Current genetic and genomic tools to further investigate these cultivars will facilitate the process of breeding and selection of hypoallergenic fruit, leading to an improvement in the health and well-being of the increasing number of people suffering from fruit allergies.

Acknowledgements

This work was supported by the Natural Science Foundation of China

(30971970) and the Special Research Fund for Public Walfare in Chinese Agriculture (Contract No. 20090344). We thank Wiley-Blackwell for their permission to use some parts from Gao and Gilissen, *Breeding of Hypoallergenic Fruits* (p.105–126), in *Breeding for Fruit Quality*, edited by Matthew A. Jenks and Penelope J. Bebeli, Copyright © 2011 John Wiley & Sons, Inc.

References

Ahrazem, O., Ibanez, M.D., Lopez-Torrejon, G., *et al.* (2005). Lipid transfer proteins and allergy to oranges. Int Arch Allergy Immunol, 137(3), 201-210.

Ahrazem, O., Jimeno, L., Lopez-Torrejon G., *et al.* (2007). Assessing allergen levels in peach and nectarine cultivars. Ann Allergy Asthma Immunol, 99, 42-47.

Alm, R., Ekefjard A., Krogh, M., *et al.* (2007). Proteomic variation is as large within as between strawberry varieties. J Proteome Res, 6, 3011-3020.

Andersen, M.B., Hall S., Dragsted, L.O. (2010). Identification of European Allergy Patterns to the Allergen Families PR-10, LTP, and Profilin from Rosaceae Fruits. Clin Rev Allergy Immunol.

Asensio, T., Crespo, J.F., Sanchez-Monge, R., *et al.* (2004). Novel plant pathogenesis-related protein family involved in food allergy. J Allergy Clin Immunol, 114(4), 896-899.

Asero, R., Mistrello, G., Amato, S., *et al.* (2006). Peach fuzz containe large amounts of lipid transfer protein: is this the cause of the high prevalence of sensitization to LTP in Mediterranean countries? Eur Annals Allergy Clin Immunol, 38(4), 118-121.

Bai, Q.Y. (2005). The allergic problems of certain foods. Food Nutr China, 7, 54-55.

Beuning, L.L., Bowen, J.H., Persson, H.A., *et al.* (2004). Characterisation of Mal d 1-related genes in Malus. Plant Mol Biol, 55(3), 369-388.

Blanco, C., Carrillo, T., Castillo, R., *et al.* (1994). Avocado hypersensitivity, Allergy, 49(6), 454-459.

Bolhaar, S., van de Weg, W. E., van Ree, R., *et al.* (2005). *In vivo* assessment with prick-to-prick testing and double-blind, placebo-controlled food challenge of allergenicity of apple cultivars. J Allergy Clin Immunol, 116(5), 1080-1086.

Bolhaar, S., van Ree, R., Bruijnzeel-Koomen, C., *et al.* (2004). Allergy to jackfruit: a novel example of Bet v 1-related food allergy. Allergy, 59(11), 1187-1192.

Bolhaar, S.T.H.P., van Ree, R., Ma, Y., Bruijnzeel-Koomen, C.A.F.M., Vieths, S., Hoffmann-Sommergruber, Knulst, A.C., Zuidmeer, L. (2005). Severe allergy to Sharon fruit caused by birch pollen. Int Arch Allergy Immunol, 136, 45-52.

Botton, A., Andreotti, C., Costa, G., *et al.* (2009a). Peach (Prunus persica L. Batsch) Allergen-encoding genes are developmentally regulated and affected by fruit load and light radiation. J Agric Food Chem, 57(2), 724-734.

Botton, A., Begheldo, M., Rasori, A., *et al.* (2002). Differential expression of two lipid transfer protein genes in reproductive organs of peach (Prunus persica L. Batsch). Plant Sci, 163(5), 993-1000.

Botton, A., Lezzer, P., Dorigoni, A., *et al.* (2008). Genetic and environmental factors affecting allergen-related gene expression in apple fruit (Malus domestica L. Borkh). J Agric Food Chem, 56(15), 6707-6716.

Botton, A., Lezzer, P., Dorigoni, A., *et al.* (2009b). Environmental factors affecting the expression of apple (Malus X domestica L. Borkh) allergen-encoding genes. J Horti Sci Biotechnol, 182-187.

Botton, A., Vegro, M., De Franceschi, F., *et al.* (2006). Different expression of Pp-LTP1 and accumulation of Pru p 3 in fruits of two Prunus persica L. Batsch genotypes. Plant Sci, 171(1), 106-113.

Breiteneder, H. (2004). Thaumatin-like proteins – a new family of pollen and fruit allergens. Allergy, 59,479-481.

Caballero, T., Martin-Esteban, M. (1998) Association between pollen hypersensitivity and edible vegetable allergy: A review. J Investig Allergo Clin Immuno, 8(1), 6-16.

Caiaffa, M.F., Cataldo, V.M., Tursi, A., *et al.* (2003). Fig and mulberry cross-allergy. Annals of Allergy Asthma Immunol, 91(5), 493-495.

Chen, L., Zhang, S. M., Illa, E., *et al.* (2008). Genomic characterization of putative allergen genes in peach/almond and their synteny with apple. BMC Genomics, 9, 543.

Chen, L.Y., Lucas, J.S, Hourihane, J.O., *et al.* (2006). Evaluation of IgE binding to proteins of hardy (Actinidia arguta), gold (Actinidia chinensis) and green (Actinidia deliciosa) kiwifruits and processed hardy kiwifruit concentrate, using sera of individuals with food allergies to green kiwifruit. Food Chem Toxico, 44, 1100-1107.

Cuesta-Herranz, J., Pastor, C., Figueredo, E., *et al.* (2003). Identification of Cucumisin (Cuc m 1), a subtilisin-like endopeptidase, as the major allergen of melon fruit. Clin Exp Allergy, 33(6), 827-833.

Diaz-Perales, A., Blanco, C., Sanchez-Monge, R., *et al.* (2003). Analysis of avocado allergen (Prs a 1) IgE-binding peptides generated by simulated gastric fluid digestion. J Allergy Clin Immunol, 112(5), 1002-1007.

Diaz-Perales, A., Collada, C., Blanco, C., *et al.* (1998). Class I chitinases with hevein-like domain, but not class II enzymes, are relevant chestnut and avocado allergens. J Allergy Clin Immunol, 102(1), 127-133.

Duque, S., Fernandez-Pellon, L., Rodriguez, F. (1999). Mango allergy in a latex-sensitized patient. Allergy, 54(9), 1004-1005.

Ebner, C., Birkner, T., Valenta, R., *et al.* (1991). Common epitopes of birch pollen and apples – studies by western and northern blot. J Allergy Clin Immunol, 88(4), 588-594.

Ebo, D.G., Ahrazem, O., Lopez-Torrejon, G., *et al.* (2007). Anaphylaxis from mandarin (Citrus reticulata) : Identification of potential responsible allergens. Intl Arch Allergy Immunol, 144(1), 39-43.

Eriksson, N.E., Formgren, H., Svenonius, E. (1982). Food hypersensitivity in patients with pollen allergy. Allery, 33, 437-443.

Fernandez-Rivas, M., Benito, C., Gonzalez-Mancebo, E., *et al.* (2008) Allergies to fruits and vegetables. Pediatric Allergy Immuno, 19(8), 675-681.

Fine, A.J. (1981). Hypersensitivity reaction to kiwi fruit (Chinese gooseberry, Actinidia Chinesis). J Allergy Clin Immunol, 68(3), 235-237.

Focke, M., Hemmer, W., Wohrl, S., *et al.* (2003). Cross-reactivity between Ficus benjamina latex and fig fruit in patients with clinical fig allergy. Clin Exp Allergy, 33(7), 971-977.

Fraser, L.G., Tsang, G.K., Datson, P.M., *et al.* (2009). A gene-rich linkage map in the dioecious species Actinidia chinensis (kiwifruit) reveals putative X/Y sex-determining chromosomes. BMC Genomics, 10,102.

Gaig, P., Bartolome, B., Lleonart, R., *et al.* (1999). Allergy to pomegranate (Punica granatum). Allergy, 54(3), 287-288.

Gandolfo, M., Baeza, M., De Barrio, M. (2001). Anaphylaxis after eating figs, Allergy, 56(5), 462-463.

Gao, Z.S. (2005a). Localization of candidate allergen genes on the apple (Malus domestica) genome and their putative allergenicity. Ph.D thesis CIP-DATA Koninklijke biobliotheek, Den Haag.

Gao, Z.S., van de Weg, W.E., Schaart, J.G., *et al.* (2005b). Genomic cloning and linkage mapping of the Mal d 1 (PR-10) gene family in apple (Malus domestica). Theo Appl Genet, 111(1), 171-183.

Gao, Z.S., van de Weg, W.E., Schaart, J.G., *et al.* (2005c). Genomic characterization and linkage mapping of the apple allergen genes Mal d 2 (thaumatin-like protein) and Mal d 4 (profilin). Theo Appl Genet, 111(6), 1087-1097.

Gao, Z.S., van de Weg, W.E., Schaart, J.G., *et al.* (2005d). Linkage map positions and allelic diversity of two Mal d 3 (non-specific lipid transfer protein) genes in the cultivated apple (Malus domestica). Theo Appl Genet, 110(3), 479-491.

Gao, Z.S., van de Weg, E.W., Matos, C.I., *et al.* (2008). Assessment of allelic diversity in intron-containing Mal d 1 genes and their association to apple allergenicity. BMC Plant Biology, 8,116.

Gao, Z.S. and Gilissen, L.J. W.J. (2011). Breeding of hypoallergenic fruits. In: Jenks, M.A., Bebeli, P.J. (eds.) Breeding for Fruit Quality. New York: John Wiley & Sons, Inc., pp. 105-126.

Garcia, B.E., Delacuesta, C.G., Santos, F., *et al.* (1989). A rare case of food allergy-monosensitivity to kiwi (Actinidia Chinensis). Allergologia Et Immunopathologia, 17(4), 217-218.

Gavrovic-Jankulovic, M., Polovic, N., Prisic, S., *et al.* (2005). Allergenic potency of kiwi fruit during fruit development. Food Agric Immunol, 16(2), 117-128.

Gavrovic-Jankulovic, M., Spasic, M., Cirkovic Velickovic, T., *et al.* (2008). Quantification of the thaumatin-like kiwi allergen by a monoclonal antibody-based ELISA. Mol Nutr Food Res, 52(6), 701-707.

Gilissen, L., Bolhaar, S., Knulst, A.C., *et al.* (2006). Production of hypoallergenic plant foods by selection, breeding and genetic modification. In: Gilissen, L. J. W., Wichers, H. J., Savelkoul, H. F. J., Bogers, R. J. (eds.). Allergy Matters: New Approaches to Allergy Prevention and Management, 97-107.

Gisbert, A.D., Martínez-Calvo, J., Llácer, G., *et al.* (2009). Development of two loquat [Eriobotrya japonica (Thunb.) Lindl.] Linkage maps based on AFLPs and SSR markers from different Rosaceae species. Mol Breed, 23, 523-538.

Hajime, K., Motoshi, W., Iwao, A. (2002). A case of oral allergy syndrome caused by intake of loquat. Allergy in Practice, 293, 723-727 (in Japanese) .

Hegde, V.L. and Venkatesh, Y.P. (2007). Anaphylaxis following ingestion of mango fruit. J Investig Allergology Clin Immunol, 17(5), 341-344.

Helsper, J., Gilissen, L., van Ree, R., *et al.* (2002). Quadrupole time-of-flight mass spectrometry: A method to study the actual expression of allergen isoforms identified by PCR cloning. J Allergy Clin Immunol, 110(1), 131-138.

Herndl, A., Marzban, G., Kolarich, D., *et al.* (2007). Mapping of Malus domestica allergens by 2-D electrophoresis and IgE-reactivity. Electrophoresis, 28(3), 437-448.

Hoffmann-Sommergruber, K. (2005). The SAFE project: Plant food allergies: Field to table strategies for reducing their incidence in Europe an EC-funded study. Allergy, 60(4), 436-442.

Hoffmann-Sommergruber, K. and Bruckmuller, M. (2009). Watermelon Contains 92% Water but It Also Contains Allergens! Int Arch Allergy Immunol, 149(4), 289-290.

Howad, W., Yamamoto, T., Dirlewanger, E., *et al.* (2005). Mapping with a few plants: Using selective mapping for microsatellite saturation of the Prunus reference map. Genetics, 171(3), 1305-1309.

Jordanwagner, D.L., Whisman, B.A., Goetz, D.W. (1993). Cross-allergenicitu among celery, cucumber, carrot and watermelom. Annals of Allergy, 71(1), 70-79.

Kader J.C. (1996). Lipid-transfer proteins in plants. Ann Rev Plant Physiol Plant Mol Biol, 47, 627-654.

Karamloo, F., Scheurer, S., Wangorsch, A., *et al.* (2001). Pyr c 1, the major allergen from pear (Pyrus communis), is a new member of the Bet v 1 allergen family. J Chromatogr B-Analytical Technologies in the Biomedical and Life Sciences, 756(1-2), 281-293.

Karlsson, A.L., Alm, R., Ekstrand, B., *et al.* (2004). Bet v 1 homologues in strawberry identified as IgE-binding proteins and presumptive allergens. Allergy, 59(12), 1277-1284.

Kitagawa, M., Moriyama, T., Ito, H., *et al.* (2006). Reduction of allergenic proteins by the effect of the ripening inhibitor (rin) mutant gene in an F-1 hybrid of the rin mutant tomato. Biosci Biotechnol Biochem, 70(5), 1227-1233.

Kleinheinz, A., Lepp, U., Hausen, B.M., *et al.* (2009). Anaphylactic reaction to (mixed) fruit juice containing dragon fruit. J Allergy Clin Immunol, 124(4), 841-842.

Kootstra, H.S., Vlleg-Boerstra, B.J., Dubois, A.E.J. (2007). Assessment of the reduced allergenic properties of the Santana apple. Annals of Allergy Asthma & Immunology, 99(6), 522-525.

Krath, B.N., Eriksen, F.D., Pedersen, B.H., *et al.* (2009). Development of hypoallergenic apples: silencing of the major allergen Mal d 1 gene in "Elstar" apple and the effect of grafting. J Horti Sci Biotechnol, 52-57.

Krebitz, M., Wagner, B., Ferreira, F., Peterbauer, C., *et al.* (2003). Plant-based heterologous expression of Mal d 2, a thaumatin-like protein and allergen of apple (Malus domestica), and its characterization as an antifungal protein. J Mol Biol, 329, 721-730.

Kwaasi, A.A.A., Harfi, H.A., Parhar, R.S., *et al.* (1999). Allergy to date fruits: characterization of antigens and allergens of fruits of the date palm (Phoenix dactylifera L.). Allergy, 54(12), 1270-1277.

Lee, M.F., Hwang, G.Y., Chen, Y.H., *et al.* (2006) Molecular cloning of Indian jujube (Zizyphus mauritiana) allergen Ziz m 1 with sequence similarity to plant class III chitinases. Mol Immunol, 43(8), 1144-1151.

Liu, J.J., Ekramoddoullah, A.K.M. (2006). The family 10 of plant pathogenesis-related proteins: Their structure, regulation, and function in response to biotic and abiotic stresses. Physiol Mol Plant Pathol, 68(1-3), 3-13.

Lorenz, A.R., Scheurer, S., Haustein, D., Vieths, S. (2001). Recombinant food allergens. J Chromatogr B, 756, 255-279.

Lucas, J.S.A., Lewis, S.A., Hourihane, J.O. (2003). Kiwi fruit allergy: A review. Pediatric Allergy and Immunology, 14(6), 420-428.

Maliepaard, C., Alston, F.H., van Arkel, G., *et al.* (1998). Aligning male and female linkage maps of apple (Malus pumila Mill.) using multi-allelic markers. Theor Appl Genet, 97(1-2), 60-73.

Marzban, G., Mansfeld, A., Hemmer, W., *et al.* (2005). Fruit cross-reactive allergens: A theme of uprising interest for consumers' health. Biofactors, 23(4), 235-241.

Matthes, A., Schmitz-Eiberger, M. (2009). Apple (Malus domestica L. Borkh.) Allergen Mal d 1: Effect of Cultivar, Cultivation System, and Storage Conditions. J Agric Food Chem, 57(22), 10548-10553.

Menu-Bouaouiche, L., Vriet, C., Peumans, W.J., *et al.* (2003). A molecular basis for the endo-beta 1,3-glucanase activity of the thaumatin-like proteins from edible fruits. Biochimie, 85(1-2), 123-131.

Musidlowska-Persson, A., Alm, R., Emanuelsson, C. (2007). Cloning and sequencing of the Bet v 1-homologous allergen Fra a 1 in strawberry (Fragaria ananassa) shows the presence of an intron and little variability in amino acid sequence. Mol Immunol, 44(6), 1245-1252.

Palacin, A., Rodriguez, J., Blanco, C., *et al.* (2008). Immunoglobulin E recognition patterns to purified kiwifruit (Actinidinia deliciosa) allergens in patients sensitized to kiwi with different clinical symptoms. Clin Exp Allergy, 38(7), 1220-1228.

Paschke, A., Kinder, H., Zunker, K., *et al.* (2001a). Characterization of cross-reacting allergens in mango fruit. Allergy, 56(3), 237-242.

Paschke, A., Kinder, H., Zunker, K., *et al.* (2001b) Characterization of allergens in mango fruit and ripening dependence of the allergenic potency. Food and Agricultural Immunology, 13(1), 51-61.

Pastorello, E.A., Ortolani, C., Farioli, L., *et al.* (1994). Allergenic cross-reactivity among peach, apricot, plum, and cherry in patients with oral allergy syndrome: An in-vivo and in-vitro study. J Allergy Clin Immunol, 94(4), 699-707.

Pastorello, E.A., Farioli, L., Pravettoni, V., *et al.* (1999). The major allergen of peach (Prunus persica) is a lipid transfer protein. J Allergy Clin Immunol, 103(3), 520-526.

Pastorello, E.A., D'Ambrosio, F.P., Pravettoni, V., *et al.* (2000). Evidence for a lipid transfer protein as the major allergen of apricot. J Allergy Clin Immunol, 105(2), 371-377.

Pastorello, E.A., Farioli, L., Pravettoni, V., *et al.* (2001). Characterization of the major allergen of plum as a lipid transfer protein. J Chromatog B-Analytical Technologies in the Biomedical and Life Sciences, 756(1-2), 95-103.

Pastorello, E.A., Farioli, L., Pravettoni, V., *et al.* (2003). Identification of grape and wine allergens as an endochitinase 4, a lipid-transfer protein, and a thaumatin. J Allergy Clin Immunol, 111(2), 350-359.

Poltl, G., Ahrazem, O., Paschinger, K., Ibanez, M.D., *et al.* (2007). Molecular and immunological characterization of the glycosylated orange allergen Cit s 1. Glycobiology, 17(2), 220-230.

Pravettoni, V., Brenna, O., Farioli, L., *et al.* (2007). Cherry allergy: How to decrease the allergenicity of cherry industrial derivatives. J Allergy Clin Immunol, 119(1), 466.

Pravettoni, V., Primavesi, L., Farioli, L., *et al.* (2009). Tomato Allergy: Detection of IgE-Binding Lipid Transfer Proteins in Tomato Derivatives and in Fresh Tomato Peel, Pulp, and Seeds. J Agric Food Chem, 57(22), 10749-10754.

Radauer, C., Hoffmann-Sommergruber, K. (2004). Profilins. In: Mills, E.N.C. and Shewry, P.R. (eds). Plant Food Allergens. Oxford: Blackwell, pp.105-124.

Radauer, C., Bublin, M., Wagner, S., *et al.* (2008). Allergens are distributed into few protein families and possess a restricted number of biochemical functions. J Allergy Clin Immunol, 121(4), 847-852.

Receveur-Brechot, V., Czjzek, M., Barre, A., *et al.* (2006). Crystal structure at 1.45-A resolution of the major allergen endo-beta-1,3-glucanase of banana as a molecular basis for the latex-fruit syndrome. Proteins-Structure Function and Bioinformatics, 63(1), 235-242.

Reche, M., Pascual, C.Y., Vicente, J., *et al.* (2001). Tomato allergy in children and young adults: cross-reactivity with latex and potato. Allergy, 56(12), 1197-1201.

Reuter, A., Fortunato, D., Garoffo, L.P., *et al.* (2005). Novel isoforms of Pru av 1 with diverging immunoglobulin E binding properties identified by a synergistic combination of molecular biology and proteomics. Proteomics, 5(1), 282-289.

Ricci, G., Dondi, A., Belotti, T., *et al.* (2010). Allergenicity of different apple cultivars assessed by means of skin prick test and sensitisation to recombinant allergens Mal d 1 and Mal d 3 in a group of Italian apple-allergic patients. Int J Food Sci Technol, 45(7), 1517-1523.

Rodriguez, J., Crespo, J.F., Burks, W., *et al.* (2000). Randomized, double-blind, crossover challenge study in 53 subjects reporting adverse reactions to melon (Cucumis melo). J Allergy Clin Immunol, 106(5), 968-972.

Sanchez-Monge, R., Blanco, C., Diaz-Perales, A., *et al.* (1999). Isolation and characterization of major banana allergens: identification as fruit class I chitinases. Clin Exp Allergy, 29(5), 673-680.

Sanchez-Monge, R., Blanco, C., Lopez-Torrejon, G., *et al.* (2006). Differential allergen sensitization patterns in chestnut allergy with or without associated latex-fruit syndrome. J Allergy Clin Immunol, 118(3), 705-710.

Sancho, A.I., van Ree, R., van Leeuwen, A., *et al.* (2008). Measurement of lipid transfer protein in 88 apple cultivars. In Arch Allergy Immunol, 146(1), 19-26.

Saraswat, A., Kumar, B. (2005). Anaphylactic reaction to apple, banana and lychee: what is common between botanically disparate plant families? Int J Dermatol, 44(12), 996-998.

Sargent, D.J., Fernandez-Fernandez, F., Ruiz-Roja, J.J., *et al.* (2009). A genetic linkage map of the cultivated strawberry (Fragaria x ananassa) and its comparison to the diploid Fragaria reference map. Mol Breed, 24(3), 293-303.

Scheurer, S., Metzner, K., Haustein, D., *et al.* (1997). Molecular cloning, expression and characterization of Pru a 1, the major cherry allergen. Mol Immunol, 34(8-9), 619-629.

Scheurer, S., Lauer, I., Foetisch, K., *et al.* (2004). Strong allergenicity of Pru av 3, the lipid transfer protein from cherry, is related to high stability against thermal processing and digestion. J Allergy Clin Immunol, 114(4), 900-907.

Sherson, D., Andersen, B., Hansen, I., *et al.* (2003). Occupational asthma due to freeze-dried raspberry. Annals of Allergy Asthma & Immunology, 90(6), 660-663.

Son, D.Y., Scheurer, S., Hoffmann, A., *et al.* (1999). Pollen-related food allergy: cloning and immunological analysis of isoforms and mutants of Mal d 1, the major apple allergen, and Bet v 1, the major birch pollen allergen. Eur J Nutr, 38(4), 201-215.

Song, J.J., Zhang, H.Y., Liu, Z.G., *et al.* (2007). Cloning of the panallergen profilin from lychee fruit and its cross-reactivity with birch pollen profilin Bet v 2. Food Agric Immunol, 18(2), 129-138.

Stafne, E.T., Clark, J.R., Weber, C.A., *et al.* (2005). Simple sequence repeat (SSR) markers for genetic mapping of raspberry and blackberry. J Ame Soci Horti Sci, 130(5), 722-728.

Tamburrini, M., Cerasuolo, I., Carratore, V., *et al.* (2005). Kiwellin, a novel protein from kiwi fruit. Purification, biochemical characterization and identification as an allergen. Protein J, 24(7-8), 423-429.

Tuft, L. and Blumenstein G.I. (1942). Studies in food allergy. Sensitization to fresh fruits: clinical and experimental observations. J Allergy, 13, 574-582

van Loon, L.G., van Strien, E.A. (1999). The family of pathogenesis-related proteins, their activities, and comparative analysis of PR-1 type protein. Physiol Mol Plant Patho, 55, 85-97.

Valenta, R., Duchene, M., Pettenburger, K., *et al.* (1991). Identification of profilin as a novel pollen allergen. IgE auto-reactivity in sensitized individuals. Science, 253, 557-560.

Vieths, S., Schoning, B. and Petersen, A. (1994). Characterization of the 18-Kda apple allergen by 2-dimensional immunoblotting and microsequencing. Int Arch Allergy Immunol, 104(4), 399-404.

Wuthrich, B., Borga, A. and Yman, L. (1997). Oral allergy syndrome to jackfruit (Artocarpus integrifolia). Allergy, 52(4), 428-431.

Yang Z.W., Chen L., Qiao L.Y., *et al.* (2010). Fruit allergy and genomic research on their allergens. J Fruit Sci, 27(2), 281-288.

Yamamoto, T., Kimura, T., Terakami, S., *et al.* (2007). Integrated reference genetic linkage maps of pear based on SSR and AFLP markers. Breed Sci, 57(4), 321-329.

Yang, Z.W., Ma, Y.T., Chen, L., *et al.* (2011). Differential transcript abundance and genotypic variation of four putative allergen-encoding gene families in melting peach. Tree Genet Genomes, 7, 903-916.

Zheng, Q., Song, J., Doncaster, K., Rowland, E., *et al.* (2007). Qualitative and quantitative evaluation of protein extraction protocols for apple and strawberry fruit suitable for two-dimensional electrophoresis and mass spectrometry analysis. J Agric Food Chem, 55, 1663-1673.

Zoccatelli, G., Pellegrina, C.D., Consolini, M., *et al.* (2007). Isolation and identification of two lipid transfer proteins in pomegranate (Punica granatum). J Agric Food Chem, 55(26), 11057-11062.

Zubini, P., Zambelli, B., Musiani, F., *et al.* (2009). The RNA Hydrolysis and the Cytokinin Binding Activities of PR-10 Proteins Are Differently Performed by Two Isoforms of the Pru p 1 Peach Major Allergen and Are Possibly Functionally Related. Plant Physiol, 150(3), 1235-1247.

Zuidmeer, L., van Leeuwen, W.A., Budde, I. K., *et al.* (2006). Allergenicity assessment of apple cultivars: Hurdles in quantifying labile fruit allergens. Int Arch Allergy Immunol, 141(3), 230-240.

Zuidmeer, L., van Leeuwen, W.A., Budde, I.K., *et al.* (2005). Lipid transfer proteins from fruit: Cloning, expression and quantification. Int Arch Allergy Immunol, 137(4), 273-281.

Zuidmeer L., Goldhahn K., Rona R., J., *et al.* (2008). The prevalence of plant food allergies: A systematic review. J Allergy Clin Immunol, 121(5), 1210-1218.

23

Creating Hypoallergenic Crops through Genetic Modification

Xing-Jun Wang*[1], Han Xia[1], Shuan-Tao Liu[2], Chun-Juan Li[3], Chuan-Zhi Zhao[1], Ai-Qin Li[1]

[1] High-Tech Research Center, Shandong Academy of Agricultural Sciences, Ji'nan, 250100, China
[2] Vegetable Research Institute, Shandong Academy of Agricultural Sciences, Ji'nan, 250100, China
[3] Shandong Peanut Research Institute, Qingdao, 266100, China
[*] E-mail: xingjunw@hotmail.com

Abstract: Food allergy is a serious human health problem with increasing prevalence world-wide. The only management of food allergy is strict avoidance of the allergenic source, which is almost impossible because food allergens are widely present in our traditional food and because trace amount of an allergen is enough to trigger an IgE mediated reaction. In addition, most food allergens are structurally very stable. Even after various harsh food processes, like boiling, allergenicity of many foods still remains. Creating hypoallergenic crops by suppressing the allergen synthesis during plants growth and development may provide an alternative way for people to avoid allergens. RNA interference (RNAi) has been proven to be a powerful tool for gene silencing in both plants and animals. During the last decade, several studies demonstrate successful removal of some major allergens from different crops by genetic modification which greatly reduced the allergenicity of these crops. This chapter summarizes the advances of studies on hypoallergenic crops creation using genetic modification mainly through the RNAi technology.

23.1 Food Allergy: Prevalence and the Lack of Efficient Management

Food allergy is a serious international health problem, especially in developed western countries. In most cases, it greatly reduces the quality of life of allergic patients and limits the choice of their daily diet. In some rare cases, food allergy can be very severe and even fatal. Many food sources, including peanut, nut, peach, apple, egg, milk, fish, shell fish, can cause allergy. For example, patients who develop allergenic reactions to peanut or peanut containing products are about 0.6% of the total population in the US (Sicherer *et al.*, 2003). In addition, the results from a large scale survey in the US indicate that the prevalence of food allergy is increasing. Recently, a study of peanut allergy revealed that the prevalence rate among school children is more than 1% in both North America and the UK (Sicherer and Sampson, 2007). Because allergy has been a public health concern for such a long time, extensive studies have been carried out for the management of food allergy. Various curative methods have been investigated for possible treatments of food allergy. These include sublingual and oral immunotherapy, anti-IgE therapy, Chinese herbal medicine treatment, mutated recombinant protein treatment for immunomodulation, peptide immunotherapy, plasmid DNA-based immunotherapy and ISS-oligodeoxynucleotide-based immunotherapy (Sicherer and Sampson, 2007). Although several studies using animal models have achieved great success and have shown promising potential for human application, their use in human allergy treatment is still limited, not least because of safety risks. The only option for food allergic patients is currently strict avoidance of allergens in their diet, which is almost impossible due to the wide occurrence of allergens in foods.

Removal of allergens from crops through genetic modification may be an alternative strategy for allergen avoidance. Previous studies, however, did not give solutions for food allergy treatment, but mainly provided a rich pool of information about allergens regarding their molecular, biochemical, immunological or clinical characteristics. During the past decades, a large number of allergens from crops, such as peanut, soybean, rice, wheat and nuts have been identified and thoroughly characterized. Screening of crop germplasms for naturally occurring mutations, or with chemically or physically induced mutations leading to less or no allergenicity, is a way to produce hypoallergenic crops, which is reviewed in Chapter 21 in this book. Alternatively, the identification of allergens and allergen-encoding genes enables biotechnological approaches, including RNA interference, to allergen gene silencing in crops. In this chapter, we are to summarize the advances in the area of creating hypoallergenic crops using biotechnological approaches with an emphasis on RNA interference.

23.2 Characterization of Major Allergens and Their Encoding Genes in Crops

Characterization of allergens at the molecular level, especially the nucleotide sequences of allergen genes from different crops, vegetables or fruits, is required for hypoallergenic crop production through biotechnology approaches including RNA interference. Extensive studies on gene cloning and characterization were carried out in two most important and widely consumed legume species, peanut and soybean. Peanut is one of the most important crops in the world as an oil and protein source. It is also a serious allergenic food, because it contains a number of allergens. So far, eight peanut allergens have been reported, Ara h 1 to Ara h 8 (Pons *et al.*, 2002; de Leon *et al.*, 2007). Ara h 1, Ara h 2 and Ara h 6 are considered to be the major allergens. More than 90% of the patient's sera show positive reaction to these three proteins (Flinterman *et al.*, 2007). Most of the peanut allergens are seed storage proteins while Ara h 5 is a profilin family protein and Ara h 8 is a pathogenesis related protein-10 (PR-10). The nucleotide sequence of allergen genes and their corresponding protein sequences have been well studied. For instance, the Ara h 1 cDNA, the genomic sequence and its promoter have been cloned and analyzed in detail (Burks *et al.*, 1995a; 1995b ; Viquez *et al.*, 2004). The Ara h 1 protein, with a molecular weight of 63.5 kDa and consisting of 626 amino acids, occurs as a multi-peptide with a total molecular weight of 100–200 kDa. The content of Ara h 1 in peanut seed is very high, accounting for 12%–16% of total protein. Ara h 1 has 23 IgE-binding sites. The conformation Ara h 1 changes after heating, but the IgE binding capacity remains (Burks *et al.*, 1995 a ; Shin *et al.*, 1998). In addition, Ara h 2 (Viquez *et al.*, 2001; Chatel *et al.*, 2003), Ara h 3 (Viquez *et al.*, 2004; Dodo *et al.*, 2004 ; Rabjohn *et al.*, 1999), Ara h 4 (Dodo *et al.*, 2004 ; Kleber-Janke *et al.*, 1999), Ara h 5, Ara h 6, Ara h 7 (Kleber-Janke *et al.*, 1999 ; Ramos *et al.*, 2006), and Ara h 8 (Mittag *et al.*, 2004) have been all cloned and characterized. GenBank accessions of the eight peanut allergen genes are listed in Table 23.1. We cloned the genes encoding the peanut allergens Ara h 1 to Ara h 8 from the Chinese cultivar Luhua-14. The predicted amino acid sequence of Ara h 1 from Luhua-14 is highly conserved, as was demonstrated in a sequence comparison with Ara h 1 genes from other cultivars. However, the amino acid sequence of Ara h 3 from Luhua-14 showed significant differences from previously reported ones. It has mutations within the IgE binding sites, which may result in reduced capacity in IgE binding. From Luhua-14, two members of Ara h 2, Ara h 2.01 and Ara h 2.02 with 12 amino acid differences, have been cloned.

In soybean, there are three major allergens, Gly m Bd 60K, Gly m Bd 30K, and Gly m Bd 28K. Compared to Gly m Bd 60K, a very abundant seed storage protein in soybean, the major allergen Gly m Bd 30K accounts for 1%–3% of the total soybean seed protein (Kalinski *et al.*, 1990; Ogawa *et al.*, 1995). The Gly m Bd 30K allergen is a seed vascular protein (P34), which belongs to the papain superfamily (Ogawa *et al.*, 1993; Sewekow *et al.*, 2008). There are about 14 human IgE epitopes in Gly m Bd 30K. The cDNA sequence including the whole coding region of Glym Bd 30K was cloned (Kalinski *et al.*, 1992). Genes encoding the other two soybean allergens, Gly m Bd 28K and Gly m Bd 60K, have also been cloned (Sebastiani *et al.*, 1990; Tsuji *et al.*, 2001). Recently, a gene encoding another soybean allergen (P39), whose function is still unknown, has been cloned and analyzed in developing and mature soybean seeds (Xiang *et al.*, 2008).

In other crops, for example in rice, 14–16 kDa proteins encoded by a multigene family are considered as major allergens (Matsuda *et al.*, 1988; Urisu *et al.*, 1991). These proteins show significant homology to alpha-amylase/trypsin inhibitors. A 33-kDa protein was also reported as a major allergen in rice seed (Nakase *et al.*, 1998), while a 60-kDa protein represented a minor allergen (Ikezewa *et al.*, 1999). cDNAs coding for these proteins have been cloned in different rice varieties (Tian *et al.*, 2009).

Wheat is also a representative of an allergenic food. Flour contains multiple allergens. A recent study reported that at least 15 allergens can be identified from wheat through IgE-binding analysis (Kimoto *et al.*, 2009). Among these allergens Tri a Bd 17K, Tri a Bd 27K and Tri a Bd 36K are the major allergens, which have α -amylase inhibitor, γ-interferon-inducible thiol reductase and peroxidase activity, respectively (Kimoto *et al.*, 1998; 2009). cDNA of these three allergens were cloned (Gautier *et al.*, 1990, Kimoto *et al.*, 2009; Sareini and Thornburg, 2002). Wheat lipid transfer proteins and omega-gliadins have also been reported as allergens: Tri a Bd 14K and Tri a Bd 19K (Palacin *et al.*, 2007; Battais *et al.*, 2005).

In apple, there are four classes of allergens reported, namely Mal d 1, Mal d 2, Mal d 3 and Mal d 4. Expression of these genes has been analyzed in different varieties of apple and under different environmental conditions (Botton *et al.*, 2008). Mal d 1 is the major allergen in apple, which belongs to the pathogenesis-related protein 10 (PR-10) family (Vieths *et al.*, 1995). From two apple cultivars, 37 genomic sequences have been cloned, representing 18 genes that are coding for this protein (Gao *et al.*, 2005). Based on the genomic structure, Mal d 1 gene family can be divided into two groups, the intron-containing and the intronless genes. Sequence comparison reveals that the sequences of Mal d 1.01 and Mal d 1.02 are highly conserved, while the sequences of Mal d 1.04, Mal d 1.05 and Mal d 1.06A, B, C are more variable (Gao *et al.*, 2008). From a genome wide screen in another Rosaceae fruit, i.e., peach, a total of 18 putative allergen genes have been identified and mapped to five linkage groups (Chen *et al.*, 2008).

In tomato, many studies for cloning and characterization of allergen genes have been carried out during the past decade (Westphal *et al.*, 2002; 2004; Willerroider *et al.*, 2003). Lyc e 1, Lyc e 2 and Lyc e 3 are major allergens in the fruit of the tomato plant.

Table 23.1 Summary of cloned allergen-coding genes and genetically-modified hypoallergenic mutants

Plant Species	Allergen	Hypoallergenic Mutants	GenBank Accession	References
Arachis hypogaea	Ara h 1	No	L34402.1	Burks *et al.*, 1995a; 1995b; Viquez *et al.*, 2003
	Ara h 2.01	Yes	EF609641	Ramos *et al.*, 2006
	Ara h 2.02	Yes	AY158467	Chatel *et al.*, 2003
	Ara h 3	No	AF093541	Dodo *et al.*, 2004; Rabjohn *et al.*, 1999
	Ara h 4	No	AF086821	Dodo *et al.*, 2004; Kleber-Janke *et al.*, 1999
	Ara h 5	No	AY726606	Kleber-Janke *et al.*, 1999; Ramos *et al.*, 2006
	Ara h 6	Yes	EF609643	Kleber-Janke *et al.*, 1999; Ramos *et al.*, 2006
	Ara h 7	No	AF091737	Kleber-Janke *et al.*, 1999; Ramos *et al.*, 2006
	Ara h 8	No	AY328088	Mittag *et al.*, 2004
Glycine max	Gly m Bd 60K	No	X17698	Sebastiani *et al.*, 1990
	Gly m Bd 30K	Yes	J05560	Herman *et al.*, 2003; Joseph *et al.*, 2006
	Gly m Bd 28K	No	AB046874	Ogawa *et al.*, 1995
Orysa sativa	14–16 kDa proteins	Yes	Q01883	Nakamura *et al.*, 1996
	33-kDa protein	No	AB017042	Usui *et al.*, 2001
	60-kDa protein	No	GQ151066	Tian *et al.*, 2009
Malus domestica	Mal d 1.01	Yes	AY827645	Gao *et al.*, 2005[a]; 2008,
	Mal d 1.02	Yes	AY827660	Gao *et al.*, 2005[a]; 2008,
	Mal d 1.03	Yes	AY822732	Gao *et al.*, 2005[a]; 2008,
	Mal d 1.04	Yes	AY827673	Gao *et al.*, 2005a; 2008,
	Mal d 2.01	No	AY792599	Gao *et al.*, 2005b
	Mal d 3.01	No	AY572516	Gao *et al.*, 2005c
	Mal d 3.02	No	AY572532	Gao *et al.*, 2005c
	Mal d 4.01	No	AY792609	Gao *et al.*, 2005b
	Mal d 4.02	No	AY792613	Gao *et al.*, 2005b
	Mal d 4.03	No	AY792617	Gao *et al.*, 2005b
Lycopersicon esculentum	Lyc e 1	Yes	AJ417553	Willerroider *et al.*, 2003
	Lyc e 2	No	AF465612	Westphal *et al.*, 2002
	Lyc e 3	Yes	AM051295	Le *et al.*, 2005
Triticum aestivum	Tri a Bd 17K	No	CAA34709	Gautier *et al.*, 1990
	Tri a Bd 27K	No	AB085212	Kimoto *et al.*, 2009
	Tri a Bd 36K	No	AF525425	Sareini and Thornburg, 2002

More information on allergen gene and protein sequences of different crop plants can be obtained from the website http://www.allergenonline.org/databasebrowse.shtml. In Table 23.1, the genes encoding for allergens in different crops, fruits and vegetables are listed.

23.2.1 RNA-Mediated Gene Silencing

RNA-mediated gene silencing was first found in transgenic plants and termed co-suppression or post-transcriptional gene silencing (PTGS). Jorgensen and colleagues aimed to transfer a gene encoding chalcone synthase (CHS) required for the biosynthesis of anthocyanin pigments to make petunia flowers more colorful. They found that the introduction of the transgene inhibited the expression of endogenous genes, which resulted in white or chimeric-colored flowers (Napoli *et al.*, 1990; Jorgensen, 1995). At that time, this phenomenon was termed co-suppression. A similar phenomenon in fungi was termed quelling (Romano and Macino, 1992; Cogoni *et al.*, 1994), while in animals it is termed RNA-interference (Lee *et al.*, 1993; Fire *et al.*, 1998). The nature of the gene silencing as described in the above mentioned examples appeared to be the double stranded RNA-mediated specific silencing of homologous genes. The mechanism of antisense silencing was later also proved to be the result of double stranded RNA degradation. Recently, Paoli and co-workers analyzed the siRNA in the transgenic petunia by high throughput sequencing. They found that in the transgenic petunia extremely abundant siRNAs associated with the co-suppression were generated (Paoli *et al.*, 2009). Since the mechanism of RNAi is not in the scope of this chapter, only a very brief introduction is provided. The double stranded RNA-mediated silencing pathway includes several key proteins which form the RNA-induced silencing complex (RISC) and different types of small RNAs, the small interference RNA and microRNA. Dicer in animal and Dicer like proteins in plants were the central players required to generate the small interference RNA from long double stranded RNA. The resulting small RNA was associated with other proteins of the silencing complex and attached to the target sequence by complementary base pairing to silence the gene expression either by mRNA degradation or by inhibition of translation. Smith and colleagues designed a construct encoding an intron-spliced RNA with a hairpin structure which greatly facilitated the application of RNAi in plants (Smith *et al.*, 2000). RNAi is currently a powerful tool for gene function analysis and inactivation of specific genes (disease-related genes or genes encoding undesirable proteins as, for instance, plant allergens). The main advantage of using RNAi is the possibility to target a specific gene or the multiple members of a gene family. RNAi-mediated partial silencing of a gene is very important for function analysis of important genes. However, this "inefficient" silencing would be a disadvantage in studies aiming at removal of undesirable proteins. In addition, cell-to-cell or tissue-tissue transmission of small interference RNAs could cause problems in tissue-specific silencing studies. The potential off-target effect could also be a serious limitation for the application of this technology (Qiu *et al.*, 2005; Ma *et al.*, 2006). miRNAs are endogeous small RNAs with a length of about 19-21 nt encoded by miRNA genes (Lee *et al.*, 1993). miRNAs play crucial roles in plant development and adaptation to the environment. The mature miRNA sequence binds to the target mRNA based on complementary pairing and guides target mRNA degradation or repression of translation. Many studies have demonstrated that replacement of the mature miRNA sequence from the miRNA precursor did not affect the miRNA biogenesis but could generate a novel artificial miRNA. This finding enables application of the miRNA precursor as a carrier for silencing of a specific gene (Niu *et al.*, 2006).

23.2.2 Creating Hypoallergenic Crops by RNA-Mediated Repression

Biotechnology-based approaches to reduce crop allergens achieved great success in different crops in recent years (Scheurer and Sonnewald, 2009; Herman, 2003). A decade ago, Japanese researchers applied an antisense RNA strategy to reduce a 16 KDa rice allergen. Transgenic plants accumulated significantly much lower amounts of the allergen compared to control plants (Nakamura and Matsuda, 1996; Tada *et al.*, 1996). Using a similar strategy, Bhalla and co-workers successfully repressed the ryegrass pollen allergen Lol p 5 (Bhalla *et al.*, 1999; 2001). Later, biotechnology-based silencing of allergen genes was applied to several important crops and in particular to peanut and soybean.

A couple of major peanut seed storage proteins are allergens. For example, Ara h 1 and Ara h 2 will cause serious allergic reactions in many individuals. In contrast to many other food allergies such as milk, soybean and wheat allergies, peanut allergy is rarely outgrown after childhood and often shows persistence over a lifetime. It is therefore crucial to reduce the content of these allergens in peanuts. Therefore, an RNA interference construct targeting the coding region of Ara h *2* was introduced into peanut by microprojectile bombardment (Chu *et al.*, 2008). One copy of the Ara h 2 silencing construct significantly suppressed Ara h 2 expression. Due to the sequence similarity between Ara h 2 and Ara h 6, some transgenic lines also showed decreased Ara h 6 expression (Chu *et al.*, 2008). Dodo and colleagues used similar approaches to repress the expression of the Ara h 2 gene. Their results showed that the gene was silenced and the accumulation of Ara h 2 was not detected in the transgenic peanut (Dodo *et al.*, 2008). In both studies, transgenic lines were tested for IgE binding with serum from peanut allergic patients. Significant reduction of IgE-binding capacity was observed (Chu *et al.*, 2008; Dodo *et al.*, 2008).

Soybean is also in focus for allergen removal studies due to its potent allergenicity and extensive use in processed food products. In contrast to other soybean allergens, Gly m Bd 30K is not so abundantly expressed in soybean seeds (Fang *et al.*, 2006). However, it has the strongest allergenic potential. This is the reason why Gly m Bd 30K is the target of low/null allergenic germplasm screening or biotech-based inactivation of the encoding gene (Joseph *et al.*, 2006; Herman *et al.*, 2003). The transgene-induced suppression of Gly m Bd 30K was carried out in a seed-specific manner. Silencing the gene led to completely removal of the allergen from transgenic seeds without any adverse effects on growth and development of the plant and in particular on the size and shape of seeds as well as on their protein and oil content (Herman *et al.*, 2003; Herman, 2003).

In tomato, studies to reduce the allergenicity using RNA interference were also successful. Repression of allergenic Lyc e 1.01 and Lyc e 1.02 profilin genes by RNA interference led to a 10-fold decrease in Lyc e 1.01 and Lyc e 1.02 accumulation in transgenic tomatoes (Le *et al.*, 2006a). The same research group identified a 9 kDa polypeptide, a non-specific lipid transfer protein, from tomato peel as an allergen, Lyc e 3 (Le *et al.*, 2006b). The authors designed dsRNA interference constructs using the N-terminal sequence of Lyc e 3. Expression of these constructs in transgenic tomatoes resulted in a 95% reduction of the allergen protein. The allergenic potential of transgenic tomatoes showed a 100-fold decrease compared with non-transgenic controls (Le *et al.*, 2006b). Reduced levels of Lyc e 3 were also found in the second generation of transgenic tomato plants (Lorenz *et al.*, 2006). Based on immunoblotting analysis in tomato plants, several other proteins such as polygalacturonase 2A, beta-fructofuranosidase, superoxide dismutase, pectinesterase and suberization-associated anionic peroxidase have been found to be potential allergens (Kondo *et al.*, 2001; Weangsripanaval *et al.*, 2003). Polygalacturonase 2A is a key enzyme involved in tomato ripening and tomato fruit softening through degradation of the cell wall. Accumulation of polygalacturonase 2A was significantly reduced and resulted in reduction of allergenicity in the fruits of tomato ripening mutants (Kitagawa *et al.*, 2006).

In apple, conventional breeding for hypoallergenicity is laborious and time consuming. The major apple allergen Mal d 1 is encoded by a large gene family, consisting of about 18 genes distributed in different clusters on three chromosomes (Gao *et al.*, 2005a; 2008). It therefore becomes a very complex matter to remove only Mal d 1 from a cultivar by conventional breeding. However, Gilissen and colleagues successfully silenced the Mal d 1 genes in apple plants using the RNAi technology (Gilissen *et al.*, 2005).

In most studies on the development of hypoallergenic crops, investigators target directly the allergen gene by RNA interference. Kitagawa and colleagues demonstrated that a mutation of the tomato ripening inhibitor, a MADS-box transcription factor, affected the accumulation of multiple proteins including at least two allergens, polygalacturonase 2A and beta-fructofuranosidase (Kitagawa *et al.*, 2006). This may open a new gate to hypoallergenic crop production through genetic modification of transcription regulatory factors other than that of specific allergens.

23.3 Concerns

Concerns on hypoallergenic plants created either in the traditional way or by biotechnology-based approaches relate to whether the reduction in a specific allergen could result in any adverse effect. For example, several allergens are seed storage proteins. It cannot be excluded that the inhibition of these proteins may affect seed viability, germination or morphology and, thereby, the economic value of the seeds. Would the elimination of a specific allergen result in the accumulation of other allergens? Or would the reduction of an allergen, for example the protein PR-10, lower the tolerance of transgenic plants to the disease? In the studies described above, aberrant plant growth and development or aberrant seed viability and germination have never been observed in hypoallergenic crops. However, compensation for allergen removal has been detected in specific hypoallergenic transgenic plants: β-conglycinin-transgenic soybean lines accumulate higher amounts of other proteins such as glycinin, proglycinin and P34 (Kinney *et al.*, 2001). Global protein analysis in Ara h 2 and Ara h 6 transgenic peanut lines showed enhanced accumulation of oleosin 1, 13-lipoxygenase and arachin (Ahy-3), while the accumulation of conarachin was decreased (Stevenson *et al.*, 2009). However, in the natural soybean mutant with an extremely low level of P34 or Gly m Bd 30K, the author did not detect any compensation effect for other proteins (Joseph *et al.*, 2006). In some studies, the inhibition of one or two proteins by RNAi or antisense RNA did not cause detectable compensatory enhanced protein synthesis, as for example in soybean (Herman *et al.*, 2003; Herman, 2003) and peanut (Dodo *et al.*, 2008). In these studies, resolution of analytical methods and the number of proteins analyzed should be considered. Ara h 2 has been suggested to act as a trypsin inhibitor which may contribute to disease resistance of plants. However, silencing of Ara h 2 did not promote fungal growth in the transgenic lines based on *in vitro Aspergillus flavus* infection (Chu *et al.*, 2008). In the case of tomato and rice, the authors did not mention if the silencing affected the synthesis of other proteins (Le *et al.*, 2006; Lorenz *et al.*, 2006; Tada *et al.*, 1996).

The stability of hypoallergenic plants with reduced or completely lack of an allergen created through RNAi silencing and related technologies is another concern. Some studies have addressed this issue over a limited time scale. For example, in rice, antisense RNA-induced silencing could be stably transmitted at least to the third generations (Tada *et al.*, 1996). Knock-out or knock-down of one or two allergens from a crop is only the first step towards the creation of hypoallergenic crops because in many crops there are multiple allergens present. Whether the targeted allergen is completely removed is also a concern. Trace amount of an allergen can elicit allergic reactions and might probably be fatal. For example, peanut varieties with a significantly reduced amount of Ara h 1 showed similar or even higher allergenicity compared with a peanut variety which has a normal concentration of the Ara h 1 protein (Krause *et al.*, 2009). In this respect, RNAi may not be the best way for silencing allergen genes because in most cases RNAi could only induce partial silencing. RNAi-induced silencing, which could spread from the target tissues to other parts of the plants, may cause problems because some allergens may also play significant biological roles in plant growth and development or plant defense.

23.4 Conclusion

Due to the lack of efficient treatments against food allergy and the difficulty of avoidance by patients in their diet, repressing the accumulation of a specific allergen through genetic modification is a promising strategy for producing hypoallergenic crops. Although there are a lot of concerns about genetically-induced repression of allergen expression, it is promising to see that some allergic patients in Asia can use hypoallergenic rice as a substitution in their daily diet. Importantly, it should be noted that simultaneous silencing of multiple allergens in the edible parts is strictly required to develop a hypoallergenic food crop. Silencing of one single allergen may not affect plant growth and development of transgenic lines. However, simultaneous knock-out several allergens may affect plant growth and development.

Acknowledgements

Our studies are supported by National Science Foundation of China (30871324; 31000720), Shandong province "Taishan Scholar" Foundation, Shandong Province Natural Science Foundation (ZR2010CZ002; ZR2010CQ008; Y2008D44)

References

Battais, F., Courcoux, P., Popineau, Y., *et al.* (2005). Food allergy to wheat: Differences in immunoglobulin E-binding proteins as a function of age or symptoms. J Cereal Sci, 42, 109-117.

Bhalla, P.L., Swoboda, I., Singh, M.B. (1999). Antisense-mediated silencing of a gene encoding a major ryegrass pollen allergen. PANS, 96,11676-11680.

Bhalla, P.L., Swoboda, I., Singh, M.B. (2001). Reduction in allergenicity of grass pollen by genetic engineering. Int Arch Allergy Immunol, 124(1-3), 51-54.

Burks, A.W., Cockrell, G., Stanley, J.S., *et al.* (1995a). Recombinant peanut allergen Ara h I expression and IgE binding in patients with peanut hypersensitivity. J Clin Invest, 96 (4), 1715-1721.

Burks, A.W., Cockrell, G., Stanley, J.S., *et al.* (1995b). Isolation, identification, and characterization of clones encoding antigens responsible for peanut hypersensitivity. Int Arch Allergy Immunol, 107(1-3), 248-250.

Bilyeu, K.D., Ren, C., Nguyen, H.T., *et al.* (2009). Association of a four-basepair insertion in the P34 gene with the low allergen trait in soybean. Plant Genet, 2, 141-148.

Botton, A., Lezzer, P., Dorigoni, A., *et al.* (2008). Genetic and environmental factors affecting allergen-related gene expression in apple fruit (Malus domestica L. Borkh). J Agric Food Chem, 56(15), 6707-16J.

Chatel, J.M., Bernard, H., Orson, F.M. (2003). Isolation and characterization of two complete Ara h 2 isoforms cDNA. Int. Arch. Allergy Immunol, 131 (1), 14-18.

Chen, L., Zhang, S., Illa, E., *et al.* (2008). Genomic characterization of putative allergen genes in peach/almond and their synteny with apple. BMC Genomics, 9, 543.

Chu, Y., Faustinelli, P., Ramos, M.L., *et al.* (2008). Reduction of IgE bBinding and nNonpromotion of aAspergillus flavus fungal growth by simultaneously silencing Ara h2 and Ara h6 in peanut. J of Agric Food Chem, 56 (23), 11225-11233.

Cogoni, C., Romano, N., Macino, G. (1994). Suppression of gene expression by homologus transgene. Antonie Van Leeuwenhoek, 65(3), 205-209.

De Paoli, E., Dorantes-Acosta, A., Zhai, J., *et al.* (2009). Distinct extremely abundant siRNAs associated with cosuppression in petunia. RNA, 15, 1965-1970.

Dodo, H.W., Konan, K.N., Chen, F.C., *et al.* (2008). Alleviating peanut allergy using genetic engineering: the silencing of the immunodominant allergen Ara h 2 leads to its significant reduction and a decrease in peanut allergenicity. Plant Biotechnol J, 6(2),135-145.

Dodo, H.W., Viquez, O.M., Maleki, S.J., *et al.* (2004). cDNA clone of putative peanut (*Arachis hypogaea* L.) trypsin inhibitor has homology with peanut allergens Ara h3 and Ara h4. J Agric Food Chem, 52(5), 1404-1409.

de Leon, M.P., Rolland, J.M., O'Hehir, R.E. (2007). The peanut allergy epidemic: Allergen molecular characterisation and prospects for specific therapy. Expert ReV Mol Med, 9(1), 1-18.

Fang, X.Q., Zhu, Y.L., Qiu, L.J. (2006). Soybean allergens and hypoallergenic germplasm enhancement. Yi Chuan, 28(8), 1043-1050.

Fire, A., Xu, S., Montgomery, M.K., *et al.* (1998). Potent and specific genetic interference by double stranded RNA in cCaenorhabditis elegans. Nature, 391, 806-811.

Flinterman, M.D., van Hoffen, E., den Jager, C., *et al.* (2007). Children with peanut allergy recognize predominantly Ara h 2 and Ara h 6, which remains stable over time. Clin Exp Allergy, 37, 1221-1228.

Gao, Z.S., van de Weg, W.E., Schaart, J.G., *et al.* (2005b). Genomic characterization and linkage mapping of the apple allergen genes Mal d 2

(Thaumatin-like protein) and Mal d 4 (Profilin). Theo Appl Genet, 111, 1087-1097.

Gao, Z. S., van de Weg, W.E., Schaart, J.G., *et al.* (2005c). Linkage map positions and allelic diversity of two Mal d 3 (non-specific lipid transfer protein) genes in the cultivated apple (*Malus domestica*). Theo Appl Genet, 110, 479-491.

Gao, Z. S., van de Weg, W.E., Matos, C.I., *et al.* (2008). Assessment of allelic diversity in intron-containing Mal d 1 genes and their association to apple allergenicity. BMC Plant Biology, 8, 116.

Gautier, M., Alary, R., Joudrier P. (1990). Cloning and characterization of a cDNA encoding the wheat (*Tritium durum* Desf.) CM16 protein. Plant Mol Biol, 14, 313-322.

Gilissen, L.J., Bolhaar, S.T., Matos, C.I., *et al.* (2005). Silencing the major apple allergen Mal d1 by using the RNA interference approach. J Allergy Clin Immunol, 115(2), 364-369.

Herman, E.M., Helm, R.M., Jung, R., *et al.* (2003). Genetic modification removes an immunodominant allergen from soybean. Plant Physiol, 132, 36-43.

Herman, E.M. (2003). Genetically modified soybeans and food allergies. J Exp Bot, 54(386), 1317-1319.

Ikezawa, Z., Tsubaki, K., Osuna, H., *et al.* (1999). Usefulness of hypoallergenic rice (AFT-R 1) and analysis of the salt insoluble rice allergen molecule. Arerugi, 48, 40-49.

Joseph, L.M., Hymowitz, T., Schmidt, M.A., *et al.* (2006). Evaluation of Glycine germplasm for nulls of the immunodominant allergen P34/Gly m Bd 30K. Crop Sci, 46, 1755-1763.

Jorgensen, R.A. (1995). Cosuppression, flower color patterns, and metastable gene expression states. Science, 268(5211), 686-691.

Kalinski, A., Melroy, D.L., Dwivedi, R.S., *et al.* (1992). A soybean vacuolar protein (P34) related to thiol proteases is synthesized as a glycoprotein precursor during seed maturation. J Biol Chem, 267 (17), 12068-12076.

Kalinski, A., Weisemann, J.M., Matthews, *et al.* (1990). Molecular cloning of a protein associated with soybean seed oil bodies that is similar to thiol proteases of the papain family. J Biol Chem, 265(23), 13843-13848.

Kinney, A., Jung, R., Herman, E. (2001). Cosuppression of the α-subunits of β-conglycinin in transgenic soybean seeds induces the formation of endoplasmic reticulum-derived protein bodies. Plant Cell, 13, 1165-1178.

Kimoto, M., Yoshikawa, M., Takahashi, K., *et al.* (1998). Identification of allergens in cereal and their hypoallergenization. I. Screening of allergens in wheat and identification of an allergen, Tri a Bd 17K. Ann Report Interdiscil Res Inst Environ Sci, 17, 53-60.

Kimoto, M., Suzuki, M., Komiyama, N., *et al.* (2009). Isolation and molecular cloning of a major wheat allergen, Tri a Bd 27K. Biosci Biotechnol Biochem, 73(1), 85-92.

Kitagawa, M., Moriyama, T., Ito, H., *et al.* (2006). Reduction of allergenic proteins by the effect of the ripening inhibitor (rin) mutant gene in an F1 hybrid of the rin mutant tomato. Biosci Biotechnol Biochem, 70(5),

1227-1233.

Kleber-Janke, T., Crameri, R., Appenzeller, U., *et al.* (1999). Selective cloning of peanut alleregns, including profiling and 2S albumins, by phage display technology. Int Arch Allergy Immunol, 119(4), 265-274.

Kondo, Y., Urisu, A., Tokuda, R. (2001). Identification and characterization of the allergens in the tomato fruit by immunoblotting. Int Arch Allergy Immunol, 126(4), 294-299.

Krause, S., Latendorf, T., Schmidt, H., *et al.* (2009). Peanut varieties with reduced Ara h1 content indicating no reduced allergenicity. Mol Nutr Food Res, 54(3), 381-387.

Lee, R.C., Feibaum, R.L., Ambros, V. (1993). The *C. elegans* heterochronic gene lin-4 encodes small RNAs with antisense complementarity to lin-14. Cell, 75(5), 843-854.

Le, L.Q., Mahler, V., Lorenz, Y., *et al.* (2006a). Reduced allergenicity of tomato fruits harvested from Lyc 1-silenced transgenic tomato plants. J Allergy Clin Immumnol, 118(5), 1176-1183.

Le, L.Q., Lorenz, Y., Scheurer, S., *et al.* (2006b). Design of tomato fruits with reduced allergenicity by dsRNAi-mediated inhibition of ns-LPT (Lyc e3) expression. Plant Biotechnol J, 4(2), 231-242.

Lorenz, Y., Enrique, E., Lequynh, L., *et al.* (2006). Skin prick tests reveal stable and heritable reduction of allergenicity potency of gene-silenced tomato fruits. J Allergy Clin Immunol, 118(3), 711-718.

Ma, Y., Creanga, A., Lum, L., *et al.* (2006). Prevalence of off-target effects in Drosophila RNA interference screens. Nature, 443, 359-363.

Mittag, D., Akkerdaas, J., Ballmer-Weber, B.K., *et al.* (2004). Ara h 8, a Bet v 1-homologous allergen from peanut, is a major allergen in patients with combined birch pollen and peanut allergy. J Allergy Clin Immunol, 114 (6), 1410-1417.

Matsuda, T., Sugiyama, M., Nakamura, R., *et al.* (1988). Purification and properties of an allergenic proteinien in rice grain. Agric Biol Chem, 52, 1465-1470.

Nakase, M., Usui, Y., Alvarez-Nakase, A.M., *et al.* (1998). Cereal allergens: Rice-seed allergens with structural similarity to wheat and barley allergens. Allergy, 53(46 Suppl), 55-57.

Nakamura, R., Matsuda, T. (1996). Rice allergenic protein and molecular-genetic approach for hypoallergenic rice. Biosci Biotechnol Biochem, 60 (8), 1215-1221.

Napoli, C., Lemieux, C., Jorgensen, R.A. (1990). Introduction of a chimeric chalcone synthase gene into petunia results in reversible co-suppression of homologous genes in trans. Plant Cell, 2(4), 279-289.

Niu, Q.W., Lin, S.S., Reyes, J.L., *et al.* (2006). Expression of artificial microRNAs in transgenic *Arabidopsis thaliana* confers virus resistance. Nat Biotechnol, 24, 1420-1428.

Ogawa, T., Tsuji, H., Bando, N., *et al.* (1993). Identification of the soybean allergenic protein, Gly m Bd 30K, with the soybean 34-kDa oil-body-associated protein. Biosci Biotechnol Biochem, 57, 1030-1033.

Ogawa, T., Bando, N., Tsuji, H., *et al.* (1995). α-Subunit of β-conglycinin, an allergenic protein recognized by IgE antibodies of soybean-sensitive patients with atopic dermatitis. Biosci Biotechnol Biochem, 59(5), 831-833.

Palacin, A., Quirce, S., Armentia, A., *et al.* (2007). Wheat lipid transfer protein is a major allergen associated with baker's asthma. J Allergy Clin Immunol, 120(5), 1132-1138.

Pons, L., Chery, C., Romano, A., *et al.* (2002). The 18 kDa peanut oleosin is a candidate allergen for IgE-mediated reactions to peanuts. Allergy, 57, 88-93.

Qiu, S., Ademal, C.A., Lane, T. (2005). A computational study of off-target effects of RNA interference. Nucleic Acids Res, 33(6), 1834-1847.

Ramos, M.L., Fleming, G., Chu, Y., *et al.* (2006). Chromosomal and phylogenetic context for conglutin genes in Arachis based on genomic sequence. Mol Genet Genomics, 275 (6), 578-592.

Romano, N., Macino, G. (1992). Quelling: Transient inactivation of gene expression in Neurospora crassa by transformation with homologous sequences. Mol Microbiol, 6(22), 3343-3353.

Rabjohn, P., Helm, E.M., Stanley, J.S., *et al.* (1999). Molecular cloning and epitope analysis of the peanut allergen Ara h3. J Clin Invest, 103 (4), 535-542.

Scheurer, S., Sonnewald, U. (2009). Genetic engineering of plant food with reduced allergenicity. Fron Biosci, 1(14), 59-71.

Sebastiani, F.L., Farrell, L.B., Schuler, M.A., *et al.* (1990). Complete sequence of a cDNA of alpha subunit of soybean beta-conglycinin. Plant Mol Biol, 15 (1), 197-201.

Sewekow, E., Keßler, L.C., Seidel-Morgenstern, A., *et al.* (2008). Isolation of soybean protein P34 from oil bodies using hydrophobic interaction chromatography. BMC Biotechnol, 8, 27.

Shin, D.S., Compadre, C.M., Maleki, S.J., *et al.* (1998). Biochemical and structural analysis of the IgE binding sites on ara h1, an abundant and highly allergenic peanut protein. J Biol Chem, 273(22), 13753-13759.

Sicherer, S.H., Muñoz-Furlong, A., Sampson, H.A. (2003). Prevalence of peanut and tree nut allergy in the United States determined by means of a random digit dial telephone survey: A 5-year follow-up study. J Allergy Clin Immunol, 112(6), 1203-1207.

Sicherer, S.H., Sampson, H.A. (2007). Peanut allergy: Emerging concepts and approaches for an apparent epidemic. J Allergy Clin Immunol, 120(3), 491-503.

Smith, N.A., Singh, S.P., Wang, M.B., *et al.* (2000). Gene expression: Total silencing by intron-spliced hairpin RNAs. Nature, 407(6802), 319-320.

Stevenson, S.E., Chu, Y., Ozias-Akins, P., *et al.* (2009). Validation of gel-free, label-free quantitative proteomics approaches: Application for seed allergen profiling. J Proteomics, 72(3), 555-566.

Tada, Y., Nakase, M., Adachi, T., *et al.* (1996). Reduction of 14–16 kDa allergenic proteins in transgenic rice plants by antisense gene. FEBS Letters, 391, 341-345.

Tian, Z., Qian, Q., Liu, Q., *et al.* (2009). Allelic diversities in rice starch biosynthesis lead to a diverse array of rice eating and cooking qualities. PNAS, 106(51), 21760-21765.

Tsuji, H., Hiemori, M., Kimoto, M., *et al.* (2001). Cloning of cDNA encoding a soybean allergen, Gly m Bd 28K. Biochim Biophys Acta, 1518(1–2), 178-182.

Urisu, A., Yamada, K., Masuda, S., *et al.* (1991). 16-kilodalton rice protein is one of the major allergens in rice grain extract and responsible for cross-allergenicity between cereal grains in the Poaceae family. Int Arch Allergy Appl Immunol, 96, 244-252.

Usui, Y., Nakase, M., Hotta, H., *et al.* (2001). A 33-Kda allergen from rice (*Oryza sativa* L. *Japonica*): cDNA cloning, expression, and identification as a novel glyoxalase I. J Biol Chem, 276, 11376-11381.

Viquez, O.M., Summer, C.G., Dodo, H.W. (2001). Isolation and molecular characterization of the first genomic clone of a major peanut allergen, Ara h2. J Allergy Clin Immunol, 107(4), 713-717.

Viquez, O. M., Konan, K.N., Dodo, H.W. (2003). Structure and organization of the genomic clone of a major peanut allergen gene, Ara h 1. Mol Immunol, 40(9), 565-571.

Viquez, O.M., Konan, K.N., Dodo, H.W. (2004). Genomic organization of peanut allergen gene, Ara h3. Mol Immunol, 41(12), 1235-1240.

Vieths, S., Janek, K., Aulepp, H., *et al.* (1995). Isolation and characterization of the 18-kDa major apple allergen and comparison with the major birch pollen allergen (Bet v I). Allergy, 50, 421-430.

Weangsripanaval, T., Nomura, N., Moriyama, T., *et al.* (2003). Identification of suberization-associated anionic peroxidase as a possible allergenic protein from tomato. Biosci Biotechnol Biochem, 67, 1299-1304.

Willerroider, M., Fuchs, H., Ballmer-Weber, B.K., *et al.* (2003). Cloning and molecular and immunological characterization of two new food allergens, Cap a2 and Lyc e 1, profilins from bell pepper (*Capsicum annuum*) and tomato (*Lycopersicon esculentum*). Int Arch Allergy Immunol, 131(4), 245-255.

Westphal, S., Foetisch, K., Lauer, I., *et al.* (2002). Cloning and characterization of β-fructofuranosidase, a newly identified glycoprotein as a minor allergen of tomato (*Lycopersicon esculentum*). J Allergy Clin Immunol, 109(1), 947.

Xiang, P., Bair, L.M., Jun, R., *et al.* (2008). P39, a novel soybean protein allergen, belongs to a plant-specific protein family and is present in protein storage vacuoles. J of Agric Food Chem, 56(6), 2266-2272.

24

Acceptance of Natural and Genetically Modified Hypoallergenic Apples by Consumers with an Oral Allergy Syndrome (OAS)

I.M. van der Meer[1]*, M.F. Schenk[1], A.R.H. Fischer[2], L.J. Frewer[2], M.P. van der Maas[1], M.J.M. Smulders[1], L.J.W.J. Gilissen[1]

[1] Plant Research International, Wageningen University and Research Center, P.O. Box 16, 6700AA Wageningen, the Netherlands

[2] Department of Marketing and Consumer Behaviour, Wageningen University, the Netherlands

* E-mail: ingrid.vandermeer@wur.nl

Abstract: Plant Research International, together with UMC-Utrecht, screened a wide collection of existing apple varieties for allergenicity, and identified the Santana as a hypoallergenic variety. This variety was introduced on the market in 2006 as "suitable for consumption by consumers with a mild apple allergy". Its packaging included a background information leaflet, a protocol regarding how to eat it (in relation to "testing" for allergic reactions) and a link to a consumer questionnaire. The results of this questionnaire showed that a high percentage of consumers had no, or very slight, complaints after eating this hypoallergenic cultivar Santana. The questionnaire also included some questions on the appreciation of hypoallergenic apple varieties if they were to be produced by genetic modification. This chapter elaborates on the acceptance by Dutch consumers of hypoallergenic apples, developed either by breeding or by genetic modification.

24.1 Apple Allergy and Hypoallergenic Foods

In many parts of the world, apple is the most frequently consumed fruit. It is considered to be healthy because it adds important nutritional value to the daily diet (Bolhaar *et al.*, 2005). However, some consumers can become allergic to apples. Apple allergy is particularly problematic for those individuals who suffer from birch pollen allergy, as approximately 70% of the birch pollen allergic patients develop an allergy to apple (Gilissen *et al.*, 2006). The phenomenon is caused by an IgE cross-reactivity between the major birch allergen Bet v 1 and the major apple allergen Mal d 1 (Breiteneder *et al.*, 2000). It is referred to as the Oral Allergy Syndrome (OAS), and it is estimated that 10 million individuals in Northwest Europe (2%–3% of the population) are affected. Although the symptoms are generally rather mild, the diet of these allergic consumers may be restricted, as they avoid eating apples and also other Rosaceae fruits, such as pears, peaches and cherries, as well as vegetables such as fresh carrot and celery, which also induce OAS.

To overcome these dietary exclusions for such a large group of consumers, we are developing hypoallergenic alternatives. Hypoallergenic foods are defined as foods that are less likely to cause allergic complaints compared to their traditional counterparts. Introduction of hypoallergenic foods may, however, complicate risk management as it requires adequate separation of production chains, as well as measures that safeguard and facilitate consumer choice in the retail environment. This includes adequate communication and end-point labeling (Mills *et al.*, 2004; Cornelisse-Vermaat *et al.*, 2008). Several food products contain multiple allergens and individual consumers may display diverse clinical response profiles in response to different allergens. Moreover, there are differences in threshold levels to a particular allergen between individual consumers. As a result, some allergic individuals may still experience an allergic reaction following consumption of a hypoallergenic food that is considered suitable for the majority of allergic consumers.

24.2 Development of Hypoallergenic Apples

Hypoallergenic apples can be developed in several ways, e.g., through selection from existing cultivars, through breeding for this particular trait, or by genetic modification to eliminate or silence the allergen gene. It has been reported by apple allergic individuals that they tolerate some apple cultivars better than others. Differences in allergenicity have indeed been demonstrated in an extensive screening of a panel of 80 existing apple cultivars using the prick-to-prick method, confirmed by double-blind placebo-controlled food challenges. Based on this screening, the cultivar Santana was selected as the least allergenic apple that is currently available. Moreover, high variation in the degree of allergenicity (ranging from very low to very high) was detected among fruits from the progeny of a cross between the cultivars Fiesta and Discovery (unpublished results). This indicates that targeted breeding could be used to develop hypoallergenic apple cultivars. However, development of a market-valuable cultivar from new genotypes takes at least 20 years as it includes tests for appearance in the field, for taste and texture of the fruits, for production and storage qualities, for consumer satisfaction and economic viability.

24.3 Santana on the Market

The hypoallergenic apple cultivar Santana has good disease resistance; hence it is also suitable for organic cultivation. Several supermarkets in the Netherlands have sold this cultivar since 2006, labeled as "Suitable for consumption for people with a mild apple allergy" (Fig. 24.1). Its packaging has included a background information leaflet, a protocol regarding consumption, in order to test for an allergic response, and a link to a consumer questionnaire. Apple allergic consumers were generally very positive about the Santana.

Fig. 24.1. Labeled hypoallergenic apple cultivar Santana as it is being sold in Dutch supermarkets. The label says in Dutch: "Suitable for consumption by people with a mild apple allergy" and "Please first read the instructions"

Development of apple plants with a significantly reduced allergen expression from existing and economically successful cultivars through GM would reduce research and development costs, as well as time needed to create new varieties (Dodo *et al.*, 2008; Gilissen *et al.*, 2008). Societal concern about GM has proven to be a major factor influencing the development and acceptance of GM products in Europe (Gaskel *et al.*, 2006). Societal attitudes towards GM are influenced by case-specific characteristics of the application, such as which organism is modified and to what purpose (Zechendorf *et al.*, 1994; Frewer *et al.*, 1997). Currently, most GM traits are directed at improving agronomic characteristics (so called "input traits"), namely herbicide and plague (insect) resistance. The second generation of GM crops includes crop traits such as drought and salt resistance. The majority of European consumers are positive about medical applications of GM, but reject agricultural applications. Medical applications are considered more necessary, because of the treatment of illness, and are thereby more acceptable than food-related applications that appeared to favor only producers. Hypoallergenic GM foods are of interest because they are on the boundary between medical and food-related applications, and the consumer may either consider them as foods, medicines, or somewhere inbetween.

24.4 Consumption of the Hypoallergenic Santana

In the seasons of July 2006 and August 2007, the Santana apple (a relatively new cultivar) was sold in several stores in the Netherlands, being advertised as hypoallergenic. To enable consumers to decide for themselves whether the Santana was suitable for them, information was provided by packaging the Santana with an explanatory leaflet that contained information, and also an eating protocol for safe consumption by apple allergic individuals. Important starting points were that Santana is only suitable for consumers having a mild apple allergy and that following the protocol for safe consumption would reduce the likelihood of possible allergic reactions. In addition, buyers were requested to participate in an Internet-based survey about the Santana apple.

The results from the questionnaire were clear and consistent over a two-year period (years 2007 and 2008). More than 400 apple allergic consumers responded to the survey. Of these, around 45% of consumers did not experience any problems, 51% reported very mild symptoms (itching), while only 4% reported severe complaints. Whenever symptoms occurred, around 75% of respondents reported them as being less intense compared to other apple cultivars. Remarkably, only 15% of the respondents actually reported following the consumption protocol. Despite the fact that the packaging clearly indicated that the Santana is only suitable for consumers with mild apple allergy, consumers with (self-reported) severe apple allergy also tried eating the Santana. In this context, it is important to note that apple allergy (and OAS in general) almost never results in a life-threatening allergic (anaphylactic) reaction. Other newly developed hypoallergenic products involving potentially life-threatening allergens, such as peanut or shell fish, will demand more explicit warnings or may not be suitable for market introduction at all.

Three factors may explain why consumption of Santana results in allergic reactions in some individuals, while not in others. Firstly, an apple contains multiple allergens (termed Mal d 1 to Mal d 4), which are members of four totally different protein families, and the Santana may not be hypoallergenic for all these allergens. Sensitization to different allergens by individual patients may cause differences in the allergic response. Actually, that is what we generally see: Apple allergic consumers in Northwest Europe react to Mal d 1, while in Southern Europe (where no birches grow) people react to Mal d 3 (Sancho *et al.*, 2008). Secondly, individual consumers have different threshold levels regarding the allergen concentration that results in an allergic reaction. Thirdly, various protein variants of each apple allergen exist (Gao *et al.*, 2005), and the abundance of particular variants varies between different apple cultivars (Marzban *et al.*, 2005; Gao *et al.*, 2008). Depending on the protein variants to which individual patients react, they may or may not react to the Santana. In principle, these three factors will complicate development of any hypoallergenic product, and a particular challenge relates to the development of products that will be tolerated by all allergic consumers.

Overall, the Santana was highly appreciated by both non-allergic and allergic consumers, except by those few allergic consumers that experienced a strong allergic reaction after consumption. Consumption of the hypoallergenic Santana by those consumers who did not experience an allergic reaction contributes to a normal and healthy diet. Consumers who reported minor allergic reactions to the Santana were generally positive about both the Santana and hypoallergenic products in general. This positive attitude may be the direct result of the experience with the Santana, but may also result from the fact that attention is being paid to the allergy problems of these consumers, reflecting a well-known psychological effect (Wickstrom *et al.*, 2000).

24.5 Acceptation of Hypoallergenic GM Food: Personal Benefits Versus Societal Benefits

Consumer preferences with regard to the development of hypoallergenic food products through other methods, including GM, are important as this will influence acceptance of these products. Our survey examined the influence of perceived benefits and risks on acceptance of GM food products. Previous studies have found a two-dimensional attitudinal structure towards GM in which "erceived benefit" and "perceived risk" were influential in determining consumer acceptance (Frewer *et al.*, 1997; Schenk *et al.*, 2008). Differences in acceptance of GM hypoallergenic products between allergic and non-allergic individuals were mainly explained by differences in the perceived benefits, while the perceived risks remained constant (Schenk *et al.*, 2008). Evidence suggests that the public perception that consumers, rather than industry, will benefit from a particular application plays an important role in GM acceptance, and it has been suggested that as long as the risks are not as large as to be completely intolerable, consumer acceptance would be driven by perceptions of personal benefit (Miles and Frewer 2001; Frewer, 2003).

In the survey, consumers expressed a clear preference for traditional breeding over breeding by GM in the development of hypoallergenic apples, which is consistent with previous research (Schenk *et al.*, 2008; Miles *et al.*, 2005). Nonetheless, acceptance of hypoallergenic GM apples was higher for apple allergic consumers, when compared to non-allergic consumers, which suggested that perceived "personal benefits" have a strong impact on GM acceptance. In comparison, we also included questions on GM apples which would require fewer pesticides during production as an example of a "societal benefit" (Deliza *et al.*, 1999). Reduced pesticide usage led to an increased acceptance of GM apples, similarly for both allergic and non-allergic consumers, although acceptance was contingent on greater levels of pesticide reduction being achieved. Non-allergic consumers may still consider hypoallergenic GM products to have a societal benefit because these products alleviate complaints among other consumers, although the results of this research did not confirm this. The results of the study suggest that the introduction of traits that have "personal benefits" to consumers, more so than "societal benefits", may facilitate acceptance of GM products, although a preference for traditional breeding techniques applied to the development of new varieties to obtain the same benefits was expressed by all consumers.

24.6 Conclusion

Hypoallergenic foods have the potential to contribute to food allergy management. The results from a large-scale sales pilot of the Santana apple showed that the Santana fully alleviated the allergic complaints in nearly half of apple allergic consumers who participated in the study. Consumers who did experience a mild allergic reaction to the Santana tended to be positive about the Santana. Acceptance of similar hypoallergenic products was high, in particular if traditional breeding was applied to develop hypoallergenic traits, and to a lesser extent if GM was involved in their development. The results support the hypothesis that consumers are more likely to accept applications of GM in the agri-food sector when the benefits are sufficiently large and personally relevant. Given the individual differences among apple allergic consumers in the response to the Santana we recommend a cautious approach regarding interventions based on hypoallergenic food.

References

Bolhaar, S.T.H.P., *et al.* (2005). *In vivo* assessment with prick-to-prick testing and double-blind, placebo-controlled food challenge of allergenicity of apple

cultivars. J Allergy Clin Immunol, 116(5), 1080-1086.
Breiteneder, H., Ebner, C. (2000). Molecular and biochemical classification of plant-derived food allergens. J Allergy Clin Immunol, 106(1), 27-36.
Cornelisse-Vermaat, J.R., Voordouw, J., Yiakoumaki, V., *et al.* (2008). Food-allergic consumers' labelling preferences: A cross-cultural comparison. Eur J Public Health, 18, 115-120.
Deliza, R., Rosenthal, A., Hedderley, D., *et al.* (1999). The importance of brand, product information and manufacturing process in the development of novel environmentally friendly vegetable oils. J Int Food Agribusiness Marketing, 10, 67-77.
Dodo, H.W., Konan, K.N., Chen, F.C., *et al.* (2008). Alleviating peanut allergy using genetic engineering: The silencing of the immunodominant allergen Ara h 2 leads to its significant reduction and a decrease in peanut allergenicity. Plant Biotechnol J, 6, 135-145.
Frewer, L., Howard, C., Shepherd, R. (1997). Public concerns in the United Kingdom and specific applications of genetic engineering: risk, benefit, and ethics. Science, Technology and Human Values, 22, 98-124.
Frewer, L. (2003). Societal issues and public attitudes towards genetically modified foods. Trends In Food Science and Technology, 14, 319-332.
Gao, Z.S., van de Weg, W. E., Schaart, J. G., *et al.* (2005). Genomic characterization and linkage mapping of the apple allergen genes Mal d 2 (thaumatin-like protein) and Mal d 4 (profilin). Theor Appl Genet, 111, 1087-1097.
Gao, Z. S., van de Weg, W. E., Schaart, J. G., *et al.* (2008). Assessment of allelic diversity in intron-containing Mal d 1 genes and their association to apple allergenicity. BMC Plant Biology, 8, 116. Open access: http://www.biomedcentral.com/1471-2229/8/116/.
Gaskell, G., *et al.* (2006). Eurobarometer 64.3: Europeans and Biotechnology in 2005: Patterns and Trends (INRA).
Gilissen, L.J.W.J., *et al.* (2005). Silencing the major apple allergen Mal d 1 by using the RNA interference approach. Journal of Allergy and Clinical Immunology, 115, 364-369.
Marzban, G., *et al.* (2005). Localisation and distribution of the major allergens in apple fruits. Plant Sci,, 169, 387-394.
Miles, S., Frewer, L.J. (2001). Investigating specific concerns about different food hazards. Nutri Food Sci, 12, 47-61.
Mills, E.N.C., *et al.* (2004). Information provision for allergic consumers - where are we going with food allergen labelling? Allergy, 59, 1262-1268.
Miles, S., *et al.* (2005). Attitudes towards low-allergen food in food allergic consumers. Nutrition and Food Science, 35, 220-228.
Schenk, M.F., *et al.* (2008). The influence of perceived benefits on acceptance of GM applications for allergy prevention. Health, Risk and Society, 10, 263-282.
Sancho, A.I., *et al.* (2008). Measurement of lipid transfer protein in 88 apple cultivars. Int Arch Allergy Immunol, 146, 19-26.

Wickstrom, G., Bendix, T. (2000). The "Hawthorne effect" –what did the original Hawthorne studies actually show? Scandinavian J Work Environ Health, 26, 363-367.

Zechendorf, B. (1994). What the public thinks about biotechnology–better than synthetic food but worse than organ-transplantation–a survey of opinion polls. Biotechnology, 12, 870-875.

25

Cow's Milk Allergens and Technologies to Control Allergenicity

Yong-Kang Luo [1*], Guan-Hao Bu [2]

[1] College of Food Science and Nutritional Engineering, China Agricultural University, Beijing, 100083, China
[2] College of Food Science and Technology, Henan University of Technology, Zhengzhou, 450001, China
[*] E-mail: luoyongkang@ cau.edu.cn

Abstract: Cow's milk is one of the most frequently occurring food allergens. Cow's milk allergy has become a social concern for human health and nutrition all over the world. Cow's milk allergy can severely affect infants' health. Different clinical symptoms of milk allergy have been observed. Cow's milk contains more than 20 proteins that can cause allergic reactions. β-lactoglobulin, α-lactalbumin and casein were found as the main allergens of cow's milk and many epitopes of these allergens were identified using sera of allergic patients or animal sera. Various approaches have been applied to reduce the allergenicity of cow's milk proteins, such as heat treatment, glycation, protease hydrolysis and lactobacillus fermentation. These treatments were able to reduce the allergenicity of milk proteins only partly. For complete elimination of milk allergenicity and for the development of hypoallergenic or non-allergenic milk products, further research on the specific effects of the various treatments is required.

25.1 Introduction

Food allergy can cause serious health problems and has become a major public health concern around the world. Most common food allergens include cow's milk, eggs, fish, crustacean, peanuts, soy, nuts and wheat (Fritsche, 2003). Cow's milk is considered as an important allergen in food allergy. In a number of epidemiological studies, cow's milk protein allergy (CMPA) is reported to be the most prevalent for infants or young children with an incidence of 2% to 6%, with serious consequences for the infants' health (Hill and Hosking, 1996; 1997).

All cow's milk proteins may be potential allergens, of which the main allergens were found to be β-lactoglobulin (β-LG), α-lactalbumin (α-LA) and casein (Fritsche, 2003; Sharma *et al.*, 2001). In addition, other proteins like bovine serum albumin and even lactoferrin (present in trace amounts) have been shown to induce milk allergies (Wal, 2001). β-LG is considered to be a dominant allergen. β-LG is normally absent from human breast milk (Sawyer and Kontopidis, 2000). It can reach the intestinal mucosa almost intact due to its resistance to pepsin hydrolysis (Wal, 2001); there, it can induce allergic reactions (Reddy *et al.*, 1988; Wal, 2001). About 82% of milk allergic patients are sensitive to β-LG (Spies, 1973). Therefore, the reduction or elimination of this major milk allergen by effective methods and technologies will be very important for milk allergic individuals.

25.2 Cow's Milk Protein Allergens

The structures of β-LG, α-LA and casein display some features that could be related to their allergenic potential.

β-LG is a major whey protein in the milk of many mammals and has been considered to be the principal milk allergen. It is predominantly present as a dimer at physiological pH, while it dissociates into its native monomers at acidic pH<3.5 or basic pH>7.5 (McKenzie and Sawyer, 1967). The β-LG monomer is a globular protein with a molecular weight of about 18 kDa and has two internal disulfide bridges between Cys66-Cys160 and Cys106-Cys119. It contains a free sulfhydryl group (-SH group of Cys121), which is supposed to be involved in significant disulfide interchange with -SH group of other whey proteins or of casein (Kleber and Hinrishs, 2007; Sawyer *et al.*, 2002). β-LG belongs to the lipocalin family and is considered a retinol binding protein. The structures of all members in this family contain a β barrel composed of eight anti-parallel β strands. This kind of molecule has a high allergenic potential (Brownlow *et al.*, 1997).

α-LA is a monomeric globular protein with a molecular weight of about 14 kDa. It contains 4 disulfide bridges within its molecule, making the molecular structure of α-LA even more stable than that of β-LG during heating (Hinrichs and Rademacher, 2005; Wal, 2001). It possesses a high affinity binding site for calcium, and this bond stabilizes its secondary structure. The complete amino acid sequence of bovine α-LA shows extensive homology with human α-LA, since 74% of the amino acid residues are identical and another 6% are chemically similar. Despite this high degree of similarity, bovine α-LA has been identified as a major cow's milk allergen (Wal, 1998; 2001).

Casein consists of α_{s1}-casein, α_{s2}-casein, β-casein and κ-casein components which are crosslinked to form well-ordered aggregates or micelles. α_{s1}- and β-casein are the major components of casein. The four casein components display some common features. They are phosphorylated proteins with a loose tertiary, highly hydrated structure. Casein is often considered poorly immunogenic due to its flexible, non-compact structure and because of its rapid and extensive degradation by proteolytic enzymes during digestion. Casein is not significantly affected by severe heat treatment but is very susceptible to all proteinases and exopeptidase (Wal, 1998; 2001).

25.3 Epitopes of Milk Proteins Allergens

CMPA is mainly an IgE mediated hypersensitivity reaction. In an allergenic response, B-cell immobilized IgE antibodies bind to specific epitope areas of the protein amino acid side chains. The three-dimensional structure is thus important in CMPA, but IgE binding studies also showed the presence of additional short linear epitopes. IgE epitope analysis of cow milk proteins is usually carried out by direct or competitive ELISA inhibition tests *in vitro* with serum from allergic patients or with animal serum.

With tryptic digests and synthetic peptides, Selo *et al.* (1999) studied allergic epitopes of β-LG using serum from 46 milk-allergic patients. The results revealed the existence of numerous epitopes that are widely scattered all along the β-LG molecule. The major epitopes were the fragments 41–60, 102–124 and 149–162, recognized by 92%, 97% and 89% of sera from allergic patients, respectively. The C-terminus 149–162 peptide forms a short α-helix and appears to be quite mobile according to the crystal structure (Brownlow *et al.*, 1997). Peptides 41–60 and 102–124 form loops on the surface of the molecule (Brownlow *et al.*, 1997) and hence are accessible to antibodies. In addition, peptides 1–8, 25–40, and 92–100 were recognized by 52%–65% of the patients' sera. Peptides 9–14 and 84–91 were recognized by 40% of sera, and the fragments 78–83 and 125–135 were very poorly recognized by human IgE. The major allergenic sites and complete amino acid sequence were shown by Selo *et al.* (1999) and Fox (2003).

Many IgE binding epitopes on bovine α-LA have also been identified using sera from allergic patients. In the study by Adams *et al.* (1991), the fragment 5–18 was shown by RAST inhibition to exhibit IgE binding capacity. In another study, the sequence 17–58 and larger peptides sharing this sequence were recognized by patient sera, which confirmed the importance of conformational epitopes in allergenicity. Moreover, the peptide 59–94 was shown to exhibit a similar or a higher IgE-binding capacity than the native corresponding fragment, suggesting the existence of sequential epitopes (Maynard *et al.*, 1997). Jarvinen *et al.* (2001) identified four IgE binding sites (1–16, 13–26, 47–58 and 93–102) on α-LA using sera from the 11 patients with cow's milk allergy (Fox, 2003).

Most patients allergic to whole casein are sensitive to all the four casein components. The intensity of the IgE responses varied greatly between the sera and the different casein components but seemed closely related to the amounts of the four casein components in milk (Bernard *et al.*, 1998). Among caseins, α_{s1}-casein is a major allergen in milk. Six major and three minor IgE-binding regions on α_{s1}-casein were identified (Chatchatee *et al.*, 2001).

25.4 Clinical Symptoms and Diagnosis

CMPA is clinically an abnormal immunological reaction to cow's milk proteins. Many milk-allergic infants suffer from gastro-intestinal or skin symptoms (approximately 50% to 60%), respiratory symptoms (30%), as well as systemic anaphylactic symptoms. In the majority of infants with CMPA, more than two symptoms may occur, and the severity of the symptoms can vary from mild to life-threatening. An anaphylactic shock is a particularly serious symptom of CMPA (Exl and Fritsche, 2001; El-Agamy, 2007)

To diagnose CMPA, *in vitro* tests (RAST, ELISA) and *in vivo* tests (skin-prick test, patch test) are frequently used, but the gold standard of CMPA diagnosis is still the double-blind placebo-controlled food challenge (DBPCFC). However, in clinical practice, especially in the case of young infants, a DBPCFC may not be practical or necessary, and a simple food challenge and elimination test might be sufficient (Exl and Fritsche, 2001).

25.5 Milk Allergy Controlling Technologies

Until now, effective methods of medical therapy to milk protein allergy are still not established. Thus, all work today is focused on the prevention of milk allergy. Breast-feeding is the best way to preventing the development of allergy in infants to milk proteins. However, when exclusive breast-feeding is not possible, hypoallergenic formulas of cow's milk are usually given during the first six months of life to prevent milk allergy. Therefore, different attempts have been made to reduce the allergenicity of cow's milk proteins, and various technological processes have been applied in order to maintain the nutritional value of cow's milk in infant formulae.

25.5.1 Heat Treatment

Heating is an important process in the manufacturing of most dairy products. Different heat treatments, such as homogenization, concentration, pasteurization, sterilization, often are used to modify the functional properties of proteins, to ensure the safety of the product and to provide higher nutritional value (Oldfield *et al.*, 2005). During the heating process, important structural and chemical changes in proteins occur, like denaturation, aggregation and the Maillard reaction with other molecules such as sugars. These alterations may have an effect on the antigenicity of milk protein allergens.

Baldo (1984) found that heat treatment at 80 ℃ and 100 ℃ for 15 min produced a drop in the IgE fixation capacity with β-LG and BSA, but no change in the case of α-LA and casein. Ehn *et al.* (2004) detected only a slight decrease in IgE binding ability of β-LG after heating a β-LG solution or milk to 74 ℃, whereas a more significant decrease was found at 90 ℃ by ELISA inhibition studies. It has also been shown that heat denaturation of β-LG exposes some new epitopes (Davis and Williams, 1998).

We investigated the effect of heat treatment on the antigenicity of α-LA and β-LG in a whey protein isolate (WPI). It was found that the antigenicity of α-LA and β-LG increased with an increase in temperature from 50 ℃ to 90 ℃, with the highest antigenicity of α-LA and β-LG detected at 90 ℃ (Fig. 25.1). However, above 90 ℃ the antigenicity of both proteins showed a remarkable decrease. When treated at 120 ℃ for 20 min, the antigenicity of α-LA decreased below the initial value of the untreated sample (Bu *et al.*, 2009b). The decrease in β-LG antigenicity can be attributed to the destruction or masking of conformational epitopes being exposed at the surface of the molecule by sulfhydryl/disulfide exchange and subsequent aggregation (Kleber and Hinrishs, 2007). In addition to

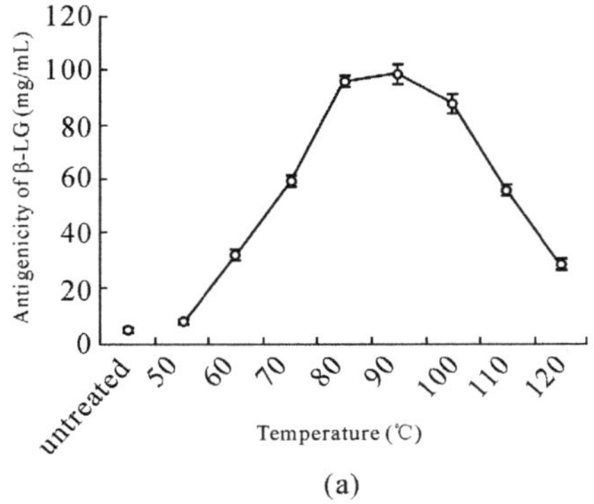

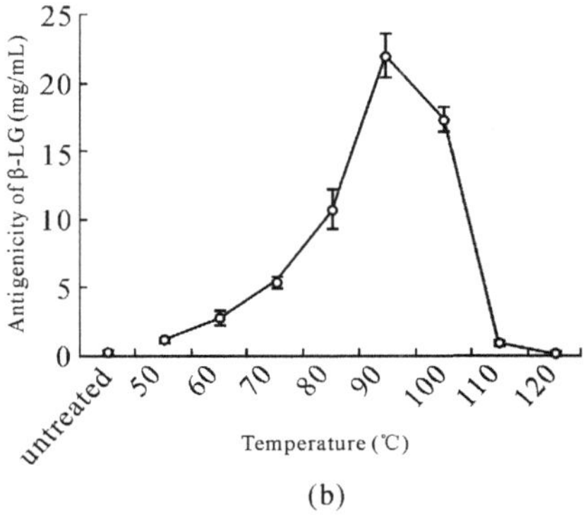

Fig. 25.1. Effect of heating temperature from 50–120 ℃ for 20 min on the antigenicity of β-LG (a) and α-LA (b) in WPI solution

disulfide-mediated aggregation, under more severe heating conditions the Maillard reaction may lead to the loss of linear epitopes with the consequence of a reduced antigenic response (Davis and Williams, 1998; Fritsche, 2003).

25.5.2 Hydrolysis

CMPA is caused by the presence of allergenic epitopes in the native proteins. Enzymatic hydrolysis of whey protein offers a practical way to destroy allergenic epitopes and reduce the antigenicity (Heyman, 1999). Asselin *et al.* (1988) showed that the antigenicity of whey protein (associated with α-LA and β-LG) could be reduced by hydrolysis with α-chymotrypsin. Pahud *et al.* (1985) indicated that the antigenicity of whey protein could be decreased by hydrolysis with trypsin. Wroblewska and Troszynska (2005) suggested that hydrolysis with Alcalase 2.4 FG could effectively reduce the antigenicity of a whey protein concentrate (WPC). However, the protein hydrolysis should not be more extensive than that required to remove the allergic response, because more extensive hydrolysis makes the product less palatable (bitter) and, in addition, free amino acids are less well absorbed than small peptides (van Beresteijn *et al.*, 1994; Ena *et al.*, 1995; Jost *et al.*, 1987). Therefore, the degree of hydrolysis must be well controlled.

In our previous research, a response surface methodology was used to study the effects of pH, temperature and enzyme-to-substrate ratio on the residual antigenicity of whey protein concentrate (WPC) hydrolysates obtained with Alcalase. It was shown that enzymatic hydrolysis with Alcalase could reduce the antigenicity of WPC for α-LA and β-LG effectively and the reduction of antigenicity could be controlled by regulation of three independent variables. Models for anti-α-LA and anti-β-LG IgG binding inhibition were established. Temperature had the greatest effect on the anti-α-LA IgG binding inhibition and pH had the greatest effect on the anti-β-LG IgG binding inhibition (Zheng *et al.*, 2008).

25.5.3 Glycation

The conjugation with reducing sugars through the Maillard reaction seems to be a promising and safe method for masking food protein allergenicity, and might also be an effective means of improving functional properties of proteins, as was shown in soy (Wilson *et al.*, 2005). Some studies have revealed that β-LG allergenicity can be modified by conjugation with different saccharides. Morgan *et al.* (1998) indicated that lactosylation of β-LG by the Maillard reaction during heating changed the epitope area. Hattori *et al.* (2000) showed that conjugation of β-LG and carboxymethyl dextran improved the emulsifying properties and reduced immunogenicity. β-LG conjugated with acidic oligosaccharides showed enhanced thermal stability and had reduced immunogenicity. It was also found that one of the mechanisms responsible for the reduced immunogenicity of β-LG-acidic oligosaccharides was the shielding of the epitopes resulting from the conjugation event (Hattori *et al.*, 2004).

Effects of the weight ratio of protein to sugar, reaction temperature and time on the antigenicity of β-LG and α-LA in conjugates of WPI with glucose using response surface methodology were also investigated in our study(Bu *et al.*, 2009a; 2010a). It was shown that conjugation of WPI with glucose could effectively reduce the antigenicity of β-LG and α-LA. This reduction of antigenicity could be controlled by regulating three independent variables (Table 25.1). The model for optimal reaction conditions of a lower antigenicity of β-LG and α-LA was established. The weight ratio of WPI to glucose had the greatest effect on the antigenicity of β-LG and α-LA. In addition, as shown in Fig. 25.2, the molecular mass of α-LA and β-LG increased in glycated WPI. It proved that some glucose was bound to whey proteins during the glycation reaction, which may have caused the masking effect on the epitopes of the whey proteins (Bu *et al.*, 2009).

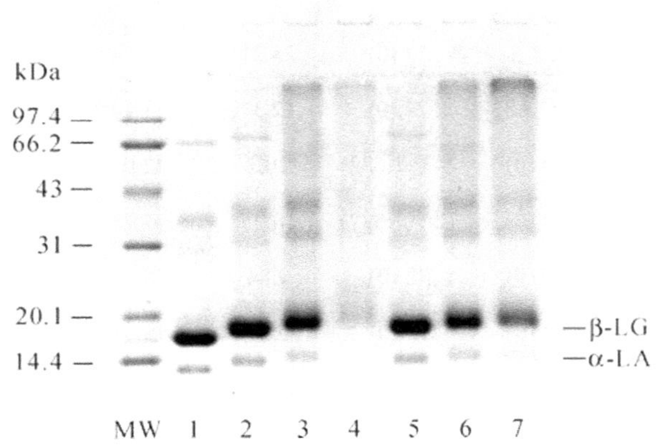

Fig. 25.2. SDS-PAGE analysis of dry-heated whey protein isolate (WPI) + glucose (G) conjugates (weight ratio 4 : 1) (Bu *et al.*, 2009a). MW: molecular weight markers; lane 1: native WPI; lane 2: WPI + G dry-heated at 40 ℃ for 72 h; lane 3: WPI + G dry-heated at 50 ℃ for 72 h; lane 4: WPI + G dry-heated at 60 ℃ for 72 h; lane 5: WPI + G dry-heated at 50 ℃ for 24 h; lane 6: WPI + G dry-heated at 50 ℃ for 72 h; lane 7: WPI + G dry-heated at 50℃ for 120 h. (With Permission of Wiley-Blackwell)

Table 25.1 Full factorial central composite design matrix for antigenicity of α-LA and β-LG in WPI-glucose conjugates

	Independence variables[a]			Dependence variables			
Assay	Weight ratio	Temperature (℃)	Time (h)	α-LA antigenicity (μg/mL)	Inhibition rate (%)[b]	β-LG anti-genicity (μg/mL)	Inhibition rate (%)[b]
1	1.72 (−1)	44 (−1)	43.5 (−1)	31.65	31.27	59.76	77.86
2	1.72 (−1)	44 (−1)	100.5 (1)	15.67	65.97	22.21	91.77
3	1.72 (−1)	56 (1)	43.5 (−1)	29.71	35.48	11.41	95.77
4	1.72 (−1)	56 (1)	100.5 (1)	7.65	83.39	21.2	92.15
5	6.28 (1)	44 (−1)	43.5 (−1)	15.98	65.3	154.94	42.59
6	6.28 (1)	44 (−1)	100.5 (1)	11.2	75.68	88.95	67.04
7	6.28 (1)	56 (1)	43.5 (−1)	4.16	90.97	47.86	82.27
8	6.28 (1)	56 (1)	100.5 (1)	1.72	96.26	109.02	59.61
9	0.17 (−1.682)	50 (0)	72 (0)	30.39	34.0	10.7	96.04
10	7.83 (1.682)	50 (0)	72 (0)	0.26	99.44	100.31	62.83
11	4 (0)	40 (−1.682)	72 (0)	28.21	38.74	106.04	60.71
12	4 (0)	60 (1.682)	72 (0)	12.37	73.14	12.63	95.32
13	4 (0)	50 (0)	24 (−1.682)	26.7	42.02	90.61	66.43
14	4 (0)	50 (0)	120 (1.682)	10.25	77.74	19.33	92.84
15	4 (0)	50 (0)	72 (0)	9.84	78.63	21.8	91.92
16	4 (0)	50 (0)	72 (0)	9.23	79.96	17.79	93.41
17	4 (0)	50 (0)	72 (0)	11.15	75.79	22.42	91.69
18	4 (0)	50 (0)	72 (0)	8.33	81.91	19.88	92.63
19	4 (0)	50 (0)	72 (0)	10.5	77.2	18.46	93.16
20	4 (0)	50 (0)	72 (0)	12.36	73.16	20.63	92.36
21	4 (0)	50 (0)	72 (0)	8.54	81.45	23.05	91.46
22	4 (0)	50 (0)	72 (0)	10.01	78.26	19.51	92.77
23	4 (0)	50 (0)	72 (0)	10.67	76.83	20.25	92.50

a) Values in parentheses are the coded levels of independence variables;

b) Inhibition Rate is compared with non-glycated WPI

25.5.4 Fermentation

Fermented foods exert a positive influence on human health or physiology, mainly due to their ability to liberate bioactive peptides from major food proteins by microorganisms' enzymatic hydrolysis. It was reported that lactic acid bacteria possess a complex proteolytic system composed of proteinases, peptidases and transport systems (Bianchi-Salvadori *et al.*, 1995; Law and Haandrikman, 1997). During fermentation by lactic acid bacteria, the hydrolysis of milk proteins may have important effects on milk digestibility and the production of bioactive peptides. Moreover, the proteolysis causes the break of some epitopes and may decrease milk allergenicity (Cross *et al.*, 2001; Bertrand-Harb *et al.*, 2003). In addition, dietary consumption of probiotics and fermented foods, such as yogurt, can alleviate some of the symptoms of atopy and reduce the development of allergies through a mechanism of immune regulation.

It has been demonstrated that fermentation could induce the degradation of some food allergens. Song *et al.* (2008) found that natural and induced fermentation significantly reduced IgE immunoreactivity up to 89% in soybean meal. Barkholt *et al.* (1998) showed that fermentation with three lactic acid bacteria and *R. microsporus* reduced the antigenicity to 10% of the antigenicity of the unfermented pea flour. It was shown that the antigenicity of whey proteins in sterilized cow's milk was reduced by over 99% as compared to raw milk after fermentation with selected lactic acid bacteria (Jedrychowski and Wroblewska, 1999). Kleber *et al.* (2006) indicated that lactic acid fermentation can attenuate β-LG antigenicity in skimmed milk and sweet whey, but did not eliminate antigenicity. Our study also showed that fermentation by lactic acid bacteria could significantly reduce the antigenicity of α-LA and β-LG in skimmed milk. Combined strains of *Lactobacillus helveticus* and *Streptococcus thermophilus* were the most effective in reducing the antigenicity of both whey proteins. Synergistic action was observed between the combined strains (Bu *et al.*, 2010b).

25.6 Conclusion

Cow's milk is one of the most common food allergens. Cow's milk protein allergy has affected human health severely, especially infants' health. In this paper, major cow's milk protein allergens and their epitopes are reviewed. It was also shown that some technological processes can be used to reduce the antigenicity and allergenicity of milk proteins by controlling and optimizing the processing conditions. A combination of different technological approaches might be of interest in the design of processing strategies. This will be helpful for developing new hypoallergenic or non-allergenic milk products.

References

Adams, S.L., Barnett, D., Walsh, B.J., *et al*. (1991). Human IgE-binding synthetic peptides of bovine β-lactoglobulin and α-lactalbumin. In vitro cross-reactivity of the allergens. Immunol Cell Biol, 69, 191-197.

Asselin, J., Amiot, J., Gauthier, S.F. (1988). Immunogenicity and allergenicity of whey protein hydrolysates. J Food Sci, 53, 1208-1210.

Baldo, B.A. (1984). Milk allergies. Austr J Dairy Technol, 39, 120-128.

Barkholt, V., Jørgensen, P.B., Sørensen, D., *et al.* (1998). Protein modification of fermentation: Effect of fermentation on the potential allergenicity of pea. Allergy, 53(Suppl 46), 106-108.

Bernard, H., Creminon, C., Yvon, M., *et al.* (1998). Specificity of the human IgE response to the different purified caseins in allergy to cow's milk proteins. Int Arch Allergy Immunol, 115, 235-244.

Bertrand-Harb, C., Ivanova, I.V., Dalgalarrondo, M., *et al.* (2003). Evolution of β-lactoglobulin and α-lactalbumin content during yoghurt fermentation. Int Dairy J, 13, 39-45.

Bianchi-Salvadori, B., Camaschella, P., Cislaghi, S. (1995). Rapid enzymatic method for biotyping and control of lactic acid bacteria used in the production of yoghurt and some cheeses. Int J Food Microbiol, 27, 253-261.

Brownlow, S., Morais Cabral, J.H., Cooper, R., *et al.* (1997). Bovine β-lactoglobu-lin at 1.8 Å resolution–still an enigmatic lipocalin. Structure, 5, 481-495.

Bu, G.H., Lu, J., Zheng, Z., *et al.* (2009a). Influence of Maillard reaction conditions on the antigenicity of bovine α-lactalbumin using response surface methodology. J Sci Food Agric, 89(14), 2428-2434.

Bu, G.H., Luo, Y.K., Lu, J., *et al.* (2010a). Reduced antigenicity of β-lactoglobulin by conjugation with glucose through controlled Maillard reaction conditions. Food Agric Immunol, 21(2), 143-156.

Bu, G.H., Luo, Y.K., Zhang, Y., *et al.* (2010b). Effect of the fermentation by lactic acid bacteria on the antigenicity of bovine whey proteins. J Sci Food Agric, 90(12), 2015-2020.

Bu, G.H., Luo, Y.K., Zheng, Z., *et al.* (2009b). Effect of heat treatment on the antigenicity of bovine α-lactalbumin and β-lactoglobulin in whey protein isolate. Food Agric Immunol, 20(3), 195-206.

Chatchatee, P., Jarvinen, K.M., Bardina, L., *et al.* (2001). Identification of IgE- and IgG-binding epitopes on α_{s1}-casein: Differences in patients with persistent and transient cow's milk allergy. J Allergy Clin Immunol, 107, 379-383.

Cross, M.L., Stevenson, L.M., Gill, H.S. (2001). Anti-allergy properties of fermented foods: an important immunoregulatory mechanism of lactic acid bacteria? Int Immunopharmacol, 1, 891-901.

Davis, P.J., Williams, S.C. (1998). Protein modification by thermal processing. Allergy, 53, 102-105.

Ehn, B.M., Ekstrand, B., Bengtsson, U., *et al.* (2004). Modification of IgE binding during heat processing of the cow's milk allergen beta-lactoglobulin. J Agric Food Chem, 52, 1398-1403.

Ena, J.M., van Beresteijn, E.C.H., Robben, A.J.P.M., *et al.* (1995). Whey protein antigenicity reduction by fungal proteinases and a pepsin/pancreatic combination. J Food Sci, 60, 104-116.

El-Agamy, E.I. (2007). The challenge of cow milk protein allergy. Small Ruminant Res, 68, 64-72.

Exl, B.M., Fritsche, R. (2001). Cow's milk protein allergy and possible means for its prevention. Nutrition, 17, 642-651.

Fox, P.F. (2003). Advanced dairy chemistry. New York: Kluwer Academic/Plenum, 1239-1254.

Fritsche, R. (2003). Role of technology in dairy allergy. Aust J Dairy Technol, 58, 89-91.

Hattori, M., Miyakawa, S., Ohama, Y., *et al.* (2004). Reduced immunogenicity of β-lactoglobulin by conjugation with acidic oligosaccharides. J Agric Food Chem, 52, 4546-4553.

Hattori, M., Nagasawa, K., Ohgata, K., *et al.* (2000). Reduced Immunogenicity of β-Lactoglobulin by Conjugation with Carboxymethyl Dextran. Bioconjugate Chem, 11, 84-93.

Heyman, M. (1999). Evaluation of the impact of food technology on the allergenicity of cow's milk proteins. Proc Nutr Soc, 58, 587-592.

Hill, D.J., Hosking, C.S. (1996). Cow milk allergy in infancy and early childhood. Clin Exp Allergy, 26, 243-246.

Hill, D.J., Hosking, C.S. (1997). Emerging disease profiles in infants and young children with food allergy. Pediatr Allergy Immunol, 8, 21-26.

Hinrichs, J., Rademacher, B. (2005). Kinetics of combined thermal and pressure-induced whey protein denaturation in bovine skim milk. Int Dairy J, 15, 315-323.

Jarvinen, K.M., Chatchatee, P., Bardina, L., *et al.*(2001). IgE and IgG binding epitopes on alpha-lactalbumin and beta -lactoglobulin in cow's milk allergy. International Archives of Allergy and Immunology, 126, 111-118.

Jedrychowski, L., Wroblewska, B. (1999). Reduction of the antigenicity of whey proteins by lactic acid fermentation. Food and Agricultural Immunology, 11, 91-99.

Jost, R., Monti, J.C., Pahud, J.J. (1987). Whey protein allergenicity and its reduction by technological means. Food Technol, 41, 118-121.

Kleber, N., Hinrichs, J. (2007). Antigenic response of β-lactoglobulin in thermally treated bovine skim milk and sweet whey. Milchwissenschaft, 62, 121-124.

Kleber, N., Weyrich, U., Hinrichs, J. (2006). Screening for lactic acid bacteria with potential to reduce antigenic response of β-lactoglobulin in bovine skim milk and sweet whey. Innovat Food Sci Emerg Tech, 7, 233-238.

Law, J., Haandrikman, A. (1997). Proteolytic enzymes of lactic acid bacteria. Int Dairy J, 7, 1-11.

Maynard, F., Jost, R., Wal, J.M. (1997). Human IgE binding capacity of tryptic peptides from bovine α-lactalbumin. Int Arch Allergy Immunol, 113, 478-488.

McKenzie, H.A., Sawyer, W.H. (1967). Effect of pH on β-lactoglobulin. Nature, 214, 1101-1104.

Morgan, F., Bouhallab, S., Mollé, D., *et al.* (1998). Lactolation of β-lactoglobulin monitored by electrospray ionization mass spectrometry. Int Dairy J, 7, 299-303.

Oldfield, D.J., Singh, H., Taylor, M.W. (2005). Kinetics of heat-induced whey protein denaturation and aggregation in skim milks with adjusted whey protein concentration. Journal of Dairy Research, 72, 369-378.

Pahud, J.J., Monti, J.C., Jost, R. (1985). Allergenicity of whey proteins: Its modification by tryptic *in vitro* hydrolysis of the protein. J Pediatr Gastroenterol Nutr, 4, 408-413.

Reddy, M., Kella, N.K.D., Kinsella, J.E. (1988). Structural and conformational basis of the resistance of β-lactoglobulin to peptic and chymotryptic digestion. J Agric Food Chem, 36, 737-741.

Sawyer, L., Barlow, P.N., Boland, M.J., *et al.* (2002). Milk protein structure-what can it tell the dairy industry? Int Dairy J, 12, 299-310.

Sawyer, L. and Kontopidis, G. (2000). The core lipocalin, bovine bete-lactoglobulin. Biochim Biophys Acta, 1482, 136-148.

Selo, I., Clement, G., Bernard, H., *et al.*(1999). Allergy to bovine β-lactoglobulin: Specificity of human IgE to tryptic peptides. Clin Exp Allergy, 29, 1055-1063.

Sharma, S., Pravindra, K., Betzel, C., *et al.* (2001). Structure and function of proteins involved in milk allergies. J Chromatogr B, 756, 183-187.

Song, Y.S., Frias, J. Martinez-Villaluenga, C., *et al.* (2008) Immunoreactivity reduction of soybean meal by fermentation, effect on amino acid composition and antigenicity of commercial soy products. Food Chem, 108, 571-581.

Spies, J., 1973. Milk allergy. J milk food technol, 36, 225-231.

van Beresteijn, E.C.H., Peeters, R.A., Kaper, J., *et al.* (1994). Molecular mass distribution, immunological properties and nutritive value of whey protein hydrolysates. J Food Protect, 57, 619-625.

Wal, J.M. (1998). Cow's milk allergens. Allergy, 53, 1013-1022.

Wal, J.M. (2001). Structure and function of milk allergens. Allergy, 56, 35-38.

Wilson, S., Blaschek, K., Gonzalez de Mejia, E. (2005). Allergenic proteins in soybean: Processing and reduction of P34 Allergenicity. Nutrition Reviews, 63, 47-58.

Wroblewska, B., Troszynska, A. (2005). Enzymatic hydrolysis of cow's whey milkproteins in the aspect of their utilization for the production of hypoallergenic formulas. Pol J Food Nutr Sci, 14, 349-357.

Zheng, H., Shen, X.Q., Bu, G.H., *et al.* (2008). Effects of pH, temperature and enzyme-to-substrate ratio on the antigenicity of whey protein hydrolysates prepared by Alcalase. Int Dairy J, 18, 1028-1033.

Part VI

Environment, Hygiene and Societal Issues

26

Prevalence of Asthma and Asthmatic Symptoms in Children in Relation to Environmental Factors

—Epidemiological Studies in School Children in Taiyuan, China

Zhuo-Hui Zhao[1*], Zhuan-Hua Wang[2], Zheng Zhang[2], D. Norbäck[3], G . Wieslander[3]

[1] Department of Environmental Health, School of Public Health, Key Laboratory of Public Health Safety, Ministry of Education, Fudan University, Shanghai 200032, China

[2] Key laboratory of Chemical Biology and Molecular Engineering of Ministry of Education, Shanxi University, Taiyuan, 030006, China

[3] Department of Occupational and Environmental Medicine, University Hospital and Uppsala University, Uppsala, Sweden

[*] E-mail: zhzhao@fudan.edu.cn.

Abstract: There has been a rapid increase in prevalence of asthma in Chinese children. However, the current level is still lower when compared to the prevalence in Western countries. Environmental factors might be associated with the increasing prevalence of children's asthma and asthmatic symptoms in China. In this study, a cross-sectional survey was performed in 10 randomly selected schools involving 1993 children (mean age 13 years old) in urban areas in Taiyuan, China. Data on children's asthma and asthmatic symptoms were collected by a questionnaire taken from the International Study on Asthma and Allergies in Childhood (ISAAC). Data on environmental exposure, including indoor and outdoor chemical air pollutants and indoor biological contamination in the settled dust, were quantitatively evaluated in the school environment. The results showed that the indoor school environment in urban areas in Taiyuan was contaminated with chemical air pollutants of outdoor origin (SO_2, NO_2, O_3 and formaldehyde), and that the air pollutants were positively associated with children's wheezing and daytime attacks of breathlessness. Different microbial components in the settled dust showed different effects regarding the prevalence of children's respiratory symptoms, for example, muramic acid, a marker of gram positive bacteria, was negatively associated with children's respiratory health, while ergosterol, a marker of fungi, showed positive associations. There was a low level of allergen contamination in the settled dust in the school environment and the detected airborne cat and dog allergens were not associated with any health parameters included in this study. In addition, environmental tobacco smoking (ETS) and emissions from new furniture in the home environment were risk factors for children's respiratory symptoms. In conclusion, chemical air pollutants in schools may adversely affect children's asthmatic symptoms while biological components resulted in more complex effects. This further research on different environmental factors and their potential interactions needs to be explored.

26.1 Introduction

Recent international research indicates that, at the present time, the increase in asthma and allergies in children is most pronounced in the more advanced developing countries (Asher *et al.*, 2006). In mainland China, a large national survey in children (0–14 years old) in 27 cities (399,193 children in 1990 and 287,329 children in 2000) showed that there was an average 64.84% increase in asthma prevalence over the 10-year period included in the surveys (Chen, 2004). The increasing prevalence of asthma or symptomatic wheezing was reported in many countries in Asia including Singapore, Thailand, Indonesia and Malaysia (Leung and Wong, 2008). The reasons remain unclear, but this trend could be attributed to changes associated with environment and lifestyle factors during the modernization process (Douwes and Pearce, 2002).

Air pollution, produced as a side effect of the industrialization and urbanization process on the mainland of China, has been blamed for the increase in asthma (Watts, 2006). However, the cause and effect relationship between air pollution and the incidence of asthma still remains controversial. Many studies have shown that air pollution aggravates existing respiratory symptoms (Chauhan *et al.*, 2003; Gent *et al.*, 2003). A recent retrospective study observed significant associations between asthma-related hospitalizations and atmospheric levels of NO_2, O_3, PM10 and PM2.5 (Ko *et al.*, 2007). In addition, air pollution produced outdoors could penetrate indoors and children who spent a large part of time in indoor environments were found to be exposed to outdoor pollution, in particular in classrooms without any air filtration system (Norback *et al.*, 2002).

Environmental biological contamination has a greater effect on asthmatic symptoms in children via inhalation pathways. Environmental allergens are the most common irritants, which usually trigger asthmatic or allergic symptoms in sensitized children. Several common kinds of environmental allergens include pets (Simpson and Custovic, 2005), house dust mites (Sporik *et al.*, 1992), various pollens, fungi and other animal allergens, the effects varying across different countries or areas. Mould contamination can easily occur in houses with higher relative humidity, or which have been flooded. In such environments, the increase in risk for children developing asthmatic symptoms is generally in the range of 1.5%–3.5% (Peat *et al.*, 1998). Ergosterol (Erg) and glucan (1-3-β-D-glucan) are two components usually used as indicators of fungal biomass (Saraf *et al.*, 1997). Muramic acid, one component in the gram positive bacteria, can be used as marker for bacteria pollution. Compared with reports indicating that fungi represent risk factors, the evidence to suggest that bacterial contamination results in asthma or allergic disease appears more controversial.

Other factors associated with asthma or asthmatic symptoms have been reported and they include the growing environment in childhood, living and dietary habits (Zhao *et al.*, 2003; Dong *et al.*, 2009). One comparative study among Hongkong children showed that the use of cotton quilts, exposure to house dampness, the use of gas as cooking fuel, and consumption of fruits and raw vegetables were potential protective factors for children who were born on the mainland of China but later migrated to Hongkong, when compared to children who were born and raised in Hongkong. The latter have a significantly higher prevalence of asthma and allergic diseases (Wong *et al.*, 2004). Environmental tobacco smoking (ETS), still common in Chinese homes, was consistently reported as a risk factor for asthma and allergic diseases (Dong *et al.*, 2007; Zhao *et al.*, 2008).

To date, Chinese epidemiological studies have placed more emphasis on investigations within the home environment compared to the school environment (Zhang *et al.*, 2002; Zhao *et al.*, 2006; 2008; Dong *et al.*, 2007). The school environment, however, potentially plays an important role in terms of children's health. This is, firstly, because children spend a lot of time in schools, and because the school environment may be poorly maintained due to a lack of maintenance routines, financial support or low awareness of risk factors.

In this article, we review environmental epidemiological studies in relation to asthma and asthmatic symptoms in children in Taiyuan, an industrialized city with metallurgy and coal industries and power plants being the major industries in the north of China. The prevalence of diagnosed asthma in children (0–14 years of age) increased from 0.51% in 1990 to 1.02% in 2000 (Chen, 2004). Air pollution is relatively higher in Taiyuan than that in many other cities due to the heavy industries and increasing traffic pollution. Environmental factors including air pollutants SO_2, NO_2, O_3 and formaldehyde, biological components of muramic acid, ergosterol and LPS (endotoxin), airborne allergens as well as the indoor climate were measured, and linked with children's asthma or asthmatic symptoms.

26.2 Study Locations, Selection Process and Methods

Ten junior high schools were randomly selected in December 2004 within the urban areas of Taiyuan (3 million inhabitants), situated 500 km southwest of Beijing (Fig. 26.1). Taiyuan used to be one of the most heavily polluted cities in the world, and Shanxi province is the major coal mining area in China, providing two-thirds of China's domestic coal production. The headmasters of the 10 selected schools were contacted, and all agreed to participate. This study was approved by the ethical committee of Uppsala University, Uppsala, Sweden.

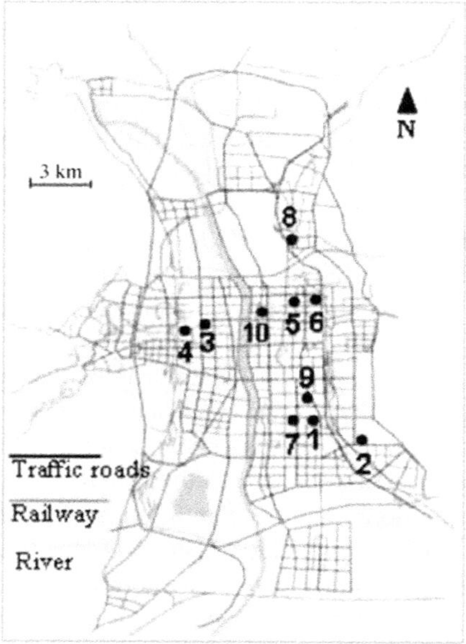

Fig. 26.1. Map of the 10 selected schools in urban areas in Taiyuan, Shanxi province, China. Modified from Taiyuan Urban Planning Committee (2007)

26.2.1 Study Population

In each of the 10 schools, 5 first-year classes were randomly selected, in different areas and floors in the school buildings. If there were fewer than 5 first-year classes, all were selected. The study population consisted of 2,209 pupils (11–15 years of age) in 46 classes; 1,993 pupils (90.2%) completed the questionnaire. There were no reports of health complaints or environmental problems associated with any of the schools before the investigation.

26.2.2 Environmental Exposure Measurements

26.2.2.1 Indoor and Outdoor Air Pollutants

Indoor levels of SO_2, NO_2, O_3, and formaldehyde were measured in the selected classrooms (maximum five per school). Outdoor levels were measured at the same representative location in each school by diffusion samplers continuously for 7 days. Indoor samplers were placed approximately 1.5–2 m above the floor. Outdoor samplers were placed 2.5–3.5 m above the ground under a well ventilated plastic cover preventing rain and snowfall. The concentrations ($\mu g/m^3$) of air pollutants were analyzed by accredited laboratories specializing in analyzing samples. Average values across the 7-day measurement period were reported. To evaluate how the indoor air was affected by outdoor air pollution, the ratios between indoor and outdoor concentrations were calculated.

26.2.2.2 Microbial Components in Settled Dust (Lipopolysaccharide (LPS), Muramic Acid (MuA) and Ergosterol (Erg)

In total, 78 dust samples were collected from 39 classrooms, while 7 classrooms could not be sampled for lack of electric outlets. Settled dust was collected by a vacuum cleaner of 400 W, equipped with a special dust collector fitted with a Millipore filter (pore size 6 lm) (ALK Abello, Copenhagen, Denmark). Two samples were collected for each classroom, one from the corridor side and the other from the window side. Vacuum cleaning was performed for 4 min for each sample, 2 min on the floor and 2 min on the desks and chairs. The filters were stored at 20℃ until analysis occurred.

The dust samples were sieved (particle diameter < 400 lm) and 1-5 mg portions of the fine dust fraction were used for analyses by GC-MS/MS. The microbial amount per sample for different microbial markers was calculated by multiplying the concentration by the total weight of the dust sample.

26.2.2.3 Indoor Allergen Exposure

Allergen levels were measured in both settled dust and airborne dust. The settled dust samples were from the same samples for microbial components analyses. Airborne dust was collected on a Petri-dish in each classroom, placed at a height of about 1.5-2 m in the window and corridor sides and kept open for 7 days continuously (Karlsson et al., 2002).

After sample extraction, two-site sandwich ELISA was applied to determine the allergens levels of cat (Fel d 1), dog (Can f 1), house dust mite (Der p 1 and Der f 1), cockroach (Bla g 2) (Indoor Biotechnologies Ltd., Manchester, UK) and horse (Equ c x) (Mabtech, Stockholm, Sweden) (Emenius *et al.*, 2001) for both settled dust and Petri-dish samples. The allergen level is expressed as ng/g dust, except for horse allergen concentration which is expressed as U/g dust, where 1 unit equals 1 ng protein of a horse hair and dander extract used as standard (Allergon, Valinge, Sweden and NIBSC, Hertfordshire, UK).

26.2.3 Questionnaire

Students were given a self-administered questionnaire to collect data on their respiratory health, parental asthma or allergy, and selected factors in the home environment. Questions on respiratory health were mainly based on the International Study of Asthma and Allergy in Childhood (ISAAC) (Asher *et al.*, 1995), the European Community Respiratory Health Survey (ECRHS) (Janson *et al.*, 2001), and previous school studies in Sweden (Smedje *et al.*, 1997) and in Shanghai, China (Mi *et al.*, 2006). They included yes/no questions on cumulative asthma, doctor-diagnosed asthma, current asthma, and allergies to furry pets or pollen. Moreover, there were questions on respiratory symptoms (without using the word asthma) including wheeze or whistling in the chest, daytime or nocturnal attacks of breathlessness in the preceding 12 months, and recent respiratory infections defined as either cold, upper respiratory infection, or middle ear infection in the preceding 3 months. Finally, there were questions on parental asthma or allergy and current home environment, including recent home painting, installation of new floor material and new furniture in the preceding 12 months, and environmental tobacco smoke (ETS) at home. The survey was performed 1 week before the classroom inspections and measurements, distributed in the school by the class teachers, and answered at home in cooperation with the parents. All the personal information from the questionnaire was kept confidential.

26.2.4 Data Analysis

A hierarchical logistic regression model was used to analyze associations between response (pupils' respiratory health on an individual level) and exposure (environmental factors on a class level or school level) with each of the environmental variables added separately, controlling for age, sex, parental asthma or allergy, and home environmental factors (new painting, new floor material, new furniture and ETS). Subsequently, the multivariate hierarchical regression model was fitted with adjustment for personal and home environmental factors and different environmental factors simultaneously. In the hierarchical model, a random intercept logit link-binomial model was applied, accounting for the hierarchical structure of the data. It was estimated by iterative generalized least square, first-order marginal quasi-likelihood followed by second order penalized quasi-likelihood. Odds ratios (OR) with 95% confidence interval (95% CIs) were applied.

Correlation analyses among different exposure factors were performed by a rank correlation test not requiring normal distribution (Kendall's tau-β). In all statistical analyses, two-tailed tests and a 5% level of significance were applied.

26.3 Prevalence of Asthma and Asthmatic Symptoms

In total, 2,209 pupils received the questionnaire and 1,993 participated (90.2% response rate). There were 49.3% girls and the average age was 13 years old (s.d. 0.64, range 11-15 years). The prevalence of asthma and asthmatic symptoms is shown in Table 26.1, stratified by parental asthma or allergy. The results demonstrated that there are lower levels of cumulative asthma (1.8%) and doctor's diagnosed asthma (1.2%), whereas the prevalence of asthmatic symptoms, such as wheezing or whistling in the chest (8.4%), or daytime attacks of breathlessness (29.8%), is higher. These symptoms are significantly higher in pupils whose parents had a history of asthma or allergy ($P<0.05$).

Table 26.1 Prevalence (%) of self-reported asthma, asthmatic symptoms and allergies among pupils in Taiyuan, China, with and without parental asthma or allergy (n=1,993)

	Total	Parental asthma or allergy		OR (95% CI)
		Yes (n=224)	No (n=1,769)	
Asthma				
Ever asthma	1.8	4.5	1.4	3.30 (1.56–6.98)**
Doctor's diagnosed asthma	1.2	1.3	1.1	1.19 (0.35–4.03)
Current asthma attacks	0.4	0.9	0.3	2.66 (0.53–13.3)
Current asthma medication	0.5	1.8	0.3	5.36 (1.50–19.2)**
Current airway symptoms in the last 12 months				
Wheezing or whistling in the chest	8.4	15.2	7.6	2.19 (1.46–3.29)***
Daytime attacks of breathlessness[a]	29.8	43.3	28.0	1.93 (1.45–2.56)***
Daytime attacks of breathlessness at rest	5.4	10.3	4.7	2.26 (1.39–3.67)**
Daytime attacks of breathlessness after exercise	27.7	38.4	26.3	1.77 (1.32–2.38)***
Nocturnal attacks of breathlessness	2.1	2.7	2.0	1.34 (0.56–3.22)
At least one airway symptom[b]	33.9	32.1	48.2	2.02 (1.51–2.70)***
A history of atopy [c]	3.8	6.7	3.4	1.92 (1.07–3.45)*
Respiratory infections (in the last 3 months)	39.3	37.1	39.6	0.90 (0.67–1.20)

a) calculated as daytime attacks of breathlessness either at rest or after exercise; b) wheezing or whistling in the chest, daytime or nocturnal attacks of breathlessness; c) self-reported furry pet or pollen allergy. * $P<0.05$; ** $P<0.01$; *** $P<0.001$

26.3.1 Environmental Exposure

26.3.1.1 Indoor and Outdoor Air Pollutants

The average indoor and outdoor concentrations of SO2, NO2, O3 and formaldehyde are listed in Table 26.2. For SO2, 5 out of 10 outdoor samples were close to saturation, and 3 were completely saturated. In the case of complete saturation, the saturation concentration (1,015 μg/m3) was used. Indoor SO2 and NO2 were approximately normally distributed, whereas O3 data were more skewed (data not shown).

Table 26.2 Indoor and outdoor levels (μg/m^3) of air pollutants in 10 schools

	No.[a]	Mean±SD	Range
		Indoor	
SO_2	34	264.8±139.0	60.0–641.1
NO_2	34	39.4±9.5	15.5–61.6
O_3	34	10.1±10.4	3.0–61.2
Formaldehyde	31	2.3±1.1	1.0–5.0
Outdoor			
SO_2	10	712.8±189.3	476.0–1015.0
NO_2	10	52.3±9.5	37.9–65.2
O_3	10	12.4±3.3	7.1–17.5
Formaldehyde	9	5.8±0.6	5.0–7.0
Indoor/Outdoor ratios			
SO_2	34	0.38±0.17	0.11–0.76
NO_2	34	0.78±0.22	0.38–1.19
O_3	34	0.91±0.93	0.18–5.1
Formaldehyde	31	0.39±0.18	0.14–0.83

a) Number of classes and schools with available measurements

26.3.1.2 Microbial Components in the Settled Dust

The concentration level and total load of microbial components, 3-OH fatty acid (3-OH FAs), MuA and Erg, three chemical markers for LPS from endotoxin in Gram-negative bacteria (G−), Gram-positive bacteria (G+) and fungi, respectively, are shown in Table 26.3. Concentration of MuA was correlated with concentrations of LPS (Tau beta 0.40, $P < 0.01$) and Erg (Tau beta 0.24, $P < 0.01$), whereas no significant correlation was found between LPS and Erg (Tau beta 0.13, $P = 0.23$).

26.3.1.3 Allergen Exposure

Allergen levels were generally very low in settled dust in classrooms in urban schools in Taiyuan (Table 26.4). In total, 5 samples (6%), all from different classrooms, contained traces of cat allergen (Fel d 1) and two samples (3%) contained traces of dog allergen (Can f 1). One of the dust samples contained horse allergen (Equ cx). One dust sample (1%) contained Der f 1 (323 ng/g dust). None contained Der p 1. Cockroach allergen (Bla g 1) was not detected in any dust sample (<100 ng/g dust). In Petri-dish samples, only cat (Fel d 1) and dog (Can f 1) allergens were analyzed and the geometric mean was 16.82 and 17.74 ng/m^2/day, respectively.

Table 26.3 Arithmetic average values of microbial components in the settled dust across 39 classrooms in Taiyuan, China

	Concentration		Amount per sample	
	M (SD)	Min–max	M (SD)	Min–max
MuA[a]	9.53 (4.60)	2.10–24.56	11.39 (8.80)	0.72–35.63
Erg [a]	0.69 (0.40)	0.15–1.78	0.71 (0.60)	0.07–2.35
LPS [b,]	18.64 (6.19)	7.92–34.69	21.01 (13.15)	1.39–54.31

M (SD): arithmetic mean (standard deviation); Min-Max: minimum-maximum. a) The concentration values are in the unit of μg/g dust and the amount values in the unit of the μg/sample; b) The concentration values are in the unit of nmol/g dust and the amount values in the unit of the nmol/sample. Total LPS concentration was calculated from the sum of the concentration of C10, C12, C14 and C16 3-OH FAs, divided by a factor of four, as each LPS molecule has four molecules of 3-OH FA (26)

Table 26.4 Allergen levels and total weight of the settled dust in 10 schools in Taiyuan, China

Settled dust	Class average level (*n*=78)	
	Median (IQR)	Min–max
Cat allergen (Fel d 1) Conc. (ng/g dust)	<100 (<100–<100)	<100–177
Dog allergen (Can f 1) Conc.(ng/g dust)	<200 (<200–<200)	<200–527
Horse allergen (Equ c x) Conc. (U/g dust)	<200 (<200–<200)	<200–<200
Cat allergen amount (ng/filter)	60 (28–83)	5–241
Dog allergen amount (ng/filter)	120 (57–165)	10–1,150
Horse allergen amount (U/filter)	110 (57–160)	10–483
Petri-dish airborne dust	GM (GSD)	Min–Max
Cat allergen (Fel d 1) (ng/m^2/day)	16.82 (6.62)	0.17–225.62
Dog allergen (Can f 1) (ng/m^2/day)	17.74 (2.67)	6.79–139.87
Total dust weight (mg/filter)	1,108 (565–1,604)	98–4,828

IQR: inter-quartile range; Min-max: minimum-maximum

26.3.2 Associations between Environmental Factors and Asthmatic Symptoms

Univariate hierarchical logistic regression was applied to determine the association between air pollutants (SO2, NO2, O3 and formaldehyde) and symptoms. Age, gender, parental asthma or allergy and home environmental factors were controlled. The results suggest that indoor air pollutants were generally positively associated with wheeze, daytime and nocturnal attacks of breathlessness. The associations between SO2, NO2 and formaldehyde and nocturnal attacks of breathlessness were statistically significant with OR 1.27 (95% CI 1.02-1.59), 1.45 (95% CI 1.00-2.08) and 1.40 (95% CI 1.02-1.92), respectively. Application of multivariate adjustment including indoor and outdoor air pollutants, indoor SO2 and wheeze, indicated that indoor formaldehyde and nocturnal attacks of breathless were significantly positively associated. Due to the lower prevalence (<1%) of current asthma attacks and asthma medication, no association analyses between these two symptoms were performed with any environmental factor (Fig. 26.2).

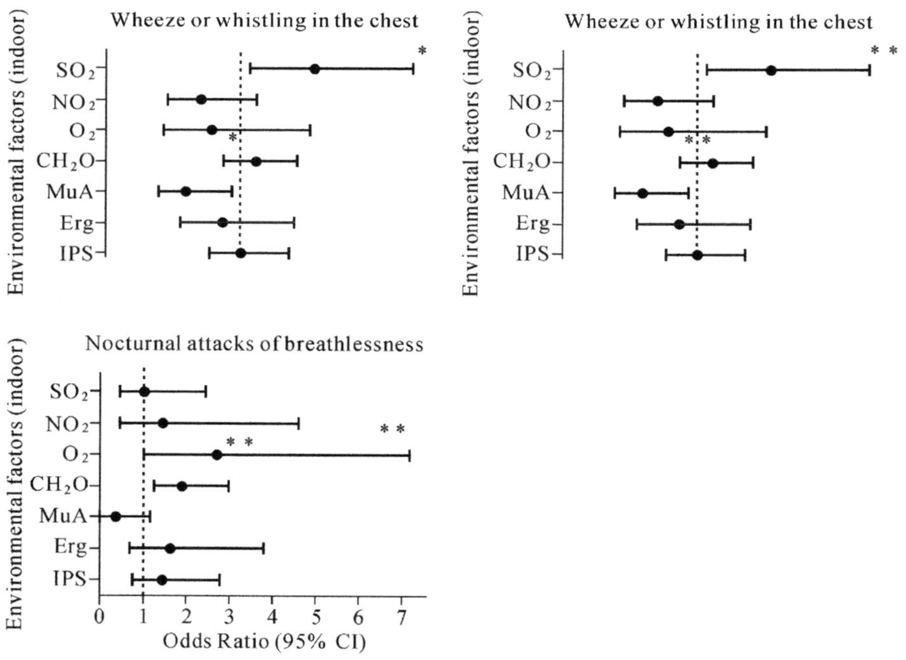

Fig. 26.2. Association between environmental factors and children's asthmatic symptoms. (OR 95% CI) †: controlling for age, gender, parental asthma or allergy, ETS and home; *: *P*<0.05; ** *P*<0.001

For microbial contaminants in the settled dust in classrooms, MuA were found to be negatively associated with wheeze (OR 0.60, 95% CI 0.39–0.94) and daytime attacks of breathlessness (OR 0.69, 95% CI 0.54–0.88). Wheezing was also negatively associated with ergosterol. MuA was more pronouncedly associated with wheeze and daytime attacks of breathlessness. LPS, however, was positively associated with daytime attacks of breathlessness (Fig. 26.2). No significant associations were found between any of the symptoms and allergen exposure. In addition, home environmental factors (including new paintwork and new furniture being introduced in the previous 12 months) were associated with children's wheeze or daytime attacks of breathlessness ($P<0.05$). For microbial components, OR was calculated for concentrations of LPS, MuA and ergosterol with an increase of 10 nmol/g, 10 Lg/g and 1 Lg/g dust, respectively. For indoor air pollutants, OR refers to a step change of 100, 10, 10, and 1 μg/m^3 for SO_2, NO_2, O_3 and formaldehyde, respectively. Air pollutants data with available measurements were applied. No association was found with outdoor air pollutants (not shown in figures).

26.4 Discussion

The indoor school environment in urban schools in Taiyuan was contaminated with chemical air pollutants of outdoor origin and the air pollutants may be risk factors for children's asthmatic symptoms. However, different microbial components in the settled dust may have different effects on children's respiratory symptoms. Low levels of allergen contamination in the settled dust in the school environment was observed in this area, and the airborne cat and dog allergens detected in the analysis were not associated with any health parameters analyzed in this study. In addition, in the home environment, ETS and emissions from new furniture seemed to exacerbate children's respiratory symptoms.

The questionnaire survey in this study had a high response rate of 90.2%, and the questions were answered with the help of parents before the school environment measurements were started. Schools were arbitrarily selected within the urban areas of Taiyuan, and first-year classes were arbitrarily selected within the schools. There were no indications of selection bias when classes participating in the questionnaire study were compared with those included in the classroom measurements. As a consequence of the data being associated with a three-level hierarchical structure (school, classroom, individual), the analysis utilized a three-level hierarchical model. The results were mostly consistent, with some variation in p-values between different models. There are no indications of selection effects on a particular statistical model, but the cross-sectional study design limits the possibility of drawing conclusions from causal relationships.

The prevalence of respiratory symptoms and airway infections was high, whereas the prevalence of diagnosed asthma and allergy to furry pets or pollen was low. The discrepancy between diagnosed asthma and asthmatic symptoms, sex differences, and the validity of the symptom reported in this school study have been discussed previously (Zhao *et al.*, 2006). A similar low prevalence of asthma among children in Taiyuan has been reported in a larger Chinese study (Chen, 2004). Moreover, a high prevalence of respiratory symptoms and airway infections has also been reported from other Asian school studies in Shanghai (Mi *et al.*, 2006) and the Republic of Korea (Kim *et al.*, 2005).

The positive association between indoor air pollutants and children's respiratory symptoms indicated that the indoor air environment in schools may have side effects on children's health. This effect was in consistence with other epidemiological studies in Chinese pupils (Mi *et al.*, 2006). Among other pollutants, SO_2, as a high level air pollutant due to local coal burning, was a risk factor for pupils' wheeze or whistling in the chest in the previous 12 months. Indoor NO_2 levels were associated with nocturnal attacks of breathlessness. The level of indoor NO_2 was relatively high (weekly mean, 39 $\mu g/m^3$; range, 16–62 $\mu g/m^3$), because all schools were located in urban areas near busy roads. It can be expected that the rapid increase in the number of cars in China will lead to a further increase in urban NO_2 levels. Indications of a slight association between levels of O_3 and respiratory symptoms in the conventional model were found, and these become more significant when the indoor O_3 level is classified into 3 categories (low, middle, high). As a consequence of the data being aggregated into weekly means, peak exposure at higher levels cannot be excluded. The data are not directly comparable with the WHO air quality guideline of 100 $\mu g/m^3$ as 8h mean value (WHO, 2005). Surprisingly, links between indoor levels of formaldehyde and wheeze and nocturnal attacks of breathlessness were observed, as well as links between outdoor formaldehyde and daytime attacks of breathlessness. In many countries, formaldehyde is considered an indoor air pollutant, but the results of this research indicate consistently higher levels outdoors (mean indoor/outdoor ratio, 0.38). One reason could be that, in this study, formaldehyde is an indicator of reactive chemistry and possibly is associated with other stronger local irritants (Sundell and Zuber, 1996; Wilkins *et al.*, 2001).

For microbial components analysis, MuA was pronouncedly protective for wheeze symptoms among children. This finding was consistent with the results from van Strien regarding MuA concentrations in mattresses of rural school children using a similar GC-MS method, even after controlling for LPS measurements (van Strien *et al.*, 2004). The results presented here support the view that possible protective effects of bacterial components other than endotoxin should also be considered. This corresponds to its interaction mechanism with TLRs in the innate immunity and regulation roles in human immunological activity (Heinrich *et al.*, 2001). Endotoxins played a more complicated role in this study. Shorter chain lengths of 3-OH FA from LPS (C10, C12 and C14) showed protective effects for wheeze or daytime attacks of breathlessness. Longer chain lengths (C14, C16 and C18) tended to be risk factors for asthmatic symptoms (data not shown). After mutual adjustment, total LPS concentration became a risk factor for daytime attacks of breathlessness. The inconsistency of LPS associations partly reflected the paradoxical nature of endotoxins (Liu, 2002; Radon, 2006), and different types of LPS could provide new evidence from an epidemiological point of view. For instance, they could reflect different Gram negative species. Ergosterol was initially protective for daytime attacks of breathlessness. Furthermore, it showed positive associations with respiratory infection by mutual adjustment. These discrepancies may be explained by the fact that Erg is present more in toxin moulds and less in harmful yeasts. From a mechanical point of view, the recognition of fungi by the innate immune system appears to be more complex than that of bacteria because fungi can exist in two forms, as hyphae or conidia (von Hertzen and Haahtela, 2006). Therefore, Erg *per se* may not be a health-relevant marker in the general sense and its use should perhaps be limited to strictly defined indoor environments.

Environmental allergens, such as cat and dog allergens, are a major problem in schools in western countries (Smedje and Norback, 2001). However, there is almost no detectable level of cat (Fel d 1) and dog (Can f 1) allergens in the settled dust in Taiyuan schools. In the airborne Petri-dish samples, these allergens were detected but neither was associated with children's asthmatic symptoms. There was a frequent cleaning of classrooms in this area, and this might explain the lower level of cumulated allergen contamination in classrooms. However, pet keeping is becoming more popular, especially in higher socio-economic families. These allergens might become a public health problem in schools in the future.

This is a cross-sectional epidemiological study focused on both environmental chemical and biological pollutants in relation to children's respiratory health. This project is the first study investigating the associations between different environmental factors in relation to children's asthmatic symptoms in this area of China. The results indicate that chemical air pollutants tend to be risk factors for children's asthmatic symptoms while biological components, such as MuA, may play protective roles after adjusting for genetic factors and home factors. Asthma and allergic diseases have complex mechanisms. These results indicate the need to investigate potential related factors from different aspects including both biological and chemical environmental factors. It is also important to improve school environment in order to protect children's health.

References

Asher, M.I., Keil, U., Anderson, H.R., *et al.* (1995). International Study of asthma and allergies in childhood (ISAAC): Rationale and methods. Eur Respir J, 8 (3), 483-491.

Asher, M.I., Montefort, S., Bjorksten, B., *et al.* (2006). Worldwide time trends in the prevalence of symptoms of asthma, allergic rhinoconjunctivitis, and eczema in childhood: ISAAC phases one and three repeat multicountry cross-sectional surveys. Lancet, 368(9537), 733-743.

Chauhan, A.J., Inskip, H.M., Linaker, C.H., *et al.* (2003). Personal exposure to nitrogen dioxide (NO_2) and the severity of virus-induced asthma in children. Lancet, 361(9373), 1939-1944.

Chen, Y.Z. (2004). Comparative analysis of the state of asthma prevalence in children from two nation-wide surveys in 1990 and 2000. Zhonghua Jie He He Hu Xi Za Zhi, 27(2), 112-116 (in Chinese).

Dong, G.H., Cao, Y., Ding, H.L., *et al.* (2007). Effects of environmental tobacco smoke on respiratory health of boys and girls from kindergarten: results from 15 districts of northern China. Indoor Air, 17(6), 475-483.

Dong, G.H., Ma, Y.N., Ding, H.L., *et al.* (2009). Pets keeping atin home, parental atopy, asthma, and asthma-related symptoms in 12,910 elementary school children from northeast China. Indoor Air, 19(2), 166-173.

Douwes, J., Pearce, N. (2002). Asthma and the westernization 'package'. Int J Epidemiol, 31(6), 1098-1102.

Emenius, G., Larsson, P.H., Wickman, M., *et al.* (2001). Dispersion of horse allergen in the ambient air, detected with sandwich ELISA. Allergy, 56(8), 771-774.

Gent, J.F., Triche, E.W., Holford, T.R., *et al.* (2003). Association of low-level ozone and fine particles with respiratory symptoms in children with asthma. JAMA, 290(14), 1859-1867.

Heinrich, J., Gehring, U., Douwes, J., *et al.* (2001). Pets and vermin are associated with high endotoxin levels in house dust. Clin Exp Allergy, 31(12), 1839-1845.

Janson, C., Anto, J., Burney, P., *et al.* (2001). The european community respiratory health survey: what are the main results so far? European community respiratory health Survey Ⅱ. Eur Respir J, 18(3), 598-611.

Karlsson, A.S., Hedren, M., Almqvist, C., *et al.* (2002). Evaluation of Petri dish sampling for assessment of cat allergen in airborne dust. Allergy, 57(2), 164-168.

Kim, J.L., Elfman, L., Mi, Y., *et al.* (2005). Current asthma and respiratory symptoms among pupils in relation to dietary factors and allergens in the school environment. Indoor Air, 15(3), 170-182.

Ko, F.W., Tam, W., Wong, T.W., *et al.* (2007). Effects of air pollution on asthma hospitalization rates in different age groups in Hong Kong. Clin Exp Allergy, 37(9), 1312-1319.

Leung, T.F., Wong, G.W., (2008). The Asian side of asthma and allergy. Current Opinion Allergy Clinical Immunology, 8(5), 384-390.

Liu, A. H., 2002. Endotoxin exposure in allergy and asthma: Reconciling a paradox. J Journal of Allergy and Clinical Immunology, 109(3), 379-392.

Mi, Y.H., Norback, D., Tao, J., *et al.* (2006). Current asthma and respiratory symptoms among pupils in Shanghai, China: Influence of building ventilation, nitrogen dioxide, ozone, and formaldehyde in classrooms. Indoor Air, 16(6), 454-464.

Norback, D., Mi, Y.H., Larsson, L., *et al.* (2002). Current asthma, respiratory infections and hypersennesitivity oft moulds in pupils in Shanghai, China, in relation to microbial components in the classrooms. The 9th International Conference on Indoor Air Quality and Climate, Monterey, California.

Peat, J.K., Dickerson, J., Li, J. (1998). Effects of damp and mould in the home on respiratory health: a review of the literature. Allergy, 53(2), 120-128.

Radon, K. (2006). The two sides of the "endotoxin coin". Occup Environ Med, 63(1), 73-78, 10.

Saraf, A., Larsson, L., Burge, H., *et al.* (1997). Quantification of ergosterol and 3-hydroxy fatty acids in settled house dust by gas chromatography-mass spectrometry: comparison with fungal culture and determination of endotoxin by a Limulus amebocyte lysate assay. Appl Environ Microbiol, 63(7), 2554-2559.

Simpson, A., Custovic, A. (2005). Pets and the development of allergic sensitization. Curr Allergy Asthma Report, 5(3), 212-220.

Smedje, G., Norback, D. (2001). Irritants and allergens at school in relation to furnishings and cleaning. Indoor Air, 11(2), 127-133.

Smedje, G., Norback, D., Edling, C. (1997). Asthma among secondary schoolchildren in relation to the school environment. Clin Exp Allergy, 27(11), 1270-1278.

Sporik, R., Chapman, M.D., Platts-Mills, T.A. (1992). House dust mite exposure as a cause of asthma. Clin Exp Allergy, 22(10), 897-906.

Sundell, J., Zuber, A. (1996). Ozone and other photochemical oxidants in ambient and indoor air--properties, sources and concentrations. Scand J Work Environ Health, 22 (Suppl 3), 5-14.

van Strien, R.T., Engel, R., Holst, O., *et al.* (2004). Microbial exposure of rural school children, as assessed by levels of N-acetyl-muramic acid in mattress dust, and its association with respiratory health. J Allergy Clin Immunol, 113(5), 860-867.

von Hertzen, L., Haahtela, T. (2006). Disconnection of man and the soil: reason for the asthma and atopy epidemic? J Allergy Clinical Immunol, 117(2), 334-344.

Watts, J. (2006). Doctors blame air pollution for China's asthma increases. Lancet, 368(9537), 719-720.

WHO, W.H.O. (2005). WHO air quality guidelines global updata.

Wilkins, C.K., Clausen, P.A., Wolkoff, P., *et al.* (2001). Formation of strong airway irritants in mixtures of isoprene/ozone and isoprene/ozone/nitrogen dioxide. Environ Health Perspect, 109(9), 937-941.

Wong, G.W., Leung, T.F., Ko, F.W., *et al.* (2004). Declining asthma prevalence in Hong Kong Chinese schoolchildren. Clinical land Experimental Allergy, 34(10), 1550-1555.

Zhang, J.J., Hu, W., Wei, F., *et al.* (2002). Children's respiratory morbidity prevalence in relation to air pollution in four Chinese cities. Environ Health Perspect, 110(9), 961-967.

Zhao, J., Ma, Y., Chen, Y.Z., *et al.* (2003). Prevalence of allergic respiratory disorders and skin prick test in Beijing urban and suburban children: A comparative study. Zhonghua Yi Xue Za Zhi, 83(21), 1879-1881.

Zhao, Z. H., Elfman, L., Wang, Z.H., *et al.* (2006). A comparative study of asthma, pollen, cat and dog allergy among pupils and allergen levels in schools in Taiyuan city, China and Uppsala, Sweden. Indoor Air, 16(6), 404-413.

Zhao, Z., Zhang, Z., Wang, Z., *et al.* (2008). Asthmatic symptoms among pupils in relation to winter indoor and outdoor air pollution in schools in Taiyuan, China. Environment Health Perspective, 116(1), 90-97.

27

The Relationship of Food Allergies and Respiratory Allergies in Urban and Rural Chinese Children

Gary W.K. Wong[1*], Yu-Zhi Chen[2], Jing Li[3], Nan-Shan Zhong[3]

[1] Department of Pediatrics, Chinese University of Hong Kong, Hong Kong, China
[2] Capital Institute of Pediatrics, Beijing, 100045, China
[3] Guangzhou Institute of Respiratory Diseases, Guangzhou, 510120, China
[*] E-mail: wingkinwong@cuhk.edu.hk

Abstract: Food allergy is considered as the first step of the "allergic march". It is frequently the first manifestation of allergies in children. It has been estimated that 1%–3% of the general population in the western world suffers from food allergies. The prevalence of food allergies has been reported to be more and more common in children and adults. As for other allergic disorders, food allergy is more common in urban and westernized countries. The manifestations can range from simple skin rash to life-threatening anaphylactic shock. The exact etiologies are largely unknown and the only preventive treatment is avoidance of causative food. However, accumulated evidence has revealed that food allergy is rather uncommon in the rural population. With the advance of molecular biology and understanding of the immunobiology of food allergies, improvement to standardized diagnostic methods can be made. When the genetic and environmental factors that lead to the development of food allergies are better understood, primary preventive strategies for this common condition can be developed.

27.1 Introduction

Many adults and children report that they possibly develop adverse reactions following exposure to certain foods. There are many different types of these adverse reactions. The most important one is related to the development of allergic reactions following food ingestion. Food allergies are most common in young children under 3 years of age. Food allergies are due to the adverse reaction to food associated with an abnormal immune reaction to the ingestion of a food or food additive. The reaction is typically related to the presence of IgE, while a minority of reactions may be related to cell mediated immune response. IgE mediated food allergy is common in many developed countries, and there is evidence of a possible increase in the prevalence of this condition (Eggesbo *et al.*, 2001; Grundy *et al.*, 2002; Roehr *et al.*, 2004; Zuberbier *et al.*, 2004; Osterballe *et al.*, 2005). However, the exact causes of food allergy remain largely unknown. The only preventive treatment for this potentially fatal condition is avoidance of food. There have been many published epidemiolgy reports from Europe and America (Zuidmeer *et al.*, 2008; Rona, 2007), but many of them do not include objective measures, which makes it difficult to draw firm conclusions regarding prevalence. The European Union funded multidisciplinary Integrated Project, EuroPrevall-INCO, was designed to evaluate the prevalence, basis and costs of food allergy in different countries using the same standardized methodology.

There have been several studies in Chinese populations which have demonstrated that asthma is common in Hong Kong when compared with cities in mainland China. Furthermore, epidemiology studies of asthma in China also confirmed that asthma and related allergic symptoms are very uncommon in rural Chinese populations (Wong *et al.*, 2004; Wang *et al.*, 2008). There are very few published studies regarding food allergies in developing countries. The EuroPrevall protocol has been adapted to studying the prevalence of food allergies in urban and rural China. Since many regions of China are developing, it may be possible to identify the possible environmental risk factors which are important in the pathogenesis of food allergies. Epidemiological studies have clearly shown that children from rural areas have a much lower prevalence of asthma and related atopic conditions (von Ehrenstein *et al.*, 2000; Riedler *et al.*, 2000; Downs *et al.*, 2001; Yu M *et al.*, 2009). Many possible factors in the rural environment are potentially modifiable. Early exposure to microbes, and the modulation of the immature immune system, may represent important contributing protective factors in the rural environment (Elliott *et al.*, 2005; Wong *et al.*, 2008). Current evidence indicates that food allergies differ between the Chinese and European population. For example, shellfish appear to be one of the most common food allergens in the Chinese population (Leung *et al.*, 2009). Peanut allergy is a relatively common food allergen in Europe and America, but it is uncommon in many Asian countries (Leung *et al.*, 2009; Sicherer *et al.*, 2007). The aim of the research within the Europrevall project is to study different groups of children living in both urban and rural environments using standardized methodologies in order to determine the prevalence of food allergies and their relationship with other atopic symptoms such as asthma, allergic rhinitis and eczema.

27.2 Methodologies and Recruitment of Subjects

The Europrevall-INCO survey was a multi-centre, cross-sectional study in random samples of children from the general populations, with a nested case-control design conducted in different parts of China. The participating centers include Hong Kong, urban and rural Beijing, Guangzhou and Shaoguan. Primary school children aged 7–10 years were targeted for study. The study participants were first screened for the presence of adverse reactions to foods. In order to obtain accurate estimates of food allergies among children from the different centers, a multi- staged approach has been adopted. The focus was on IgE-mediated allergies to a panel of foods most commonly reported to result in Type I allergic reactions. These foods were hen's egg, cow's milk, peanut, soy, hazelnut, walnut, celery, kiwi, apple, peach, sesame, mustard, wheat, fish, shrimp, buckwheat, corn, carrot, tomato, melon, banana, lentils, sunflower seeds and poppy seeds. In addition, mango and crab were also tested.

The random sample of school children was obtained through the school system as attendance at school is compulsory by law. A complete list of the primary schools was obtained from the Education Departments from different centers. Schools were then selected by computer randomization. Of the schools agreeing to participate in the study, all schoolchildren from primary one to five were invited. The participants were first screened by a standardized one-page questionnaire to check for possible adverse reactions to food. Then, a case-control sample was recruited for completion of a face-to-face questionnaire, skin-prick test, and blood collection for serum specific IgE measurement. Cases (with reported adverse food reactions) and controls (those without adverse food reactions) were then recruited for the second stage of the study. The case-control questionnaires asked for details regarding possible environmental risk factors and the presence of other allergic symptoms such as asthma and allergic rhinitis.

27.3 Skin-Prick Testing

All subjects underwent a skin-prick test against a panel of food allergens and aero-allergens along with positive and negative controls (Table 27.1). Standardized allergen extracts and control solutions were obtained from ALK-Abelló (Madrid, Spain). Skin prick tests were performed using the ISAAC Phase II protocol. A drop of each allergen extract was placed on the skin of the volar side of the forearms and pricked through using ALK lancets (Horsholm, Denmark). After 15 minutes, the weal reaction was measured as the mean of the longest diameter and the length of the perpendicular line through its middle. Serum IgE to 25 foods were determined as described previously. The sera are first tested on 5 mixtures (2 commercially available and 3 custom-made). Samples testing positive (0.35 kU_A/l) to a mix are tested on the individual foods of that mix. Additional serum specific IgE measurements against mango and crab were performed. Serum specific IgE against the common environmental allergens including house dust mite, cat, birch, grass, mugwort, *Parietaria* pollen, and total IgE were also determined.

Table 27.1 Allergen extracts for skin-prick test used in the study

Food Allergens		Aeroallergens
Shrimp	Melon	Dustmite mix
Crab	Cat	Grass mix
Peanut	Wheat	Mugwort
Cow's milk	Tomato	Cockroach
Hen's egg	Beef	
Soy	Orange	
Fish	Apple	
Mussel	Date palm profilin	
Hazelnut	Mango	
Walnut	Peach	

27.4 Results

A total of 27,543 children (Hong Kong 6,194, urban Beijing 5,948, rural Beijing 4,724, Guangzhou 5,542, Shaoguan 5,135) agreed to enroll in the study. The prevalence of asthma diagnosis and wheeze (at any time) was highest in Hong Kong (27%) when compared to prevalence in urban Beijing (6%, $P<0.001$), rural Beijing (2.5%, $P<0.001$), urban Guangzhou (6.6%), and rural Shaoguan (2.5%). In both Beijing and Guangzhou, the prevalence was significantly higher in the urban regions than that in the rural counterparts. The prevalence of self-reported food allergies was also highest in Hong Kong, with 3.9% of the children reported to have had more than 4 episodes of adverse reactions in the past, as compared to 2.0% in urban Beijing, 0.9% in rural Beijing. A random subsample of 1,875 children from the 5 sites was recruited for skin prick testing. The preliminary results from Hong Kong, as shown in Fig. 27.1, suggest that shrimp is the most common food associated with adverse reactions. The frequency of reported reactions to milk and egg were less than those reported in America and Western Europe.

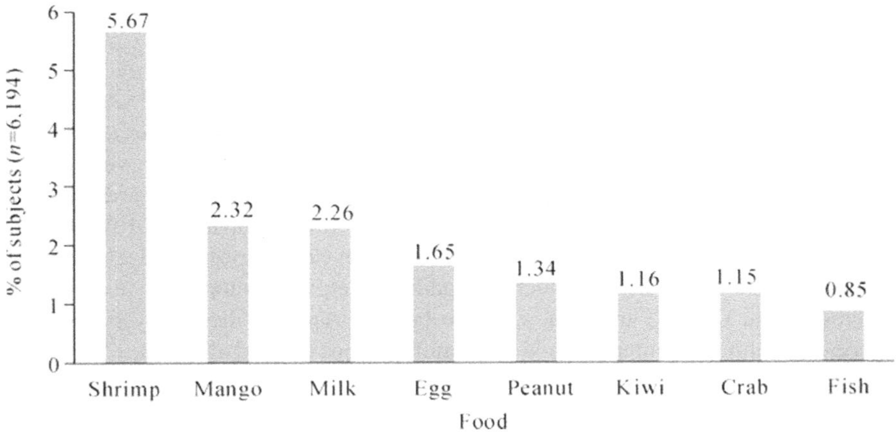

Fig. 27.1. Reported foods resulting in adverse reactions for children recruited from Hong Kong, China

The pattern of sensitization of children recruited from Hong Kong, China as determined by SPT was shown in Fig. 27.2. The most common food of sensitization was shrimp while sensitization was very uncommon against peanut and egg. These results were similar to the parents' reports of adverse reaction. For study participants with self-reported food allergies, the prevalence of wheezing and asthma attack as the predominant symptom was similar in all centers ranging from 4.1% to 5%. Increasing numbers of self-reported food reactions were associated with a higher prevalence of asthma diagnosis and allergic rhinitis (*P* for trend, <0.001).

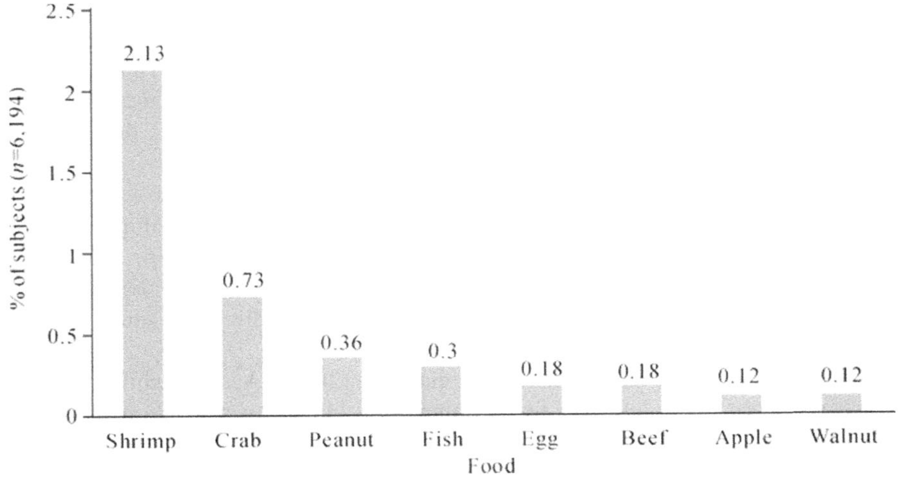

Fig. 27.2. Pattern of food allergen sensitization (SPT) in children recruited from Hong Kong, China

27.5 Conclusion

These results indicated that living in a rural environment is not only protective against the development of asthma and wheeze, but is also associated with a lower prevalence of food allergies. Although the prevalence of food allergies is very low in the rural environment, symptom manifestations of food allergies were similar no matter whether the children are from the urban area or rural area. Further analyses, including the analyses of the sera as well as comparison of the various environmental factors, are needed in order to reveal the possible factors which may be associated with protection against allergy development in the rural environment. Genetic analyses will also be important, as there may be possible gene-environmental interaction in the manifestations of food allergies.

References

Downs S.H., Marks G.B., Mitakakis T.Z., *et al.* (2001). Having lived on a farm and protection against allergic diseases in Australia. Clin Exp Allergy, 31, 570-575.

Eggesbo, M. (2001). The prevalence of CMA/CMPI in young children. Allergy, 56, 393-402.

Elliott, A.M., Mpairwe, H., Quigley, M.A., *et al.* (2005). Helminth infection during pregnancy and development of infantile eczema. J Am Med Assoc, 294, 2032-2034.

Genuneit, J., Cantelmo, J.L., Weinmayr, G., *et al.* (2009). A multi-centre study of candidate genes for wheeze and allergy: The international study of asthma and allergies in childhood phase 2. Clin Exp Allergy, 39, 1875-1888.

Grundy, J., Matthews, S., Bateman, B., *et al.* (2002). Rising prevalence of allergy to peanut in children: data from 2 sequential cohorts. J Allergy Clin Immunol, 110, 784-789.

Leung, T.F., Yung, E., Wong, Y.S., *et al.* (2009). Parent-reported adverse food reactions in Hong Kong Chinese pre-schoolers: Epidemiology, clinical spectrum and risk factors. Pediatr Allergy Immunol, 20, 339-346.

Osterballe, M., Hansen, T.K., Mortz, C.G., *et al.* (2005). The prevalence of food hypersensitivity in an unselected population of children and adults. Pediatr Allergy Immunol, 16, 567-573.

Riedler, J., Eder, W., Oberfeld, G., Schreuer, M. (2000). Austrian children living on a farm have less hay fever asthma and allergic sensitization. Clin Exp Allergy, 30, 194-200.

Roehr, C.C., Edenharter, G., Reimann, S., *et al.* (2004). Food allergy and non-allergic food hypersensitivity in children and adolescents. Clin Exp Allergy, 34, 1534-1541.

Rona, R.J., Keil, T., Summers, C., *et al.* (2007). The prevalence of food allergy: A

meta-analysis. J Allergy Clinical Immunol, 120, 638-646.
Sicherer, S.H., Sampson, H.A. (2007). Peanut allergy: emerging concepts and approaches for an apparent epidemic. J Allergy Clin Immunol, 120, 491-503.
von Ehrenstein, O.S., von Mutius, E., Illi, S., *et al.* (2000). Reduced risk of hay fever and asthma among children of farmers. Clin Exp Allergy, 30, 187-193.
Wang, H.Y., Wong, G.W., Chen, Y.Z., *et al.* (2008). Prevalence of asthma among Chinese adolescents living in Canada and in China. Canadian Med Assoc J, 18(179), 1133-1142.
Wong, G.W., Chow, C.M. (2008). Childhood asthma epidemiology: Insights from comparative studies of rural and urban populations. Pediatr Pulmonol, 43, 107-116.
Wong, G.W., Ko, F.W., Hui, D.S., *et al.* (2004). Factors associated with difference in prevalence of asthma in children from three cities in China: Multicentre epidemiological survey. Brit Med J, 329, 486-489.
Yu, M., Jing, Z., Han, Z.R., Chen, Y.Z., *et al.* (2009). Very low prevalence of asthma and allergies in schoolchildren from rural Beijing, China. Pediatr Pulmonol, 44, 793-799.
Zuberbier, T., Edenharter, G., Worm, M., *et al.* (2004). Prevalence of adverse reactions to food in Germany: A population study. Allergy, 59, 338-345.
Zuidmeer, L., Goldhahn, K., Rona, R.J., *et al.* (2008). The prevalence of plant food allergies: A systematic review. J Allergy Clin Immunol, 121, 1210-1218.

28

Socio-Economic Issues Associated with Food Allergy

L.J. Frewer[1,2]*, J. Voordouw[2], M.F. Schenk[3], M.C. van Putten[2], B. Gremmen[4], G. Antonides[5], J. Cornelisse-Vermaat[2]

[1] Food and Society, Center for Rural Economy, School of Agriculture, Food and Rural Development, Newcastle University, Agriculture Building NE17RU, Newcastle upon Tyne, UK

[2] Marketing and Consumer Behavior Group, Social Sciences Department, Wageningen University, Hollandseweg 1, 6706 KN Wageningen, the Netherlands

[3] Plant Research International, Wageningen UR, P.O. Box 47, 6700 AA Wageningen, the Netherlands

[4] Methodical Ethics and Technology Assessment (META), Social Sciences Department, Wageningen University, Hollandseweg 1, 6706 KN Wageningen, the Netherlands

[5] Economics of Consumers and Households Group, Social Sciences Department, Wageningen University, Hollandseweg 1, 6706 KN Wageningen University, the Netherlands

* E-mail: lynn.frewer@newcastle.ac.uk

Abstract: Food allergies affect a substantial proportion of the population, with prevalence estimates ranging from 1% to 11% of the population suffering from some complaints. Dietary exclusion of allergenic foods, ingredients and derived products, if necessary, represents the primary risk management strategy at the present time. Given the potentially profound consequences of experiencing an allergic reaction, food allergy has the potential to negatively affect the socio-economic functioning of those food allergic patients, as well as their families. The results of research focused on the socio-economic impact of food allergy suggest that, whilst food allergy has the potential to have a profound impact on the well-being of consumers, not all food allergy management activities are equally preferred by food allergy sufferers. The introduction of novel hypoallergenic products produced using potentially controversial food technologies is not universally accepted by food allergic consumers. There is considerable consumer demand for accurate allergen labeling of food products. Questions arise as to whether current regulatory frameworks (for example, the General Food Law in Europe) are adequate in terms of optimizing consumer protection. Potential future areas of research will be discussed, in particular those where international collaboration is needed to attain the research objectives.

28.1 Introduction

Food allergies affect a substantial proportion of the population. Prevalence estimates range from 1% to 11% of the population suffering from some complaints (Blok *et al*., 2007; Flokstra-de Blok *et al*., 2009; Rona *et al*., 2007). The seriousness and type of symptoms, experienced after consuming problematic food and food ingredients, can vary from uncomfortable (e.g., itching, rashes) to potentially fatal (e.g., anaphylactic shock) (Buttriss, 2002). There is also evidence to suggest that the negative impact on the quality of life of food allergic patients is quite profound (Flokstra-de Blok *et al*., 2009; Sicherer *et al*., 2001) and there is potential for food allergy to negatively affect the economic functioning of food allergic patients and their families (Fox *et al*., 2009; Voordouw *et al*., 2010). As a consequence, the development and implementation of effective risk management strategies are extremely important in terms of optimizing the health and well-being of food allergy sufferers.

Given that, at the present time, risk management is focused on total exclusion of allergens from the diet (Primeau *et al*., 2000), the continuous need for alertness (for example, when shopping for food, in restaurants or when buying food outside of the home, or at social events) can have a negative effect on the social lives of patients and their families, or even prevent social interactions or attendance at social events involving food consumption (Cornelisse-Vermaat *et al*., 2008a). The adoption of "exclusion diets" may increase the time spent in food shopping, intensify the use of health services on the part of food allergic patients, and restrict the "opportunity costs for food allergic individuals and their carers. Other negative effects may relate to days spent absent from work or from school, or reduced employment options. Both direct and indirect economic costs may negative impacts on the socio-economic status of food allergic individuals, and their households (Fox *et al*., 2009; Voordouw *et al*., 2010). Societal costs (e.g., costs associated with health care systems and costs accruing to industry relating to employee work absences) may result.

The aim of this chapter is to briefly review the socio-economic impact of food allergy, and to assess the potential of risk mitigation strategies which can be applied to reduce the negative impact of food allergy on the quality of life and economic functioning of food allergy sufferers. These strategies fall into two major categories. The first category focuses on facilitating consumer avoidance of problematic foods and ingredients through more effective labeling and communication about allergenic foods and ingredients, and their location in the food chain (Mills *et al.*, 2004; Cornelisse-Vermaat *et al.*, 2008b). The second focuses on the introduction of novel hypoallergenic foods and ingredients as potential substitutes for allergenic foods and ingredients, either through development of novel hypoallergenic foods and ingredients and their integration into the entire human food chain (van Putten *et al.*, 2007) or through the development of hypoallergenic food products sold specifically to food allergic consumers (Schenk *et al.*, 2008). At the end of this chapter, implications for harmonized international legislation and policy will be discussed.

28.2 Economic Impact of Food Allergy

The potential economic impact of food allergy may be considered at the level of society (for example, in terms of costs to the food industry or to health services), or just at the level of individuals suffering from food allergies and their families. In the case of the latter, costs may be direct, indirect or intangible, for example as assessed in terms of impact on welfare (Miles *et al.*, 2005). At the present time, information regarding costs to health services is not available. In terms of costs to industry, factors driving responses to allergen risk can be summarized into 6 main areas: research and development, farming and suppliers, supply chain logistics, manufacturing processes, catering practices, and corporate/managerial issues. These authors note that the food industry may need to manage and control the presence of allergenic ingredients in food production chains through careful selection and dedicated specifications and quality control of increased labor costs, related to increased administrative requirements, training of staff, additional cleaning and so forth, investments in new production lines and staff costs needed to guarantee allergen free products, improved ingredient traceability systems, product losses due to increased waste as reworking (i.e., recycling ingredients and mixes) becomes impractical, and costs of recalls if an accidental allergen contamination is identified. Research is needed to quantify the actual costs of allergen control to the food industry. However, preliminary analysis suggests that these are likely to be substantial.

The costs of food allergy to individuals and their families are likely to depend not only on the costs of health services, but also on other relevant factors, such as lost opportunity costs through increased time spent in shopping or lost opportunity costs resulting from absence from work or education. Validated questionnaires have been developed to assess the direct, indirect and intangible costs of food allergy at the level of the household. Different versions of questionnaires have been developed, applicable to food allergic adults and the parents of food allergic children. Data have been collected in several European countries (the Netherlands, Poland, Spain and the UK) for both food allergic cases (food allergic adults and parents of food allergic children) and controls (equivalents in the population who do not suffer from food allergies). In the analysis, age, gender, education level, income, working hours, household composition, severity of food allergy and type of food allergy were also assessed as potentially influential factors in terms of economic functioning. The total direct and indirect costs across all countries show higher costs for controls compared to cases. This applies to both costs for adults and children. When only looking at health care cost items, cases have a higher cost than controls. In particular, the cost items related to medical care, consultation with health professionals and travel costs to see health care professionals, showed significantly higher costs for cases compared to controls in some, although not in all, countries. The cost of living (costs for food, either dishes prepared at home or outside of the home, the total expenses of holidays, the cost of special aids to prepare safe food, and the cost of domestic help) were significantly higher in the Netherlands, Spain and the UK, but not in Poland. In addition, the analysis of the impact of food allergy on welfare suggests that welfare is negatively impacted by food allergy.

28.3 Allergen Traceability and Consumer Information

The principle of traceability, and the anticipated positive impact on consumer confidence, has been identified in various legislatures (e.g., the "General Food Law" following the publication of Regulation (EC) No. 178/2002). However, in order to investigate whether traceability does have a positive effect on consumer confidence, knowledge regarding consumers' perceptions and needs with regard to traceability information in general, and how this information can be communicated most effectively, needs to be identified. This is particularly important in the area of food safety, where the failure to implement effective traceability for potentially risky foods and ingredients can have negative influenceon consumer health, and confidence in the food chain (van Rijswijk *et al.*, 2008). Failure to implement effective communication with consumers about potential allergens can also be problematic in terms of socio-economic impact. For example, if allergens are unidentified (and unlabeled) in a specific food chain, there may be negative consequences for consumer health. A decline in consumer confidence in the brand may also decrease, which may extend beyond the issue of allergenicity to other areas of food safety. This might result in negative impacts on food safety.

Research has addressed whether information provided through current labeling practices meets the need of food allergic consumers (Cornelisse-Vermaat *et al.*, 2008a; Voordouw *et al.*, 2009). Food allergic participants in 2 European member states (Greece and the Netherlands) were interviewed whilst shopping in a supermarket. Participants were asked to purchase 15 potentially problematic food items as if for consumption in their own household. Their "information search" behavior was observed, and participants were questioned about their preferences for food allergen information provision. Participants in the study reported experiencing problems associated with current food allergy information provision. A great deal of concern about information legibility and consistency in presentation was identified, including international harmonization of information, and the use of different languages, which could be problematic if food allergic consumers are traveling. The use of "precautionary ("may contain" labeling was also raised as a potential issue. It was reported that inappropriate or highly varied use of labeling, application of precautionary labeling, and the lack of harmonization in labeling practices across countries was perceived to cause (un)necessary dietary restrictions for food allergic consumers. Optimal allergen information delivery strategies needed to be developed and tested, and international labeling regulations amended if appropriate. For example, an important result of changes in European food safety legislation is that the food industry has tended to introduce "precautionary" or "may contain" labeling if there is uncertainty about whether products contain potential allergens or not. Participants now reported experiencing precautionary labeling more frequently now than in the past, prominent examples including "may contain traces of nuts" or "produced in a factory where nuts are processed". Participants claimed that they believed the warnings were inaccurate and unnecessarily restricted their choices. Moreover, the participants indicated that, in order to protect itself, the food industry was increasingly applying precautionary warnings to products unnecessarily. Food allergic consumers reported that the use of these warnings might eventually result in a decrease in the credibility of the warning, the product and the brand.

Additional preferences for harmonised labeling were also described by food allergic consumers. They expressed the view that it was important to ensure the legibility of allergen labeling (e.g., through use of legible font sizes and font contrasts). Identification of allergen labels (e.g., through positioning of labels on the packaging in the same position, and through indication of recipe changes, in order to identify when a previously "safe" product would potentially contain allergens as a consequence of changed ingredients) was also important. A particular problem lies in the use of different languages to identify allergens, and for this reason the application of symbolic labeling was favored by many food allergic consumers. With respect to the latter in particular, participants in the studies indicated a need for international harmonization of symbols. For example, many reported uncertainty regarding whether a symbol representing a particular ingredient represented its presence or its absence in the product.

Given the potential shortcomings of existing allergen labeling strategies, it is relevant to investigate the feasibility of introducing novel information and technology (ICT) approaches to the provision of consumer information, for example using smart card technology to deliver individualized information to consumers which could include allergy information as well as other information of interest, e.g., information about food production methods, such as organic production, or nutritional information, such as low in fat or salt, or food miles used in product delivery from the producer to retailers. Specific examples might include hand-scanners linked to data bases which link specific products, via their bar-codes, to their composite ingredients. Of course, the accuracy of such methods is dependent on the efficacy of the underlying traceability systems themselves, and as such may deliver the same certainty regarding the location of allergenic ingredients in products as labels. Precautionary labeling would not be reduced through application of ICT methods. However, the ease of identifying problematic foods (e.g., through application of an auditory warning when the product is scanned) may facilitate the ease of product identification.

It is important to note that multiple information methods may be required to deliver consumer information to interested consumers, at least in the short to medium term. The adoption of ICT approaches may not be feasible in all retail environments, as a high level of technical sophistication is required. As already stated, the effectiveness of traceability systems may not yet be sophisticated enough to discard traditional labeling strategies, in particular "precautionary labeling". In addition, the use of precautionary labeling may reflect actual manufacturing practices, in that many production lines are not able to guarantee segregation between allergenic and non-allergenic ingredients. Social exclusion of population groups who are unable to use, or cautious of using, ICT approaches may also be important, implying that more traditional information delivery systems need to be maintained in addition to the adoption of novel ICT approaches. Despite these caveats, participants reported that the advantages offered by emerging ICT information delivery systems would potentially deliver many advantages to food allergic consumers in terms of easier identification of potential allergens (Voordouw *et al.*, 2010).

It is important to note that novel approaches to delivering information to food allergic consumers are unlikely to be adopted if they are not accepted by key stakeholders in the food chain. In an analysis of stakeholder views, 8 different information scenarios were developed, including novel ICT approaches to information delivery, and their feasibility of application in European food chains discussed with different food chain actors in five European countries (Germany, Greece, the Netherlands, Spain, and the UK). The food chain actors tended to prefer the use of the standardized label, although participants in the study were also positive about the feasibility of introducing novel ICT approaches to traceability and information delivery. Concerns were raised about accountability, (who will be ultimately responsible for the upkeep of food allergy data bases, for example), and the additional costs for the services to the food industry (Cornelisse-Vermaat *et al.*, 2008).

28.4 Development of Novel Hypoallergenic Foods and Ingredients

Allergy prevention measures currently under development may involve the application of potentially controversial technologies. For example, in Europe, the potential benefits of genetic modification of foods may be offset by consumer concerns about risks, uncertain impacts on human health and the environment, and the perceived unnaturalness of the food products developed through application of the technology (Frewer *et al.*, 1997). What's relevant to the present discussion is the potential for genetic modification to be applied to reduce or remove the allergens in problematic foods. Such developments should take societal concerns about genetic modification into account, and address the needs of consumers in general and food allergic consumers in particular, especially if GM foods and ingredients are to be introduced into the total food chain. Schenk *et al.* (2008) examined the attitude of allergic and non-allergic consumers towards applications of genetic modification for allergy prevention in 1 food allergy application (apple) and 2 hay fever applications (birch, grass). Study participants suffering from self-reported allergy perceived greater "benefits" associated with the birch application as compared to non-sufferers correlating with self-reported negative impact on the quality of life. No differences were found between sufferers and non-sufferers for the food allergy application, possibly because the perceived benefits (being able to eat a food to which the participant was previously allergic) did not outweigh perceived risks and disadvantages. Non-food allergic consumers rejected the inclusion of hypoallergenic apples, and related ingredients, in the food chain in general. In this example, the introduction of a novel hypoallergenic food did not appear to solve the problem with food allergy. In a subsequent study, Schenk *et al.* (2011) compared differences in consumer acceptance of genetically modified hypoallergenic apples (created using genes from another plant species) with hypoallergenic apples created by Cisgenisis technology (where the genetic modification involves a gene from the same plant species) and hypoallergenic apples created using conventional technology. An additional manipulation was in terms of sustainability. Novel apples were "created" using the same techniques, which resulted in reduced use of pesticides. The results confirmed the conclusions of the previous study. Hypoallergenicity was not a benefit perceived to be advantageous by participants in the study, independent of the production method applied. However, participants who scored high on environmental concern were in favor of the more sustainably produced apples. Again, preferences were expressed for those produced using conventional production methods, but genetic modification was also viewed more positively if applied towards improved sustainability.

28.5 Feasibility of Developing Specific Markets for Hypoallergenic Foods

It seems unlikely that hypoallergenic foods and ingredients will be used as a replacement for traditional, allergenic equivalents throughout the food chain. This is because consumers appear to be unenthusiastic about such "global" replacement (in particular if genetic modification is used as part of the process of food and ingredient development), because there is some uncertainty regarding the level of safety delivered (some food allergic consumers have allergic reactions to even hypoallergenic foods, which is unacceptable, particularly in the case of very severe reactions) and because global monitoring of ingredient replacement is unfeasible (van Putten, 2010). An alternative to introducing novel hypoallergenic foods into the food chain might be to develop hypoallergenic food products which could be specifically "targeted" at food allergic consumers. Stakeholder support for such an approach was found to be lukewarm (Cornelisse *et al.*, 2008). The majority of stakeholders were not enthusiastic about the inclusion of a special food aisle in the supermarket, as it would lead to "stigmatization" of food allergic consumers. Non-specialized food producers supplying products aimed at the general population would not be encouraged to produce allergen-free products, which could lead to a decrease in food safety more generally. Food allergic consumers would be forced to eat special products, resulting in higher shopping costs (as special foods are more expensive) and reduced product choice. Another argument against a special food aisle was that it would prevent consumers from making spontaneous purchases in the course of their routine shopping activities, which would have a negative impact on the interests of the retailer. Stakeholders were, however, much more positive about the possibility of retailing such products via an Internet-based retail outlet.

The results of these various research studies suggest that, whilst food allergy has the potential to have a profound impact on the well-being of consumers, not all potential solutions to the problem of allergy are equally preferred by food allergy sufferers. For example, whilst food allergy may have a negative impact on the economic functioning of the household, the introduction of novel hypoallergenic products produced using potentially controversial food technologies is not universally accepted by food allergic consumers. There is considerable consumer demand for accurate allergen labeling of food products. However, the introduction of effective allergen labeling will be contingent on the introduction of effective traceability systems at a global level, and not just on the clarity of the allergen information provided on foods and ingredients. Questions arise as to whether current regulatory frameworks (e.g., the General Food Law in Europe) are adequate in terms of optimizing consumer protection. Current European legislation concerning allergens and their labeling, in particular in relation to the need to optimize consumer protection and improve the quality of life for food allergic consumers, has been identified. Adequate communication concerning the presence of (potentially) allergenic ingredients is, for susceptible individuals, essential if intake of these hazardous ingredients is to be avoided. At the present time, European food labeling legislation requires pre-packaged foodstuffs to be labeled with a list of ingredients, including a reference to allergens. Despite prioritization of the labeling lists, some food products and ingredients are excluded from existing labeling requirements, and it is possible that these exceptions may represent important hazards for food allergic individuals. In order to reach a higher level of consumer protection, some possibilities for improving the European legislative framework can be identified (Hendriks *et al.*, 2011)

In particular, food producers voluntarily use precautionary labeling to warn allergic consumers about the possible unintentional presence of allergens in products. An embedding of precautionary labeling in European legislation may facilitate consumer protection. Food producers may apply precautionary labeling more widely than necessary (Hefle *et al.*, 2007; Taylor and Hefle, 2001), resulting in a burden for food allergic consumers because it restricts the foods included in their diets unnecessarily (Voordouw *et al.*, 2009).

Precautionary labeling should only be used when elimination of cross contamination is not possible within a specific food chain. European legislation should explicitly define precautionary labeling, and the presentation and meaning should be clarified. The warning should appear in the same "field of vision" as the list of ingredients, preferably in the same location, for example at the bottom of the ingredient list. Food allergic consumers would prefer replacing precautionary labeling with a negative or positive allergen indicator. This would lead to practical problems for the food industry, since there is no legal definition or requirement for allergen "free" products. New possibilities for precautionary labeling may arise when threshold levels and detection limits are established for different allergens (Hendricks *et al.*, 2011). In addition, legislation needs to address issues relating to eligibility of font sizes and harmonization of symbolic labeling of allergens. Whilst pragmatic problems may arise in attempting the international harmonization of regulations regarding labeling practices, this does not mean that the issue should not be discussed at an international level.

28.6 Conclusion

In conclusion, research has been initiated, which addresses the socio-economic impacts of allergy. The negative impact of food allergy on the quality of life and welfare of food allergic patients is clear-cut. The evidence regarding the economic impact on the food industry requires further assessment. Some negative impacts of food allergy on the economic functioning of the individuals affected by food allergy, and their households, is less clear-cut, and it is probably dependent on local factors such as preference for traditional diets (and of course the inclusion of allergenic foods and ingredients in these diets), individual and national differences in preferences for processed foods and eating outside of home, and differences in national funding and the structure of national healthcare systems. Nonetheless, it is arguable that changes in the quality of life and economic functioning may be used to assess interventions (e.g., in legislation or communication) designed to reduce the negative effects of food allergy on the lives of food allergic consumers.

References

Blok, B.M.J., Vlieg-Boerstra, B.J., Oude Elberink, J.N.G., *et al.*, (2007). A framework for measuring the social impact of food allergy across Europe. Allergy, 62(7), 733-737.

Flokstra-de Blok, B.M.J., DunnGalvin, A., *et al.*, (2009). Development and validation of a self-administered Food Allergy Quality of Life Questionnaire for childrenChildren. Clin Exp Allergy, 39(1), 127-137.

Buttriss, J. (2002). Adverse reactions to food: The report of a British nutrition foundation task force. Oxford, United Kingdom: Blackwell Science Ltd.

Cornelisse-Vermaat, J.R., Voordouw, J., Theodoridis, G., *et al.* (2008a). Food-allergic consumers' labelling preferences. Similarities and differences in Greece and the Netherlands. Eur J Public Health, 18(2), 115-120.

Cornelisse-Vermaat, R.A., Pfaff, S., Frewer, L.J., *et al.* (2008b). The information needs and labelling preferences of food allergic consumers. The views of stakeholders regarding information scenarios. Trends Food Sci Technol, 19(12), 669-676.

EU, 2002. Regulation (EC) No. 178/2002 of the European Parliament and of the Council of 28 January 2002 laying down the general principles and requirements of food law, establishing the European Food Safety Authority and laying down procedures in matters of food safety. Official Journal of the European Communities, L31(45), 1-24.

Fox, M., Mugford, M., Voordouw, J., *et al.* (2009). Social and economic costs of food allergies in Europe: the development of a socioeconomic impact questionnaire. Health Services Research, 44(5 Pt. 1), 1662-1678.

Frewer, L.J., Howard, C., Shepherd, R. (1997). Public concerns about general and specific applications of genetic engineering: Risk, benefit and ethics. Science, Technology and Human Values, 22(1), 98-124. Hefle, S.L., Taylor, S.L. 2004. Food allergy and the food industry. Curr Allergy Asthma Reports, 4(1), 55-59.

Hefle, S.L., Furlong, T.J., Niemann, L., *et al.* (2007). Consumer attitudes and risks associated with packaged foods having advisory labeling regarding the presence of peanuts. J Allergy Clin Immunol, 120, 171-176.

Hendriks, M.J., Frewer, L.J., van der Meulen, B.M.J. (2011). Allergens in law; European legislation assessed against the preferences of food allergic consumers. European Food and Feed Law Review, 2, 74-87.

Miles, S., Fordham, R., Mills, C., *et al.* (2005). A framework for measuring costs to society of IgE-mediated food allergy. Allergy, 60(8), 996-1003.

Mills, E.N.C, Valovirta, E., Madsen, C., *et al.* (2004). Information provision for allergic consumers – where are we going with food allergen labelling? Allergy, 59(12), 1262-1269.

Primeau, M.N., Kagan, R, Joseph, L., *et al.* (2000). The psychological burden of peanut allergy as perceived by adults with peanut allergy and the parents of peanut-allergic children. Clin Exp Allergy, 30(8), 1135-1143.

Rona, R.J., Keil, T., Summers, C., *et al.* (2007). The prevalence of food allergy: A meta-analysis. J Allergy Clin Immunol, 120(3), 638-646.

Schenk, M., Fischer, A., Jacobsen, E., *et al.* (2008). Patient and non-patient attitudes to different mitigation strategies for birch pollen allergies and Ooral allergy syndrome. Health Risk and Society, 10(3), 263-282.

Schenk, M.F., Maas, M. P., Smulders, M.J.M., *et al.* (2011) Consumer attitudes towards hypoallergenic apples that alleviate mild apple allergy. Food Quality Preference, 22(1): 83-91.

Sicherer, S.H., Noone, S.A., Munoz-Furlong, A. (2001). The impact of childhood food allergy on quality of life. Ann Allergy Asthma Immunol, 87(6), 461-464.

Taylor, S.L., Hefle, S.L. (2001). Will genetically modified foods be allergenic? J Allergy Clin Immunol, 107, 765-771.

van Putten, M.C., Frewer, L.J., Gilissen, L.J.W.J., *et al.* (2007). Novel Foods and Ffood Aallergies. The Iissues. Trends Food Sci Technol, 17(6), 289-299.

van Putten, M., Frewer, L., Gilissen, L., *et al.* (2010). Stakeholder and consumer views regarding novel hypoallergenic foods. Brit Food J, 112(9), 949-961.

van Rijswijk, W. Frewer, L.J., Menozzi, D., *et al.* (2008). Consumer perceptions of traceability: A cross-national comparison of the associated benefits. Food Qual Prefer, 19(5), 452-464.

Voordouw, J., Cornelisse-Vermaat, J., Yiakoumaki, V., et al. (2009). Food allergic consumers' preferences for labelling practices: A qualitative study in a real shopping environment. Int J Consumer Studies, 33(1), 94-102.

Voordouw, J., Fox, M., Cornelisse-Vermaat, M., *et al.* (2010). Household costs associated with food allergy: an exploratory study. Brit Food J, 112, 1205-1215.

29

The Association between Helminth Infections and Atopic Diseases

L.A.H. Hermsen

Faculty of Earth and Life Science, VU University of Amsterdam, 1012 ZA Amsterdam, the Netherlands
E-mail: l.hermsen@vumc.nl

Abstract: Worldwide, allergies are increasingly becoming a public health problem. Several studies have indicated a relation between allergies and a "western" lifestyle. These observations have contributed to the formulation of the hygiene hypothesis, which suggests that low levels of exposure to infections in early childhood increase the risk of developing allergic diseases, which is mainly due to insufficient stimulation of the Th1 immune response. This is in favor of the Th2 response and results in a higher frequency of chronic diseases like allergies, which are characterized by a specific hyperactive Th2 immune response. Helminth infections and allergens both have similar Th2 immune responses. However, it has been noticed that helminth endemic areas are often associated with a lower prevalence of allergic diseases. Therefore, helminth infections are thought to play an important role in the process of allergic disorders, and indicate a relation between helminth infections and allergens. However, studies showed controversial results with both positive and negative associations between helminths and allergens. This indicates a dual or controversial influence of helminth infections on the immune system. Different studies indicate that helminth infections interfere with the predictive value of diagnostic tests for atopy. Generally, atopy is defined by measuring the total level of IgE by the presence of specific IgE for allergens or by the evidence of hypersensitivity by skin prick test (SPT). However, in developing countries, high serum IgE could also indicate chronic helminth infections. Therefore, the use of serum IgE might lead to conflicting results in atopic diagnostics, especially in areas where helminth infections are endemic. In this chapter, the controversial effects of helminth infections on the occurrence of allergies and on the related consequences for atopy diagnostics will be elaborated, with an emphasis on the situations in countries at different stages of economic development.

29.1 Introduction

At the present time, atopy and atopic diseases are becoming more and more prevalent. Some studies indicate an association between allergies and a more western lifestyle (Yemaneberhan *et al.*, 1997), because an increase in allergic diseases has occurred primarily in "westernized" societies. It has been proposed that high exposure to infections in early childhood reduces the risk of developing allergies. This theory has led to the formulation of the hygiene hypothesis. Allergic diseases and helminth infections both show a similar immune response. Nevertheless, it has been noticed that populations in helminth endemic areas often show little sign of allergic disorders. Therefore, helminth infections are thought to play an important role in the process of allergic disorder, and indicate an association between allergic diseases and helminth infections. However, studies showed controversial results with both positive and negative association and different immunological explanations. Another important issue relates to the observation that helminth infections are thought to play a role in the predictive value of diagnostic tests for atopy, because in developing countries it appears that IgE is not only an expression of atopy. High serum levels of IgE could also indicate chronic helminth infections (Medeiros *et al.*, 2006).

The aim of this chapter is to analyze the relationship between atopy and atopic disease and their interaction with geohelminth infections. To shed more light on this relationship, a literature study was performed. In addition, a cross-sectional study with longitudinal outcome was applied to analyze the complex interaction between atopic diseases and helminth infections, and the related consequences for atopic diagnostics. Overall, some possible explanations for the controversial results found in different studies are provided, because further understanding of immunologic mechanisms is required for the development of preventive strategies.

29.2 Atopic Diseases

29.2.1 Prevalence

In the second half of the 20th century, one can observe an increase in atopy and atopic diseases. The most common atopic diseases are atopic dermatitis, allergic rhinitis (hay fever) and asthma. These three manifestations of allergic disease are a major cause of morbidity in developed countries, where they often occur as a chronic disease in childhood (Cooper *et al.*, 2006). There has been an increase in the prevalence of allergic diseases. A study by Steerenberg found an increase of 50% every 10 years, over the past 20 to 30 years (Steerenberg *et al.*, 2001). At this very moment, more than 130 million people suffer from asthma and these numbers are still rising (Yazdanbakhsh *et al.*, 2002). Furthermore, the prevalence of hay fever has increased rapidly in recent years. Some studies indicate genetics play an important role in atopic disease, as the risk of an individual being atopic is 50% when one of their parents is atopic and 60% when both parents are atopic (Steerenberg *et al.*, 2001). Environmental and lifestyle factors are other important determinants of atopic disease, which could be responsible for the recent increase in atopic disease.

In China, the prevalence of asthma seems to be lower than that in western countries (ISAAC, 1998). When looked within China, the prevalence of asthma vastly differs among regions (Leung *et al.*, 1997). For example, the prevalence of asthma and atopic symptoms seems to be higher in children living in Hong Kong, compared to children living in mainland China, for example in Beijing or Guangzhou (Wong *et al.*, 2004). Furthermore, the atopy rate was higher in Hong Kong (41.2%), compared to Beijing (23.9%) or Guangzhou (30.8%) (Wong *et al.*, 2001). Overall, it seems that in China the severity of asthma and other atopic diseases is tending to increase, but this trend is not comparable to the rate of increase observed in the western world.

29.2.2 Immune Response

A theoretical explanation for atopic diseases relates to the balance between Th1 and Th2 cells. T-cells are also known as lymphocytes, a special group of white blood cells. T-cells are divided into T helper Th1, T helper Th2 and cytotoxic Tc cells. T helper cells are involved in activating and directing other immune cells and are considered most important in atopy. Th1 cells are involved in cell mediated immunity and are associated with bacterial and viral infections and autoimmune diseases. These cells maintain the control of invading micro-organisms. Th1 cells recognize class II MHC on macrophages and produce soluble factors, so called cytokines. Th1 cells produce mainly Il-2 and IFN-y cytokines, which inhibit Th2 response and recruit cytotoxic lymphocytes and macrophages. They activate phagocytic cells to switch on their intracellular antimicrobial mechanism. Th2 cells dominate in case of parasitic infections and allergic diseases. Th2 cells recognize class II MHC on B-cells and help them to produce antibodies. Cytokines, like Il-4, Il-5, Il-10 and Il-13, are produced by Th2 cells. These cytokines inhibit the Th1 response and initiate the production of IgE and recruitment of mast cells and eosinophils. Allergic sensitization shows an imbalance between the Th1 and Th2 response, in favor of the Th2 response. This eventually causes allergic diseases.

29.2.3 Hygiene Hypothesis

Given that allergic diseases have occurred primarily in "westernized" societies, and are less common in developing countries, it may be possible to deduct causative factors by comparing conditions in the two cases. Furthermore, allergies are more prevalent in urban than in rural settings (Yazdanbakhsh *et al.*, 2002; Yemaneberhan *et al.*, 1997). The increased prevalence of allergies and asthma is thought to be primarily due to changes in the environment as a result of modernization, especially in developed countries. A comparison of environmental conditions in developed and developing countries has provided insight into possible causes. In developing countries children are often more exposed to infectious agents, because of larger family sizes, rural homes in close proximity to livestock, less antibiotic use, poor sanitation and a higher burden of parasitic infections (Wills-Karp *et al.*, 2001). In developed countries, these environmental factors are less likely to apply. All these observations have contributed to the formulation of the hygiene hypothesis, which suggests that high exposure to infections in early childhood reduces the risk of developing allergies (Strachan *et al.*, 1989). There is an immunological framework for the hygiene hypothesis which shows an imbalance between type 1 and type 2 immune responses (Yazdanbakhsh *et al.*, 2001). During pregnancy, foetal immunity is dominated by the humoral responsiveness (Th2 response). This is because the Th2 response protects the fetus from rejection. It is of great importance to establish a balance between Th1 and Th2 responses in early childhood, up to the age of 5 years. Bacterial and viral infections are highly prevalent, especially during childhood. Eventually, these infections will adequately change the Th2 phenotype to a Th1 phenotype. This is precisely what the hygiene hypothesis proposes. Yazdanbakhsh (2002) explained the hygiene hypothesis in the following way: "Limited exposure to childhood infections results in an insufficient stimulation of the Th1 response. There is an imbalance between Th1 and Th2 in favor of the Th2 response, with expansion of Th2 cells". This eventually results in higher frequency of chronic diseases such as allergies, which are characterized by a specific hyperactive immune response.

29.3 Helminth Infections

29.3.1 Prevalence

It has been noticed that helminth endemic areas are associated with lower prevalence of allergic diseases. From this, it is postulated that helminth infections are thought to play an important role in the process of allergic disorders. In developing countries, geohelminths like *Ascaris lumbricoides*, *Trichuris trichiura* and hookworm (*Ancylostoma duodenale* and *Necator americanus*) are very common. They are capable of causing chronic infections (Cooper *et al.*, 2004). The estimated global prevalence of *Ascaris lumbricoides, Trichuris trichiura* and hookworm are 1.5 billion, 1.3 billion and 900 million, respectively (Cooper *et al.*, 2006). Worldwide, the geohelminths infect over 2 billion people (Moncayo *et al.*, 2006).

In China, geohelminth infections are also prevalent. Two nationwide Chinese surveys were completed during the last 15 years. In 1995, an overall prevalence of 47.0%, 18.8%, and 17.2% for *Ascaris lumbricoides, Trichuris trichiura* and hookworm infections, respectively, were identified (Xu *et al.*, 1995). The number of infections due to *Ascaris lumbricoides*, *Trichuris trichiura* and hookworm was estimated as 531 million, 212 million, and 194 million, respectively (Xu *et al.*, 1995). Egg counts showed light infection in 75%–95% of the study participants living in China (Xu *et al.*, 1995). In 2003, a total of 356,629 individuals from 31 Provinces, Autonomous Regions and Municipalities (P/A/M) were included in the survey. They recorded 26 different helminth species (Coordinating Office of the National Survey on the Human Parasitic Diseases, 2005). Furthermore, they estimated that approximately 129 million individuals in China were infected with at least one helminth species (Coordinating Office of the National Survey on the Human Parasitic Diseases, 2005). Of these infections, approximately 85.93 million were infected with *Ascaris lumbricoides*, 29.09 million with *Trichuris trichiura* and 39.3 million with hookworm (Coordinating Office of the National Survey on the Human Parasitic Diseases, 2005). The survey described an overall infection rate of 19.56% for soil-transmitted helminths. More specifically, the survey described an average rate of 12.72% for *Ascaris* infection, 4.63% for *Trichuris* infection and 6.12% for hookworm infection. The standardized rate of *Ascaris, Trichuris* and hookworm infections have decreased by 71.29%, 73.6% and 60.72%, since 1990 (Ministry of Health, China, 2008). Overall, the survey showed a decrease of 407 million infections between 1995 and 2003. This is probably due to the control strategies implemented in China (Ministry of Health, China, 2008). Variation in reduction in different areas was also reported. The infection rate tended to be higher in South China, when compared to North China. It has been noticed that there are some exceptions. For example, the infection rate in Jiangsu, Guangdong, Shanghai and Zhejiang was lower than other central and south P/A/M, mainly due to control strategies. Overall, the prevalence of soil transmitted helminths remained very high in central and western China (Zhou *et al.*, 2008; Li *et al.*, 2010).

29.3.2 Life Cycles

Geohelminths are also known as intestinal helminths or soil-transmitted helminths. These nematode parasites have a direct life cycle, which involves no intermediate host. They are transmitted by fecal contamination of soil, food and water supplies. *Ascaris lumbricoides* (roundworm) and *Trichuris trichiura* (whipworm) are transmitted through ingestion of embryonated eggs from the environment. The life cycle of Ascariasis involves infective eggs being swallowed by the human host, which eventually invade the intestinal mucosa. They are carried via the portal and systemic circulation to the lungs. Subsequently, the larvae mature in the lungs, penetrate the alveolar walls and ascend the bronchial tree to the throat, where they are swallowed. They then enter the small intestine, where larvae develop into adult worms and start to produce eggs. Eventually, these eggs are passed with the feces again. This process can take about within 2–3 months. *Trichuris trichiura* is a whipworm with a purely enteric life cycle, which causes an infection of the large intestine in the human host. The highest prevalence and intensity of both types of geohelminths is seen between the age of 5 and 15 years, especially in helminth endemic areas.

Hookworm infection is transmitted through skin contact with free living larvae in the soil. Larvae penetrate the skin and are carried to the veins of the heart. The hookworm has a phase of larval migration through the lungs, as for Ascariasis. Eventually, the larvae reach the small intestine, where they mature into adult worms. Adult worms live in the lumen of the small intestine, where they attach to the intestinal wall and produce eggs, which are again passed in the stool. The hookworm has a peak prevalence and intensity in adulthood. Overall, helminth infections are a major cause of morbidity through the effect on nutritional, growth and general health status (Moncayo *et al.*, 2006).

29.3.3 Association between Helminth Infection and Allergic Diseases

Helminth infections and allergic diseases both have a similar Th2 immune response, which activates B-cells and in turn leads to the production of IgE antibodies. When the antibodies are released into the circulation, they will attach to mast cells or eosinophilia. More IgE antibodies can attach to one mast cell, where they are capable of connecting to each other. This so-called "cross linking" between antibodies causes mast cell degranulation, in which mediators, like histamines, are released and cause inflammatory reactions. There are many studies that indicate an association between helminth infections and allergic diseases. Some studies have demonstrated a negative association between helminth infection and allergies, with reduced risk of allergic diseases when chronically infected with helminths, while some other studies found a positive association (Table 29.1).

Table 29.1 Studies that have examined the relationship between helminths and allergies: Correlations between helminth infections and allergies (van Riet *et al.*, 2007) (With permission of Elsevier)

Study area	Population (age range)	Helminth	Outcome	References
Allergy: Human Association Studies Reported negative associations				
Venezuela	children	Geohelminths	Reduced skin reactivity to environmental and ascaris antigen	Hagel *et al.* (1993)
Brazil	175 subjects	*Schistosoma mansoni*	Reduced skin response to aeroallergens	Araujo *et al.* (2000)
Gabon	520 children (5–14)	*Schistosoma haematobium*	Reduced skin reactivity to mite	Van den Biggelaar *et al.* (2000, 2001)
Ethiopia	604 adults (> 16)	Hookworm	Reduced risk of wheeze	Nyan *et al.* (2001)
Gambia	448 adults (> 15)	Geohelminhths	Reduced from skin reactivity	Scrivener *et al.* (2001) Nyan *et al.* (2001)
Ecuador	2865 children 5–19 years	Geohelminths	Reduced skin reactivity to allergens	Cooper *et al.* (2003a)
Brazil	84 asthma patients (6–35)	*Schistosoma mansoni*	Reduced course of asthma	Medeiros *et al.* (2003)
Ethiopia	563 children (1–4)	*Ascaris humbricoides*, hookworm	Reduced wheezing	Dagoye *et al.* (2003)
Uganda	62 infants	Maternal helminthiasis (filariasis and hookworm) at delivery	Protection against infantile eczema	Elliott *et al.* (2005)
Reported no association				
Ecuador	4,433 children (5–18)	Geohelminths	No association with allergic symptoms	Cooper *et al.* (2003b)
Ethiopia	7,649 (> 5)	Geohelminths	No association with weeze/asthma	Davey *et al.* (2005)
Reported positive association				
The Netherlands	1,379 children (4–12)	*Toxocara* spp.[a]	Increase in allergic manifestations	Buijs *et al.* (1997)
China	2,164 children (8–18)	Ascaris lumbricoides	Increased sensitization to aerollergens, increased risk of asthma	Palmer *et al.* (2002)
South Africa	359 children (6–14)	Ascaris lumbricoides (ascaris-specific IgE)	Increased SPT positivity to aeroallergens	Obihara *et al.* (2006)

[a] Intestinal parasite of cats and dogs, human is a non-compatible host

29.3.4 Immunological Explanation

The role of the immune system in combating infections (helminth infections) and the contribution to the development of allergies is still controversial. There are two possible immunological explanations for the negative association between helminth infections and allergic diseases. The first explanation is that helminths induce polyclonal IgE or IgG4. These specific antibodies saturate IgE receptors on mast cells or basophiles, which prevent the occurrence of an effector phase (release of histamines), by blocking degranulation (Yazdanbakhsh *et al.*, 2001). The second explanation is the induction of an anti-inflammatory network (Yazdanbakhsh *et al.*, 2001). Chronic helminth infections could induce a regulatory response, where T regulatory cells are activated and produce Il-10 and TGF-B. These anti-inflammatory cytokines are capable of down regulating the immune system (Yazdanbakhsh *et al.*, 2001) and cause suppression of immune responses to environmental allergens. However, there is also evidence for a positive association between helminths and the skin prick test, especially in areas with low geohelminth prevalence. The possible immunological explanation for a positive association between helminth infections and allergic diseases might be that helminths induce a large amount of polyclonal IgE, which could result in enhanced allergic reactivity at low intensity of infection (Cooper *et al.*, 2004; 2006). In China, it has been reported that infection with *Ascaris lumbricoides* was closely associated with an increased risk of childhood asthma, airway responsiveness to methacholine, and sensitization to some aeroallergens (Palmer *et al.*, 2002). Overall, they found a positive association, which could be explained by the mild infections mostly seen in China. These mild infections could result in enhanced allergic reactivity.

29.3.5 Controversial Results

Some studies have reported controversial results regarding the association between helminth infection and allergies. A possible explanation could be that the studies were executed in a wide variation of countries. Furthermore, endemicity of helminth infections is an important determinant of the effect of helminth infection on atopy. There is enhanced allergic reactivity in populations with low prevalence of infection (<10%) and suppression of allergic reactivity in populations where the infections are highly prevalent (>50%) (Cooper *et al.*, 2006). Other explanations could relate to the intensity and pattern of transmission, which are both thought to be important determinants of the host's immune reaction towards the parasite (Cooper *et al.*, 2002). Acute infections are primary infections with infrequent or short exposure because the infection periods are interrupted. When infections show continuous infection periods and maintenance of high parasite burdens, the acute infection will develop into a more chronic infection stage. Clinical allergic reaction appears to be rare among children who are chronically infected by helminths (Cooper *et al.*, 2006). This suggests that either host, or parasite, or both can down-regulate the allergic inflammatory response. This favors survival of the parasite, which is capable of avoiding the host's pathology and parasite elimination. The last possible explanation is a wide variation in helminth species, where different species show different effects to allergic responses. Some studies found differences between hookworm and both trichuriasis and ascariasis (Wordemann *et al.*, 2008).

29.3.6 Predictive Value of IgE, the Role of Helminth Infections

Helminth infections are also thought to play a role in the predictive value of diagnostic tests for atopy. As already mentioned, helminth infections and allergic diseases both have similar immune responses. Allergic disease is strongly associated with atopy, which is an immediate hypersensitivity reaction against environmental allergens. There are several diagnostic tests to measure the immunological (atopic) reaction. In most studies, atopy is defined by the total levels of IgE, the presence of specific IgE (sIgE) for specific allergens, or evidence of hypersensitivity by the skin prick test (SPT) (Cooper *et al.*, 2004). However, not every atopic individual will develop an allergic disease. Clinical symptoms in individuals do not always show elevated levels of serum IgE levels for allergens (Rorke *et al.*, 2006). Furthermore, they are often influenced by other confounding factors like inflammation, organ function, the presence of infection, stress, hormonal influences and environmental factors (Söderström *et al.*, 2003). For a long time it was assumed that the increased production of specific IgE and total IgE were the most important diagnostic factors for atopy and atopic diseases. At the moment, it appears that atopy and IgE levels are two different kinds of phenotypes (Mutius *et al.*, 2004). In developing countries, IgE is not only an expression of atopy but it can also indicate chronic helminth infections (Medeiros *et al.*, 2006). Levin *et al.* (2008) demonstrated that helminth infection is a major determining factor for IgE levels in some populations. There are conflicting results regarding the diagnostic value of serum IgE in measuring atopy, especially when helminth infections are endemic.

A cross sectional study with longitudinal outcome (in which the author participated) analyzed the relationship between atopy and atopic disease and their interaction with geohelminth infections in Cuban schoolchildren. The aim of the research was to determine the correlation of serum IgE and SPT with atopy and atopic disease. This study identified a significant correlation between serum IgE and SPT reactivity, with a better correlation between specific IgE and SPT reactivity, compared to total IgE. The association between total IgE and SPT reactivity indicates a possible role for helminth infections in the diagnostic value of total IgE for atopy. No role was found for helminth infections in determining specific IgE levels. The results indicate specific IgE to be a better diagnostic tool for atopy, compared to total IgE. Some other studies describe the predictive value of IgE in measuring atopy and the role for helminth infections in this process. The predictive value of IgE showed differences between countries, Gabon (Biggelaar *et al.*, 2001), Europe (Carosso *et al.*, 2007), Cuba (Hermsen *et al.*, 2008), Rural Tropics (Cooper *et al.*, 2008), South Africa (Levin *et al.*, 2008).

Overall, the influence of helminth infections on IgE levels suggests that IgE levels need to be used with special care when used as a diagnostic tool for measuring atopy, especially in helminth endemic areas.

29.4 Conclusion

Research into the role of helminth infection on both atopy and atopic diseases is not conclusive. Helminth endemicity and prevalence may both represent important factors, but there is uncertainty regarding their relevant importance to the development or otherwise of allergic diseases. More research in areas with different endemicity of helminths is required to compare mechanisms of geohelminth mediated immunomodulation of atopy and atopic disease between these areas. Furthermore, a question remains as to whether the negative association remains or disappears after anti-helminth treatment.

References

Buijs, J., Borsboom, G., Renting, M., *et al.* (1997). Relationship between allergic manifestations and Toxocara seropositivity: A cross-sectional study among elementary school children. Eur Respir J, 10, 1467-1475.

Carosso, A., Bugiani, M., Migliore, E., *et al.* (2007). Reference values of total serum IgE and their significance in the diagnosis of allergy in young european adults. Int Arch Allergy Immunol, 142, 230-238.

Cooper, P.J. (2002). Can intestinal helminth infections (geohelminths) affect the development and expression of asthma and allergic disease? Clin Exp Immunol, 128, 398-404.

Cooper, P.J., Chico, M.E., Griffin, G.E., *et al.* (2003). Allergy symptoms, atopy, and geohelminth infections in a rural area of Ecuador. Am J Respir Crit Care Med, 168, 313-317.

Cooper, P.J., Chico, M.E., Rodriques, L.C., *et al.* (2003). Reduced risk of atopy among school-aged children infected with geohelminth parasites in a rural area in the tropics. J Allergy Clin Immunol, 111(5), 995-1000.

Cooper, P.J. (2004). Intestinal worms and human allergy. Parasite Immunology, 26, 455-467. Int J Biochem Cell Biol, 38, 1031-1035.

Cooper, P.J., Barreto, M.L., Rodrigues, L.C. (2006). Human allergy and geohelminth infections: a review of the literature and a proposed conceptual model to guide the investigation of possible causal association. Brit Med Bulletin, 79-80, 203-218.

Cooper, P.J., Alexander, N., Moncayo, A.L., *et al.* (2008). Environmental determinants of total IgE among school children living in the rural Tropics: Importance of geohelminth infection and effect of anthelminth treatment. BMC Immunology, 9, 33.

Coordinating Office of the National Survey on the Important Human Parasitic Diseases (2005). A national survey on current status of the important parasitic disease in human population. Zhongguo Ji Sheng Chong Xue Yu Ji Sheng Chong Bing Za Zhi, 23 (5 Suppl), 332-340.

Dagoye, D., Bekele, Z., Woldemichael, K., *et al.* (2003). Wheezing, allergy, and parasite infection in children in urban and rural ethiopia. Am J of Resp and Critical Med, 167, 1369-1373.

Davey, G., Venn, A., Belete, H., *et al.* (2005). Wheeze, allergic sensitization and geohelminth infection in Butajira, Ethiopia. Clin Exp Allergy, 35, 301-307.

Elliott, A.M., Mpairwe, H., Quigley, M.A., *et al.* (2005). Helminth infection during pregnancy and development of infantile eczema. JAMA, 294, 2032-2034.

Hagel, I., Lynch, N.R., Di Prisco, M.C., *et al.* (1993). Ascaris reinfection of slum children: Relation with the IgE response. Clin Exp Immunol, 94, 80-83.

Hermsen, L.A.H. (2008) Correlations of skin prick test and IgE levels with atopy and atopic disease in Cuban schoolchildren living in helminth endemic areas. MSc thesis.

International Study of Asthma and Allergies in Childhood (ISAAC) Steering Committee (1998). Worldwide variation in prevalence of symptoms of asthma, allergic rhinoconjunctivitis, and atopic eczema. Lancet, 351, 1125-1132.

Leung, R., Wong, G., Lau, J., *et al.* (1997). Prevalence of asthma and allergy in Hong Kong schoolchildren: An ISAAC study. Eur Respir J, 10, 354-360.

Levin, M.E., Le Souëf, P.N., Motala, C. (2008). Total IgE in urban Black South African teenagers: The influence of atopy and helminth infection. Pediatr Allergy Immunol, 22.

Li, T., He, S.Y., Zhao, H., *et al.* (2010). Major trends in human parasitic diseases in China. Trends in Parasitdogy, 26, 264-270.

Medeiros, D., Silva, A.R., Rizzo, J.A., *et al.* (2006). Total IgE level in respiratory allergy: Study of patients at high risk for helminthic infection. Jornal de Pediatria.

Ministry of Health, People's Democratic Republic of China (2008). Current status and control of schistosomiasis, soil transmitted helminths, foodborne trematodes and echinococcosis in China. Personal communication.

Moncayo, A.L., Cooper, P.J. (2006). Geohelminth infections: Impact on allergic diseases. The International Journal of Biochemistry Cell Biology, 38(7): 1031-1035.

Mutius, E. (2004). Influences in allergy: Epidemiology and the environment. American Academy of Allergy, Asthma and Immunology.

Nyan, O.A., Walraven, G.E.L., Banya, W.A.S., *et al.* (2001). Atopy, intestinal helminth infection and total serum IgE in rural and urban adult Gambian communities. Clin Exp Allergy, 31, 1672-1678.

Obihara, C.C., Beyers, N., Gie, R.P., *et al.* (2006). Respiratory atopic disease, ascaris-immunoglobulin E and tuberculin testing in urban South African children. Clin Exp Allergy, 36, 640-648.

Palmer, L.J., Celedon, J.C., Weiss, S.T., *et al.* (2002). Ascaris lumbricoides infection is associated with increased risk of childhood asthma and atopy in rural China. Am J of Respir Crit Care Med, 165, 1489-1493.

Rorke, S., Holgate, S.T. (2006). The atopy phenotype revisited. Journal of the American Academy of Dermatology, 54(1), 172-179.

Scrivener, S., Yemaneberhan, H., Zebenigus, M., *et al.* (2001). Independent effects of intestinal parasite infection and domestic allergen exposure on risk of wheeze in Ethiopia: A nested case control study. Lancet, 358, 1493-1499.

Söderström, L., Kober, A., Ahlstedt, S., *et al.* (2003). A further evaluation of the clinical use of specific IgE antibody testing in allergic diseases. Allergy, 58, 921-928.

Steerenberg, P.A., Loveren, van H., Vandebriel, R.J., *et al.* (2001). The prevalence of asthma and allergy increases; a world wide problem. RIVM rapport, 640250004.

Strachan, D.P. (1989). Hayfever, hygiene and household size. Brit Med J, 299, 1259-1260.

van den Biggelaar, A.H.J.R., van den, R., Rodrigues, L.C., *et al.* (2000). Decreased atopy in children infected with Schistosoma haematobium: A role for parasite-induced interleukin-10. Lancet, 356, 1723-1727.

van den Biggelaar, van den A.H., Lopuhaa, C., *et al.* (2001). The prevalence of parasite infestation and House Dust Mite sensitization in Gabonese schoolchildren. Int. Arch Allergy Immunol, 126(3), 231-238.

van Riet, E., Hartgers, F.C., Yazdanbakhsh, M. (2007). Chronic helminth infections induce immunomodulation: Consequences and mechanisms. Immunobiology, 212, 475-490.

Wills-Karp, M., Santeliz, J., Karp, C.L. (2001). The germless theory of allergic disease: Revisiting the hygiene hypothesis. Nature Rev Immunol, 1, 69-75.

Wong, G.W.K., Hui, D.S.C., Chan, H.H., *et al.* (2001). Prevalence of respiratory and atopic disorders in Chinese schoolchildren. Clinical and Eexperimental Aallergy, 31, 1225-1231.

Wong, G.W.K., Ko, F.W.S., Hui, D.S.C., *et al.* (2004). Factors associated with difference in prevalence of asthma in children from three cities in China: Multicentre epidemiological survey. BMJ, 329 (7464), 486.

Wordemann, M., Diaz, R.J., Heredia, L.M., *et al.* (2008). Association of atopy, asthma allergic rhinoconjunctivitis, atopic dermatitis and intestinal helminth infections in Cuban children. Trop Med Int Health, 13(2),180-186.

Xu, L.O., Yu, S.H., Jiang, Z.X., *et al.* (1995). Soiltransmitted helminthiases:

Nationwide survey in China. Bull WHO, 73, 507-513.
Yazdanbakhsh, M., *et al.* (2002). Allergy, parasites and the hygiene hypotheses. Science, 296, 490.
Yazdanbakhsh, M., van den Biggelaar, A., Maizels, R.M., *et al.*(2001). Th2 responses without atopy: Immunoregulation in chronic helminth infections and reduced allergic diseases. Trends Immunol, 22(7).
Yemaneberhan, H., Bekele, Z., Venn, A., *et al.* (1997). Prevalence of wheeze and asthma and relation to atopy in urban and rural Ethiopia. Lancet, 350(9071), 85-90.
Zhou, P., Chen, N., Zhang, R.L., *et al.* (2008). Food-borne parasitic zoonoses in China: Perspective for control. Trends in Parasitology, 24,190-196.

Index